AF371256

Environmental Impact Assessment by Green Processes

Environmental Impact Assessment by Green Processes

Editors

Pasquale Avino
Massimiliano Errico
Aristide Giuliano
Hamid Salehi

MDPI • Basel • Beijing • Wuhan • Barcelona • Belgrade • Manchester • Tokyo • Cluj • Tianjin

Editors

Pasquale Avino
University of Molise
Italy

Massimiliano Errico
University of Southern
Denmark
Denmark

Aristide Giuliano
Trisaia Research Centre
Italy

Hamid Salehi
University of Greenwich
UK

Editorial Office
MDPI
St. Alban-Anlage 66
4052 Basel, Switzerland

This is a reprint of articles from the Special Issue published online in the open access journal *International Journal of Environmental Research and Public Health* (ISSN 1660-4601) (available at: https://www.mdpi.com/journal/ijerph/special_issues/green_processes).

For citation purposes, cite each article independently as indicated on the article page online and as indicated below:

LastName, A.A.; LastName, B.B.; LastName, C.C. Article Title. *Journal Name* **Year**, *Volume Number*, Page Range.

ISBN 978-3-0365-5895-0 (Hbk)
ISBN 978-3-0365-5896-7 (PDF)

Contents

About the Editors

Pasquale Avino

Pasquale Avino received his Ph.D. in Chemical Sciences at the University of Rome "La Sapienza", and in 1997 he was appointed as Post-Doc at the University of California, Irvine (USA), with Prof. F.S. Rowland (Nobel Prize in Chemistry in 1995) where he was interested in air pollution studies carrying out sampling in remote areas (South-East area of the Pacific Ocean) and at high altitudes (mountains and transoceanic air flights). From 1998 to 2018 he was Researcher at the ISPESL/INAIL institute and in 2018 he moved to University of Molise where he is currently Associate Professor in Analytical Chemistry and Environmental Chemistry. He is author or co-author of more than 190 peer-review papers according to the Scopus data base, he has coordinated several research projects in the environmental field (in 2022 he was recipient of a Horizon project on indoor air pollution). For his studies he received the NASA Award, the "Environment and Health" Sapio National Award and the Ecological Chemistry Medal of the Moldova Chemical Society for his contribution for the development of ecological chemistry.

Massimiliano Errico

Massimiliano Errico is Associate Professor in Chemical Engineering Biotechnology and Environmental Technology at the University of Southern Denmark. His research activity is focused on energy saving in distillation systems for biofuels production. In the specific field of distillation sequences with thermal coupling and thermal integration. He is the author of 95 publications in international journals and he joined to 5 international projects.

Aristide Giuliano

Aristide Giuliano is a researcher of the Italian Research Centre ENEA–Italian Agency for New Technologies, Energy and Sustainable Economic Development, Department of Energy Technologies and Renewable Sources. After achievement of PhD in Chemical Engineering at University of Salerno (Italy), he won the competition of the Italian Association of Chemical Engineering (AIDIC) as "Best PhD Thesis in Italy 2016". In June 2022, he achieved the National Scientific qualification as associate in the Italian higher education system for the disciplinary field of 09/D3—Chemical plants and technologies. He is an 'IEA Bioenergy—Task 42 Biorefining in a Circular Economy' member, he studies the process design and optimization of multi-product lignocellulosic biorefineries to identify the bio-based high-added value compounds ready for commercialization. He is the author of 36 publications in international journals and he is the inventor of 2 international patents.

Hamid Salehi

Hamid Salehi is currently a Senior Lecturer in Chemical Engineering in the Faculty of Engineering and Science after joining the University of Greenwich as a Research Fellow with the Wolfson Centre in April 2018. He was a Research Associate for one year at the Biomass Technology Centre, of the Swedish University of Agricultural Science. His work was focused on the development of a new method for quantifying the flow properties of biomass particulate solids. Dr Salehi carries over 8 years of work/research experience in various disciplines of particulate material technology, food and chemical engineering. He is the author of more than 15 scientific peer reviewed journal papers in high impact journals.

International Journal of
*Environmental Research
and Public Health*

MDPI

Editorial

Environmental Impact Assessment by Green Processes

Aristide Giuliano [1], Massimiliano Errico [2], Hamid Salehi [3] and Pasquale Avino [4,*]

[1]   ENEA–Italian Agency for New Technologies, Energy and Sustainable Economic Development,
     Department of Energetic Technologies, Trisaia Research Centre, I-75026 Rotondella, Italy
[2]   Department of Green Technology, University of Southern Denmark, Campusvej 55, 5230 Odense, Denmark
[3]   Wolfson Centre for Bulk Solids Handling Technology, Faculty of Engineering & Science,
     University of Greenwich, London SE10 9LS, UK
[4]   Department of Agricultural, Environmental and Food Sciences (DiAAA), University of Molise, Via de Sanctis,
     86100 Campobasso, Italy
*   Correspondence: avino@unimol.it; Tel.: +39-338-4672816

Citation: Giuliano, A.; Errico, M.; Salehi, H.; Avino, P. Environmental Impact Assessment by Green Processes. *Int. J. Environ. Res. Public Health* **2022**, *19*, 15575. https://doi.org/10.3390/ijerph192315575

Received: 11 November 2022
Accepted: 21 November 2022
Published: 24 November 2022

Publisher's Note: MDPI stays neutral with regard to jurisdictional claims in published maps and institutional affiliations.

Global primary energy consumption has been steadily increasing since the Industrial Revolution, and it is showing no sign of slowing down in the coming years. This trend is accompanied by increasing concentrations of pollutants in the Earth's biosystems and general concerns regarding the health and environmental impacts that will ensue. Air quality, water purity, atmospheric CO_2 concentration, etc., are some examples of environmental parameters that are degrading due to human activities. Pollutant abatement systems, biobased processes, novel environmental assessment tools, and pollutant monitoring equipment and methods can help us to reach environmental neutrality of human activities.

In this background, the Special Issue of the *International Journal of Environmental Research and Public Health* on the Environmental Impact Assessment by Green Processes offers insights into the sustainable conversion processes and environmental assessment using novel monitoring methods and tools useful for policymakers. A significant collection of contributions and studies is presented. Overall, 15 manuscripts were published in this Special Issue after evaluation by the Guest Editors and by many experts involved in the peer-review process.

The collected articles cover a wide range of macro-themes:

-   The environmentally friendly conversion processes;
-   Novel monitoring systems and technologies;
-   Environmental assessment tools and qualifying parameters that are useful for political decision-makers.

De Bari et al. [1] individuated wastewater treatment as a high-environmental-impact process in biorefinery systems, while Giuliano et al. [2] showed that direct DME synthesis can be used as an environmentally friendly strategy for carbon dioxide recycling. Two papers collected in the first group, "the environmentally friendly conversion processes", studied wastewater from two different points of view. Catizzone et al. [3] developed innovative processes for the purification of wastewater from biomass conversion thermochemical processes. Two commercial activated carbons and two residual biochars obtained through pyrolysis and gasification processes were assessed as potential adsorbents for the purification of wastewater produced in a syngas wet scrubber unit of a biomass gasification plant. Phenol solution was used as a model solution for the investigations. The results indicated the superiority of activated carbons due to the higher pore volume, including biomass-derived char. The phenol adsorption capacity increases from about 65 m g^{-1} for gasification biochar to about 270 mg g^{-1} for commercial activated carbon. Camilleri-Rumbau et al. [4] studied the experimental setup of membrane-based manure and digestate purification processes in a review. The effects of the feed characteristics, membrane operating conditions (pressure, cross-flow velocity, temperature), pH, flocculation–coagulation and membrane cleaning on fouling and membrane performance were presented. Membrane fouling represents one of the major drawbacks when using membrane technologies during farm effluent processing. However, by understanding the related mechanisms and fouling composition and establishing efficient membrane pretreatment and cleaning strategies, membrane technologies can lead to outstanding performance in terms of volume reduction and nutrient recovery.

Four papers were collected in the second group concerning novel monitoring technologies. Lotrecchiano et al. [5] aimed to analyze the air quality data of the monitoring network of the regional agency for environmental protection of the Campania region (Italy), integrated with an innovative monitoring station based on IoT technology to highlight criticalities in the levels of pollution. In-depth analyses showed that two events related to Saharan dust occurred, which led to an increase in the measured PM10 values. These Saharan phenomena were the cause of very high values for PM_{10} (until 260 µg m^{-3}). From air quality to the toxicity of heavy metals in an aquatic ecosystem, Adnan et al. [6] presented the acute and chronic toxicity based on luminescence inhibition assay using newly isolated *Photobacterium* sp. NAA-MIE as the indicator. Two different mathematical approaches, Toxicity Unit (TU) and Mixture Toxicity Index (MTI), were used to describe it. The toxicity result of this strain may be effective for taking the toxicity levels of a mixture of pollutants that occur together in ecosystems into theoretical consideration and in the same way contribute extrapolation studies under acute to chronic exposures for mixture metal toxicity in deriving safe limits and standards aimed at protecting organisms in the environment. Sui and Lv [7] applied the environmental Kuznets curve hypothesis and a decoupling analysis to examine the relationship between crop production and agricultural carbon emissions during 2000–2018, and it further provided a decomposition analysis of the changes in agricultural carbon emissions using the log mean Divisia index (LMDI) method. Overall, agricultural economic growth played a significant role in the increase in agricultural carbon emissions, while agricultural carbon emission intensity was the main factor behind the decline in agricultural carbon emissions in China, especially in the years 2003 and 2008, when turning points toward a downward trend in agricultural carbon emissions and strong decoupling states appeared. Different from strong decoupling, both agricultural carbon emission intensity and agricultural labor forces acted as positive driving factors of agricultural carbon emissions, while the agricultural structure and agricultural economic growth acted as inhibitory driving factors of agricultural carbon emissions. Celades et al.'s [8] sampling methodology and mathematical data treatment were developed, which enable us to determine not only total suspended particulates emitted at channeled sources but also the PM_{10}, $PM_{2.5}$, and PM_1 mass fractions and emission factors, using a seven-stage cascade impactor. The proposed methodology was applied to different stages of the ceramic process, including ambient temperature (milling, shaping, glazing) and medium–high-temperature (spray-drying, drying, firing, and frit melting) stages. In total, more than 100 measurements were performed (pilot scale and industrial scale), which leads to a measurement time of 1500 h.

Regarding the environmental assessment tools and qualifying parameters useful for political decision-makers, nine papers were submitted to the Special Issue. In particular, Lin et al. [9] introduced the concept of innovative human capital by developing a new index that measures human capital based on the number of patents for every one million R&D full-time equivalent staff. On the other hand, Wang et al. [10] studied the impact of the Emissions Trading System (ETS) on the Green Total Factor Productivity (GTFP) based on the spatial difference-in-differences approach. The paper resulted in the ETS significantly improving the GTFP of the pilot cities, producing a spatial spillover effect. The results were robust to the placebo test; further analysis showed that the policy effect was mainly driven by improving energy efficiency, promoting green innovation, and optimizing the industrial structure. The third environmental parameter was proposed by Liu et al. [11], using the carbon tax as the main policy tool to promote low-carbon economic development. This had a positive impact on reducing corporate carbon dioxide emissions, promoting the development of more energy-saving emission reduction technologies, and exploring more renewable resources. The research also showed that the levy of a carbon tax will adversely affect economic development, residents' income, and social welfare. However, introducing a suitable carbon tax recycling mechanism when formulating carbon tax policies can reduce the impact of the carbon tax on related industries. The same research group, Liu et al. [12], reconsidered the carbon tax recovery policy supplemented by technological progress in the clean power sector, showing that it can promote economic growth, improve social welfare, and reduce the intensity of carbon dioxide emissions. The introduction of clean power technology advances makes up for the negative impact of economic growth and social welfare losses, promoting the sustainable growth of the green economy. Advances in clean power technology promote the transformation of the power structure. The advancement of clean power technology drives the production of the clean power sector and replaces thermal power generation. Mehmood Ali Shah et al. [13], differently to previous studies, adopted a mediation model and unfolded not only the role of green human resource practices in the psychological climate and green organizational culture but also clarified the mediating role of the green psychological climate and green organizational culture in sustainable environmental efficiency. The work recommended that a green psychological climate and green organizational culture can be strengthened to achieve more environmental benefits of green human resource management practices in organizations. The results urged researchers to reconsider green human resource management policies for more clarification on the present needs of organizations for developing a green environment. Finally, four works focused on manufacturing activities' environmental compliance. Liu et al. [14] used the random-effects Tobit model and the double hurdle model to empirically conduct a robustness test. The robust conclusion was that environmental compliance had a significant U-shaped relationship with enterprise innovation, which means that environmental compliance will inhibit enterprise innovation on the left of the inflection point of environmental compliance, while environmental compliance on the right of the inflection point will promote enterprise innovation. How CEO tournament incentives induce top executives to invest more in green innovation was studied by Ullah et al. [15]. The main results supported tournament theory, which proposed that better incentives induced top executives' efforts to win the tournament incentives, and such efforts were subject to fiercer competition among employees, which improved firms' social and financial performance. Sunday Adebayo et al. [16] re-assessed the environmental Kuznets curve, taking into consideration the role of hydroelectricity consumption and urbanization. As a result, regulations that decrease the usage of hydroelectricity will hurt economic growth. Any hydroelectricity deficit will also hinder economic progress. Furthermore, a decrease in output will hurt the hydroelectricity demand. Any shock to one of the series of interests will be felt in the others, and the feedback flow will keep the chain going. As a result, expansionary hydroelectricity plans would bring benefits. Hatice Sezgin et al. [17] focused on human development in reducing CO_2 emissions in the G7 and/or BRICS (Brazil–Russia–India–China–South Africa) economies. According to the G8 countries' joint declarations, efforts were consistently taken to significantly reduce greenhouse gas emissions and to support sustainable energy and human development. However, as most scholars indicate, there was still a gap between political decisions, on the one hand, and the position of different researchers and analysts regarding the need to further and intensify environmental stringency policies, on the other hand. Our findings indicated that environmental stringency policies and human development were important for environmental sustainability. However, environmental stringency policies can negatively affect economic growth and employment by raising costs at the beginning. However, the countries offset the negative economic effects of environmental stringency policies through innovation, considering the Porter hypothesis and empirical findings over time. On the other hand, improvements in human development were also effective for environmental sustainability.

Author Contributions: Conceptualization, A.G., M.E., H.S. and P.A.; writing—original draft preparation, A.G.; writing—review and editing, P.A.; supervision, P.A. All authors have read and agreed to the published version of the manuscript.

Funding: This research received no external funding.

Acknowledgments: The Guest Editors wish to thank all the authors who contributed to the success of this Special Issue.

Conflicts of Interest: The authors declare no conflict of interest.

References

1. De Bari, I.; Giuliano, A.; Petrone, M.T.; Stoppiello, G.; Fatta, V.; Giardi, C.; Razza, F.; Novelli, A. From Cardoon Lignocellulosic Biomass to Bio-1,4 Butanediol: An Integrated Biorefinery Model. *Processes* **2020**, *8*, 1585. [CrossRef]
2. Giuliano, A.; Catizzone, E.; Freda, C. Process Simulation and Environmental Aspects of Dimethyl Ether Production from Digestate-Derived Syngas. *Int. J. Environ. Res. Public Health* **2021**, *18*, 807. [CrossRef] [PubMed]
3. Catizzone, E.; Sposato, C.; Romanelli, A.; Barisano, D.; Cornacchia, G.; Marsico, L.; Cozza, D.; Migliori, M. Purification of Wastewater from Biomass-Derived Syngas Scrubber Using Biochar and Activated Carbons. *Int. J. Environ. Res. Public Health* **2021**, *18*, 4247. [CrossRef]
4. Camilleri-Rumbau, M.S.; Briceño, K.; Fjerbæk Søtoft, L.; Christensen, K.V.; Roda-Serrat, M.C.; Errico, M.; Norddahl, B. Treatment of Manure and Digestate Liquid Fractions Using Membranes: Opportunities and Challenges. *Int. J. Environ. Res. Public Health* **2021**, *18*, 3107. [CrossRef] [PubMed]
5. Lotrecchiano, N.; Capozzi, V.; Sofia, D. An Innovative Approach to Determining the Contribution of Saharan Dust to Pollution. *Int. J. Environ. Res. Public Health* **2021**, *18*, 6100. [CrossRef]
6. Adnan, N.A.; Halmi, M.I.E.; Abd Gani, S.S.; Zaidan, U.H.; Abd Shukor, M.Y. Comparison of Joint Effect of Acute and Chronic Toxicity for Combined Assessment of Heavy Metals on *Photobacterium* Sp.NAA-MIE. *Int. J. Environ. Res. Public Health* **2021**, *18*, 6644. [CrossRef] [PubMed]
7. Sui, J.; Lv, W. Crop Production and Agricultural Carbon Emissions: Relationship Diagnosis and Decomposition Analysis. *Int. J. Environ. Res. Public Health* **2021**, *18*, 8219. [CrossRef] [PubMed]
8. Celades, I.; Sanfelix, V.; López-Lilao, A.; Gomar, S.; Escrig, A.; Monfort, E.; Querol, X. Channeled PM_{10}, $PM_{2.5}$ and PM_1 Emission Factors Associated with the Ceramic Process and Abatement Technologies. *Int. J. Environ. Res. Public Health* **2022**, *19*, 9652. [CrossRef] [PubMed]
9. Lin, X.; Zhao, Y.; Ahmad, M.; Ahmed, Z.; Rjoub, H.; Adebayo, T.S. Linking Innovative Human Capital, Economic Growth, and CO_2 Emissions: An Empirical Study Based on Chinese Provincial Panel Data. *Int. J. Environ. Res. Public Health* **2021**, *18*, 8503. [CrossRef] [PubMed]
10. Wang, S.; Chen, G.; Han, X. An Analysis of the Impact of the Emissions Trading System on the Green Total Factor Productivity Based on the Spatial Difference-in-Differences Approach: The Case of China. *Int. J. Environ. Res. Public Health* **2021**, *18*, 9040. [CrossRef] [PubMed]
11. Liu, W.; Li, Y.; Liu, T.; Liu, M.; Wei, H. How to Promote Low-Carbon Economic Development? A Comprehensive Assessment of Carbon Tax Policy in China. *Int. J. Environ. Res. Public Health* **2021**, *18*, 10699. [CrossRef] [PubMed]
12. Liu, W.; Liu, M.; Liu, T.; Li, Y.; Hao, Y. Does a Recycling Carbon Tax with Technological Progress in Clean Electricity Drive the Green Economy? *Int. J. Environ. Res. Public Health* **2022**, *19*, 1708. [CrossRef] [PubMed]
13. Shah, S.M.A.; Jiang, Y.; Wu, H.; Ahmed, Z.; Ullah, I.; Adebayo, T.S. Linking Green Human Resource Practices and Environmental Economics Performance: The Role of Green Economic Organizational Culture and Green Psychological Climate. *Int. J. Environ. Res. Public Health* **2021**, *18*, 10953. [CrossRef] [PubMed]
14. Liu, M.; Liu, Y.; Zhao, Y. Environmental Compliance and Enterprise Innovation: Empirical Evidence from Chinese Manufacturing Enterprises. *Int. J. Environ. Res. Public Health* **2021**, *18*, 1924. [CrossRef] [PubMed]
15. Ullah, S.; Khan, F.U.; Cismaş, L.-M.; Usman, M.; Miculescu, A. Do Tournament Incentives Matter for CEOs to Be Environmentally Responsible? Evidence from Chinese Listed Companies. *Int. J. Environ. Res. Public Health* **2022**, *19*, 470. [CrossRef] [PubMed]
16. Adebayo, T.S.; Agboola, M.O.; Rjoub, H.; Adeshola, I.; Agyekum, E.B.; Kumar, N.M. Linking Economic Growth, Urbanization, and Environmental Degradation in China: What Is the Role of Hydroelectricity Consumption? *Int. J. Environ. Res. Public Health* **2021**, *18*, 6975. [CrossRef] [PubMed]
17. Sezgin, F.H.; Bayar, Y.; Herta, L.; Gavriletea, M.D. Do Environmental Stringency Policies and Human Development Reduce CO_2 Emissions? Evidence from G7 and BRICS Economies. *Int. J. Environ. Res. Public Health* **2021**, *18*, 6727. [CrossRef] [PubMed]

International Journal of
*Environmental Research
and Public Health*

Article

Channeled PM$_{10}$, PM$_{2.5}$ and PM$_1$ Emission Factors Associated with the Ceramic Process and Abatement Technologies

Irina Celades [1,*], Vicenta Sanfelix [1], Ana López-Lilao [1], Salvador Gomar [1], Alberto Escrig [1], Eliseo Monfort [1] and Xavier Querol [2]

1 Institute of Ceramic Technology (ITC-AICE), University Jaume I, Campus Universitario Riu Sec, Av. Vicent Sos Baynat s/n, 12006 Castellón, Spain
2 Institute of Environmental Assessment and Water Research (IDAEA-CSIC), C/Jordi Girona 18, 08034 Barcelona, Spain
* Correspondence: irina.celades@itc.uji.es; Tel.: +34-964-34-24-24

Abstract: A sampling methodology and a mathematical data treatment were developed that enable to determine not only total suspended particulates (TSP) emitted at channeled sources but also the PM$_{10}$, PM$_{2.5}$, and PM$_1$ mass fractions (w$_{10}$, w$_{2.5}$, and w$_1$) and emission factors (E.F.), using a seven-stage cascade impactor. Moreover, a chemical analysis was performed to identify the elements present in these emissions. The proposed methodology was applied to different stages of the ceramic process, including ambient temperature (milling, shaping, glazing) and medium–high-temperature (spray-drying, drying, firing, and frit melting) stages. In total, more than 100 measurements were performed (pilot scale and industrial scale), which leads to a measurement time of 1500 h. Related to the mass fractions, in general, the mean values of w$_{10}$ after the fabric filters operated at high performance are high and with little dispersion (75–85%), and it is also observed that they are practically independent of the stage considered, i.e., they are not significantly dependent on the initial PSD of the stream to be treated. In the case of the fine fraction w$_{2.5}$, the behavior is more complex (w$_{2.5}$: 30–60%), probably because the only variable is not the cleaning system, but also the nature of the processed material. Regarding abatement measures, the use of high-efficiency cleaning systems considerably reduces the emission factors obtained for fractions PM$_{10}$, PM$_{2.5}$, and PM$_1$. In reference to chemical analysis, the presence of ZrO$_2$ and Ni in the spray-drying and pressing stages, the significant concentration of ZrO$_2$ in the glazing stage, the presence of Pb, As, and Zn in the firing stage, and the presence of Zn, Pb, Cd, and As compounds in the frits manufacturing should all be highlighted. Nevertheless, it should be pointed out that the use of some compounds, such as cadmium and lead, has been very limited in the last years and, therefore, presumably, the presence of these elements in the emissions should have been also reduced in the same way.

Keywords: channeled emission; emission factor; particulate matter; abatement technology; ceramic industry; PM$_{10}$, PM$_{2.5}$, and PM$_1$

Citation: Celades, I.; Sanfelix, V.; López-Lilao, A.; Gomar, S.; Escrig, A.; Monfort, E.; Querol, X. Channeled PM$_{10}$, PM$_{2.5}$ and PM$_1$ Emission Factors Associated with the Ceramic Process and Abatement Technologies. *Int. J. Environ. Res. Public Health* **2022**, *19*, 9652. https://doi.org/10.3390/ijerph19159652

Academic Editors: Yinchang Feng and Paul B. Tchounwou

Received: 19 April 2022
Accepted: 29 July 2022
Published: 5 August 2022

Publisher's Note: MDPI stays neutral with regard to jurisdictional claims in published maps and institutional affiliations.

1. Introduction

Air pollution has been recognized as the single biggest environmental threat to human health, based on its notable contribution to disease burden [1]. In this sense, European air quality regulations related to particulate matter (hereinafter PM) have been established and significantly modified in the last decades. In the 1990s, only total suspended particles (TSP) were regulated. However, since Directive 1999/30/EC entered into force, limit values for PM$_{10}$ (particulate matter which passes through a size selective inlet with a 50% efficiency cut-off at 10 µm aerodynamic diameter) have been established, specifically, an annual limit value of 40 µg PM$_{10}$ µg/m^3 with a maximum of 35 days of exceedances of the daily limit value of 50 µg/m^3. This PM$_{10}$ can be divided into two different categories: coarse fraction, which is mainly deposited in the tracheobronchial region (2.5–10 µm), and fine fraction (<2.5 µm, including ultrafine particles (<0.1 µm)), which can penetrate deep into the lungs and translocate to the other parts of the body [2]. In this regard, in the last years, epidemiologic studies [3–6] have evidenced the negative effect of fine fraction on health. For this reason, in 2008, Directive 2008/50/EC added $PM_{2.5}$ fraction (particulate matter which passes through a size-selective inlet with a 50% efficiency cut-off at 2.5 µm aerodynamic diameter) and set an annual target value of 25 µg/m^3.

In the same line, the recent publication in 2021 of the document WHO Global Air Quality Guidance [1] promotes the necessity to reduce the limit values because of their impact on health. In this regard, this document defines quantitative health-based recommendations for air quality, expressed as either long- or short-term concentrations of different key air pollutants. Specifically, Air Quality Annual Guidance levels of 15 µg/m^3 and 5 µg/m^3 are recommended for PM_{10} and $PM_{2.5}$, respectively. These guidelines are not legally binding standards; however, they provide countries with an evidence-informed tool which they can use to inform legislation and policy.

The main contributors of PM are traffic, natural phenomena, combustion in agriculture, domestic fuel burning, and industry [7]. In fact, the emissions into the air are one of the main environmental impacts from industrial activities. With regard to industrial emissions, as with air quality, regulations are becoming increasingly restrictive in terms of permitted concentrations and have broadened the parameters of interest [8]. This behavior is driven following the enforcement of Industrial Emissions Directive (IED, Directive 2010/75/UE) and Integrated Pollution Prevention and Control (IPPC, Directive 1996/61/EC), where Emission Limit Values associated with Best Available Techniques (BAT-AELs) have been established according to BAT Reference Documents (BREFs). As an example, for the ceramics industry (CER-BREF [9]), a generic BAT-AELs for dust is 30–50 mg/Nm^3. Nevertheless, it should be highlighted that in the BREFs updated after 2012 (cement, wood, ferrous metals, non-ferrous metals, large combustion plants, glass, and waste incineration [10–16]), the limits established for dust are becoming much more restrictive (1–20 mg/Nm^3).

In fact, in the discussion and approval of the recent BREFs, PM_{10} and $PM_{2.5}$ are included as parameters to be monitored, but when deriving BAT-AELs, the limit is only established for TSP including PM_{10} and $PM_{2.5}$ [17]. The PM_{10} parameter has appeared for the first time for emissions, additionally to the usual TSP, in tools derived from the IED, knows as Pollutant Release and Transfer Register in Europe (E-PRTR) and Spain (PRTR-Spain [18]).

On the other hand, $PM_{2.5}$ determination can be deemed essential not only because of its potential impact on health, but also because it allows for the detection of anthropogenic particulate pollutants, excluding crustal particulate interference [19]. However, this parameter alone does not seem to be adequate to quantify the impact of some industries with significant primary particulate emissions (such as those of ceramics [19]). For this reason, it is necessary to obtain accurate information on both fractions (PM_{10} and $PM_{2.5}$) in order to identify the contribution of different particulate matter sources and, therefore, to establish specific measures that allow for the improvement of air quality [20].

In Spain, according to the values declared in PRTR-Spain, more than 8381.4 tons of industrial primary PM_{10} were emitted into the atmosphere in 2020, of which more than 10% corresponds to ceramic industries [18]. In fact, air quality studies performed in ceramic areas have evidenced the influence of ceramic industry on air quality not only by the presence of high PM concentration, but also for the significant levels of different heavy metals [21–24]. The contribution of ceramic and related industries to PM_{10} and heavy metals which may be considered as tracers of the ceramic industry [24], information which is available in PRTR, is shown in Figure 1. In Table 1, the main characteristics of ceramic process emissions are shown.

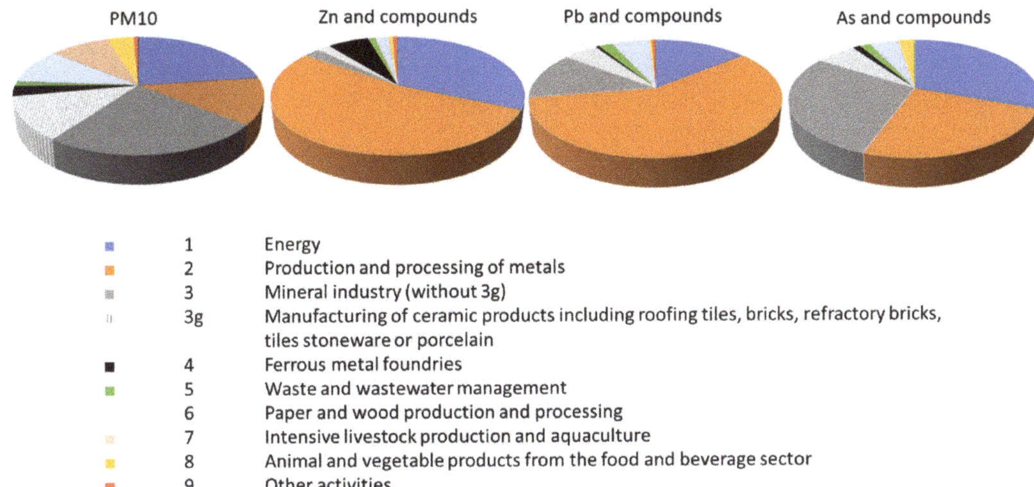

Figure 1. Contribution of ceramic industries (3g) to some air quality indicators in Spain, 2020 [18].

Table 1. Ceramic process emissions.

Process	Stage	Emission	Flow	Type	Pollutant
Tiles	Storage and handling of raw materials	Variable	Continuous	Diffuse	PM
	Milling (dry) Milling (wet)	Variable	Continuous/ Discontinuous	Ambient channeled	PM
	Spray-dried	Constant	Continuous	[1] Hot channeled	PM and gases
	Pressing	Variable	Continuous	Ambient channeled	PM
	Dry	Constant	Continuous	[1] Hot channeled	PM and gases
	Glaze preparation	Variable	Discontinuous	Ambient channeled	PM
	Glazing	Variable	Continuous	Ambient channeled	PM
	Firing	Constant	Continuous	[1] Hot channeled	PM and gases
Frits	Milling	Variable	Continuous/ Discontinuous	Ambient channeled	PM
	Frits melting	Constant	Continuous/ Discontinuous	[1] Hot channeled	PM and gases

[1] Hot channeled refers to those medium–high-temperature processes.

As it can be drawn for Table 1, the ceramic tile production process may generate both channeled and diffuse emissions:

- Diffuse emissions are those which pass to the atmosphere without being channeled. In the ceramic industry, they are mainly related to bulk material storage, handling, and transport. They can also occur in some operations such as milling, grinding, and trucking.
- Channeled emissions are those which pass to the atmosphere through a pipe. They can be divided into medium–high- and ambient-temperature emissions. The main measures proposed in the CER-BREF [9] to reduce this type of emissions are (1) primary

measures: related to reducing the use of raw materials that could contain hazardous components; and (2) secondary measures: different abatement technologies are available as wet scrubber systems, Venturi type, fabric filters, and electrostatic precipitators.

Although the impact of the ceramic industry on air quality is well known [21–24], PM emission factors related with channeled emissions from ceramic process are not available in the EMEP/EEA air pollutant emission inventory guidebook 2019 [25]. There are some reference documents [26] which are based on US-EPA (Environmental Protection Agency) documents and previous studies [27–46] in which we can check on emission factors for spray-drying, drying, glazing, and firing resulting in a global emission factor between 2.4–11.1 kg TSP/ton depending on the implemented mitigation measures. Nevertheless, these emission factors are only available for TSP. In order to extend the information in public inventories such as E-PRTR and harmonize the key control parameters between air quality and industrial emissions (TSP, PM_{10}, and $PM_{2.5}$), the rationale of the present study was, firstly, to develop a sampling methodology based on the previous study performed by Erlich et al. [47,48] and other previous studies performed at industrial scale [49–55]; secondly, to determine PM_{10}, $PM_{2.5}$, and PM_1 emission factors associated with different stages that take place during the ceramic tiles manufacturing process.

To this aim, this study was performed in a wide variety of facilities located in the ceramic production area of Castelló (Figure 2). This area extends from the coastal flat (mainly occupied by residential areas and orange tree plantations) to the mountain chain of La Cruz. This area is the largest ceramic-tile-producing zone in the EU, accounting for a turnover in 2020 of approximately EUR 3842M [56] and EUR 1200M [57] for ceramic tiles and frit and pigments, respectively. As a consequence of the high concentration of ceramic and related industries in a small area, an Air Quality Plan [58] was elaborated in 2008 to implement high-efficiency PM emissions abatement technologies in ceramic facilities and to replace impurity-bearing raw materials.

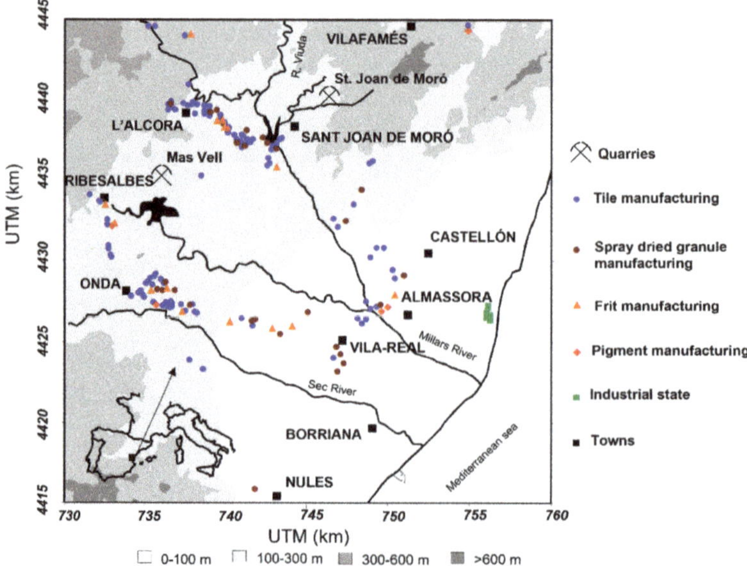

Figure 2. Map of the Castelló ceramic cluster [59].

2. Methodology

The method to perform the PM monitoring campaigns is based on the inertial separation of the PM target fractions and its subsequent gravimetric determination. The applied methodology was performed in accordance with the reference standards [60,61] and the specific previous studies focused on ceramic emissions [62,63], and its application allowed us to obtain the chemical characterization of the ceramic PM channeled emissions.

2.1. Physical Characterization of the PM Emitted

The physical characterization allowed for the determination of particle size distribution (PSD), mass fractions (w_x: w_{10}, $w_{2.5}$, and w_1) and specific emission factors (EF_{10}, $EF_{2.5}$, and EF_1) for each process stage. Measurement campaigns were carried out at several ceramic companies which manufacture wall and floor tiles and frits (Table 2). All measurements were carried out under real operating conditions. The sampling period was chosen in such a way that sufficient mass is collected to permit weighing with the required accuracy without overloading the stages. Since the total PM concentrations were usually low at the tested industrial plants, very long sampling times must be provided for the reasons mentioned.

Table 2. Industrial scenarios.

Industrial Process	Stage Process
Ceramic tile	Milling Spray-drying Pressing Glaze preparation and glazing Drying Firing
Ceramic frit	Frit melting

It should be highlighted that, in some stages, measures were performed before and after the cleaning system (cyclone, wet scrubber, fabric filter, and electrostatic precipitator).

With this aim, experimental measures were taken at industrial scale using a cascade impactor (Anderson Impactor type Mark III). This impactor is designed to meet the specifications reported by VDI 2066 [64,65] and to fractionate suspended particles into different sizes categories according to their inertia.

The cut-size associated with each impactor stage depends on flow and temperature of the airstream. To calculate the PSD and mass fractions of interest (w_{10}, $w_{2.5}$, and w_1) easily and accurately, the results need to be adjusted to a distribution. This distribution could be the log-normal one, which is the most usual to treat PSD data because, from the mathematical point of view, it ensures that all obtained values are positive and, therefore, they have a physical meaning. From the literature review, different types of distributions has been identified, which yield very good results in this field, such as the Rosin–Rammler–Sperling–Bennet distribution (RRSB) [47,66]. For this reason, the results obtained with log-normal distribution in the present study were compared with those obtained by RRSB distribution.

The procedure applied to calculate the cumulative log-normal and RRSB distribution is described in Figure 3.

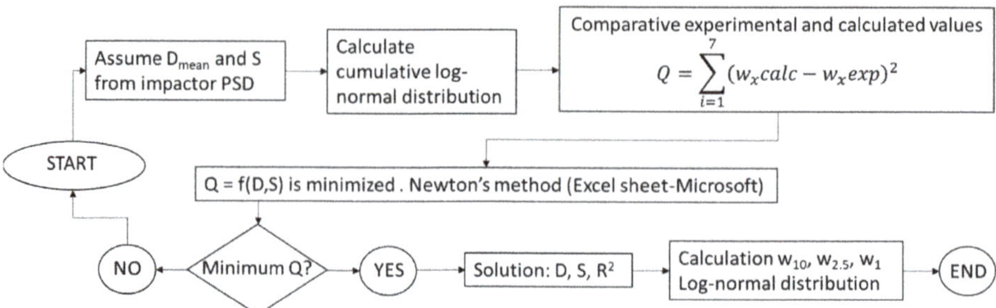

Figure 3. Flow chart of the mathematical treatment used to calculate PSD and w_x. D: geometric diameter (µm); S: geometric deviation; R: correlation index.

The detailed assessment of the proposed methodology was carried out in previous studies [62,63]. The evaluation criteria used was the compatibility index (CI) (EN ISO/IEC 17043 [67]), which allows us to know if two results associated with their respective uncertainties are comparable. The results compared were the total concentration measurement determined by using a cascade impactor and one obtained with the standard method (EN 13284-1 [68]). The CI was satisfactory in most of the cases studied, so it was considered that there was a good correlation between the compared concentrations. Therefore, both distributions can be considered appropriate for the objectives of the present study.

Despite the favorable results obtained in this study, and following the recommendations of the standards (EN ISO 23210 [69]), the use of a cascade impactor for the quantification of the total PM concentration is not recommended in those cases where the objective of the measurement is to ensure compliance with the established regulations.

2.2. Chemical Characterization

The objective of the chemical characterization was to obtain the chemical profile of the mass fractions w_{10} and $w_{2.5}$ from the emissions associated with the ceramic process.

The selection of the analysis technique for the determination of the mass concentration of specific elements in the particulate matter emissions of the studied industrial processes was dependent on the amount of sample required to perform the analysis. In fact, it was the main drawback of samplings carried out at industrial scale.

This situation is more critical because of the extensive implementation of Best Available Techniques in the ceramic plants, which significantly reduces the emissions and requires very long sampling times, which makes it difficult to comply with the technical criteria established in the sampling standards. In these cases, the measurements were carried on a pilot scale (emission simulator; Figure 4), whose use was evaluated in previous studies [62,63].

The PM emissions generator allows for the regulation of the flow rate and the amount of solid material fed into an airstream with an air velocity similar to industrial installations (10–15 m/s) which transfer the dust to the sampling area, obtaining a range of PM concentrations. The feasibility of this system was deemed essential to obtain enough amount of sample for chemical analysis, upholding the technical requirements of the sampling standards. In this system, the powdered material used was provided by ceramic industries from the waste captured by the fabric filters installed to abate PM emissions generated in each stage of the process (Table 2).

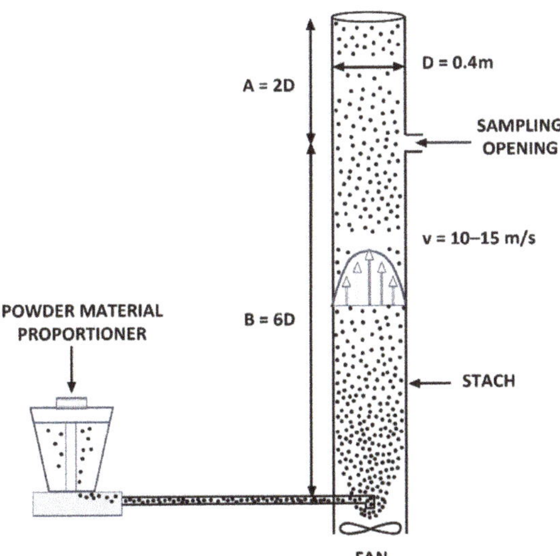

Figure 4. PM emissions generator diagram.

The sampling by means of the PM emissions generator is based on the assumption that the content of PST and fractions w_{10}, $w_{2.5}$, and w_1 of the material collected by the different cleaning systems (fabric filter and electrostatic precipitator) is similar to the composition of the w_{10}, $w_{2.5}$, and w_1 of the emissions generated after the abatement system. This assumption is based on the fact that the temperature of the gases as they pass through the cleaning systems is of the same order as the emission temperature, and therefore, a priori, it is not expected that condensation processes, which could modify the chemical composition of the issued PM, take place.

The device used for collecting the sample was a Tecora cyclone, designed to meet the specifications reported by USEPA in the Method 201A [70] and to measure PM_{10} and $PM_{2.5}$ in stack emission [62,63]. This device allowed us to obtain the required amount of sample to subsequently perform the chemical analysis of the PM_{10} and $PM_{2.5}$ captured. The cyclone is required since the cascade impactor has different stages and, therefore, it would be very difficult to obtain enough mass of sample for the subsequent analysis using this device.

The sampling of particulate matter was carried out by means of quartz glass filters (QF20 Schleicher and Schuell). Once the PM concentrations were obtained by weighting the filters using standard procedures, one-half of each of them was digested and analyzed following the method by Querol et al. [19,71]. This method is based on an acid attack using low-pressure Teflon bombs. The solution obtained was then centrifuged and analyzed by (a) inductively coupled plasma–atomic emission spectrometry (ICP-AES) for major elements, and (b) inductively coupled plasma–mass spectrometry (ICP-MS). A quarter of each filter was used to analyze boron by the Azomethine-H method. Finally, the last quarter was sometimes used for the morphological characterization.

2.3. Summary of Sampling Campaigns

In total, more than 100 measurements were performed (pilot scale and industrial scale), which led to a measurement time of 1500 h (Table 3).

Table 3. Sampling campaigns description.

	Physical Characterization	**Chemical Characterization**
Scenario	Industrial	Pilot scale (PM emission generator)
Device	Cascade impactor	$PM_{10}/PM_{2.5}$ cyclone
Sampling campaigns	Number of samplings: 47 Sampling hours: 1150	Number of samplings: 95 Sampling hours: 470

3. Results

3.1. Physical Characterization of the PM Emissions

3.1.1. Determination of PSD and w_x

The determination of w_x was obtained from the PSD, applying the mathematical treatment described in Figure 3. In this regard, two mathematical methods were evaluated: log-normal and RRSB (Section 2.1). It can be observed that both methods are comparable in all cases (Figure 5), so the log-normal model was applied, since it is the commonly used one in the surveyed literature.

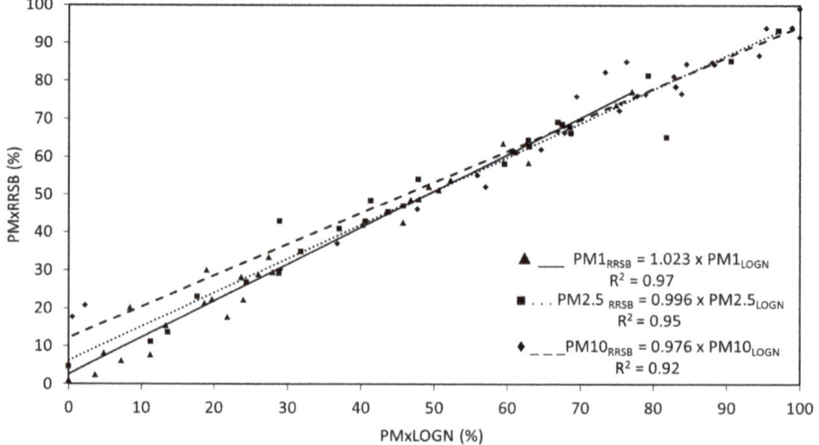

Figure 5. Granulometric fractions w_{10}, $w_{2.5}$, and w_1 calculated by the log-normal method and RRSB method.

The average fractions w_{10}, $w_{2.5}$, and w_1 obtained from the PSD are shown in Tables 4 and 5. More detailed information about the average PSD is shown in the Supplementary Material (Figures S1–S3 and Tables S1 and S2). The average process stage PSD was calculated from the sum of the mass of the particles (within the same size range) of each of the individual samplings.

Table 4. w_{10}, $w_{2.5}$, and w_1 obtained during milling, pressing, and glazing (ambient emissions).

Process Stage	Cleaning System	Number of Samplings (n)	T_{gases} (°C)	Average Values			
				C_{TI} (mg/Nm^3)	w_{10} (%)	$w_{2.5}$ (%)	w_1 (%)
Milling	Fabric filter	4	15–30	<5	74.8	53.4	38.0
Pressing	None	1	15–30	109 ± 33	21.0	2.1	0.23
	Fabric filter	1		<5	75.3	28.9	5.3
Glaze preparation and glazing	None	2	18–40	132 ± 71	51.8	20.1	7.5
	Fabric filter	3		<5	74.5	41.7	22.9

Table 5. w_{10}, $w_{2.5}$, and w_1 obtained during drying, spray-drying, firing, and frits melting (medium- and high-temperature emissions).

Process Stage	Cleaning System	Number of Samplings (n)	T_{gases} (°C)	Average Values			
				C_{TI} (mg/Nm^3)	w_{10} (%)	$w_{2.5}$ (%)	w_1 (%)
Spray-drying	Cyclone	1	75–120	>1000	73.4	41.3	21.8
	Fabric filter	2		<5	75.4	50.0	33.2
	Cyclone + wet scrubber	2	60–65	75 ± 34	97.7	80.6	39.2
Drying	None	1	110–120	<5	84.5	66.9	52.2
Firing	None	2	160–210	11 ± 5	99.4	93.9	75.9
	Fabric filter + reagent	2	140–160	<5	81.6	59.2	41.7
Frits melting	None	3	110–260	415 ± 318	74.9	59.1	43.9
	Fabric filter	1	110–210	<5	83.1	61.1	43.5
	Electrostatic precipitator	1		<5	88.0	67.5	49.2

In order to compare the results, graphs representing the cumulative probability (cumulative mass expressed in %) versus particle diameter were produced (Figures S1–S3). This type of graph is easy to interpret and yields a straight line whenever the characterized particles come from a single source or when they have similar sizes, even if they come from several sources [72–74].

The average w_x (expressed in %), as a function of the particle size, was grouped attending to the following criteria, to be easily understood:

- Emissions generated in ambient-temperature processes (<50 °C): milling, pressing, and glaze preparation and glazing (Table 4).
- Emissions generated in medium–high-temperature processes (60 °C to >150 °C): drying, spray-drying, firing, and frits melting (Table 5).

It is remarkable the long sampling times (>40 h) required to determine the w_x and PSD in those stack emissions with low particle concentrations (dryers or emissions after treatment).

3.1.2. Determination of EF

This section sets out the specific emission factors obtained for the PM$_{10}$, PM$_{2.5}$, and PM$_1$ for the different stages of the ceramic process, expressed as mgPMx/m^2 or mgPMx/kg depending on the characteristics of the processed product. For this purpose, a specific weight of 21 kg of spray-dried granulate/m^2 was considered and the specific flow rates were obtained from previous studies in the ceramic industry [58,59].

The sources studied were divided into two groups, based on process temperature, as was performed in Section 3.1.2, since the literature reports and results obtained from the present study evidenced that this characteristic notably influences the size and composition of the PM emitted from these sources. Emission factors are shown in Tables 6 and 7.

Table 6. Emission factors for milling, pressing, and glaze preparation and glazing (ambient-temperature processes).

	Process Stage	Cleaning System	^{1} Q (Nm^3/kg)	Units	Average Values		
					EF$_{PM10}$	EF$_{PM2.5}$	EF$_{PM1}$
Samplings at ambient temperature sources (T < 40 °C)	Milling	Fabric filter	4	mg/kg	2	2	1
	Pressing	None Fabric filter	4	mg/m^2	1923 8	192 3	21 1
	Glaze preparation and glazing	None Fabric filter	4	mg/m^2	7183 48	2212 31	724 29

^{1} Specific flow rate obtained from Monfort et al., 2013 [59] and Conselleria de Medi Ambient, Aigua, Urbanisme i Habitatge, 2008 [58].

Table 7. Emission factors for spray-drying, drying, firing, and frits melting (medium- and high-temperature processes).

	Process Stage	Cleaning System	^{1} Q (Nm^3/kg)	Units	Average Values		
					EF$_{PM10}$	EF$_{PM2.5}$	EF$_{PM1}$
Samplings at medium–high-temperature ceramic sources	Spray-drying	None	4	mg/kg	4147	2734	1137
		Fabric filter			3	2	1
		Wet scrubber			297	244	99
	Drying	None	3	mg/m^2	155	123	96
	Firing	None	4	mg/m^3	901	857	690
		Fabric filter + solid reagent			11	8	5
	Frit melting	None	4.4	mg/kg frit	1376	1045	643
		Fabric filter			1	1	1
		Electrostatic precipitator			19	15	11

^{1} Specific flow rate obtained from Monfort et al., 2013 [59] and Conselleria de Medi Ambient, Aigua, Urbanisme i Habitatge, 2008 [58].

3.2. Chemical Composition of PM Emissions

In this section, the average chemical profiles are shown (Table 8), including major and trace elements of PM emission from ceramic process stages. This average profile was obtained from at least three individual valid samplings for each process stage. The major elements (expressed as oxides) are those whose percentage in composition is higher than 1%, and the trace elements are those where the concentration is higher than 100 mg/kg.

Table 8. PM emission composition (major and trace elements) associated with different ceramic process stages.

Process Stage	Major Elements	Trace Elements
Spray-drying Drying Pressing	ZrO_2, ZnO, BaO, PbO	Hf, Cr
Glazing	B_2O_3, BaO, PbO, ZnO, ZrO_2	Cu, Cr, Cd, Sn, Hf
Firing	Na_2O, K_2O, ZnO, PbO	S, Tl, As, Cr, Rb, Cs, Cu
Frits melting	SiO_2, Al_2O_3, CaO, MgO, K_2O, Na_2O, BaO, ZnO	S, Sr, Cs

In stages such as spray-drying, pressing, and drying, where the processed product is quite similar (in terms of chemical composition) and has not suffered any significant physical–chemical transformations (low–medium process temperature), it is considered that the chemical profile is common for the different stages. More detailed information about the identified major and trace elements can be obtained from the Supplementary Material (Figures S4–S9).

From these results (Table 8), and taking into account previous air quality studies performed in the ceramic area of Castellón, the main tracers and other legislative elements (Ni and Cd) were selected to study the segregation of these elements in fractions PM_{10} (Figure 6) and $PM_{2.5}$ (Figure 7), associated with the different stages of ceramic process considered.

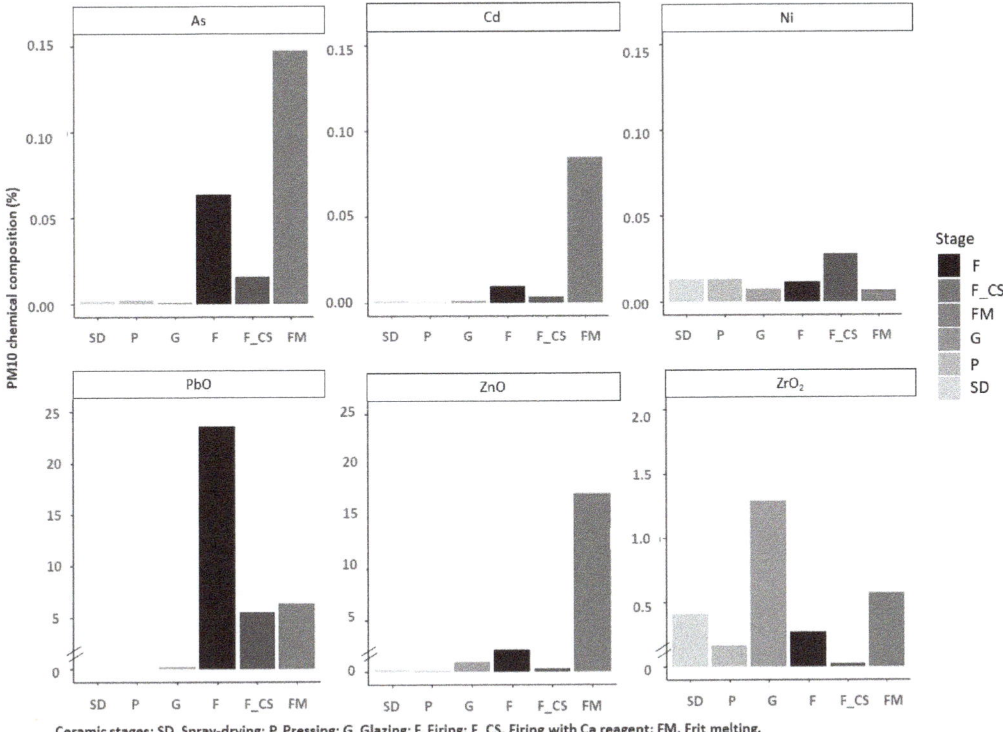

Figure 6. PM_{10} composition in the different stages of the ceramic process.

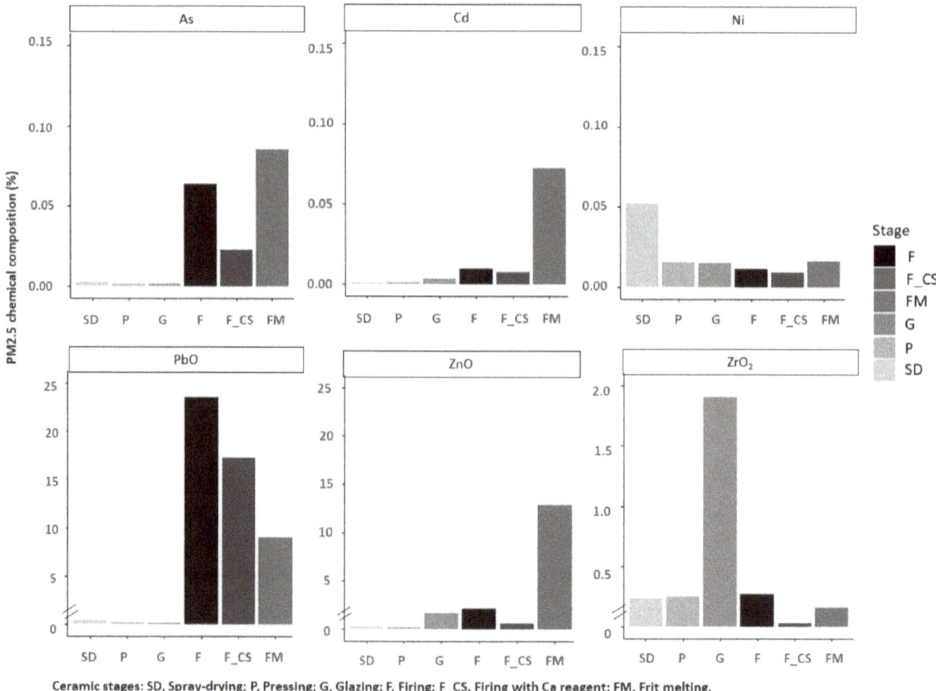

Figure 7. $PM_{2.5}$ composition in the different stages of ceramic process.

4. Discussion

The Discussion section follows the same structure as the Results section.

4.1. Physical Characterization of the Emitted PM

4.1.1. Determination of PSD and w_x

Regarding the comparison between the log-normal and RRSB distributions, it can be seen that both methods are comparable since in all cases (w_{10}, $w_{2.5}$, and w_1) a trend line with a slope coefficient close to 1 and a regression coefficient higher than 0.90 was obtained. In addition, none of the methods have a clear tendency to either overestimate or underestimate the results. It can therefore be concluded, considering this evaluation, that both methods can be indistinctly used, so the log-normal model was applied, since it is the most commonly used one in the surveyed literature.

In reference to the PSD obtained in the present study, the fit parameters to a log-normal distribution, obtained by applying the calculation procedure described in Figure 3 for the calculation of w_x, showed good agreement ($R^2 > 0.90$).

From the results obtained without cleaning system, wide PSD can be observed. In the case of ambient- and medium-temperature process stages, it can be due to the presence of coarse particles, in the form of granulates and agglomerates, and fine particles. The fine particles are associated with individual particles of the processed material and particles generated by the breakage of agglomerates/aggregates (Figure 8). In high-temperature processes, wide PSD is also observed as a result of different origins for PM, coarser material generated by carryover of batch particles and finer PM from volatilization–condensation processes (Figure 9).

Figure 8. SEM photograph of the PM emitted by the cyclones after spray-drying.

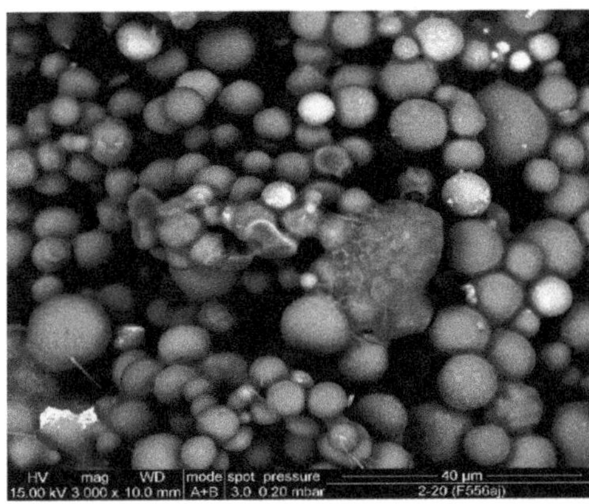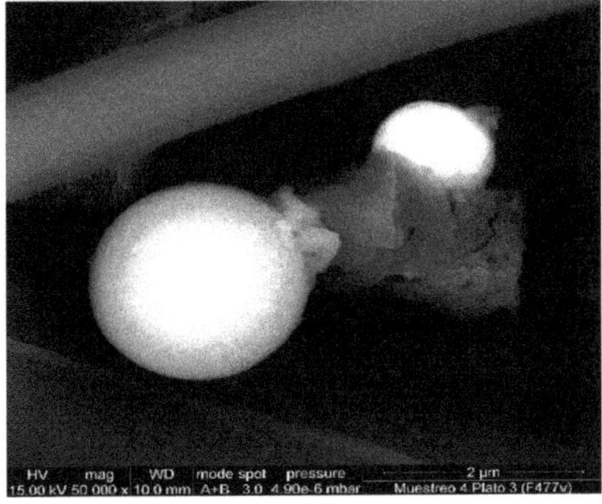

Figure 9. SEM photograph of the PM emitted in the frit melting stage.

Regarding the influence of the cleaning systems on the particle size of the emitted PM, it should be noted that, in most of the processes studied (except the firing stage with solid reagent injected in the exhaust stream to remove gaseous pollutants), the finest fractions are enriched after the cleaning system. Moreover, the PSDs were less wide, presumably due to the fact that the coarser fraction was highly efficiently reduced. Finally, it was considered interesting to make some specific comments on some of the stages and cleaning systems studied (Table 9).

Table 9. Comments about w_x associated with different ceramic process stages and cleaning systems.

Process Stage	Cleaning System	Granulometry
Spray-drying	Cyclone + wet scrubbing system	The increase of w_x is due to the breakage of the agglomerates by wetting.
Firing	Fabric filter + solid reagent	The reduction of w_x is a direct consequence of injecting solid reagent to remove gaseous pollutants.
Frit melting	Fabric filter	The increase of the w_x emitted post-cleaning can be explained by the thermal origin (volatilization–condensation) of the particles when the temperature of the exhaust gases is reduced (<200 °C) before entering into the cleaning system.

In general, the mean values of w_{10} after the fabric filters operated at high performance are high and with little dispersion (75–85%), and it is also observed that they are practically independent of the stage considered, i.e., they are not significantly dependent on the initial PSD of the stream to be treated.

In the fine fraction $w_{2.5}$, the behavior is more complex ($w_{2.5}$: 30–60%), probably because the main variable is not the cleaning system, but also the nature of the processed material.

4.1.2. Determination of EF

In general, the use of high-efficiency cleaning systems considerably reduces the emission factors obtained for fractions w_{10}, $w_{2.5}$, and w_1. Nevertheless, the type of cleaning system and other operational parameters (such as the material processed and/or process temperature) can have an influence on the obtained results:

- In medium–high-temperature stages, concretely in the spray-drying and frits melting, the emission factors obtained differ by a factor of 10 in the case of frit melting and by a factor of 100 in the case of spray-drying, depending on the Best Available Technique (BAT) implemented. The lowest values correspond to the technological scenario corresponding to fabric filters.
- In those stages where the processed materials are similar, but the process temperature is significantly different, such as the milling and spray-drying stages, the average emission factors obtained were very similar when a fabric filter was used as an abatement technology (EF_{PM10}: 2–3 mg/kg, $EF_{PM2.5}$: 2 mg/kg, and EF_{PM1}: 1 mg/kg).
- On the contrary, when the material processed is different and the process temperature is similar, such as the pressing and glazing stages, the emission factors obtained differ considerably (EF_{PM10}: 8–48 mg/m^2, $EF_{PM2.5}$: 3–31 mg/m^2, and EF_{PM1}: 1–29 mg/m^2).

The obtention of specific emission factors for different particle size and stage processes, including the influence of the abatement system, is considered of great practical interest for the ceramic industry, technological providers, public authorities, and research groups for performing emission inventories (e.g., E-PRTR), deriving new BAT-AELs (Emission Limit Values associated with Best Available Techniques), air pollution diagnosis studies, environmental impact studies from a lifecycle analysis perspective, and air quality assessment studies, among others.

Despite the potential use of the ceramic specific emission factors mentioned above, it is remarkable that these are not currently available in those reference emission factors guides such as the AP-42/EPA [26] and EMEP-EEA [25] and in the BAT Reference Document applicable to the ceramic and related industries (CER BREF [9], GLS BREF [12], and WGC BREF [17]).

Table 10 shows a compilation of emission factors obtained in other similar studies [47] and in the present study. In general, they are coherent PM emissions from combustion processes (e.g., firing and fusing) finer than those generated from mechanical treatments (e.g., press and milling). In one case, significant deviations were detected between comparable processes, such as isostatic pressing of minerals; this could be due to the different sizes of the material processed.

4.2. Chemical Characterization of the PM Emissions

Finally, regarding the chemical analysis, the following conclusions may be drawn:

- In the spray-drying and pressing stages there was an important presence of ZrO_2 and Ni, which can be attributed to the presence of these components in the raw materials and also in the ceramic sludges. It should be highlighted that the ceramic sludges, which are reused in the spray-drying stage, are generated during different cleaning operations (glaze preparation and glazing stage) and, therefore, they have a similar composition to the raw materials used in these last-mentioned processes.
- In the case of glazing, the concentration of ZrO_2, which is mainly used to achieve opacity, is significant. Nevertheless, this composition can be very variable because it

depends on the final aesthetic requirements of the ceramic tiles produced, and this compound is not present in all glaze compositions.

- Regarding the firing stage (emissions before the abatement system), the presence of PbO, As, and Zn compounds was associated with the raw materials used in body and glaze composition.
- Finally, in the case of the melting stage for frits manufacturing, the presence of Zn, PbO, Cd, and As compounds was linked to raw material compositions. Nevertheless, it should be highlighted that the use of cadmium and lead in the frits composition has been very limited in the last years, and, therefore, presumably the presence of these elements in the emissions should have been reduced in the same way.

Table 10. Compilation of emission factors.

Source	Industrial Process	Subsector, Basic Input Material, Fuel	Cleaning System	TSP (mg/Nm^3)	Mean Values (%)		
					EF$_{PM10}$	EF$_{PM2.5}$	EF$_{PM1}$
Findings of this research	Ceramic tile manufacturing	Ceramic tile manufacturing, milling, batch	Fabric filter	<5	74.8	53.4	38.8
Previous studies [47]	Treatment natural stone, sand	Crusher plant, limestone, dolomite	Fabric filter	1.2	69.2	14.2	5.0
Findings of this research	Treatment natural stone, sand	Preparation of ceramic raw materials, loam, clay, porosity material	Fabric filter	0.8	80.4	34.4	16.5
Findings of this research	Ceramic tile manufacturing	Ceramic tile manufacturing, isostatic compression press, spray-dried powder	Fabric filter	<5	75.3	28.9	5.3
Previous studies [47]	Manufacture of porcelain/press	Isostatic compression press, porcelain substance	Fabric filter	0.1	94.9	57.4	38.3
Findings of this research	Ceramic tile manufacturing	Ceramic tile manufacturing, firing, continuous, natural gas	None	11	99.4	93.9	75.9
Previous studies [47]	Tunnel oven ceramic industry	Oven without additive, loam, clay, gas	None	5.3	93.9	85.0	79.7
		Oven with additive, loam, clay, gas, lime	None	3.4	95.4	88.6	84.9
Findings of this research	Glass industry	Ceramic frit manufacturing	Fabric filter	<5	83.1	61.1	43.5
Ehrlich et al. 2007 [47]	Glass industry	Manufacture of goblets and beakers, bath, cullet, batch, natural gas	Fabric filter	0.8	93.4	53.3	37.7

Another aspect studied was the possible segregation of the components and elements of interest (As, Cd, Ni, PbO, ZnO, and ZrO$_2$) in the PM$_{10}$ and PM$_{2.5}$ fractions for each of the process stages. It was observed that the enrichment in one or another fraction is associated with the emission mechanism of the component and/or element evaluated and with the granulometry of the source material.

- For example, in the spray-drying stage, there is an enrichment in Ni in the PM$_{2.5}$ fraction. The presence of this element is associated with the use of ceramic pigments whose granulometry is usually very fine.
- Regarding the emission mechanism, the fluxing nature of lead compounds means that this is present mostly by volatilization from the melt; hence, an enrichment of this component is observed in the finest fraction, specifically in the frit melting and firing stages after the cleaning system.

- In the melting stage of ceramic frits, both As, a trace element associated with the natural raw materials that introduce boron into the composition of ceramic frits, and ZrO_2, a raw material for frits and atomized granules, are enriched in the PM_{10} fraction, probably because the emission mechanism in both cases is mechanical in nature.

5. Conclusions

The conclusions of this study are presented in accordance with the structure followed in the previous sections.

5.1. Physical Characterization

5.1.1. Assessment of Methodology Used to Determine w_x and PSD

- The cascade impactor is suitable for determining and studying both the PSD and the w_{10}, $w_{2.5}$, and w_1 fractions. The use of the impactor is not recommended due to the uncertainty that may be associated with the filters weighted.
- The two mathematical processing methods of the PSD data (log-normal distribution and Rosin–Rammler–Sperling–Bennet) exhibit similar results for the determination of the w_{10}, $w_{2.5}$, and w_1 fractions. The log-normal adjustments with R^2 greater than 0.90 were obtained.

5.1.2. Determination of PSD and w_x

- The PSDs obtained are relatively wide, due to the different mechanisms of origin of the particulate matter, mechanical, and/or volatilization–condensation.
- The average values obtained for w_{10} and $w_{2.5}$ after fabric filters, operated at high performance, are in the range of 75–85% and 30–60%, respectively. In the case of the fine fractions, the wide range is due to the influence not only of the cleaning system but also of the nature of the processed material.

5.1.3. Determination of the EF

- The highest efficiency, for the ceramic stages studied, is reached when fabric filters are applied.
- In those cases where the materials processed are similar, such as the milling and spray-drying stages, the process temperature does not significantly affect the EFs obtained.

5.2. Chemical Characterization of the PM Emissions

- The use of the pilot-scale emission simulator made it possible to significantly reduce the sampling times and consequently to obtain a large amount of samples of the different granulometric fractions, which allowed a complete chemical characterization.
- The chemical analysis showed that ZrO_2, ZnO, BaO, PbO, B_2O_3, Hf, Cr, Cu, Cd, and Sn were the main components in the emissions related to ambient-temperature processes (spray-drying, pressing, and glazing) while SiO_2, Al_2O_3, CaO, MgO, Na_2O, K_2O, BaO, ZnO, PbO, S, Tl, As, Cr, Rb, Cs, Cu, and Sr were the main components of the medium–high-temperature processes emissions (firing and frits melting).
- The emission mechanism (mechanical and/or volatilization–condensation) and the particle size of the source material are the parameters that most influence the potential segregation of the components and elements evaluated (As, Cd, Ni, PbO, ZnO, and ZrO_2) in the PM_{10} and $PM_{2.5}$ fractions.

6. Future Research Lines

From the results obtained in the present study, a series of future research lines to complement some of the results achieved are proposed:

- To develop an emission simulator in order to modify the temperature of the stream and the introduction of gases, and thus study the effect of the temperature and the composition of the gas stream on the characteristics of the particulate matter.

- To use more complete particle size distribution determination systems, even complementing several systems simultaneously, which allow us to obtain information in a wider range of particle sizes. In this sense, it would be especially interesting to address the study of the submicron and ultrafine fractions in emissions from high-temperature processes, both in concentrations and their variation and possible correlation with emission mechanisms and scrubbing systems. These fractions are of increasing environmental interest, and therefore receive greater legislative attention due to their possible effects on health.
- It would be extremely interesting to initiate a line in collaboration with analytical chemists to clarify some aspects that could not be determined in this work, such as the detailed study of the behavior of some elements and their compounds, in particular of the most volatile elements, such as boron, lead, arsenic, thallium, etc., in processes and emissions at higher temperatures, in order to deepen the knowledge of the volatilization–condensation mechanisms and minimize possible emissions. To this end, it is necessary to develop methodologies that integrate equipment allowing for the characterization of very small samples.
- To carry out a specific study of high-temperature process steps in order to determine the influence of raw material composition, process variables, scrubbing systems, etc., on the characteristics of particulate matter emissions.
- To complement the results of this study with air quality studies in the area, for which it is necessary to determine the behavior of particulate matter and gases emitted by different sources in the atmospheric environment, studying the mechanisms of interaction between different primary and secondary pollutants.

Supplementary Materials: The following supporting information can be downloaded at https://www.mdpi.com/article/10.3390/ijerph19159652/s1: Figure S1. PSD of emissions generated in ambient temperature processes; Figure S2. PSD of emissions generated in medium-temperature processes; Figure S3. PSD of emissions generated in high-temperature processes; Figure S4. PM_{10} and $PM_{2.5}$ composition of emissions generated during spray-drying emissions; Figure S5. PM_{10} and $PM_{2.5}$ composition of emissions generated during pressing; Figure S6. PM_{10} and $PM_{2.5}$ composition of emissions generated during drying; Figure S7. PM_{10} and $PM_{2.5}$ composition of emissions generated during firing (without cleaning system); Figure S8. PM_{10} and $PM_{2.5}$ composition of emissions generated during firing (after cleaning system); Figure S9. PM_{10} and $PM_{2.5}$ composition of emissions generated during frit melting. Table S1: Individual samplings of ambient temperature processes; Table S2. Individual samplings of medium- and high-temperature processes.

Author Contributions: Conceptualization, I.C., E.M., X.Q., V.S. and S.G.; methodology, I.C., X.Q. and A.E.; software, A.E.; validation, I.C., V.S. and S.G.; formal analysis, I.C., E.M., X.Q. and V.S.; investigation, I.C.; resources, I.C. and V.S.; data curation, I.C. and A.E.; writing—original draft preparation, I.C., V.S. and A.L.-L.; writing—review and editing, I.C., V.S., X.Q. and A.L.-L.; visualization, I.C. and E.M.; supervision, I.C. and E.M.; project administration, I.C. and E.M.; funding acquisition, I.C. and E.M. All authors have read and agreed to the published version of the manuscript.

Funding: This study has been funded by the Ministry of Science and Technology in the framework of the National Plan for Scientific Research, Development and Technological Innovation, reference REN2003-08916-C02-01.

Institutional Review Board Statement: Not applicable.

Informed Consent Statement: Not applicable.

Acknowledgments: The authors would like to thank IDAEA-CSIC for their committed cooperation in the sample's chemical analysis and also all the ceramic companies in which the measurements were carried out for their continuous support.

Conflicts of Interest: The authors declare no conflict of interest. The funders had no role in the design of the study; in the collection, analyses, or interpretation of data; in the writing of the manuscript, or in the decision to publish the results.

References

1. World Health Organization. WHO Global Air Quality Guidelines: Particulate Matter (PM2.5 and PM10), Ozone, Nitrogen Dioxide, Sulfur Dioxide and Carbon Monoxide; Licence: CC BY-NC-SA 3.0 IGO. 2021. Available online: https://apps.who.int/iris/handle/10665/345329 (accessed on 28 May 2020).
2. EPA. Particle Pollution and Your Patients' Health. 2021. Available online: https://www.epa.gov/pmcourse/particle-pollution-exposure (accessed on 4 July 2022).
3. Dockery, D.W.; Pope, C.A.; Xu, X.; Spengler, J.D.; Ware, J.H.; Fay, M.E.; Ferris, B.G., Jr.; Speizer, F.E. An association between air pollution and mortality in six US cities. *N. Engl. J. Med.* **1993**, *329*, 1753–1759. [CrossRef]
4. Abbey, D.E.; Nishino, N.; McDonnell, W.F.; Burchette, R.J.; Knutsen, S.F.; Lawrence Beeson, W.; Yang, J.X. Long-term inhalable particles and other air pollutants related to mortality in nonsmokers. *Am. J. Respir. Crit. Care Med.* **1999**, *159*, 373–382. [CrossRef] [PubMed]
5. Hoek, G.; Brunekreef, B.; Goldbohm, S.; Fischer, P.; van den Brandt, P.A. Association between mortality and indicators of traffic-related air pollution in the Netherlands: A cohort study. *Lancet* **2002**, *360*, 1203–1209. [CrossRef]
6. Pope, C.A., III; Burnett, R.T.; Thun, M.J.; Calle, E.E.; Krewski, D.; Ito, K.; Thurston, G.D. Lung cancer, cardiopulmonary mortality, and long-term exposure to fine particulate air pollution. *JAMA* **2002**, *287*, 1132–1141. [CrossRef] [PubMed]
7. Karagulian, F.; Belis, C.A.; Dora, C.F.C.; Prüss-Ustün, A.M.; Bonjour, S.; Adair-Rohani, H.; Amann, M. Contributions to cities' ambient particulate matter (PM): A systematic review of local source contributions at global level. *Atmos. Environ.* **2015**, *120*, 475–483. [CrossRef]
8. Mallol, G.; Monfort, E.; Busani, G.; Lezaun, J. *Depuración de los Gases de Combustión de la Industria Cerámica. Guía Técnica*, 2nd ed.; AICE, Ed.; Instituto de Tecnología Cerámica: Castellón, Spain, 2001; ISBN 84-923176-5-5.
9. Reference Document on Best Available Techniques in the Ceramic Manufacturing Industry. Available online: https://eippcb.jrc.ec.europa.eu/reference/BREF/cer_bref_0807.pdf (accessed on 28 May 2020).
10. Reference Document on Best Available Techniques for the Ferrous Metals Processing Industry. Available online: https://eippcb.jrc.ec.europa.eu/sites/default/files/2019-11/FMP_D1_web.pdf (accessed on 28 May 2020).
11. Reference Document on Best Available Techniques for Large Combustion Plants. Available online: https://eippcb.jrc.ec.europa.eu/sites/default/files/2019-11/JRC_107769_LCPBref_2017.pdf (accessed on 28 May 2020).
12. Reference Document on Best Available Techniques for the Manufacture of Glass. Available online: https://eippcb.jrc.ec.europa.eu/sites/default/files/2019-11/GLS_Adopted_03_2012_0.pdf (accessed on 28 May 2020).
13. Reference Document on Best Available Techniques for the Non-Ferrous Metals Industries. Available online: https://eippcb.jrc.ec.europa.eu/sites/default/files/2020-01/JRC107041_NFM_bref2017.pdf (accessed on 28 May 2020).
14. Reference Document on Best Available Techniques for the Production of Cement, Lime and Magnesium Oxide. Available online: https://eippcb.jrc.ec.europa.eu/sites/default/files/2019-11/CLM_Published_def_0.pdf (accessed on 28 May 2020).
15. Reference Document on Best Available Techniques for Waste Incineration. Available online: https://eippcb.jrc.ec.europa.eu/sites/default/files/2020-01/JRC118637_WI_Bref_2019_published_0.pdf (accessed on 28 May 2020).
16. Reference Document on Best Available Techniques for the Production of Wood-based Panels. Available online: https://eippcb.jrc.ec.europa.eu/sites/default/files/2019-11/WBPbref2016_0.pdf (accessed on 28 May 2020).
17. Best Available Techniques (BAT) Reference Document for Common Waste Gas Management and Treatment Systems in the Chemical Sector. Available online: https://eippcb.jrc.ec.europa.eu/sites/default/files/2022-03/WGC_Final_Draft_09Mar2022-B-W-Watermark.pdf (accessed on 7 April 2022).
18. EPER (The European Pollutant Emission Register). Available online: http://www.prtr-es.es (accessed on 29 September 2021).
19. Querol, X.; Alastuey, A.; Rodriguez, S.; Plana, F.; Mantilla, E.; Ruiz, C.R. Monitoring of PM10 and PM2. 5 around primary particulate anthropogenic emission sources. *Atmos. Environ.* **2001**, *35*, 845–858. [CrossRef]
20. Querol, X. *Bases Científico-Técnicas para un Plan Nacional de Mejora de la Calidad del Aire (No. 504 507)*; Editorial CSIC: Madrid, Spain, 2012.
21. Vicente, A.B.; Sanfeliu, T.; Jordan, M.M. Assessment of PM10 pollution episodes in a ceramic cluster (NE Spain): Proposal of a new quality index for PM10, As, Cd, Ni and Pb. *J. Environ. Manag.* **2012**, *108*, 92–101. [CrossRef]
22. Minguillón, M.C.; Monfort, E.; Escrig, A.; Celades, I.; Cuerra, L.; Busani, G.; Sterni, A.; Querol, X. Air quality comparison between two European ceramic tile clusters. *Atmos. Environ.* **2013**, *74*, 311–319. [CrossRef]
23. Minguillón, M.C.; Monfort, E.; Querol, X.; Alastuey, A.; Celades, I.; Miró, J.V. Effect of ceramic industrial particulate emission control on key components of ambient PM10. *J. Environ. Manag.* **2009**, *90*, 2558–2567. [CrossRef]
24. Querol, X.; Viana, M.; Alastuey, A.; Amato, F.; Moreno, T.; Castillo, S.; Pey, J.; de la Rosa, J.; Sánchez de la Campa, A.; Artíñano, B.; et al. Source origin of trace elements in PM from regional background, urban and industrial sites of Spain. *Atmos. Environ.* **2007**, *41*, 7219–7231. [CrossRef]
25. *EMEP/EEA Air Pollutant Emission Inventory Guidebook 2019*; Technical Guidance to Prepare National Emission Inventories; EEA Report No 13/2019; EMEP/EEA: Copenhagen, Denmark, 2019.
26. US EPA. Ceramic Products Manufacturing. 1995. Available online: https://www.epa.gov/air-emissions-factors-and-quantification/ap-42-compilation-air-emissions-factors#5thed (accessed on 29 September 2021).
27. *Particulate Emission Testing for Florida Tile Corporation, Lawrenceburg, Kentucky, 7–8 March 1989*; Air Systems Testing, Inc.: Marietta, GA, USA, 1989.

28. *Particulate Emission Testing for Florida Tile Corporation, Lawrenceburg, Kentucky, 19 April 1989*; Air Systems Testing, Inc.: Marietta, GA, USA, 1989.

29. *Metropolitan Ceramics, Canton, Ohio, Tunnel Kiln #3 Exhaust Stack, Particulate, SO₂, NOx, Hydrofluoric Acid Emission Evaluation, Conducted—17–18 November 1993*; Envisage Environmental Incorporated: Richfield, OH, USA, 1993.

30. *Metropolitan Ceramics, Inc., Canton, Ohio, TK1, TK2, TK3 Exhausts, Particulate, Sulfur Dioxides, & Fluorides Emission Evaluation, Conducted—30 March–14 April 1994*; Metropolitan Ceramics, Inc.: Canton, OH, USA, 1994.

31. *Envisage Environmental Incorporated, Richfield, OH, May 9, 1994. 11Source Evaluation Results, U.S. Ceramic Tile Company, East Sparta, Ohio, 11 August 1993*; Envisage Environmental Incorporated: Richfield, OH, USA, 1993.

32. *Particulate Emissions Test for American Olean Tile Company, Fayette, AL, Crushing and Screening Line #1, 15 October 1991*; Pensacola POC, Inc.: Pensacola, FL, USA, 1991.

33. *Particulate Emissions Test for American Olean Tile Company, Fayette, AL, Crushing and Screening Line #2, 16 October 1991*; Pensacola POC, Inc.: Pensacola, FL, USA, 1991.

34. *Exhaust Emission Sampling for Norton Company, Soddy-Daisy, TN, 19–20 April 1994*; Armstrong Environmental, Inc.: Dallas, TX, USA, 1994.

35. *Particulate Emission Evaluation for Steward, Inc., Chattanooga, TN, 30 March 1993*; FBT Engineering and Environmental Services: Chattanooga, TN, USA, 1993.

36. *Report to American Standard on Stack Particulate Samples Collected at Tiffin, OH (Test Date 18 August 1992)*; Affiliated Environmental Services, Inc.: Sandusky, OH, USA, 1992.

37. *Report to American Standard on Stack Particulate Samples Collected at Tiffin, OH (Test Date 19 August 1992)*; Affiliated Environmental Services, Inc.: Sandusky, OH, USA, 1992.

38. *Report to American Standard on Stack Particulate Samples Collected at Tiffin, OH (Test Date 8 February 1994)*; Affiliated Environmental Services, Inc.: Sandusky, OH, USA, 1994.

39. *Emission Test Report—Plant A, Roller Kiln, May 1994*; Document No. 4602-01-02; Confidential Business Information Files; Contract No 68-D2-0159; Assignment No. 2-01; U.S. Environmental Protection Agency: Research Triangle Park, NC, USA, 1995.

40. *Emission Test Report (Excerpts)—Plant A, Roller Kiln, February 1992*; Document No. 4602-01-02; Confidential Business Information Files; Contract No 68-D2-0159; Assignment No. 2-01; U.S. Environmental Protection Agency: Research Triangle Park, NC, USA, 1995.

41. *Emission Test Report—Plant A, Spray Dryer, October 1994*; Document No. 4602-01-02; Confidential Business Information Files; Contract No 68-D2-0159; Assignment No. 2-01; U.S. Environmental Protection Agency: Research Triangle Park, NC, USA, 1995.

42. *Emission Test Report (Excerpts)—Plant A, Spray Dryer, April 1994*; Document No. 4602-01-02; Confidential Business Information Files; Contract No 68-D2-0159; Assignment No. 2-01; U.S. Environmental Protection Agency: Research Triangle Park, NC, USA, 1995.

43. *Emission Test Report (Excerpts)—Plant A, Spray Dryer, January 1993*; Document No. 4602-01-02; Confidential Business Information Files; Contract No 68-D2-0159; Assignment No. 2-01; U.S. Environmental Protection Agency: Research Triangle Park, NC, USA, 1995.

44. *Emission Test Report (Excerpts)—Plant A, Spray Dryer, February 1992*; Document No. 4602-01-02; Confidential Business Information Files; Contract No 68-D2-0159; Assignment No. 2-01; U.S. Environmental Protection Agency: Research Triangle Park, NC, USA, 1995.

45. *Lead and Particulate Emissions Testing, Spray Booth 2A Stack*; Stationary Source Sampling Report Reference No. 6445; Entropy Environmentalists, Inc.: Research Triangle Park, NC, USA, 1989.

46. Palmonari, C.; Timellini, G. Pollutant emission factors for the ceramic floor and wall tile industry. *J. Air Pollut. Control. Assoc.* **1982**, *32*, 1095–1100. [CrossRef]

47. Ehrlich, C.; Noll, G.; Kalkoff, W.-D.; Baumbach, G.; Dreiseidler, A. PM10, PM2. 5 and PM1. 0—Emissions from industrial plants—results from measurement programmes in Germany. *Atmos. Environ.* **2007**, *41*, 6236–6254. [CrossRef]

48. Ehrlich, C.; Noll, G.; Wusterhausen, E.; Kalkoff, W.D.; Remus, R.; Lehmann, C. Respirable Crystalline Silica (RCS) emissions from industrial plants–Results from measurement programmes in Germany. *Atmos. Environ.* **2013**, *68*, 278–285. [CrossRef]

49. Antonsson, E.; Cordes, J.; Stoffels, B.; Wildanger, D. The European Standard Reference Method systematically underestimates particulate matter in stack emissions. *Atmos. Environ.* **2021**, *12*, 100133. [CrossRef]

50. Wada, M.; Tsukada, M.; Namiki, N.; Szymanski, W.W.; Noda, N.; Makino, H.; Kanaoka, C.; Kamiya, H. A two-stage virtual impactor for in-stack sampling of PM2. 5 and PM10 in flue gas of stationary sources. *Aerosol. Air Qual. Res.* **2016**, *16*, 36–45. [CrossRef]

51. John, A.C.; Kuhlbusch, T.A.J.; Fissan, H.; Bröker, G.; Geueke, J. Development of a PM 10/PM 2.5 cascade impactor and in-stack measurements. *Aerosol. Sci. Technol.* **2003**, *37*, 694–702. [CrossRef]

52. Sánchez de la Campa, A.M.; Moreno, T.; de la Rosa, J.; Alastuey, A.; Querol, X. Size distribution and chemical composition of metalliferous stack emissions in the San Roque petroleum refinery complex, southern Spain. *J. Hazard. Mater.* **2011**, *190*, 713–722. [CrossRef]

53. González-Castanedo, Y.; Moreno, T.; Fernández-Camacho, R.; Sánchez de la Campa, A.M.; Alastuey, A.; Querol, X.; de la Rosa, J. Size distribution and chemical composition of particulate matter stack emissions in and around a copper smelter. *Atmos. Environ.* **2014**, *98*, 271–282. [CrossRef]

54. Gupta, R.K.; Majumdar, D.; Trivedi, J.V.; Bhanarkar, A.D. Particulate matter and elemental emissions from a cement kiln. *Fuel Processing Technol.* **2012**, *104*, 343–351. [CrossRef]

55. Akinshipe, O.; Kornelius, G. Quantification of atmospheric emissions and energy metrics from simulated clamp kiln technology in the clay brick industry. *Environ. Pollut.* **2018**, *236*, 580–590. [CrossRef]

56. ASCER. Available online: https://www.ascer.es/sectorDatos.aspx?lang=es-ES (accessed on 29 September 2021).
57. ANFFEC. Available online: https://www.anffecc.com/es/cifras-del-sector (accessed on 29 September 2021).
58. *Cerámica de Castellón, Zona*; Plan de Mejora de la Calidad del Aire de la Zona ES 1003; Mihares-Penaygolosa (A. Costera) y Aglomeración ES 1015; Conselleria de Medi Ambient, Aigua, Urbanisme i Habitatge: Castelló, Spain, 2008.
59. Monfort, E.; Sanfelix, V.; Minguillón, M.C.; Celades, I. *Mitigation Strategies: Castellón, Spain*; Future Science Ltd.: London, UK, 2013; pp. 150–160.
60. *ISO 9096:2003*; Stationary Sources Emissions—Manual Determination of Mass Concetration of Particulatte Matter. International Organization for Standardization: Geneva, Switzerland, 2003.
61. *ISO 12141:2002*; Stationary Source Emissions—Determination of Mass Concentration of Particulate Matter (Dust) at Low Concentrations—Manual Gravimetric Method. International Organization for Standardization: Geneva, Switzerland, 2002.
62. Monfort, E.; Gazzula, M.F.; Celades, I.; Gómez, P. Critical factors in measuring solid particulate emissions in the ceramics industry. *Ceram. Forum Int. Ber. DKG* **2001**, *78*, E40–E44.
63. Monfort, E.; Celades, I.; Mestre, S.; Sanz, V.; Querol, X. PMX data processing in ceramic tile manufacturing emissions. In *Key Engineering Materials*; Trans Tech Publications Ltd.: Bäch, Switzerland, 2004; Volume 264, pp. 2453–2456.
64. Pilat, M.J. *Operations Manual Pilat (University of Washington) Mark 3 and Mark 5 Source Test Cascade Impactor*; University of Washington: Seattle, WA, USA, 1998.
65. *VDI 2066 Blatt 10:2004*; Particulatte Matter Meassurement. Dust Meassurement in Flowing Gases. Measurement of PM10 and PM2.5 Emissions at Stationary Sources by Impaction Method. VDI/DIN-Kommission Reinhaltung der Luft (KRdL)—Normenausschuss: Düsseldorf, Germany, 2004.
66. Batel, W. Schaubildliche Darstellung und Kennzeichnung von Korngrößenverteilungen. In *Einführung in die Korngrößenmeßtechnik*; Springer: Berlin/Heidelberg, Germany, 1964; pp. 4–27. ISBN 978-3-642-53251-1.
67. *EN ISO/IEC 17043:2010*; Conformity Assessment—General Requirements for Proficiency Testing. European Standard: Pilsen, Czech Republic, 2010.
68. *EN 13284-1:2018*; Stationary Source Emissions—Determination of Low Range Mass Concentration of Dust—Part 1: Manual Gravimetric Method. European Standard: Pilsen, Czech Republic, 2017.
69. *EN ISO 23210:2010*; Stationary Source Emissions—Determination of PM10/PM2.5 Mass Concentration in Flue Gas—Measurement at Low Concentrations by Use of Impactors (ISO 23210:2009). ISO: Geneva, Switzerland, 2009.
70. *EPA Method 201*; Emission Determination of PM10 Emissions. Exhaust Gas Recycle Procedure. U.S. Environmental Protection Agency: Washington, DC, USA, 2017.
71. Querol, X.; Minguillón, M.C.; Alastuey, A.; Monfort, E.; Mantilla, E.; Sanz, M.J.; Sanz, F.; Roig, A.; Renau, A.; Felis, C.; et al. Impact of the implementation of PM abatement technology on the ambient air levels of metals in a highly industrialised area. *Atmos. Environ.* **2007**, *41*, 1026–1040. [CrossRef]
72. *IS K 0302*; Measuring Method for Particle—Size Distribution of Dust in Flue Gas. Japanese Industrial Standards Committee: Tokyo, Japan, 1989.
73. Hinds, W.C. *Aerosol Technology: Properties, Behaviour, and Measurement of Airborne Particles*, 2nd ed.; John Wiley & Sons: Hoboken, NJ, USA, 1999.
74. Vincent, J.H. *Aerosol Sampling: Science, Standards, Instrumentation and Applications*; John Wiley & Sons: Hoboken, NJ, USA, 2007.

International Journal of
*Environmental Research
and Public Health*

MDPI

Article

Does a Recycling Carbon Tax with Technological Progress in Clean Electricity Drive the Green Economy?

Weijiang Liu [1,2], Min Liu [2,*], Tingting Liu [2], Yangyang Li [2] and Yizhe Hao [2]

1 Center for Quantitative Economics, Jilin University, Changchun 130012, China; liuwj@jlu.edu.cn
2 Business School, Jilin University, Changchun 130012, China; liutingting19@mails.jlu.edu.cn (T.L.);
   liyangyang20@mails.jlu.edu.cn (Y.L.); haoyz20@mails.jlu.edu.cn (Y.H.)
* Correspondence: minliu19@mails.jlu.edu.cn

Abstract: The environmental issue is a significant challenge that China faces in leading the development of the green economy. In this context, reducing CO_2 emissions is the key to combatting this problem. Taking the 2017 social accounting matrix (SAM) as the database and combing macroeconomic parameters from previous studies, this article constructed the environmentally computable general equilibrium (CGE) model as an analytical model to analyze the economic–environmental–energy impacts of recycling carbon tax with technological progress in clean electricity. We found that when the rate of clean electricity technological progress reaches 10%, the carbon recycling tax that reduces corporate income taxes will achieve a triple dividend of the carbon tax, namely, promoting economic development, reducing carbon emissions, and improving social welfare. In the meantime, on the basis of carbon tax policies that raise the price of fossil energy, clean electricity technological progress will help accelerate the transformation of electricity structure, reduce the proportion of thermal power generation, and better promote emission reduction. In addition, due to the high carbon emission coefficient, coal contributes significantly to carbon emission reduction. Therefore, China should implement a carbon tax recycling policy supplemented by the progress of clean power technology as soon as possible to better promote green economy development.

Keywords: carbon tax recycling policy; green economy; technological progress; CGE model; triple dividend; carbon emissions

Citation: Liu, W.; Liu, M.; Liu, T.; Li, Y.; Hao, Y. Does a Recycling Carbon Tax with Technological Progress in Clean Electricity Drive the Green Economy? *Int. J. Environ. Res. Public Health* **2022**, *19*, 1708. https://doi.org/10.3390/ijerph19031708

Academic Editors: Pasquale Avino, Massimiliano Errico, Aristide Giuliano and Hamid Salehi

Received: 9 December 2021
Accepted: 31 January 2022
Published: 2 February 2022

Publisher's Note: MDPI stays neutral with regard to jurisdictional claims in published maps and institutional affiliations.

1. Introduction

Green and sustainable development is a core driver of economic development worldwide. However, the growing amount of carbon emissions poses a tremendous threat to it. The current status of China's carbon emissions has remained underwhelming. In 2020, China's primary energy demand increased by 2.1%, and total carbon dioxide emissions increased by 0.6%. By contrast, the proportion increased. Specifically, fossil energy consumption accounted for 85.2% of total energy consumption. The high proportion of fossil energy has led to higher CO_2 emissions. As the largest carbon emission producer in the world, China is actively committed to adopting more effective policies and measures to reduce carbon emissions [1]. On 22 September 2020, President Xi Jinping proposed at the general debate of the 75th United Nations General Assembly that "China will increase its nationally determined contributions (NDC), striving to reach the peak of carbon dioxide emissions by 2030, further, try to achieve carbon neutrality by 2060". Later, at the Climate Ambition Summit, he announced that China's carbon dioxide emissions per unit of GDP will drop by more than 65% compared to 2005 by 2030, the proportion of non-fossil energy to primary energy consumption will reach about 25%, and wind and solar power generation capacity will reach more than 1.2 billion kilowatts. In response, it is necessary to study how China can promote carbon reduction, achieve carbon emission commitments, and drive the development of a green economy.

In the process of promoting the development of a green economy, the power industry, as an industry with a relatively large volume of emissions, should be of particular concern. Coal-fired power generation is the main source of power generation in most parts of the world [2]. As for China, in 2020, nuclear power generation accounted for 4.7% of produced power; hydropower generation accounted for 17%; and renewable energy power generation accounted for 11%. Based on the above data, clean powers' proportions are relatively low. Coal-fired power generation was still the primary source of power generation, accounting for 63% of the power produced (https://www.bp.com/, (accessed on 23 January 2022)). China's coal reserves are relatively sufficient, and coal-fired power generation has the advantages of controllability and stability. Therefore, thermal power generation relies primarily on coal combustion [3]. It is undeniable that high carbon emissions accompany the combustion process. According to the IEA, the global electricity and heat sector contributes 42% to CO_2 emissions, with China contributing a larger share in comparison (https://www.iea.org/, (accessed on 23 January 2022)). At the same time, in the light of the CEADs database statistics, the industries involved in the production and supply of electricity, steam, and hot water produce the highest carbon dioxide emissions (https://www.ceads.net/, (accessed on 23 January 2022)). China's power industry has enormous potential for emission reduction. The green development and transformation of China's fuel power generation industry have become a general trend.

Countries worldwide have been exploring corresponding measures to achieve green economic development, including carbon pricing policies, which can advance progress towards the achievement of climate goals [4]. Carbon emission reduction incentives based on carbon pricing policies help the competitive market play a role, reducing carbon emissions further. Specifically, carbon tax and trading policies are two effective carbon pricing policies [5,6]. The carbon tax policy is a price control strategy levied on fossil energy's carbon content or emissions, whereas the carbon trading policy is a total quantity control strategy [7,8]. In addition, the carbon tax is simpler and faster than the Emission Trading Scheme (ETS) for developing countries within the short to medium term [9]. At the same time, a more rational carbon tax corrected the intra-sector distortions created by ETS [10]. However, the operation of a carbon tax policy causes the price of related energy to rise, affecting corporate investment and household consumption, which is not conducive to economic growth or the improvement of social welfare. In this regard, a properly designed carbon tax recycling policy is the key to achieving the double dividend (promoting economic growth and reducing carbon emissions) [11]. To be more specific, a carbon tax recovery policy refers to the imposition of a carbon tax while returning it to residents or businesses in various forms, such as reducing existing distortionary tax rates (lower corporate income tax, resident income tax, and others), affecting socio-economic variables such as corporate income and resident income [12]. Furthermore, as crucial factors of production, the efficiency and technology of energies are also effective measures that can be used to avoid the adverse effects of carbon taxes. Technological progress can curb carbon emission intensity while not reducing long-term economic growth [13]. Combined with the above analysis, both carbon tax recovery and technological progress are effective and important tools to maintain economic growth under carbon reduction targets.

Nevertheless, existing studies mainly focus on individual research on carbon tax policy and the impact of technological progress on carbon emission reduction, whereas there are few studies on the combination of carbon tax recycling policy and technological progress. Moreover, the analysis does not focus on technological progress in the clean electricity sector, applying the same rate of technological progress to all sectors and making it difficult to distinguish the impact of technological progress on specific industries. In this regard, this article constructs an environmental CGE model that includes the subdivision of power sectors, combining the carbon tax recycling policy with the progress of clean electricity technology for research, intending to explore the carbon tax's emission reduction and economic growth effect. Furthermore, we explore the possible carbon tax policy's social welfare effects on this basis. Our study is conducive to realizing the recovery of the green economy and achieving China's Nationally Determined Contribution (NDC), thus promoting China's ecological civilization construction and green economy development.

The structure of this article is as follows: Section 2 compares relevant literature, Section 3 describes the model construction and data source, and Section 4 analyzes the results of the policy combination simulation. The last section summarizes the conclusions drawn by the model and puts forward relatively reasonable policy recommendations to provide relevant policy references for national policymakers.

2. Literature Review

2.1. Research on Electricity and Carbon Emissions

Since the Industrial Revolution, the rapid increase in greenhouse gas emissions has caused severe environmental and health problems [14]. To date, the issue of climate change caused by a large amount of greenhouse gas emissions is still receiving widespread attention from countries all over the world. Among them, CO_2 is a kind of greenhouse gas that has attracted much attention. Focusing on the power industry that emits more carbon dioxide, many scholars have studied the relationship between electricity and carbon dioxide emissions. The consumption of electricity has a detrimental effect on carbon emissions. In particular, electricity generated from fossil fuels causes damage to carbon emissions and is the primary source of carbon emission reduction. Renewable power generation can weaken the adverse impact of power generation on carbon emissions [15,16]. Combining the STIRPAT model and the panel threshold model, Lin and Li found that electricity usage affects carbon emissions negatively, especially when clean electricity accounts for a relatively high proportion of the energy used [17]. Wong and Zhang conducted a natural experiment on Australia's carbon pricing mechanism (CPM), finding that the carbon tax has little impact on areas where renewable energy power generation accounts for a relatively high proportion of the energy produced. After abolishing the carbon tax, the market behavior, which had begun to shift from coal-fired power generation to other energy sources, was reversed, showing that coal-fired power has higher carbon emissions [18]. Yang and Song introduced clean coal-fired power generation to study the effects of emission reductions in the coal-fired power generation sector [19]. Haxhimusa and Liebensteiner evaluated the impact of electricity demand reductions on emission reductions based on the analysis of COVID-19 shocks to electricity demand using an applied econometric model in an instrumental variables framework [20]. Fang et al. analyzed the characteristics of carbon emissions and carbon intensity in the power sector from a spatial perspective using the Moran index. Additionally, based on the multi-regional input-output table, inter-provincial embodied carbon transfer in the power sector was explored [21]. By using a system dynamics approach, Mostafaei et al. discovered a new model for using renewable electricity to reduce CO_2 emissions [22]. The empirical results of the above literature show that electricity impacts carbon emissions. Therefore, it is important that we dig deeper into the issue of carbon emission reduction within the power sector as a starting point.

2.2. Research on Carbon Tax and Carbon Tax Recycling Policy

For a long time, the carbon tax policy has been a policy tool used to reduce greenhouse gas emissions and weaken the negative impact of climate change. In 2020, the carbon tax policy took effect or was in the process of being implemented in 30 countries and regions [23]. Many scholars have made corresponding assessments on the effect of the carbon tax policy's implementation, mainly reflected in three aspects. First of all, the existing literature has tested the emission reduction effect of the carbon tax and proved the effectiveness of carbon tax for carbon emission reduction. Secondly, as for the macroeconomic impact of carbon taxes, Yamazaki's research found that although income-neutral carbon taxes do not harm employment, carbon taxes have caused employment to shift from carbon-intensive industries to cleaning service industries [24]. Furthermore, carbon taxes affect the GDP negatively and affect economic growth [25,26]. Finally, regarding the effects of the carbon tax on residents' welfare, Khastar et al. demonstrated that the Finnish carbon tax had a specific negative impact on social welfare [27].

There is a need to recycle carbon taxes in the face of the negative impact of carbon taxes on the economy and social welfare. Pearce showed that the carbon tax policy has a "double dividend" effect. Specifically, for one thing, a carbon tax can affect carbon emissions and improve environmental quality; for another thing, a carbon tax can offset corporate income tax to increase corporates' investment and enhance the efficiency of economic operations. Implementing appropriate carbon tax recovery policies can effectively increase residents' and enterprises' income, further stimulate consumption and investment, and promote economic growth [28]. The realization of the double dividend depends on the recycling method of the carbon tax revenue. Specifically, there are two types of carbon tax recovery: the income-neutral principle and the income-positive principle [29]. The principle of income neutrality introduces carbon taxes while canceling or reducing some taxes to maintain income neutrality [30], while under the principle of income positivity, the government recycles the additional carbon tax received back to the household or corporate sectors to increase the corresponding income. Ojha et al. used a recursive dynamic CGE model to study the issue of carbon tax recycling in India. They considered that based on carbon emission reduction, the recycling of carbon tax would benefit economic growth and weaken the inequality of income distribution [31]. Li et al. simulated the impact of a carbon tax on employment under various carbon tax recovery scenarios [12]. Generally speaking, recycling carbon taxes reduces the tax burden and the negative impact on the economy, and different carbon tax recovery schemes will have different results [32]. The ultimate acceptance and successful implementation of a carbon tax depend on how the carbon tax revenue is utilized [33]. Most scholars study whether the carbon tax's double dividend will be realized. At the same time, there are few considerations of social welfare, causing the absence of rational exploration and discovery of the carbon tax's triple dividend.

2.3. Research on Technological Progress and Carbon Emission Reduction

Carbon tax recycling policy has indeed contributed to carbon emission reduction. At present, achieving emission reduction targets while ensuring sustainable economic development is an important challenge for the government [34]. Technological progress is undoubtedly the most promising solution to the challenge [13]. Chen et al. found that technological advances reduced carbon emissions during the study period in China [35]. Considering that technological progress is an essential factor in promoting carbon emission reduction, Wu et al. introduced technological progress into the CGE model to predict the carbon emission situation of the global economy and evaluate the effect of different technological advancements on carbon emission reduction [36]. After simulation and prediction analysis, Guo et al. once again posited that the continuous improvement of total factor productivity is one of the critical conditions that needs to be met for China to achieve the goal of energy regulation [37]. Using the best technologies to reduce unnecessary energy consumption and improve energy efficiency deserves attention [38]. However, the relationship between technological progress and carbon dioxide emissions is still complex and requires in-depth research [39]. Wang et al. comprehensively studied the relationship between technological progress and CO_2 emissions and found heterogeneity in the effects of technological progress on different economic sectors and emission agents [40]. As far as technological progress in the power industry is concerned, there is still room for growth. For example, the intermittent and random nature of renewable energy power generation has brought massive challenges to the operation and planning of the power system [41]. Sofia et al. approved that the electricity market must be reformed to encourage the development of zero-emission technologies [42]. Nonetheless, few documents have targeted their analysis on the effects of technological progress in the clean power sector, lacking a reasonable expectation of technological advancement in a specific clean power sector.

2.4. Research on CGE Model Involving Electricity

The literature mentioned above uses various theories and empirical models for research. Nevertheless, the CGE model is widely used in policy simulation. As a tool for policy simulation analysis, CGE can study issues such as carbon taxes, carbon trading rights, and energy efficiency improvements. In addition, many scholars have studied power-related issues using the CGE model. Based on the static CGE model, He et al. analyzed the impact of coal price adjustments on the power industry and electricity price adjustments on China's macroeconomics [43]. Meng constructed a CGE model to examine the impact of Australia's carbon tax policy on the power sector [44]. Lin and Jia used the improved CGE model to analyze only the relevant effect of the power industry's carbon trading market, indicating that the annual decline factor can reach 0.5% when allocating carbon allowances in the power sector [45]. Mardones and Brevis constructed a social accounting matrix (SAMEA) with environmental accounts and found differences between power sectors. If the power sector is not highly decomposed, the simulation effects of energy and environmental policies are biased [46]. Cui et al. used CGE models with different nesting structures and power sector substitution flexibility to study reducing the impact of renewable power reduction policies on economic development and the environment. Their studies have shown that reducing renewable electricity cuts can achieve various green benefits, such as cutting CO_2 emissions generated from the power sector and improving the real GDP [47]. Zhang et al. assessed the impact of three policies, including technological progress, on power generation, carbon emissions, and prices, analyzing how nuclear power generation could be promoted [48]. Nong built a CGE model of the electricity environment to study South Africa's carbon tax policy [49]. Based on these works, we selected the CGE model for research.

From the above research review, most of the existing articles analyze and simulate scenarios where only one variable changes, and it is impossible to further compare the results of the combined effects of multiple strategies. Although previous studies have recognized the negative impact of clean electricity on carbon emissions, none have analyzed whether its technological progress will further promote carbon emission reduction. To our knowledge, this study is the only of its kind attempting to fill this gap in understanding. Motivated by these limitations, this article mainly put forth the following contributions. First, this article simulated a combination of a carbon tax recycling policy and clean power technology, exploring the effectiveness of this interaction; second, this article introduces technological progress of the clean electricity sector and divides the power sector in detail. In this way, we analyzed the critical role of the combination of technological progress in the clean power sector and a carbon tax recycling policy in a targeted manner to identify the heterogeneity of different power sectors; finally, under the principle of tax neutrality, based on the double-dividend theory of carbon tax, this article conducts further excavations to explore carbon tax's triple-dividend effect.

3. Methodology and Data

3.1. CGE Model Structure

The CGE model builds on the Walras general equilibrium theory and portrays the real economic system. Precisely, the model consists of a production module, a trading module (domestic product demand and distribution), an institutional module, an equilibrium module, a social welfare module, and a carbon emission and carbon tax module, covering three types of institutions, namely the resident, the corporate, and the government. Figure 1 presents a general framework of the model.

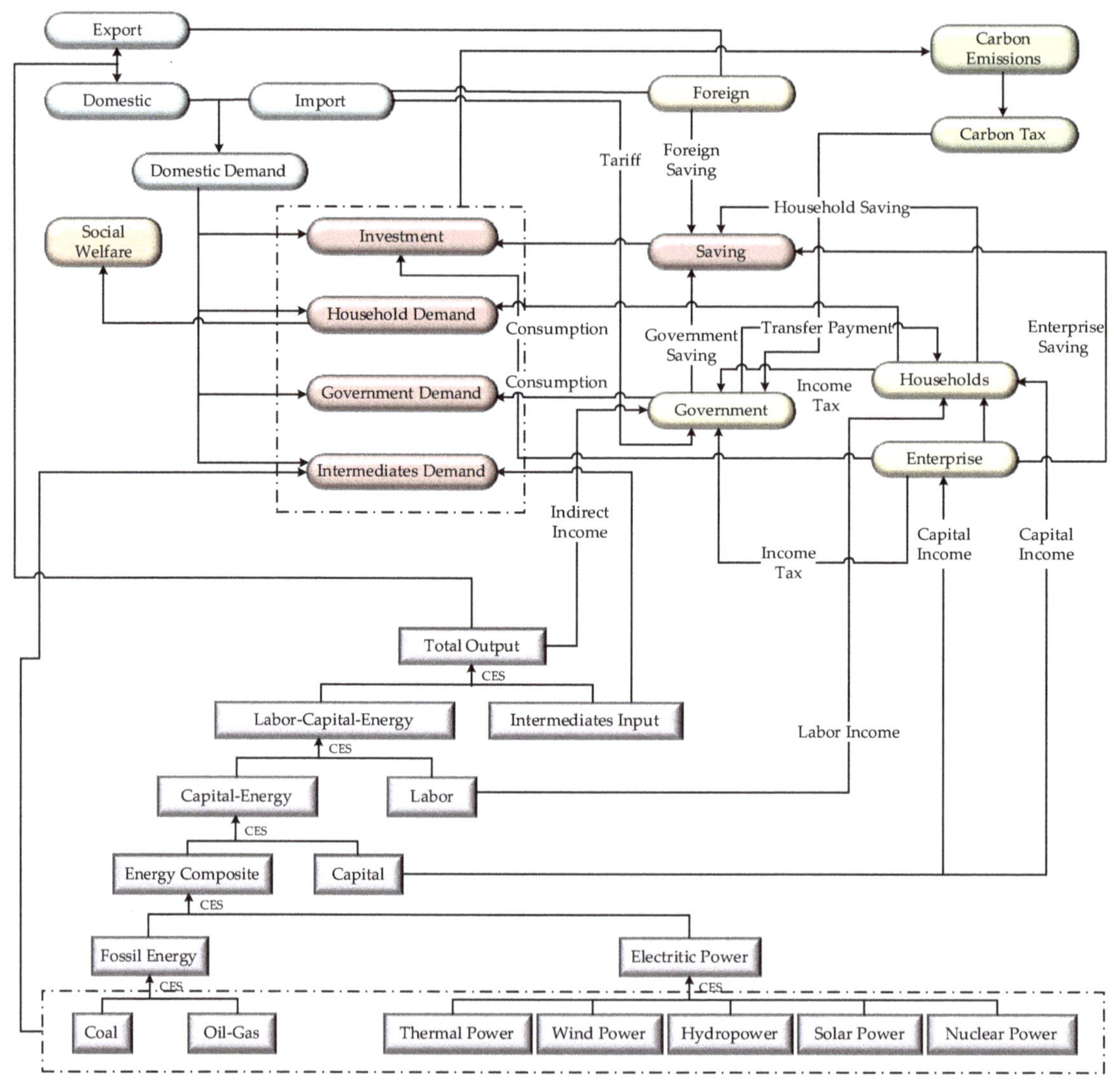

Figure 1. The framework of the CGE model.

3.1.1. Production Module

The production module of the CGE model in this paper includes five nested layers, all linked by a constant elasticity of substitution (CES) function, except for the intermediate inputs, which use the Leontief function. In the first layer, fossil energy is subdivided into coal and oil–natural gas, and electricity is divided into five types according to different generation technologies, namely thermal, hydro, nuclear, solar, and wind power. The second layer is the synthesis between fossil energy and electricity. The third layer is the synthesis of capital and energy complexes. The fourth layer is the synthesis between capital, energy complexes, and labor, i.e., the value-added component of the synthesis. The fifth layer synthesizes capital, energy, and labor complexes with intermediate inputs.

3.1.2. Trade Module

The allocation of domestic products in the trade module takes the form of a CET function, which describes the distribution strategy of commodities produced by domestic production activities between domestic production and domestic sales and exports. The demand for domestic products takes the form of an Armington function, which further compounds domestic production and domestic sales with imported goods to form domestic market commodities.

3.1.3. Institutional Module

The institutional module mainly includes residents, enterprises, and government modules. We set the corresponding income and expenditure functions. In the resident module, residents' income primarily comes from labor income, capital income, and transfer payments from enterprises and the government. The income of residents is used for consumption and saving, and this paper used a linear function to describe the consumption behavior of residents. In the enterprise module, enterprise income is mainly derived from capital income, and enterprise savings are derived from enterprise after-tax income minus enterprise transfers to residents. In the government module, the government income consists of indirect taxes, customs duties, residential and corporate income taxes, and carbon taxes. These revenues are in turn used for government consumption as well as savings, etc.

3.1.4. Balance Module

Four balances are set up in the equilibrium module, namely factor market equilibrium, product market equilibrium, savings–investment equilibrium, and balance-of-payments equilibrium. In this paper, we adopted the neoclassical closure rule, which assumes that the prices of labor and capital are endogenous and that full employment is achieved in the labor and capital markets. Therefore, in factor markets, the aggregate supply of labor and capital is equal to the aggregate demand for that factor. For product markets, the aggregate supply of products equals the aggregate demand. For saving and investment, when the economy reaches equilibrium, both aggregate investments equal aggregate savings. For the balance of payment equilibrium, we chose the closure rule with the exchange rate as the endogenous variable and foreign savings as the exogenous variable. Meanwhile, we set up the nominal GDP and real GDP equations. Nominal GDP is defined as the sum of total capital input, total labor input, and indirect tax revenue, while real GDP consists of consumptions, investments, and net exports.

3.1.5. Social Welfare Module

In the social welfare module, we introduced Hicks equivalence changes to measure the change in social welfare following a policy shock [50]. Specifically, the Hicks equivalent change calculates the change in utility before and after the policy's implementation, using the commodity's price before the policy shock as the standard. As shown in Equation (1):

$$EV = E\left(U^s, PQ^b\right) - E\left(U^b, PQ^b\right) = \sum_i PQ_i^b \cdot HD_i^s - \sum_i PQ_i^b \cdot HD_i^b \tag{1}$$

where, EV measures the change in the welfare of the population, $E\left(U^s, PQ^b\right)$ and $E\left(U^b, PQ^b\right)$ denote the expenditure required to achieve population welfare U^s and U^b, respectively, at PQ_i^b consumer prices before the policy shock. HD_i^s and HD_i^b represent the populations' consumption after and before implementing the policy, respectively. When EV is positive, it means that the population's welfare improved after the policy's implementation, and contrarily, if it is negative, the population's welfare has been compromised.

3.1.6. The Carbon Emission and Carbon Tax Module

This module calculated CO_2 emissions, further introducing a carbon tax. To analyze the carbon tax policy effect, we introduced a carbon tax shock module:

$$CTAX_i = tc \cdot \sum_j E_{i,j} \cdot \theta_j, \tag{2}$$

$$CTAX_j = tc \cdot \sum_i E_{i,j} \cdot \theta_j, \tag{3}$$

$$TCTAX = \sum_j CTAX_j, \tag{4}$$

$$t_{cj} = \frac{CTAX_j}{PQ_j \cdot QQ_j}, \tag{5}$$

where, $j = coal, oil - gas$. Among the above formulas, θ_j indicates the CO_2 emission factor per unit of energy from fossil fuels (coal and oil–gas), tc denotes the amount of carbon tax levied per ton of CO_2 emissions, t_{cj} indicates the corresponding carbon tax rate of fossil energy source j. $CTAX_i$ and $CTAX_j$ denote the amount of carbon tax levied on the sector i and the intermediate input component of fossil energy j, respectively. $TCTAX$ represents the total amount of carbon tax. This paper levied a carbon tax on intermediate energy inputs in the production process, not on the final demand sector.

3.2. Data

The data basis for the CGE model is the SAM table. The SAM table describes the supply and use flows between accounts in the System of National Accounts (SNA) and their equilibrium relationships, comprehensively portraying the economic linkages among sectors. Based on the 2017 China Input-Output tables (http://data.stats.gov.cn/, (accessed on 23 January 2022)), taxation, capital flows, and other relevant data, we constructed the SAM table in this paper. Among them, indirect taxes, as well as taxes such as income tax, were derived from the 2018 China Financial Yearbook (https://www.epsnet.com.cn/, (accessed on 23 January 2022)), and the capital and balance of payment transfer data are from the 2018 China Statistical Yearbook (http://www.stats.gov.cn/tjsj/ndsj/, (accessed on 23 January 2022)). In the SAM table, we set up six main account types: production activity accounts, commodity accounts, factor accounts, institutional accounts, investment accounts, and foreign accounts. According to the research needs, the CGE model constructed in this paper contains 13 industries and 13 commodities; the factor accounts include labor and capital, and the institutional accounts cover households, corporates, and the government. Specifically, in terms of methods of relevant research, combining relevant data from the 2019 China Electricity Yearbook (https://navi.cnki.net/knavi/yearbooks/, (accessed on 23 January 2022)) and the 2017 China Industrial Statistics Yearbook (https://data.cnki.net/Yearbook/, (accessed on 23 January 2022)), we divided the electricity sector [46,51]. Electricity production consists of five power generation technologies [37]: thermal, hydro, nuclear, wind, and solar power, including four clean power generation sectors: hydro, nuclear, wind, and solar power. Of these, coal and oil–gas only have intermediate inputs to thermal power, and there are no intermediate inputs to the clean power sector.

Calibration and setting of parameters in the model mainly include the carbon emission factor and the elasticity of substitution factor. The CO_2 emission factor for fossil energy is calculated based on the ratio of CO_2 emissions to actual energy consumption, where the CO_2 emissions data were obtained from International Energy Statistics (https://www.eia.gov/, (accessed on 23 January 2022)). As for the setting of the relevant elasticity coefficients required for the model, we refer to the relevant research literature [52–54], setting the production function elasticity of substitution coefficient, the CET function relevant elasticity of substitution, and the Armington function elasticity of substitution.

The share parameters were calibrated using the variable base year data and the elasticity of substitution calculations.

4. Simulation Analyses

Carbon tax policy is an effective tool for reducing carbon emissions, but it may adversely affect economic development and social welfare. This article analyzes the economic–environmental–energy effects of carbon tax policies and carbon tax recycling policies with and without clean power technology progress under specific carbon emission reduction targets and explores the possible triple dividends of carbon taxes. According to related simulation studies and empirical analysis, we set the simulated clean power sector's technological progress rate as 1%, 5%, and 10%, finding changing trends in various variables as technology advances [55,56]. We conducted relevant analysis against the benchmark scenario (no carbon tax and carbon tax recovery policy). The five scenarios are as follows:

Scenario 1: Assuming that the annual carbon dioxide emissions are reduced by 5% compared to the benchmark scenario, a carbon tax is levied on the intermediate energy input in the production process, while no carbon tax is levied on the final demand sector.

Scenario 2: Assuming that the annual carbon dioxide emissions are reduced by 5% compared to the benchmark scenario, a carbon tax is levied on the intermediate energy input in the production process, while no carbon tax is levied on the final demand sector. Under the premise of ensuring the neutrality of government revenue, the resident income tax rate will be reduced.

Scenario 3: Based on Scenario 2, we set different rates of technological progress for the clean power sector.

Scenario 4: Assuming that the annual carbon dioxide emissions are reduced by 5% compared to the benchmark scenario, a carbon tax is levied on the intermediate energy input in the production process, while no carbon tax is levied on the final demand sector. Unlike Scenario 2, under the premise of ensuring the neutrality of government revenue, the corporate income tax rate will be reduced.

Scenario 5: Based on Scenario 4, we set different rates of technological progress for the clean power sector.

4.1. The Impact of Policy Scenarios on Macroeconomic Variables

Based on the scenario shocks set in this article, the prices of input factors and relevant products were affected accordingly, and the entire economic system was further exerted through enterprise modules, resident modules, and other modules. Taking into account the three goals of economic growth, carbon emission reduction, and social welfare improvement, Table 1 presents the percentage changes in nominal GDP, real GDP, total carbon dioxide emission intensity, residential income, and enterprise income relative to the baseline scenario under different policy scenarios, showing the resulting social welfare.

Table 1. The impact of macroeconomic variables.

Scenarios	RTP	GDP (%)	RGDP (%)	TCOEI (%)	EV	YTH (%)	YTE (%)
Scenario 1	0%	−0.0377	−0.0493	−4.9641	−424.3465	0.0162	−0.0753
Scenario 2	0%	−0.0500	−0.0505	−4.9525	851.6128	−0.0144	−0.1022
	1%	−0.0457	−0.0334	−4.9566	865.6442	−0.0126	−0.0889
Scenario 3	5%	−0.0289	0.0335	−4.9725	922.0437	−0.0052	−0.0371
	10%	−0.0093	0.1144	−4.9912	993.0338	0.0034	0.0240
Scenario 4	0%	−0.0303	−0.0441	−4.9712	−530.8357	−0.0103	−0.0730
	1%	−0.0266	−0.0273	−4.9747	−470.2937	−0.0086	−0.0606
Scenario 5	5%	−0.0125	0.0389	−4.9881	−232.3547	−0.0018	−0.0127
	10%	0.0041	0.1189	−5.0039	55.9304	0.0062	0.0439

Note: RTP stands for the rate of technological progress for the clean power sectors; GDP stands for nominal GDP; RGDP stands for real GDP; TCOEI stands for total carbon dioxide emission intensity; EV stands for social welfare; YTH stands for the resident income; YTE stands for the corporate income.

4.1.1. Nominal GDP and Real GDP

It is clear from Table 1 that in Scenarios 1, 2, and 4, the nominal and real GDP decline. As carbon tax implementation leads to a fall in total output, the capital required for production falls, the price of capital decreases, and capital revenues fall. The labor price is the benchmark price, which keeps labor income constant. Indirect taxes do not account for a large proportion of the GDP, so the nominal GDP declines. For the real GDP, fossil energy price also rises due to implementing the carbon tax policy. First, the rising cost of fossil energy leads to an increase in the price of related commodities and a decline in overall consumer demand; secondly, the decrease in the demand for fossil energy by various sectors causes a decrease in enterprise output and investment; finally, the domestic levy of carbon taxes increases domestic commodity-related prices. Further, net exports are reduced. In summary, the real GDP falls.

As technological advances in clean electricity continue to rise, the capital required for production increases, and capital prices and capital revenues increase, leading to a gradual trend of increasing nominal GDP under both carbon tax recovery scenarios. Regarding the actual GDP, in Scenarios 2 and 3, when the carbon tax is recycled to residents to reduce residents' income tax, residents' income tax is lowered, which boosts the relevant consumer demand of residents. Likewise, the carbon tax is recycled to corporates, the corporate income tax is reduced, and the corporates' savings rise. According to the neoclassical closure rule used in this paper, investment is determined by savings. Thus, corporate investment rises further, promoting economic growth. In addition, the progress of clean power technology has further mobilized the enthusiasm of clean power companies in production, and the actual GDP continues to rise.

4.1.2. Resident Income

Resident income comprises labor income, capital income, and transfer payments from enterprises and the government. The model uses labor price as the benchmark price for related simulations as far as labor income is concerned. Therefore, the labor income of residents remains unchanged. In terms of capital income, the carbon tax policy raises the price of fossil energy, especially the cost of production for resource-intensive enterprises, causing a fall in total output and a decline in the demand for capital. Correspondingly, capital prices and total capital income are reduced. Residents receive capital income according to a set proportional coefficient, and therefore, their capital income decreases accordingly. In Scenario 1, due to the implementation of the carbon tax policy, government revenues increase, which increases the transfer payments to residents in turn, so overall residents' income can be improved. As for Scenario 2, the carbon tax is transferred to residents at a reduced income tax rate, which declines residents' income tax. However, the government income decreases relative to Scenario 1, resulting in a decrease in government transfers to residents. In general, residents' income decreased. In Scenario 3, with the advancement of clean power technology, the clean power output increases significantly, and the demand for capital rises, which corresponds to the rise in capital prices. The corporate capital income and transfer payments to residents increase. In addition, residents' capital income also increases, making residents' income gradually rise compared with Scenario 2. Similarly, in Scenario 4, the government's transfer payments to residents decrease, and residents' revenues decline. In Scenario 5, the progress of clean power technology causes residents' income to gradually rise compared with Scenario 4.

4.1.3. Corporate Income

The corporate income is derived from the income generated by the capital elements input. Total capital income is allocated to the business on a proportionate share base. Like the relevant analysis of residents' income, in Scenarios 1, 2, and 4, the carbon tax levy reduces capital income and corporate income. For Scenarios 3 and 5, advances in clean power technology further increase the demand for capital in the production process. Capital prices gradually rise, capital income increases, and corporate income also shows a gradual upward trend. In the end, under the advancement of 10% clean electricity technology for both scenarios where the carbon tax is recycled to residents and businesses, business revenues rise by 0.0240% and 0.0439% relative to the baseline scenario, respectively.

4.1.4. Social Welfare

As for Scenario 1, on the one hand, the increase in residents' income leads to an increase in residents' demand. On the other hand, the levy of carbon taxes increases the production costs of enterprises, further increasing product prices and reducing consumer demand, which ultimately causes negative social welfare. Regarding Scenario 2, the carbon tax is recycled for residents. Although the decline in capital income and government transfer payments to residents causes a decrease in residents' income, the income tax rate of residents decreases, which eventually increases residents' demand, and social welfare becomes positive. Regarding Scenario 3, as the rate of clean power technology progress gradually increases, capital income rises, corporate transfer payments to residents increase, residents' consumption demand further increases, and social welfare continues to improve.

For Scenario 4, the carbon tax is recycled to corporates, residents' income is decreased, and consumption is reduced. Compared to Scenario 1, the social welfare is smaller. For Scenario 5, the carbon tax is recycled to corporates, and as the clean electricity technology advances, corporate further increases the production in the clean electricity sector, which in turn increases the corresponding consumption of the residents. Moreover, due to the increase in firms' income, production in other sectors also rises to various degrees, increasing the corresponding consumption. As for residents, the increase in residents' income also drives up residents' consumption. Therefore, social welfare continues to rise as well. Eventually, social welfare becomes positive when clean electricity technology advances by 10%.

4.1.5. CO_2 Emission Intensity

The main reason for the reduction in carbon dioxide emission intensity is that the price of fossil energy has risen, the use of fossil energy has fallen, and more clean energy sources have replaced fossil energy with higher carbon content. Accordingly, there is a corresponding reduction in CO_2 emissions. In particular, the clean power sector technology further promotes fossil energy conversion, the share of clean power usage rises, and the industrial structure is adjusted and upgraded. Under the conditions of a certain decline in carbon dioxide, the nominal GDP has a rising trend, and the intensity of carbon dioxide emissions continues to decline. As shown in Table 1, with the progress of clean power technology, the reduction in carbon dioxide emission intensity gradually increases, which is beneficial to achieving China's carbon emission reduction targets.

From the results in Table 1, we can see that the carbon tax recovery policy and technological progress are important conditions in achieving the triple dividend of the carbon tax. The carbon tax recycled to residents directly improves social welfare, and with the advancement of clean power technology, social welfare is further improved. While carbon tax recovery to companies initially hurts the welfare of residents, social welfare improves when clean electricity technology advances to a certain level. Finally, at the 10% level of clean electricity technology, carbon tax recycling to corporates realizes the triple dividend of the carbon tax. Compared to the baseline scenario, GDP rises, carbon reduction intensity decreases significantly, and social welfare improves.

4.2. The Impact of Policy Scenarios on Carbon Dioxide Emissions

Based on the impact of carbon tax policy and clean power technology progress on CO_2 emissions, we simulated the relationship between carbon tax recovery with clean power technology advancement and a carbon tax rate under the 5% emission reduction constraint, analyzing the emission reduction contribution of different fossil energy sources.

We only selected the 5% clean power technology progress rate as a representative value. Table 2 displays the carbon tax rate and fossil energy emission reduction contribution under different policy simulation combinations.

Table 2. The carbon tax rate and contribution of fossil energy emission reduction.

Scenarios		1	2	3	4	5
Emission reduction	Coal	97.5820	97.6405	96.8750	98.1246	97.2784
contribution (%)	Oil–gas	2.4180	2.3595	3.1250	1.8754	2.7216
Ad valorem tax rate	Coal	0.0505	0.0508	0.0435	0.0517	0.0443
	Oil–gas	0.0129	0.0130	0.0111	0.0133	0.0113
Carbon tax rate (CNY/ton)		15.3563	15.4551	13.2062	15.7485	13.4477

Under specific CO_2 emission reduction targets, the carbon tax rate was determined. Regardless of the carbon tax recovery policy, with the technological advancement of the clean power sector, the carbon tax rate declines. This results from technological advances in clean power, where more fossil energy is replaced by clean power, with less fossil energy consumption and lower carbon emissions per unit. As a result, the carbon tax rate declines under a set carbon reduction target. We found that technological advances in appropriate clean electricity lower the carbon tax rate, which is beneficial for promoting carbon tax policies.

As seen in Table 2, the ad valorem tax rate on coal is relatively high. This is because coal has a higher carbon content and a higher carbon dioxide emission coefficient than oil–gas. Coal is more impacted by carbon tax policies and bears more taxes. Similarly, with the advancement of clean power technology, ad valorem tax rates for coal and oil–gas fall.

In terms of emission reduction contribution, it is not difficult to find that coal emission reduction's contribution is enormous, while the oil–gas emission reduction contribution is relatively low. Being a major source of CO_2 emissions in China, coal has a high carbon emission coefficient, which is the key to carbon emission reduction. Therefore, the use of coal resources should be reduced. Additionally, coal-to-electricity technology should be promoted, with relatively clean energy sources as an alternative.

4.3. The Impact of Policy Scenarios on Electricity Energy

Refining the power sector helped us study the power energy changes in detail during the simulation period. In this regard, the focus was to analyze the impact of carbon tax recovery policies on power consumption under the progress of clean power technology and the impact of departmental thermal power consumption; thus, we demonstrated the transformation of the power structure further and analyzed the effect of the policy.

4.3.1. Electricity Consumption

The collection of carbon tax policy affects the price of fossil energy and increases the production cost of enterprises. In this regard, the consumption demand of various energy sources will also change. In addition, technological progress in the clean power sector has a specific impact on various energy sources, especially electric energy. Selecting the 5% clean electricity technology progress rate as a representative value, we considered the changes in electricity consumption under carbon tax policies and different carbon tax recovery policies, as shown in Figure 2.

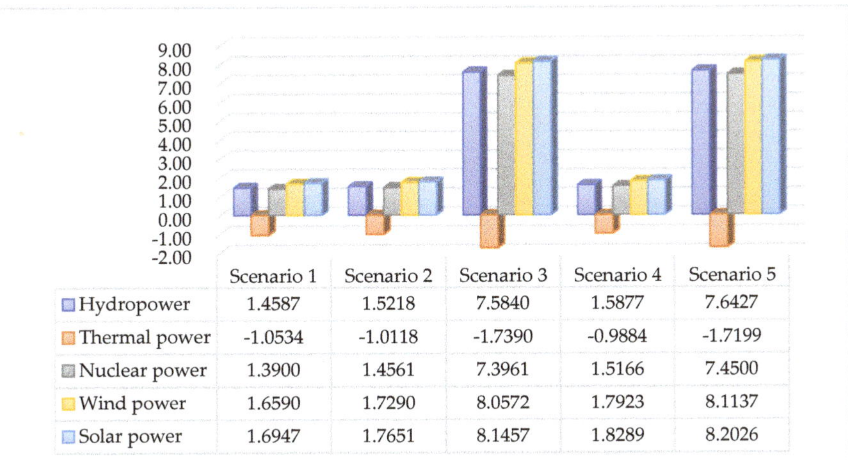

Figure 2. Percentage changes in electricity consumption (2017).

After implementing the carbon tax policy, coal and oil–natural gas were directly impacted. Corporates continue to seek alternative energy sources. Electricity, as an alternative energy source, is also indirectly affected. As a secondary energy source, thermal power indirectly emits carbon dioxide, and the levy of a carbon tax reduces the energy consumption of thermal power. Hydropower, nuclear power, wind power, and solar power generation, on the one hand, as clean energy, can replace fossil energy and thermal power and reduce carbon dioxide emissions. As a result, the consumption of clean power energy increases. On the other hand, as the rate of technological progress in the clean power sector increases, the clean power sector has increased production to replace thermal power generation. Thermal power consumption continues to decline, while clean power energy consumption increases significantly. The progress of clean power technology has a greater impact on the power industry. At the same time, the above results show that different power sectors have different responses to the impact of carbon taxes and technological progress, confirming the importance of subdividing the power sector.

4.3.2. Sectoral Thermal Power Consumption

In China, coal is the primary raw material for thermal power generation. We focused on selecting thermal power sources with higher carbon dioxide emissions, analyzing the impact of carbon tax recovery policies and clean power technology progress on thermal power consumption in various sectors.

Undoubtedly, from Figure 3, it can be seen that the consumption of thermal power in the clean power sector has increased, while the consumption of thermal power in other sectors has decreased. Among them, the coal sector has the largest decline in thermal power consumption, followed by oil–gas and construction industries. This is because the levy of a carbon tax increases the price of fossil energy and indirectly increases the cost of thermal power generation. The carbon tax has a relatively small impact on clean power, and China's clean power accounts for a relatively small proportion. Enterprises' demand for clean power increases, the clean power sector continues to grow, and the consumption of thermal power in the clean power sector increases. This change is noticeable, especially after the advancement of clean power technology, because the clean power sector develops more rapidly.

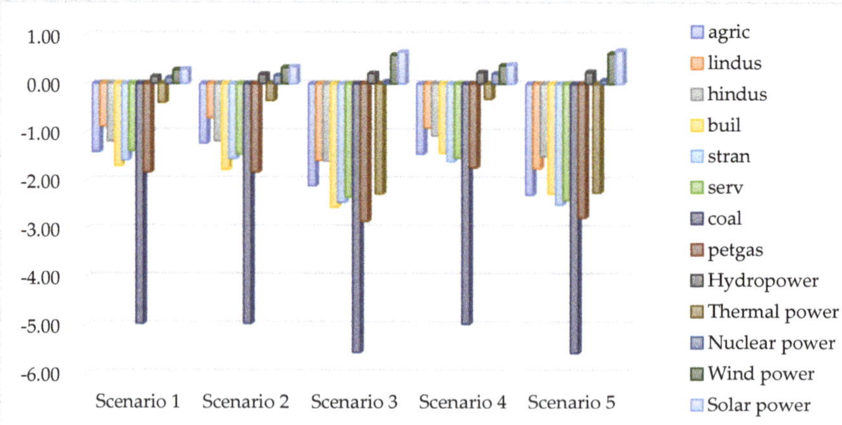

Figure 3. Percentage changes in sectoral thermal power consumption (2017).

4.3.3. Electricity Structure

The levy of a carbon tax increases the price of fossil energy, while the output of fossil energy and thermal power energy declines. Coupled with the progress of clean power technology, the output of clean power energy increases. In this regard, the power structure has been improved continuously.

As shown in Figure 4, thermal power is the main source of power generation. Progress in clean power technology significantly changes the power structure. With technological progress, the production efficiency of the clean power sector increases, and the output increases. In contrast, thermal power production declines. Therefore, the proportion of thermal power declines. The advancement of clean power technology is significant for changing the power structure and reducing carbon emissions. The proportion of thermal power is still large. Thus, reasonable measures need to be taken to adjust the power structure to reduce the proportion of thermal power continuously.

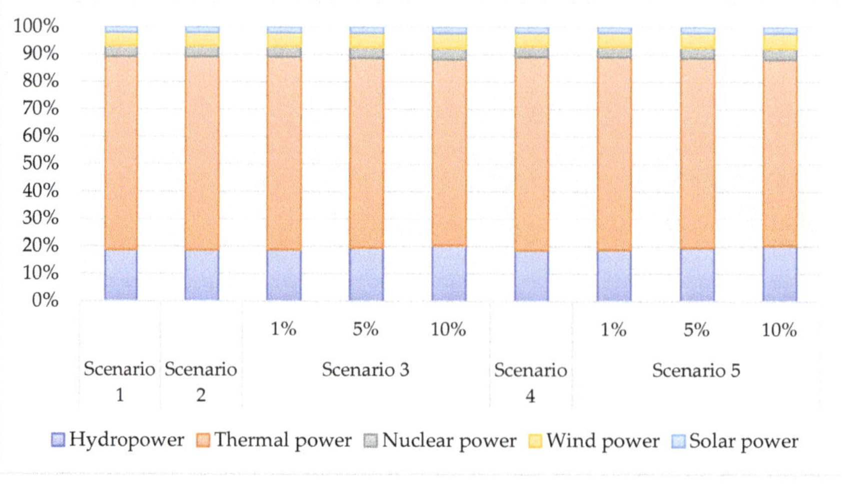

Figure 4. Changes in the power structure (2017).

5. Conclusions and Policy Suggestions

This paper constructed a computable general equilibrium model of the power sector segmentation, taking into account the two tools for reducing carbon dioxide emissions, technological progress, and carbon tax, studying the economic–environment–energy effects of the carbon tax recovery policy with clean power technology progress. Through the above research, we have drawn several conclusions. First, a carbon tax recovery policy supplemented by technological progress in the clean power sector can promote economic growth, improve social welfare, and reduce the intensity of carbon dioxide emissions, realizing the triple dividend of the carbon tax. Carbon tax alone can reduce carbon emissions and adjust the energy structure, but the carbon tax policy will adversely affect economic development and social welfare [25–27]. A carbon tax recovery policy can remedy this problem. At the same time, the introduction of clean power technology advances make up for the negative impact of economic growth and social welfare losses, promoting the sustainable growth of the green economy. Second, advances in clean power technology promote the transformation of the power structure. The advancement of clean power technology drives the production of the clean power sector and replaces thermal power generation. Without difficulty, we found that the proportion of thermal power production declines, while the proportion of other clean power increases, which is conducive to promoting carbon emission reduction. Lastly, increasing the use of clean energy is the key to reducing carbon emissions. By analyzing the contribution of carbon dioxide emission reduction, we found that the contribution of coal emission reduction is relatively significant, and the ad valorem tax rate is high. The above is because coal has a higher carbon content and emits more carbon dioxide.

Based on the conclusions of the above research, we put forward the following policy recommendations for the future development of China's economy. Firstly, we should pay attention to the role of carbon tax policy and carbon tax recycling. As an essential environmental fiscal policy, the carbon tax is an important measure needed to achieve carbon peak and carbon-neutral goals. However, China has not yet implemented a carbon tax policy and is still in the exploratory stage. The carbon tax policy can effectively supplement China's existing ETS [57]. On this basis, we should adopt reasonable carbon tax recovery methods to reverse the adverse effects of carbon tax policies continuously. Secondly, we need to continue to accelerate technological progress in the clean power sector and deepen the technical research and development of China's clean power sector and accelerate the transformation of electricity, that is, to promote the transformation of electricity from high-carbon to low-carbon sources, in a fossil-energy-based to clean-energy-based transformation. The substitution of thermal power and clean power is worthy of attention. We should vigorously develop clean power and reduce carbon dioxide emissions. Finally, we must promote clean coal and coal-to-electricity policies and accelerate the implementation of energy decarbonization policies. Coal significantly contributes to carbon emissions; therefore, it is essential that we accelerate and promote the transition from coal to electricity. Additionally, the state should increase the promotion of alternative renewable fuels [58].

This article provides a scientific basis for realizing carbon emission reduction targets and provides a particular reference for policymakers weighing economic development, environmental quality, and social welfare goals. However, there is still room for exploration in this article. First, a dynamic CGE model should be introduced to further compare the long and short-term effects of a carbon tax recovery policy along with the development of clean power technology on the power industry. Second, China's economic system is relatively complex. To avoid one-sided research, more policy tools should be taken into consideration while exploring a more effective combination of carbon emission reduction and stable economic growth policies, contributing to the sustainable development of China's green economy.

Author Contributions: W.L. designed and conceived this article; M.L. built the model, executed the program, and wrote most of the content of this article; T.L. processed the data and built the SAM table; Y.L. checked the data and proofread the text of this article; and Y.H. revised the chart of this essay. The published version of the manuscript has been read and agreed upon by all authors. All authors have read and agreed to the published version of the manuscript.

Funding: This study was financially supported by the research on "Factor Allocation and Industrial Upgrading Policies for Stable Economic Growth under the New Normal", Major Project of Key Research Bases of Humanities and Social Sciences, Ministry of Education, China (16JJD790015).

Institutional Review Board Statement: Not applicable.

Informed Consent Statement: Not applicable.

Data Availability Statement: China Financial Yearbook (https://www.epsnet.com.cn/, (accessed on 23 January 2022)), China Statistical Yearbook (http://www.stats.gov.cn/tjsj/ndsj/, (accessed on 23 January 2022)), China Electricity Yearbook (https://navi.cnki.net/knavi/yearbooks/, (accessed on 23 January 2022)), China Industrial Statistics Yearbook (https://data.cnki.net/Yearbook/, (accessed on 23 January 2022)), China Input-Output Tables (http://data.stats.gov.cn/, (accessed on 23 January 2022)), International Energy Statistics (https://www.eia.gov/, (accessed on 23 January 2022)).

Conflicts of Interest: The authors declare no conflict of interest.

References

1. Wu, S.; Li, S.; Lei, Y.; Li, L. Temporal Changes in China's Production and Consumption-Based CO_2 Emissions and the Factors Contributing to Changes. *Energy Econ.* **2020**, *89*, 104770. [CrossRef]
2. Amster, E.; Lew Levy, C. Impact of Coal-Fired Power Plant Emissions on Children's Health: A Systematic Review of the Epidemiological Literature. *Int. J. Environ. Res. Public Health* **2019**, *16*, 2008. [CrossRef] [PubMed]
3. Ye, P.; Xia, S.; Xiong, Y.; Liu, C.; Li, F.; Liang, J.; Zhang, H. Did an Ultra-Low Emissions Policy on Coal-Fueled Thermal Power Reduce the Harmful Emissions? Evidence from Three Typical Air Pollutants Abatement in China. *Int. J. Environ. Res. Public Health* **2020**, *17*, 8555. [CrossRef] [PubMed]
4. Pradhan, B.K.; Ghosh, J. COVID-19 and the Paris Agreement Target: A CGE Analysis of Alternative Economic Recovery Scenarios for India. *Energy Econ.* **2021**, *103*, 105539. [CrossRef]
5. Wang, H.; Chen, Z.P.; Wu, X.Y.; Niea, X. Can a Carbon Trading System Promote the Transformation of a Low-Carbon Economy under the Framework of the Porter Hypothesis? -Empirical Analysis Based on the PSM-DID Method. *Energy Policy* **2019**, *129*, 930–938. [CrossRef]
6. Zhou, Y.; Fang, W.; Li, M.; Liu, W. Exploring the Impacts of a Low-Carbon Policy Instrument: A Case of Carbon Tax on Transportation in China. *Resour. Conserv. Recycl.* **2018**, *139*, 307–314. [CrossRef]
7. Jia, Z.J.; Lin, B.Q. Rethinking the Choice of Carbon Tax and Carbon Trading in China. *Technol. Forecast. Soc. Change* **2020**, *159*, 120187. [CrossRef]
8. Zhao, Y.H.; Li, H.; Xiao, Y.L.; Liu, Y.; Cao, Y.; Zhang, Z.H.; Wang, S.; Zhang, Y.F.; Ahmad, A. Scenario Analysis of the Carbon Pricing Policy in China's Power Sector through 2050: Based on an Improved CGE Model. *Ecol. Indic.* **2018**, *85*, 352–366. [CrossRef]
9. Dissanayake, S.; Mahadevan, R.; Asafu-Adjaye, J. Evaluating the Efficiency of Carbon Emissions Policies in a Large Emitting Developing Country. *Energy Policy* **2020**, *136*, 111080. [CrossRef]
10. Zhang, P.; Yin, G.; Duan, M. Distortion Effects of Emissions Trading System on Intra-Sector Competition and Carbon Leakage: A Case Study of China. *Energy Policy* **2020**, *137*, 111126. [CrossRef]
11. Sun, Y.; Mao, X.; Yin, X.; Liu, G.; Zhang, J.; Zhao, Y. Optimizing Carbon Tax Rates and Revenue Recycling Schemes: Model Development, and a Case Study for the Bohai Bay Area, China. *J. Clean. Prod.* **2021**, *296*, 126519. [CrossRef]
12. Li, X.; Yao, X.; Guo, Z.; Li, J. Employing the CGE model to analyze the impact of carbon tax revenue recycling schemes on employment in coal resource-based areas: Evidence from Shanxi. *Sci. Total Environ.* **2020**, *720*, 137192. [CrossRef] [PubMed]
13. Li, P.; Ouyang, Y.F. Quantifying the Role of Technical Progress towards China's 2030 Carbon Intensity Target. *J. Environ. Plan. Manage* **2021**, *64*, 379–398. [CrossRef]
14. Kutlu, L. Greenhouse Gas Emission Efficiencies of World Countries. *Int. J. Environ. Res. Public Health* **2020**, *17*, 8771. [CrossRef]
15. Belaïd, F.; Zrelli, M.H. Renewable and Non-renewable Electricity Consumption, Environmental Degradation and Economic Development: Evidence from Mediterranean Countries. *Energy Policy* **2019**, *133*, 110929. [CrossRef]
16. Ehigiamusoe, K.U. A Disaggregated Approach to Analyzing the Effect of Electricity on Carbon Emissions: Evidence from African Countries. *Energy Rep.* **2020**, *6*, 1286–1296. [CrossRef]
17. Lin, B.; Li, Z. Is More Use of Electricity Leading to Less Carbon Emission Growth? An Analysis with a Panel Threshold Model. *Energy Policy* **2020**, *137*, 111121. [CrossRef]
18. Wong, J.B.; Zhang, Q. Impact of Carbon Tax on Electricity Prices and Behaviour. *Financ. Res. Lett.* **2021**, *44*, 102098. [CrossRef]

19. Yang, W.; Song, J. Simulating Optimal Development of Clean Coal-Fired Power Generation for Collaborative Reduction of Air Pollutant and CO_2 Emissions. *Sustain. Prod. Consum.* **2021**, *28*, 811–823. [CrossRef]
20. Haxhimusa, A.; Liebensteiner, M. Effects of electricity demand reductions under a carbon pricing regime on emissions: Lessons from COVID-19. *Energy Policy* **2021**, *156*, 112392. [CrossRef]
21. Fan, F.; Wang, Y.; Liu, Q. China's carbon emissions from the electricity sector: Spatial characteristics and interregional transfer. *Integr. Environ. Asses.* **2022**, *18*, 258–273. [CrossRef]
22. Mostafaeipour, A.; Bidokhti, A.; Fakhrzad, M.B.; Sadegheih, A.; Mehrjerdi, Y.Z. A new model for the use of renewable electricity to reduce carbon dioxide emissions. *Energy* **2022**, *238*, 121602. [CrossRef]
23. Metcalf, G.E. Carbon Taxes in Theory and Practice. *Annu. Rev. Resour. Econ.* **2021**, *13*, 245–265. [CrossRef]
24. Yamazaki, A. Jobs and Climate Policy: Evidence from British Columbia's Revenue-Neutral Carbon Tax. *J. Environ. Econ. Manag.* **2017**, *83*, 197–216. [CrossRef]
25. Lin, B.; Jia, Z. The Energy, Environmental and Economic Impacts of Carbon Tax Rate and Taxation Industry: A CGE Based Study in China. *Energy* **2018**, *159*, 558–568. [CrossRef]
26. Hu, H.; Dong, W.; Zhou, Q. A Comparative Study on the Environmental and Economic Effects of a Resource Tax and Carbon Tax in China: Analysis Based on the Computable General Equilibrium Model. *Energy Policy* **2021**, *156*, 112460. [CrossRef]
27. Khastar, M.; Aslani, A.; Nejati, M. How Does Carbon Tax Affect Social Welfare and Emission Reduction in Finland? *Energy Rep.* **2020**, *6*, 736–744. [CrossRef]
28. Pearce, D. The Role of Carbon Taxes in Adjusting to Global Warming. *Econ. J.* **1991**, *101*, 938–948. [CrossRef]
29. Pohit, S.; Pal, B.; Ojha, V.; Roy, J. *GHG Emissions and Economic Growth: A Computable General Equilibrium Model Based Analysis for India*; Springer: Berlin/Heidelberg, Germany, 2014.
30. Metcalf, G.E. On the Economics of a Carbon Tax for the United States. *Brook. Pap. Econ. Act.* **2019**, *2019*, 405–484. [CrossRef]
31. Ojha, V.P.; Pohit, S.; Ghosh, J. Recycling Carbon Tax for Inclusive Green Growth: A CGE Analysis of India. *Energy Policy* **2020**, *144*, 111708. [CrossRef]
32. Kirchner, M.; Sommer, M.; Kratena, K.; Kletzan-Slamanig, D.; Kettner-Marx, C. CO_2 Taxes, Equity and the Double Dividend—Macroeconomic Model Simulations for Austria. *Energy Policy* **2019**, *126*, 295–314. [CrossRef]
33. Steenkamp, L.-A. A Classification Framework for Carbon Tax Revenue Use. *Clim. Policy* **2021**, *21*, 897–911. [CrossRef]
34. Wills, W.; La Rovere, E.L.; Grottera, C.; Naspolini, G.F.; Le Treut, G.; Ghersi, F.; Lefèvre, J.; Dubeux, C.B.S. Economic and Social Effectiveness of Carbon Pricing Schemes to Meet Brazilian NDC Targets. *Clim. Policy.* **2021**, *22*, 1–16. [CrossRef]
35. Chen, J.; Gao, M.; Ma, K.; Song, M. Different effects of technological progress on China's carbon emissions based on sustainable development. *Bus. Strategy Environ.* **2020**, *29*, 481–492. [CrossRef]
36. Wu, L.; Liu, C.; Ma, X.; Liu, G.; Miao, C.; Wang, Z. Global Carbon Reduction and Economic Growth under Autonomous Economies. *J. Clean. Prod.* **2019**, *224*, 719–728. [CrossRef]
37. Guo, Z.Q.; Zhang, X.P.; Ding, Y.H.; Zhao, X.N. A Forecasting Analysis on China'S Energy Use and Carbon Emissions Based on A Dynamic Computable General Equilibrium Model. *Emerg. Mark. Financ. Trade* **2021**, *57*, 727–739. [CrossRef]
38. Settimo, G.; Avino, P. The Dichotomy Between Indoor Air Quality and Energy Efficiency in Light of the Onset of the COVID-19 Pandemic. *Atmosphere* **2021**, *12*, 791. [CrossRef]
39. Chen, J.; Gao, M.; Mangla, S.K.; Song, M.; Wen, J. Effects of Technological Changes on China's Carbon Emissions. *Technol. Forecast. Soc. Change* **2020**, *153*, 119938. [CrossRef]
40. Wang, S.; Zeng, J.; Liu, X. Examining the Multiple Impacts of Technological Progress on CO_2 Emissions in China: A Panel Quantile Regression Approach. *Renew. Sust. Energy Rev.* **2019**, *103*, 140–150. [CrossRef]
41. Zeng, Y.; Zhang, R.; Wang, D.; Mu, Y.; Jia, H. A Regional Power Grid Operation and Planning Method Considering Renewable Energy Generation and Load Control. *Appl. Energy* **2019**, *237*, 304–313. [CrossRef]
42. Sofia, D.; Gioiella, F.; Lotrecchiano, N.; Giuliano, A. Cost-Benefit Analysis to Support Decarbonization Scenario for 2030: A Case Study in Italy. *Energy Policy* **2020**, *137*, 111137. [CrossRef]
43. He, Y.X.; Zhang, S.L.; Yang, L.Y.; Wang, Y.J.; Wang, J. Economic Analysis of Coal Price-Electricity Price Adjustment in China Based on the CGE Model. *Energy Policy* **2010**, *38*, 6629–6637. [CrossRef]
44. Meng, S. How May a Carbon Tax Transform Australian Electricity Industry? A CGE Analysis. *Appl. Econ.* **2014**, *46*, 796–812. [CrossRef]
45. Lin, B.; Jia, Z. What will China's Carbon Emission Trading Market Affect with only Electricity Sector Involvement? A CGE Based Study. *Energy Econ.* **2019**, *78*, 301–311. [CrossRef]
46. Mardones, C.; Brevis, C. Constructing a SAMEA to Analyze Energy and Environmental Policies in Chile. *Econ. Syst. Res.* **2020**, *33*, 1–27. [CrossRef]
47. Cui, Q.; Liu, Y.; Ali, T.; Gao, J.; Chen, H. Economic and Climate Impacts of Reducing China's Renewable Electricity Curtailment: A Comparison Between CGE Models with Alternative Nesting Structures of Electricity. *Energy Econ.* **2020**, *91*, 104892. [CrossRef]
48. Zhang, T.; Ma, Y.; Li, A. Scenario Analysis and Assessment of China's Nuclear Power Policy Based on the Paris Agreement: A Dynamic CGE Model. *Energy* **2021**, *228*, 120541. [CrossRef]
49. Nong, D. Development of the electricity-environmental policy CGE model (GTAP-E-PowerS): A case of the carbon tax in South Africa. *Energy Policy* **2020**, *140*, 111375. [CrossRef]

50. Mas-Colell, A.; Whinston, M.D.; Green, J.R. *Microeconomic Theory*; Oxford University Press: New York, NY, USA, 1995; Volume 1.
51. Lindner, S.; Legault, J.; Guan, D. Disaggregating the Electricity Sector of China's Input-Output Table for Improved Environmental Life-Cycle Assessment. *Econ. Syst. Res.* **2013**, *25*, 300–320. [CrossRef]
52. Guo, Z.Q.; Zhang, X.P.; Zheng, Y.H.; Rao, R. Exploring the Impacts of a Carbon Tax on the Chinese Economy Using a CGE Model with a Detailed Disaggregation of Energy Sectors. *Energy Econ.* **2014**, *45*, 455–462. [CrossRef]
53. Guo, Z.Q.; Zhang, X.P.; Feng, S.D.; Zhang, H.N. The Impacts of Reducing Renewable Energy Subsidies on China's Energy Transition by Using a Hybrid Dynamic Computable General Equilibrium Model. *Front. Energy Res.* **2020**, *8*, 25. [CrossRef]
54. Zhai, M.; Huang, G.; Liu, L.; Guo, Z.; Su, S. Segmented Carbon Tax May Significantly Affect the Regional and National Economy and Environment-a CGE-Based Analysis for Guangdong Province. *Energy* **2021**, *231*, 120958. [CrossRef]
55. Fujimori, S.; Masui, T.; Matsuoka, Y. Development of a global computable general equilibrium model coupled with detailed energy end-use technology. *Appl. Energy* **2014**, *128*, 296–306. [CrossRef]
56. Cheng, B.; Dai, H.; Wang, P.; Xie, Y.; Chen, L.; Zhao, D.; Masui, T. Impacts of low-carbon power policy on carbon mitigation in Guangdong Province, China. *Energy Policy* **2016**, *88*, 515–527. [CrossRef]
57. Lin, B.; Jia, Z. Can Carbon Tax Complement Emission Trading Scheme? The Impact of Carbon Tax on Economy, Energy and Environment in China. *Clim. Chang. Econ.* **2020**, *11*, 2041002. [CrossRef]
58. Sofia, D.; Gioiella, F.; Lotrecchiano, N.; Giuliano, A. Mitigation Strategies for Reducing Air Pollution. *Environ. Sci. Pollut. Res.* **2020**, *27*, 19226–19235. [CrossRef]

International Journal of
*Environmental Research
and Public Health*

Article

Do Tournament Incentives Matter for CEOs to Be Environmentally Responsible? Evidence from Chinese Listed Companies

Sajid Ullah [1], Farman Ullah Khan [2,*], Laura-Mariana Cismaş [3,*], Muhammad Usman [4,*] and Andra Miculescu [3]

[1] School of Economics and Management, Xi'an University of Technology, Xi'an 710048, China; sajidkhan6695@stu.xaut.edu.cn
[2] School of Management, Xi'an Jiaotong University, Xi'an 710049, China
[3] Faculty of Economics and Business Administration, West University of Timisoara, 300006 Timisoara, Romania; andra.miculescu@e-uvt.ro
[4] School of Accounting, Nanjing Audit University, Nanjing 210017, China
* Correspondence: farman@stu.xjtu.edu.cn (F.U.K.); laura.cismas@e-uvt.ro (L.-M.C.); 320335@nau.edu.cn (M.U.)

Abstract: Relying on tournament theory and environmental management research, we examine how CEO tournament incentives induce top executives to invest more in green innovation. Using a sample of Chinese listed companies from 2010 to 2016, we find evidence that CEO tournament incentives are positively associated with green innovation. In addition, we find that a positive relationship between CEO tournament incentives and green innovation is stronger in state-owned enterprises than in non-state-owned enterprises. These results support tournament theory, which proposes that better incentives induce top executives' efforts to win the tournament incentives, and such efforts are subject to fiercer competition among employees, which improves firms' social and financial performance. Moreover, our findings have implications for policy makers and regulators who wish to enhance environmental legitimacy by providing tournament incentives to top executives.

Keywords: green innovation; CEO tournament incentive; tournament theory; state-owned enterprises; China

Citation: Ullah, S.; Khan, F.U.; Cismaş, L.-M.; Usman, M.; Miculescu, A. Do Tournament Incentives Matter for CEOs to Be Environmentally Responsible? Evidence from Chinese Listed Companies. *Int. J. Environ. Res. Public Health* **2022**, *19*, 470. https://doi.org/10.3390/ijerph19010470

Academic Editors: Pasquale Avino, Massimiliano Errico, Hamid Salehi and Aristide Giuliano

Received: 23 November 2021
Accepted: 30 December 2021
Published: 1 January 2022

Publisher's Note: MDPI stays neutral with regard to jurisdictional claims in published maps and institutional affiliations.

1. Introduction

The recent stratospheric emphasis on incentivizing executives for achieving social goals has caught scholars' attention. Companies concerned about environmental degradation are more likely to correlate executive compensation to environmental performance, recognizing that management should be rewarded for the higher risk associated with long-term goals (i.e., environmental performance) [1]. Indeed, some firms connect CEO compensation to environmental performance, recognizing that if they connect CEO compensation and environmental performance, they can effectively reduce environmental degradation in polluting industries [1]. As developing countries' environmental quality are extremely low, moreover, these countries are likely to face green innovation adoption challenges [2]; therefore, researchers are looking for drivers to promote green innovation in underdeveloped countries [3]. Because investment in green innovation is highly risky, and the payback period is long [4,5], the agency costs attached to such innovation are likely to be high, and designing an incentive package to induce innovation is specifically demanding [6].

Prior researchers in the field of environmental management sought to determine the factors that drive green innovation. In doing so, it was established that internal drivers are the most fascinating factors affecting sustainability performance [7–9]. The top management is a crucial internal factor for adopting green initiatives to improve social and environmental performance [10–13] because the top managers are accountable for firms' ethical actions, such as ensuring the implementation of sustainability initiatives [14–16]. Being a powerful member of a top management team, a chief executive officer (CEO) pledges with his/her strong will to focus on all stakeholders, including the customers, shareholders, society, and environment at large [17,18]. Some studies asserted that green practices in firms cannot be entrenched without the top management's (CEOs) support [19,20].

Therefore, research on top management has bolstered an effort to disclose what determines and motivates executives to be environmentally responsible. Because of their individual qualities, which provide them with a high level of satisfaction in such endeavors, top executives engage in environmentally friendly activities [21,22]. The top executives tend to deploy firm resources for the enhancement of their own narrow financial benefits at the cost of the company's long-term objectives, and, hence, are hesitant to invest in environmental projects that help organizations in the long run [1,23]. In this vein, scholars have largely focused on executives' traits and their effect on green performance, such as a CEO's power, [24], political affiliation [25], duality [26], gender [27], hubris [18], education and tenure [28] as well as compensation [1]. Based on upper echelons theory, research has found a link between CEO gender and corporate social performance. Many academics contend that female CEOs are more concerned with CSR-related topics than male CEOs [29,30]. According to the authors, women are more society- and environment-oriented, whereas male managers are more self-oriented, based on the concept of social role theory. The differences in the CEOs' careers have varied social outcomes [31]. Because young managers must deliver good observable results to the market, the authors suggest, they are more inclined to adopt measures that focus on short-term observable outcomes and less likely to increase social- or environmental-related field operations. Unlike younger executives, elderly CEOs are not under as much market pressure as their younger counterparts; therefore, they are more motivated to care about corporate social performance. Several studies using upper echelons theory have found a correlation between CEO education and corporate social performance. Huang [32] conducted research in which the educational specialization of CEOs was shown to be related to the CSR performance of companies. Similarly, Lewis, Walls, and Dowell [28] discovered that managers with MBA degrees are more likely than other enterprises to disclose environmental information, but peers with law degrees are less likely. However, our research goes beyond the existing literature on the demographics of CEOs by exploring whether tournament incentives motivate chief executives to spend more on environmentally friendly innovations.

The connection of CEO remuneration to social and environmental performance (e.g., greenhouse gas reduction targets, staff training in green practices, and adherence to moral principles in developing countries)—is a modern phenomenon in corporate governance. Many companies have bound compensation structure to sustainability performance; for instance, Xcel Energy implies that 75 percent of its incentives are still based on profits per share growth, but the other 25% are based on ecological footprint and greenhouse gas emissions reductions. Likewise, Intel connects CEO pay to corporate environmental targets, such as product energy efficiency, carbon emissions and power consumption savings, as well as improved environmental leadership reputation [33]. Superior social and environmental performance is often the result of long-term efforts that demand a long-term approach [34,35]. Firms may gain trust and acquire the social license to operate by actively interacting with local communities and contributing to their well-being over time. Similarly, businesses may reduce their environmental impacts by investing in renewable technology, sustainable energy, and other green initiatives that will take time to yield tangible results. We anticipate that incentivizing managers based on social and environmental performance goals will attract them to focus on longer-term initiatives, leading them to embrace a longer time horizon. As a result, they are less likely to ignore valuable long-term stakeholder initiatives, improving the firm's worth. Key stakeholders' pressure might encourage CEOs to practice quality management in order to maintain their market reputation. As a result, the CEO's quality management efforts are strengthened by fulfilling the needs of the major stakeholders. Moreover, winning an industry award improves a CEO's reputation while also increasing a firm's credibility in the eyes of its key stakeholders [36,37]. Higher levels of reputation, as well as primary and secondary stakeholder pressures, are linked to higher levels of quality management, which leads to environmental innovation [38]. As a result, when managers are properly rewarded, they are more likely to take steps to reduce their environmental impact. They could, for example, reduce pesticide use, minimize energy, implement a recycling system, involve their employees in community clean-ups and environmental protection efforts, upgrade their facilities to prevent oil spills and other environmental disasters, construct green buildings, or switch to renewable energy and clean fuels. We expect firms' emissions to diminish as a result of such measures [33].

The recent trend in CEO tournament incentives suggests that tournament incentives are the pay gap between CEO and other executive pay and likely a powerful steward that may prompt executives' performance [39–43]. It is obvious that participants in the competition deploy their full potential if pay is based on a tournament scheme [44]. Specifically, the tournament theory posits that the variance in pay between CEOs and other executives leads to a fiercer competition that exerts more efforts to improve firms' performances for winning the contest [45,46]. Thus, the motivation of this study takes its philosophical roots from tournament theory, which seeks to understand the mechanism by which incentive packages induce green performance. As top executives are assessed on the basis of their relative performance, rather than absolute performance, our study predicts that tournament incentives will motivate CEOs to be more environmentally responsible in response to huge incentives gaps.

Our study contributes in two significant aspects. First, this study extends from prior research on the internal determinants of green performance [8,9,12,13,47] by exploring chief executive tournament incentives as an important determinant of green innovation performance. We submit that CEO tournament incentives positively affect firms' green innovation, suggesting that a compensation gap drives CEOs to be green by investing more in green innovation. Second, the situation becomes particularly vital when we demonstrate this within the Chinese context. China is a transition economy designed to pool resources of land, capital, and human resource in emerging markets. Simultaneously, the state plays a monopolistic role, in terms of asset, policy, and law execution [48], and runs SOEs to exercise financial and social activities [49]. As the institutional environment of China is distinct in nature [50,51], former researchers recommended that effective corporate governance must acknowledge the institutional environment wherein corporate entities maneuver [52,53]. However, in the case of China, most of the listed firms are state-owned enterprises, and the state as a major shareholder gives more priority to social, environmental, and political goals [54,55]. Thus, we contribute by exploring whether the impact of chief executive tournament incentives on green innovation differs among state firms and non-state firms. Our study also finds that the positive impact of tournament incentive is more noticeable in state enterprises than it is in non-state enterprises.

2. Literature Review

2.1. CEO Tournament Incentives and Green Innovation

A firm's participation in social activities is influenced by a variety of factors at firm level, and, among them, the top executives' incentives play a central role in motivating management efforts. The impact of executive pay disparity on a company's social and financial consequences has sparked open discussion [56,57]. According to certain academics, CEO compensation is positively related to environmental performance [1,58]. A stream of studies suggests that pay gap can be an effective motivator and stimulate pay rivalry, which motivates executives to exert effort for involvement in long term projects—such as R&D projects [59–62]. More related to social and environmental performance, previous research highlights that executives are awarded higher incentives for social and environmental endeavors that have many significant implications including personal reputation, career enhancement, stakeholder satisfaction, and organizational legitimacy [63–66]. The previous studies suggested that environmental process innovation, business analytics, and environmental product innovation have a significant impact on the green competitive advantage of the companies [67,68]. Corporate social responsibility is exemplified by environmental innovation. Significant eco-innovation may have a favorable influence on a firm's reputation and political links, resulting in intangible assets and the ability to raise capital. As a result, substantive eco-innovation is linked to the long-term success of businesses, and businesses should invest in substantive eco-innovation in order to obtain resources [69].

Conversely, other authors documented a negative and weak relationship between executive compensation and environmental and social performance [70–73], as investment in social objectives could be a financial burden on a firm's financial performance. Furthermore, stakeholder mismatching happens when stakeholders who receive an advantage from environmental reputation (e.g., the general community) are not the same individuals who assess a firm's performance (e.g., stockholders) [74]. In the executive pay structure, the same paradox exists. The large pay gaps between executives and other employees leads to a negative impression that discourages the collective efforts of employees, which is detrimental to innovation [75,76].

However, in this research, we look at how CEO tournament incentives influence green innovation in China, where the regulatory structure embodies the intriguing features of an emerging and transition economy [77,78], and where the general public is becoming completely conscious of the worsening air quality and the dangers posed by industrial pollution [79,80]. As the environmental standards are lower in China compared to developed countries, and environmental rules and their enforcement are notoriously weak [81,82], we posit that such tournament incentives can encourage top managers to harvest green practices.

The interesting logic behind tournament incentives is that a CEO is sitting on the top ladder rung of the firm, where there are no chances of further promotion; thus tournament incentives can be considered as a catalyst to trigger a CEO's effort in order to invest in social projects [55], while other senior executives exert effort for promotion-based and performance-based incentives [83,84]. If senior executives receive a promotion to the CEO position, they receive modest pay and incentives, and such compensation schemes enhance their motivation level for winning prizes and corporate performance [45,85]. Firms that believe that an organization's competitive environment encourages talent will prefer high compensation, as assumed by tournament theory [86,87]. According to this theory, compensation based on position is an effective reward method for motivating employees [45]. Previous researchers demonstrate that higher tournament incentives tend to increase wide-spread firm activities, including research and development that translates into firm performance [88,89]. The perceived value of tournament incentives is that they motivate managers to put in more effort, which leads to more innovative performance. Consistent with this conjecture, Shen and Zhang [90] find evidence in U.S. public firms that tournament incentives strengthen the efficiency of firms' innovations. This implies that managers usually respond actively to tournament incentives to achieve better results.

A group of studies has shown that one manifestation of the greater effort made by competing against other executives is risk-taking [91–93]. For instance, in professional sports competitions, there is only one winner who receives a high prize; nonetheless, all other players adopt an aggressive risk-taking approach since they have nothing to lose and are hoping to gain great results by taking risks (Grund et al. 2013). Similarly, a higher tournament incentive brings over investment in risky innovative projects without calculating the potential benefits that might have a drastic effect on a firm's innovative capability [94]. Goel and Thakor [95] also find that executives can increase the probability of winning a tournament by plunging into excessive risky projects. The evidence on how tournament incentives affect executives' risk-taking behaviors and innovations, however, remains relatively scant in organizational governance literature [90].

Drawing on tournament theory, we hypothesize that CEO tournament incentives are a driving force that triggers green innovation because investment in green projects relies on the CEOs' decisions, and these incentives will be a stepping stone towards the long-term stability of an organization by achieving higher environmental performance. Overall, this discussion leads us to the following hypothesis:

Hypothesis 1 (H1). *Tournament incentive is positively related to green innovation.*

2.2. The Moderating Role of Ownership Structure

Next, we predict a higher correlation between a CEO tournament incentive and green innovation in state-owned businesses than in non-state-owned companies. Ownership type is an environmental catalyst for green innovation that influences the tendency towards going green. The government, as a major shareholder, tightly controls the financial and non-financial resources for innovation [96].

In a developing market (such as China), the government seems to be the most important institution, wielding enormous influence over finite resources and enacting policies to determine enterprises' goals in a competitive market [97]. The government is considered a significant stakeholder to secure the uncertain risk involved in green innovation because, if the state does not provide enough support in policies, it is tough for businesses to generate long-term growth [98]. China is a transition economy characterized by public and private ownership, where the regulatory framework illustrates the features of an emerging market [77]. In particular, the Chinese government motivate firms towards environmental legitimacy [99], and the government acts as a stewards of SOEs to implement its social and economic agenda [49]. SOEs are often large, whereas private businesses are typically small. As a result, disparities in CSR that appear to be caused by ownership may instead be due, instead, to the size of the company. Furthermore, CSR assessments differ from one another since various items are used in different research. Faced with market transition, increased competition, and crucial trade frictions, Chinese private firms failed to compete with SOEs, foreign-owned enterprises, and multinational enterprises, especially due to their tiny size, inexperienced management, and poor technology [100].

According to institutional theory, companies also improve their environmental legitimacy because of external stakeholder pressure. In such a scenario, a firm's establishment of a connection with the government is one significant approach to create legitimacy [101]. Major Chinese state-owned firms solely rely on the government for receiving opportunities involved in new product development; thus, such enterprises avail themselves the benefits of approving patents and the financial and other resources required for innovation. As managers of state-owned enterprises have easy access to government support, policy information and other guidelines, this may also affect the firms' likelihoods towards investment in environmental innovation, which further provides a signal of organizational legitimacy [102,103]. Because CEOs build social ties with the government in a way that strengthens recognition as a government representative, they obtain more government support and resources for combating environmental risk. Farag et al. [104] found that SOEs are more motivated to invest in socially responsible activities than NSOEs and are more conscious about environmental protection. Such evidence echoes that of Usman et al. [105] and Huang et al. [106], who identified Chinese listed firms that politically led CEOs to promote more green innovation in SOEs than NSOE. These CEOs established diplomatic relations with the government, which provides an edge to access information, environmental policies, and state support. In response, firms may face more pressure for environmental legitimacy.

Taken together, top executives in state enterprises are more motivated to make decisions in favor of the state, which places them in a better position to win tournaments or promotion rewards [107]. Ethical leadership of state enterprises shows a commitment to enhance the reputation of a firm using the investment in CSR because of state regulatory due diligence [108]. The executives in SOEs are assessed on the basis of social and financial performance; thus, social activist CEOs strive to engage more in green innovation for image-building in order to obtain government privileges and win incentives, as CEOs of state-owned firms do not face these financial barriers. Furthermore, CEOs of Chinese state-run firms are usually supervised, hired, and promoted through political interference, and their pay is compressed in alignment with state interests. As a result, SOE board members and management are highly incentivized to make decisions that benefit the state by putting social goals ahead of economic objectives [61,109]. Moreover, the Chinese government offers grants to state firm top executives [110] in return for legitimate business actions [111]. Consequently, sound tournament incentives will induce the state-owned enterprise CEOs to invest more in green innovation. We draw the hypothesis that the influence of CEO pay incentives on green innovation would be more evident in state enterprises than in non-state enterprises. Thus, we formulate the following hypothesis:

Hypothesis 2 (H2). *The positive impact of CEO tournament incentives on green innovation is more pronounced in SOEs than in non-SOEs.*

3. Methodology

3.1. Data Sources and Sample

This study used a panel dataset of Chinese A-share companies listed on the Shanghai and Shenzhen stock exchanges for the period ranging from 2010 to 2016. The data relating to green innovation is obtained from the National Intellectual Property Administration (NIPA) database. The NIPA database provides summary details for all information about innovation type, title, application date, applying firm or person, and a short summary of the patent. We also used the enterprises' annual reports and China Stock Market and Accounting Research (CSMAR) to collect data on the independent variable (tournament incentive) and other control variables. The sample distribution by industry is given in Appendix A. The study's final sample is comprised of 6515 observations excluding firm-year observations for which the required information on explanatory variables were missing.

3.2. Measures

3.2.1. Dependent Variable

Our study's dependent variable is green innovation (GI), representing environmentally related patents. Consistent with previous studies on green innovation and the availability of data in the context of China, we selected a summary of green patents using several terminologies related to the environment, such as green, less carbon, environmental protection, clean, emission reduction, sustainable, recycling, economical, ecology, and environmental performance [112,113]. Eventually, we accumulated the patent number at the firm level and obtained each firms' green innovation.

3.2.2. Independent Variable

Our main independent variable is the tournament incentive (T_Incentives). Following previous studies [55,88,114,115], we measure CEO tournament incentives (T_Incentives) as the difference between CEO pay and the other executives of a firm. We applied the following distinctive formula to calculate the tournament incentive:

$$T\_Incentives = Log\ (CEO\ Pay - Average\ Executives\ Pay)$$

In addition, we estimate the impact of CEO tournament incentives on green innovation in SOEs and non-SOEs. Hence, we measured SOE as a dummy variable that equals '1' when the controlling shareholder of Chinese firms is the local or central government and '0' otherwise.

3.2.3. Control Variables

In line with previous studies on green innovation and CEO attributes [1,18,55,89,116,117], we control for a number of firm-level characteristics. CEO features include CEO age (CEO_Age), calculated in number of years. CEO duality is a dummy variable for which we assign one when a CEO also serves as a board chairman and zero otherwise. CEO ownership (CEO_Shares) is defined as the percentage of shares held by the CEO. This study also adds board ownership (Board_Shares), or the percentage of shares owned by a firm's board members, and board size (Board_Size), measured as the number of directors on the board because board members also contribute to important strategic choices. A company's financial attributes can also influence its societal engagements [50]. Thus, we include firm size (Size), measured as a log of total assets; firm size (Growth), calculated as the increase in a firm's total assets; financial leverage (Leverage), which is the ratio of the total debt divided by the total assets; and firm performance (ROA), explains accounting-based performance measured as the net profit divided by the total assets. Finally, we include year and industry dummies in each regression analysis to capture the fixed effects of time and industry that may have an impact on the relationship between CEO tournament incentives and green innovation.

3.3. Statistical Model

To test the hypothesis of whether a firm's tournament incentives inclined CEOs towards green innovation, we estimate Equation (1). To determine whether CEOs in state-owned enterprises are more motivated for green innovation as compared to non-state-owned enterprises, we apply Equation (1) for both the state and non-state firms' samples. Consistent with previous studies [50,55,118], we apply an ordinary least squares (OLS) model as the baseline approach for estimating Equation (1). The model is as follows:

$$GI_{it} = \beta_0 + \beta_1 \, T\_Incentives_{it} + \sum_{i=1}^{n} \beta_n \, Controls_{it} + \varepsilon_{it} \tag{1}$$

In the above model, GI indicates green innovation, representing the dependent variable. T_Incentives is our main independent variable, and controls refers to all control variables. Table 1 reports a detailed description of each variable.

Table 1. Variables definitions and details.

Variable	Details
GI	Green innovation is defined as the number of environment related patents (such as patents related to environmental protection, low carbon, eco-energy saving, emission reduction, sustainability, recycling, pollution, water resources, and biodiversity).
T_Incentives	Log of the CEO compensation gap and the average pay of other executives.
SOE	Defined as the dummy variable which equals one if the local or central government is the ultimate owner and zero otherwise.
CEO_Shares	Indicates the percentage of shares owned by a CEO.
CEO_Age	Equals age of the CEO in years.
CEO_Duality	A dummy variable equals one if the CEO is also the chairperson of the board and zero otherwise.
Board_Shares	The percentage of shares held by a firm's board of directors.
Board_Size	Defined as number of directors on the board.
Size	Indicates the natural log of total assets.
Growth	Calculated as change in total assets.
Leverage	Indicates ratio of total debt to total assets.
ROA	Shows return on assets, which is measured by dividing the net profit by total assets.

3.4. Descriptive Statistics and Correlations

Table 2 illustrates the summary statistics for all variables used in the analysis. The results show that the mean GI score is 2.553, which indicates that, on average, Chinese firms' secure three environment-related patents that reflects their green performance. A higher mean score of green innovation shows higher environmental performance. The mean of the tournament incentives (T_Incentives) is 402,665 Yuan (CNY), showing that CEOs receive an average 402,665 Yuan (CNY) annually beyond the pay that other executives receive. The mean value of SOEs is 0.467, which declares that 47 percent of the firms in the sample are state-owned enterprises where state owners are the major shareholders.

Table 2. Descriptive statistics.

Variable	Mean	Std. Dev.	Min	Max
GI	2.553	5.23	0	39
T_Incentives	402,665	545,034	−915,000	1,500,000
SOE	0.467	0.498	0	1
CEO_Shares	0.041	0.107	0	0.805
CEO_Age	48.588	6.229	25	77
CEO_Duality	0.246	0.431	0	1
Board_Shares	0.103	0.186	0	0.892
Board_Size	10.211	2.557	5	26
Size	3.090	0.058	2.704	3.350
Growth	7.665	1.336	1.609	13.223
Leverage	0.456	0.562	0.007	46.159
ROA	0.039	0.545	−48.316	22.005

Table 3 presents the correlation for selected variables using a variance inflation factor (VIF) test and a Pearson's correlation. The results show that the correlation between the green innovation and tournament incentive is positive and significant. We also find SOEs to be positively associated with green innovation; thus, the results provide preliminary support for our propositions. Additionally, the correlation between all independent variables is within the acceptable limits, which shows that there is no problem of multicollinearity.

Table 3. VIF test and Pearson's correlation.

Variables	VIF	(1)	(2)	(3)	(4)	(5)	(6)	(7)	(8)	(9)	(10)	(11)	(12)
(1) GI	–	1.000											
(2) T_Incentives	1.13	0.064 *	1.000										
		(0.000)											
(3) CEO_Shares	2.47	−0.040 *	−0.081 *	1.000									
		(0.001)	(0.000)										
(4) CEO_Age	1.13	0.061 *	0.140 *	−0.004	1.000								
		(0.000)	(0.000)	(0.711)									
(5) CEO_Duality	1.51	−0.029 *	0.021 *	0.475 *	0.129 *	1.000							
		(0.013)	(0.039)	(0.000)	(0.000)								
(6) Board_Share	2.56	−0.055 *	−0.132 *	0.698 *	−0.118 *	0.257 *	1.000						
		(0.000)	(0.000)	(0.000)	(0.000)	(0.000)							
(7) Board_Size	1.15	0.082 *	0.117 *	−0.133 *	0.043 *	−0.104 *	−0.168 *	1.000					
		(0.000)	(0.000)	(0.000)	(0.000)	(0.000)	(0.000)						
(8) Growth	1.04	−0.017	0.012	0.050 *	−0.028 *	0.048 *	0.064 *	0.049 *	1.000				
		(0.152)	(0.228)	(0.000)	(0.004)	(0.000)	(0.000)	(0.000)					
(9) Size	1.35	0.002	−0.008	−0.002	−0.009	−0.004	−0.005	0.019 *	0.179 *	1.000			
		(0.893)	(0.452)	(0.810)	(0.344)	(0.715)	(0.621)	(0.049)	(0.000)				
(10) Leverage	1.22	0.065 *	0.012	−0.102 *	−0.012	−0.042 *	−0.150 *	0.066 *	−0.031 *	0.004	1.000		
		(0.000)	(0.243)	(0.000)	(0.238)	(0.000)	(0.000)	(0.000)	(0.001)	(0.713)			
(11) ROA	1.01	−0.002	0.029 *	0.013	0.017	−0.004	0.018	−0.020 *	0.063 *	0.000	−0.738 *	1.000	
		(0.834)	(0.004)	(0.210)	(0.076)	(0.678)	(0.056)	(0.040)	(0.000)	(0.996)	(0.000)		
(12) SOE	1.60	0.078 *	0.068 *	−0.339 *	0.144 *	−0.296 *	−0.493 *	0.217 *	−0.079 *	0.009	0.113 *	−0.011	1.000
		(0.000)	(0.000)	(0.000)	(0.000)	(0.000)	(0.000)	(0.000)	(0.000)	(0.345)	(0.000)	(0.270)	

* shows significance level at 5%.

4. Results

Table 4 displays the OLS regression results for the proposed hypotheses. Hypothesis 1 expects a significant and positive relation between CEO tournament incentives and green innovation. Model 1 of Table 4 lends support to the first hypothesis of this study by showing that the coefficient of CEOs tournament incentives (T_Incentives) is significantly positive ($\beta = 0.165$, $p < 0.05$). This result confirms that tournament incentives encourage CEOs to invest more in the firm's green innovation to boost environmental performance. Moreover, the result supports tournament theory, which posits that the compensation gaps enhance competitiveness among CEOs and other top executives, thereby encouraging top executives to be greener by investing in environmentally related projects. This finding is consistent with latest study of Ali et al. [55], who documented a positive influence of CEO tournament incentives on CSR.

Table 4. Main evidence of CEOs tournament incentives and green innovation.

Variables	OLS Regressions			Cluster-OLS Regressions		
	Full Sample	SOEs	N-SOEs	Full Sample	SOEs	N-SOEs
T_Incentives	0.165 **	0.697 ***	0.203 ***	0.613 ***	0.711 ***	0.681 ***
	(2.465)	(5.128)	(2.666)	(4.798)	(3.362)	(3.087)
CEO_Shares	0.145	−15.515	0.330	1.183	1.095	−15.582
	(0.185)	(−1.181)	(0.458)	(0.972)	(0.931)	(−1.163)
CEO_Age	0.001	0.052	0.007	0.019	−0.016	0.053 *
	(0.136)	(0.466)	(0.684)	(1.328)	(−1.245)	(1.732)
CEO_Duality	−0.100 **	−0.365 *	−0.073 *	−0.574 **	−0.352	−0.376
	(−2.148)	(−1.741)	(−1.813)	(−2.267)	(−1.402)	(−0.619)
Board_Shares	0.609	−6.760	0.286	−1.747 ***	−0.995	−6.893
	(1.394)	(−0.955)	(0.686)	(−2.599)	(−1.573)	(−0.980)
Board_Size	0.011 *	0.059 **	0.058 *	0.076 *	0.099 **	0.061
	(1.765)	(2.378)	(1.939)	(1.839)	(2.203)	(0.867)
Growth	−0.000	−0.151	−0.009	0.011	0.033	−0.050
	(−0.009)	(−1.486)	(−0.214)	(0.260)	(0.950)	(−0.562)
Size	1.691 ***	0.023 **	1.193 ***	2.121 **	1.473	3.473 **
	(31.997)	(2.335)	(17.381)	(2.236)	(1.491)	(2.278)
Leverage	−0.699 ***	−3.480 ***	−0.743 ***	0.011	−0.017	−6.774 **
	(−3.177)	(−6.586)	(−3.157)	(0.080)	(−0.217)	(−2.206)
ROA	0.348 *	7.119 ***	0.231 *	1.044 ***	5.500 **	3.931 ***
	(1.687)	(2.910)	(1.784)	(2.776)	(2.718)	(3.105)
Constant	−14.609 ***	−13.616 ***	−10.849 ***	−9.094 ***	−5.516 **	−13.571 ***
	(−13.463)	(−6.270)	(−8.743)	(−4.606)	(−2.368)	(−4.185)
Observations	6515	2740	3775	6515	2740	3775
R-squared	0.223	0.142	0.149	0.100	0.081	0.140
Industry & Year	Yes	Yes	Yes	Yes	Yes	Yes
Chi^2 value = 9.600						
p-value = 0.001						

Note: Table 4 reports result for the proposed hypotheses. T_Incentives indicate tournament incentives of CEOs. T-statistics are documented in parentheses. ***, **, and * denote significance level at 1, 5, and 10%, respectively. This table also reports the value of Chi^2 for the sake of sub-sample comparison.

Table 4 also presents the findings for the sub-samples of state-owned and non-state-owned firms. Model 2 shows the finding of SOEs sub-sample, where the coefficient of the tournament incentive is 0.697, significant at 1%. The finding implies that top executives of state-owned firms receive tournament incentives so that they are more ingrained in environmental performance and earn legitimacy. The reason is that the CEOs of state-owned firms receive pressure from the government and public to meet political and social targets. The result also supports institutional theory [119], whichpostulates that organizations seeking legitimacy succumb to institutional pressure, thus influencing their behavior. Alternatively, Model 3 reports the coefficient of CEO incentives for non-SOEs which is 0.203, significant at 1%, indicating that CEO tournament incentives in non-state firms also have a positive impact on green innovation.

However, comparing the findings of state and non-state enterprises indicates that a state enterprise's executive incentives have a greater and more favorable impact on green innovation than non-state firms, which acknowledges our second hypothesis. The coefficient value of Model 2 is significantly greater than that of Model 3. In order to compare the beta values for every model, we use an independent regression technique. Because of the significant influence of tournament incentives in state firms, the Chi^2 value (9.601), which has a *p*-value = 0.001, shows that green innovation is more visible in state firms than in non-state firms.

Other robustness checks were also used to verify the authenticity of our conclusions. For the baseline regression findings, we attempted ordinary least squares (OLS) estimation, but the OLS criterion that observations should be independent could not be met in our analysis, as we use more than one observation in each of the selected companies, which could lead to biased estimates. To overcome this limitation, we used OLS regression with clustering of the standard error by firm (Cluster-OLS). Table 4 highlights the results for the cluster OLS estimations. These findings again confirm that tournament incentives positively affect corporate green innovation, and this relationship is stronger in state-owned firms.

Table 4 also highlights results for firm-level control variables. We find that except for CEO ownership, age, board ownership, and growth, all other variables have significant impacts on green practices. In particular, the coefficient of CEO duality indicates that it may generate a negative impact on environmental innovation. This is because a CEO may grow more powerful with dual job status, and, hence, may utilize corporate resources for self-interest rather than for societal interest. Conversely, Board_Size is positively associated with GI because a larger board contains the diversified knowledge and experience that often leads to firms going green. Similarly, firm size (Size) and profitability (ROA) have positive relationships with green innovation, as larger and more profitable firms receive more attention from stakeholders with regards to environmental concerns. In addition, such firms are considered resourceful to invest in green innovation. Finally, the coefficient of *leverage* suggests that with higher level of leverage, firms will be less likely to invest in green innovation.

5. Endogeneity

Our OLS regression results on the link between CEO tournament incentives and green innovation may be biased due to potential endogeneity issues. To solve this problem, our study uses three alternative statistical methodologies following previous research studies [55,113,120].

First, this study uses fixed-effects estimations to address the issue of unobserved factors (The Hausman test has been used to check the probability that a fixed-effects model or random effect is suitable to use. In doing so, the value of the Hausman test (247.66 at $p < 0.01$) confirms that the fixed-effects estimation is suitable over the random effect model.). For instance, one of the major issues in the estimation of panel datasets is the possibility of unobservable time-invariant characteristics. In such a case, a fixed-effects model has an advantage over other OLS estimations of explicit models' characteristics, which are quite stable, thereby reducing cross-sectional fluctuations.

Table 5 presents the findings of the fixed-effect model and affirms that T_Incentives significantly and positively impacts a firm's environmental performance. In addition, the findings of the sub-samples (SOEs and non-SOEs) suggest that tournament incentives in state firms increases the CEOs' likelihood, more than in non-state firms, for investing in environmentally responsible projects. However, the regression models using fixed effects provide results that are consistent with the baseline analysis, confirming that our conclusion is unaffected by the omitted-variable bias.

Second, due to self-selection bias, our first OLS findings may be inaccurate, i.e., the features of a more environmentally responsible and a less environmentally responsible firm may differ, resulting in different outcomes for firms to be greener. In such a case, firms are considered greener because of firms-level characteristics, rather than due to increased CEO incentives. To address this issue, this study creates a dummy variable that equals one if a firm secures environmentally related patents more than the industry median, and zero otherwise, and compares the firms on the basis of the control variables by applying a propensity score matching (PSM) technique.

Table 5. Endogeneity test: fixed-effect model.

Variables	Full Sample	SOEs	N-SOEs
T_Incentives	0.264 ***	0.292 ***	0.236 *
	(3.079)	(2.699)	(1.657)
CEO_Shares	−0.582	1.623	−0.771
	(−0.555)	(0.060)	(−0.741)
CEO_Age	0.030 **	0.053 **	0.010
	(2.298)	(2.454)	(0.605)
CEO_Duality	−0.206	−0.474	−0.006
	(−1.036)	(−1.358)	(−0.024)
Board_Shares	0.522	20.592 *	0.060
	(0.506)	(1.758)	(0.059)
Board_Size	0.011	0.015	−0.006
	(0.454)	(0.393)	(−0.178)
Growth	−0.092 ***	−0.052	−0.121 ***
	(−3.001)	(−1.065)	(−3.182)
Size	0.510 ***	0.334 ***	0.798 ***
	(4.252)	(2.815)	(4.663)
Leverage	0.020	−2.086 *	−0.775
	(0.059)	(−1.902)	(−1.157)
ROA	0.116	−2.168	0.040
	(0.767)	(−1.024)	(0.227)
Constant	−1.128 ***	−6.453 ***	−1.494 ***
	(−2.770)	(−2.711)	(−2.819)
Observations	6515	2740	3775
R-squared	0.089	0.110	0.081
Year	Yes	Yes	Yes

Note: Table 5 presents findings of fixed-effect model, which addresses issues of omitted factors. T-statistics are documented in parentheses. ***, **, and * denote significance level at 1, 5, and 10%, respectively.

The findings of PSM models are highlighted in Table 6. However, consistent with our main evidence in Table 4, we find in all three models of Table 6, that T_Incentives are positively and significantly related to green innovation. In addition, we observe in Model 2 and Model 3 that the coefficient of T_Incentives remains greater for the SOEs' sample, confirming that tournament incentives induce the SOEs' CEOs more for investing in green innovation than the CEOs of NSOEs. Overall, these results suggest that our main findings are not driven by the firm-level characteristics.

Next, we apply an instrumental variable method known as Two-Stage least squares regression (2SLS) to address endogeneity issues. In this technique, the criteria that an "instrumental variable should be correlated with the independent variable but not with the dependent variable" must be fulfilled. Following this condition, our study uses the tournament incentives' industry average as an instrumental variable because it is highly correlated with our independent variable (T_Incentives) but has no correlation with our dependent variable (GI). The results of the 2SLS method are reported in Table 7. Again, we find that these results are consistent with our initial findings, which makes our main evidence more robust.

Table 6. Propensity score matching (PSM) results.

Variables	Full Sample	SOEs	N-SOEs
T_Incentives	0.1444 ***	0.1548 ***	0.0830 ***
	(7.75)	(5.15)	(3.26)
CEO_Shares	0.4137 *	−1.4760	0.1738
	(1.80)	(−0.49)	(0.71)
CEO_Age	0.0059 **	0.0123 ***	0.0016
	(2.22)	(2.71)	(0.48)
CEO_Duality	−2334 ***	−2532 ***	−1213 **
	(−5.35)	(−3.25)	(−2.18)
Board_Shares	−5153 ***	0.1516	−1162
	(−4.19)	(.10)	(−0.84)
Board_Size	0.0258 ***	0.0247 ***	0.1222
	(3.96)	(2.76)	(1.22)
Size	0.0056	−0271	0.02551 **
	(0.47)	(−1.22)	(1.97)
Growth	0.0023	0.0035	0.2197 ***
	(1.12)	(1.12)	(9.63)
Leverage	−3348 ***	−4381 ***	−1596 **
	(−6.33)	(−4.26)	(−2.28)
ROA	−0939	−0375	−2039
	(−0.52)	(−0.07)	(−0.51)
Constant	−2.7491 ***	−3.159 ***	−3.3937 ***
	(−10.59)	(−7.57)	(−9.44)
Observations	6515	2740	3757
R-squared	0.0315	0.0214	0.0412
Industry & Year	Yes	yes	Yes

Note: Table 6 highlight results to validate our main hypotheses using PSM approach. T-statistics are documented in parentheses. ***, **, and * denote significance level at 1, 5, and 10%, respectively.

Table 7. Two Stage least squares (2SLS) results.

Variables	Full Sample	SOEs	N-SOEs
T_Incentives	1.002 ***	1.285 ***	0.368 ***
	(7.712)	(5.384)	(2.618)
CEO_Shares	1.692	−9.922	0.436
	(1.472)	(−0.544)	(0.445)
CEO_Age	0.011	0.039	−0.016
	(0.763)	(1.436)	(−1.124)
CEO_Duality	−0.830 ***	−0.761 *	−0.198
	(−3.806)	(−1.767)	(−0.880)
Board_Shares	−1.733 ***	−14.071	0.538
	(−2.713)	(−1.443)	(0.942)
Board_Size	0.074 **	0.031	0.077 *
	(2.210)	(0.601)	(1.885)
Growth	−0.101	−0.225	−0.068
	(−1.321)	(−1.636)	(−0.867)
Size	0.022 **	0.028 **	1.361 ***
	(2.488)	(2.449)	(4.505)
Leverage	−2.633 ***	−7.014 ***	−0.847 ***
	(−8.400)	(−8.112)	(−3.021)
ROA	2.912 *	8.587 ***	1.370
	(1.781)	(2.715)	(0.807)
Constant	−11.560 ***	−18.606 ***	−12.490 ***
	(−6.512)	(−5.784)	(−6.681)
Observations	4596	2044	2547
R-squared	0.105	0.155	0.157
Industry & Year	Yes	Yes	Yes

Note: Table 7 presents results of instrumental variable approach (2SLS). T-statistics are documented in parentheses. ***, **, and * denote significance level at 1, 5, and 10%, respectively.

6. Discussion and Conclusions

We consider the importance of a top executive's role in environmental performance, as green practices cannot be initiated without top management support [20]. This research takes a holistic view to examine whether CEO tournament incentives result in improved environmental performance and fills the gap identified by [55]. Our expectations are founded on tournament theory, which states that promotions, awards, and incentives incentivize managers to accomplish company goals. These types of incentives create a competitive spirit among senior managers, who promote both financial and social performance within organizations [45].

When the executives choose to concentrate on consumers, workers, the environment, and other stakeholders, rather than shareholders, they have a great deal of flexibility [121]. It is established that compensation for non-financial performance is an approach by which a firm's internal governance may affect its corporate environmental performance. Theoretically, a well-calculated prize range can stimulate prolific performance in a competition for incentives. It is argued that competing participants tend to perform better when the determination of compensation is carried out, keeping in mind the tournament viewpoint [44]. This indicates that if there is a pay disparity, a competition between the CEO and other managers will develop, resulting in better organizational effectiveness in terms of financial and non-financial performance, which greatly enhances the firm's position in the marketplace. In line with this conventional notion, we argue that executives tournament incentives push them to spend more on environmentally friendly innovations, since such innovations may promote the reputation of an organization, stakeholders' trust, brand image, and reduce tensions among the many stakeholders. Furthermore, as the CEOs positions rise, they are more inclined to focus on the company's long-term success.

6.1. Theoretical Implications

This study used a sample of Chinese listed firms over the time period from 2010 to 2016 and confirms that CEOs are more inclined to invest in green innovation as a result of the tournament incentive. Our findings support tournament theory, suggesting that when compensation is determined on the basis of tournaments, the competing participants perform better. Compensation gaps encourage top decision makers to reach decisions with social, economic, and strategic consequences in mind. The pay disparity between the tournament winner and the runner-up acts as a strong motivator for senior managers to put greater efforts into winning the tournament. Organizations may gain from the aggregate enthusiasm of managers because pay incentives drive individual efforts to perform well [56,122]. Hence, our findings imply that firms' green practices and their environmental legitimacy can be improved at the expense of incentivizing their executives. Top executives are considered powerful figures; thus the structuring of handsome compensation schemes for CEOs is advisable to stimulate a CEO's likelihood to deploy efforts and resources towards green initiatives.

In addition, we find that when receiving tournament incentives in SOEs, executives are more engaged in green practices than in non-SOEs. These findings contribute to institutional theory, which advocates that, in a particular business environment, organizations need to imitate social means as the existence of the firm depends on external social approval [119]. Moreover, social and environmental behavior is contextualized by institutional pressures [123]. The SOEs face more stringent regulatory and institutional pressure (such as government and media) as compared to non-SOEs [50], which may induce SOEs further to adopt environmentally friendly policies. Neo-institutional theory also suggests that the institutional structure influences the decision making of a firm [124] and permits discovering and comparing executives' non-financial motives (like green investment) in both formal and informal institutional contexts [125].

6.2. Limitations of the Study

Although, the study objectives are attained, certain limitations are essential to explain. These limitations should be considered for future potential research, as well as for the interpretation and generalizability of the results. First, our study sample is confined by the unique structure of the Chinese market, where most of the firms are controlled by the government. In this way, the generalization of the findings is constrained, as the Chinese market varies from those of advanced nations. Therefore, a future study could be carried out to verify the same arguments in developed countries. Second, patents are not the only way to safeguard green innovations; lead times, industrial capacity, and complicated standards can all have an impact [126,127]. Therefore, future research work can be conducted to capture the other dimensions of green innovations that will further determine the applicability of results. Finally, the study results assert that top executive tournament incentives push more state-owned firms towards environmental innovation as compared to non-government firms, but this relationship may be aroused through mediated factors, such as societal expectations or government pressure. The reason is that private firms take social objectives as a cost, not as holy obligations, while, the sense of responsibility to achieve societal targets and government supervision drives state organizations to pursue environmental stewardship. This concern must be addressed in future work through an exploration of the role of social- and state-level pressures on CEOs' excellent performance in state-level organizations and how these influence their incentives and environmental innovation.

Author Contributions: Conceptualization, S.U.; investigation, S.U. and F.U.K.; methodology, M.U.; visualization, S.U. and F.U.K.; writing—original draft, S.U.; funding acquisition, L.-M.C.; writing—review and editing, S.U., F.U.K., L.-M.C., M.U. and A.M. All authors have read and agreed to the published version of the manuscript.

Funding: We have not received any external funding for this manuscript.

Institutional Review Board Statement: Not applicable.

Informed Consent Statement: Not applicable.

Data Availability Statement: The data that support the findings of this study are available on reasonable request from the corresponding author.

Conflicts of Interest: All the authors declare no conflict of interest.

Appendix A

Table A1. Sample Distribution by Industry ($N = 6515$).

Industry Name	Full Sample	SOE Obs	N-SOE Obs
Agriculture, forestry, animal husbandry, and fishery	75	29	46
Mining industry	279	97	182
Manufacturing industry	4181	1593	2588
Electricity, thermal, gas and water production, and supply industry	218	163	55
Construction business	225	102	123
Wholesale and retail business	295	178	117
Transportation, warehousing, and postal service	206	173	33
Accommodation and catering	72	28	44
Information transmission, software and information technology services	428	132	296
Real estate	172	96	76
Culture, sports and entertainment	33	10	23
Leasing and business services	41	13	28
Scientific research and technology services	46	18	28
Water conservancy, environment and public facilities management	34	11	21
Education	58	24	34
Health and social work	115	46	69
Comprehensive	37	25	12

References

1. Berrone, P.; Gomez-Mejia, L.R. Environmental Performance and Executive Compensation: An Integrated Agency-Institutional Perspective. *Acad. Manag. J.* **2009**, *52*, 103–126. [CrossRef]
2. Ullah, S.; Ahmad, N.; Khan, F.; Badulescu, A.; Badulescu, D. Mapping Interactions among Green Innovations Barriers in Manufacturing Industry Using Hybrid Methodology: Insights from a Developing Country. *Int. J. Environ. Res. Public Health* **2021**, *18*, 7885. [CrossRef] [PubMed]
3. Ullah, S.; Khan, F.U.; Ahmad, N. Promoting sustainability through green innovation adoption: A case of manufacturing industry. *Environ. Sci. Pollut. Res.* **2021**, 1–21. [CrossRef]
4. Oh, W.-Y.; Chang, Y.K.; Cheng, Z. When CEO Career Horizon Problems Matter for Corporate Social Responsibility: The Moderating Roles of Industry-Level Discretion and Blockholder Ownership. *J. Bus. Ethic.* **2016**, *133*, 279–291. [CrossRef]
5. Oltra, V. Environmental innovation and industrial dynamics: The contributions of evolutionary economics. *Cah. GREThA* **2008**, *28*, 77–89.
6. Holmström, B. *Agency Costs and Innovation*; Saltsjobaden: Stockholm, Sweden, 1989.
7. Pinto, L.; Allui, A. An analysis of drivers and barriers for sustainability supply chain management practices. *J. Asia Entrep. Sustain.* **2016**, *12*, 197.
8. Horbach, J.; Rammer, C.; Rennings, K. Determinants of eco-innovations by type of environmental impact—The role of regulatory push/pull, technology push and market pull. *Ecol. Econ.* **2012**, *78*, 112–122. [CrossRef]
9. Bossle, M.B.; de Barcellos, M.D.; Vieira, L.M.; Sauvée, L. The drivers for adoption of eco-innovation. *J. Clean. Prod.* **2016**, *113*, 861–872. [CrossRef]
10. Closs, D.J.; Jacobs, M.A.; Swink, M.; Webb, G.S. Toward a theory of competencies for the management of product complexity: Six case studies. *J. Oper. Manag.* **2007**, *26*, 590–610. [CrossRef]
11. Sarkis, J.; Setthasakko, W. Barriers to implementing corporate environmental responsibility in Thailand. *Int. J. Org. Anal.* **2009**, *17*, 169–183.
12. Qi, G.; Shen, L.; Zeng, S.; Jorge, O.J. The drivers for contractors' green innovation: An industry perspective. *J. Clean. Prod.* **2010**, *18*, 1358–1365. [CrossRef]
13. Bhanot, N.; Rao, P.V.; Deshmukh, S.G. An integrated approach for analysing the enablers and barriers of sustainable manufacturing. *J. Clean. Prod.* **2017**, *142*, 4412–4439. [CrossRef]
14. Bai, C.; Sarkis, J.; Dou, Y. Corporate sustainability development in China: Review and analysis. *Ind. Manag. Data Syst.* **2015**, *115*, 5–40. [CrossRef]
15. Giunipero, L.C.; Hooker, R.E.; Denslow, D. Purchasing and supply management sustainability: Drivers and barriers. *J. Purch. Supply Manag.* **2012**, *18*, 258–269. [CrossRef]
16. Walker, H.; Di Sisto, L.; McBain, D. Drivers and barriers to environmental supply chain management practices: Lessons from the public and private sectors. *J. Purch. Supply Manag.* **2008**, *14*, 69–85. [CrossRef]
17. Waldman, D.A.; de Luque, M.S.; Washburn, N.; House, R.J.; Adetoun, B.; Barrasa, A.; Bobina, M.; Bodur, M.; Chen, Y.-J.; Debbarma, S.; et al. Cultural and leadership predictors of corporate social responsibility values of top management: A GLOBE study of 15 countries. *J. Int. Bus. Stud.* **2006**, *37*, 823–837. [CrossRef]
18. Arena, C.; Michelon, G.; Trojanowski, G. Big Egos Can Be Green: A Study of CEO Hubris and Environmental Innovation. *Br. J. Manag.* **2018**, *29*, 316–336. [CrossRef]
19. Fabrizi, M.; Mallin, C.; Michelon, G. The Role of CEO's Personal Incentives in Driving Corporate Social Responsibility. *J. Bus. Ethic.* **2014**, *124*, 311–326. [CrossRef]
20. Khan, S.A.R.; Zhang, Y.; Anees, M.; Golpîra, H.; Lahmar, A.; Qianli, D. Green supply chain management, economic growth and environment: A GMM based evidence. *J. Clean. Prod.* **2018**, *185*, 588–599. [CrossRef]
21. Gerstner, W.-C.; König, A.; Enders, A.; Hambrick, D.C. CEO Narcissism, Audience Engagement, and Organizational Adoption of Technological Discontinuities. *Adm. Sci. Q.* **2013**, *58*, 257–291. [CrossRef]
22. Weidenbaum, M.; Jensen, M. Introduction to the transaction. In *The Modern Corporation & Private Property*; Tenth Printing; Berle, A., Means, G., Eds.; Transaction Publishers: New Brunswick, NJ, USA, 2009.
23. de Villiers, C.; Naiker, V.; Van Staden, C.J. The Effect of Board Characteristics on Firm Environmental Performance. *J. Manag.* **2011**, *37*, 1636–1663. [CrossRef]
24. Walls, J.L.; Berrone, P. The Power of One to Make a Difference: How Informal and Formal CEO Power Affect Environmental Sustainability. *J. Bus. Ethic.* **2015**, *145*, 293–308. [CrossRef]
25. Shahab, Y.; Ye, C. Corporate social responsibility disclosure and corporate governance: Empirical insights on neo-institutional framework from China. *Int. J. Discl. Gov.* **2018**, *15*, 87–103. [CrossRef]
26. Garcia-Sanchez, I.-M.; Cuadrado-Ballesteros, B.; Sepulveda, C. Does media pressure moderate CSR disclosures by external directors? *Manag. Decis.* **2014**, *52*, 1014–1045. [CrossRef]
27. Han, S.; Cui, W.; Chen, J.; Fu, Y. Female CEOs and Corporate Innovation Behaviors—Research on the Regulating Effect of Gender Culture. *Sustainsbility* **2019**, *11*, 682. [CrossRef]
28. Lewis, B.W.; Walls, J.L.; Dowell, G.W.S. Difference in degrees: CEO characteristics and firm environmental disclosure. *Strat. Manag. J.* **2014**, *35*, 712–722. [CrossRef]

29. Kassinis, G.; Panayiotou, A.; Dimou, A.; Katsifaraki, G. Gender and Environmental Sustainability: A Longitudinal Analysis. *Corp. Soc. Responsib. Environ. Manag.* **2016**, *23*, 399–412. [CrossRef]

30. Cook, A.; Glass, C. Women on corporate boards: Do they advance corporate social responsibility? *Hum. Relat.* **2018**, *71*, 897–924. [CrossRef]

31. Delmas, M.A.; Toffel, M.W. Organizational responses to environmental demands: Opening the black box. *Strateg. Manag. J.* **2008**, *29*, 1027–1055. [CrossRef]

32. Huang, S.K. The Impact of CEO Characteristics on Corporate Sustainable Development. *Corp. Soc. Responsib. Environ. Manag.* **2013**, *20*, 234–244. [CrossRef]

33. Flammer, C.; Hong, B.; Minor, D. Corporate governance and the rise of integrating corporate social responsibility criteria in executive compensation: Effectiveness and implications for firm outcomes. *Strat. Manag. J.* **2019**, *40*, 1097–1122. [CrossRef]

34. Eccles, R.G.; Ioannou, I.; Serafeim, G. The Impact of Corporate Sustainability on Organizational Processes and Performance. *Manag. Sci.* **2014**, *60*, 2835–2857. [CrossRef]

35. Flammer, C.; Bansal, P. Does a long-term orientation create value? Evidence from a regression discontinuity. *Strat. Manag. J.* **2017**, *38*, 1827–1847. [CrossRef]

36. Hayibor, S.; Agle, B.R.; Sears, G.J.; Sonnenfeld, J.A.; Ward, A. Value Congruence and Charismatic Leadership in CEO–Top Manager Relationships: An Empirical Investigation. *J. Bus. Ethic.* **2011**, *102*, 237–254. [CrossRef]

37. Heugens, P.P.; van Riel, C.B.M.; van den Bosch, F.A. Reputation Management Capabilities as Decision Rules*. *J. Manag. Stud.* **2004**, *41*, 1349–1377. [CrossRef]

38. Konadu, R.; Owusu-Agyei, S.; Lartey, T.A.; Danso, A.; Adomako, S.; Amankwah-Amoah, J. CEOs' reputation, quality management and environmental innovation: The roles of stakeholder pressure and resource commitment. *Bus. Strat. Environ.* **2020**, *29*, 2310–2323. [CrossRef]

39. Pissaris, S.; Heavey, A.; Golden, P. Executive Pay Matters: Looking Beyond the CEO to Explore Implications of Pay Disparity on Non-CEO Executive Turnover and Firm Performance. *Hum. Resour. Manag.* **2017**, *56*, 307–327. [CrossRef]

40. Park, J.; Kim, S. Pay Dispersion and Organizational Performance in Korea: Curvilinearity and the Moderating Role of Congruence with Organizational Culture. *Int. J. Hum. Resour. Manag.* **2015**, *28*, 1291–1308. [CrossRef]

41. Patel, P.C.; Li, M.; Triana, M.D.C.; Park, H.D. Pay dispersion among the top management team and outside directors: Its impact on firm risk and firm performance. *Hum. Resour. Manag.* **2018**, *57*, 177–192. [CrossRef]

42. Elsayed, N.; Elbardan, H. Investigating the associations between executive compensation and firm performance. *J. Appl. Account. Res.* **2018**, *19*, 245–270. [CrossRef]

43. Elkins, H. Measuring Compensation System Structure: The Interrelation Between Equitable Pay and Firm Performance. Available online: https://dx.doi.org/10.2139/ssrn.3198893 (accessed on 15 March 2018).

44. Hannan, R.L.; Krishnan, R.; Newman, A. The Effects of Disseminating Relative Performance Feedback in Tournament and Individual Performance Compensation Plans. *Account. Rev.* **2008**, *83*, 893–913. [CrossRef]

45. LaZear, E.P.; Rosen, S. Rank-Order Tournaments as Optimum Labor Contracts. *J. Political Econ.* **1981**, *89*, 841–864. [CrossRef]

46. Nalebuff, B.J.; Stiglitz, J.E. Prizes and Incentives: Towards a General Theory of Compensation and Competition. *Bell J. Econ.* **1983**, *14*, 21. [CrossRef]

47. Weng, M.-H.; Lin, C.-Y. Determinants of green innovation adoption for small and medium-size enterprises (SMES). *Afr. J. Bus. Manag.* **2011**, *5*, 9154–9163.

48. Nee, V.; Opper, S.; Wong, S. Developmental State and Corporate Governance in China. *Manag. Organ. Rev.* **2007**, *3*, 19–53. [CrossRef]

49. Wong, S.C. Improving corporate governance in SOEs: An integrated approach. *Corp. Gov. Int.* **2004**, *7*, 19–29.

50. Khan, F.U.; Zhang, J.; Usman, M.; Badulescu, A.; Sial, M.S. Ownership Reduction in State-Owned Enterprises and Corporate Social Responsibility: Perspective from Secondary Privatization in China. *Sustainability* **2019**, *11*, 1008. [CrossRef]

51. Guariglia, A.; Yang, J. A balancing act: Managing financial constraints and agency costs to minimize investment inefficiency in the Chinese market. *J. Corp. Financ.* **2016**, *36*, 111–130. [CrossRef]

52. Davis, G.F. New Directions in Corporate Governance. *Annu. Rev. Sociol.* **2005**, *31*, 143–162. [CrossRef]

53. He, L.; Fang, J. Subnational institutional contingencies and executive pay dispersion. *Asia Pac. J. Manag.* **2016**, *33*, 371–410. [CrossRef]

54. Chang, E.C.; Wong, S.M. Governance with multiple objectives: Evidence from top executive turnover in China. *J. Corp. Financ.* **2009**, *15*, 230–244. [CrossRef]

55. Ali, S.; Zhang, J.; Usman, M.; Khan, M.K.; Khan, F.U.; Siddique, M.A. Do tournament incentives motivate chief executive officers to be socially responsible? *Manag. Audit. J.* **2020**, *35*, 597–619. [CrossRef]

56. O'Reilly, C.A.; Main, B.; Crystal, G.S. CEO Compensation as Tournament and Social Comparison: A Tale of Two Theories. *Adm. Sci. Q.* **1988**, *33*, 257. [CrossRef]

57. Ridge, J.W.; Aime, F.; White, M.A. When much more of a difference makes a difference: Social comparison and tournaments in the CEO's top team. *Strateg. Manag. J.* **2015**, *36*, 618–636. [CrossRef]

58. Russo, M.V.; Harrison, N.S. Organizational Design and Environmental Performance: Clues from the Electronics Industry. *Acad. Manag. J.* **2005**, *48*, 582–593. [CrossRef]

59. Behrens, J.; Patzelt, H. Incentives, Resources and Combinations of Innovation Radicalness and Innovation Speed. *Br. J. Manag.* **2017**, *29*, 691–711. [CrossRef]

60. Henderson, A.D.; Fredrickson, J.W. Top management team coordination needs and the CEO pay gap: A competitive test of economic and behavioral views. *Acad. Manag. J.* **2001**, *44*, 96–117.

61. Firth, M.; Fung, P.M.; Rui, O.M. Corporate performance and CEO compensation in China. *J. Corp. Financ.* **2006**, *12*, 693–714. [CrossRef]

62. Mehran, H. Executive compensation structure, ownership, and firm performance. *J. Financ. Econ.* **1995**, *38*, 163–184. [CrossRef]

63. Zou, H.L.; Zeng, S.X.; Lin, H.; Xie, X.M. Top executives' compensation, industrial competition, and corporate environmental performance. *Manag. Dec.* **2015**, *53*, 2036–2059. [CrossRef]

64. Bearman, P.S.; Galaskiewicz, J. Social Organization of an Urban Grants Economy: A Study of Business Philanthropy and Nonprofit Organizations. *Soc. Forces* **1988**, *66*, 846. [CrossRef]

65. Barnea, A.; Rubin, A. Corporate Social Responsibility as a Conflict Between Shareholders. *J. Bus. Ethic.* **2010**, *97*, 71–86. [CrossRef]

66. Maas, K. Do Corporate Social Performance Targets in Executive Compensation Contribute to Corporate Social Performance? *J. Bus. Ethic.* **2018**, *148*, 573–585. [CrossRef]

67. Skordoulis, M.; Ntanos, S.; Kyriakopoulos, G.L.; Arabatzis, G.; Galatsidas, S.; Chalikias, M. Environmental Innovation, Open Innovation Dynamics and Competitive Advantage of Medium and Large-Sized Firms. *J. Open Innov. Technol. Mark. Complex.* **2020**, *6*, 195. [CrossRef]

68. Zameer, H.; Wang, Y.; Yasmeen, H.; Mubarak, S. Green innovation as a mediator in the impact of business analytics and environmental orientation on green competitive advantage. *Manag. Decis.* **2020**. ahead-of-print. [CrossRef]

69. Liao, Z. Is environmental innovation conducive to corporate financing? The moderating role of advertising expenditures. *Bus. Strat. Environ.* **2020**, *29*, 954–961. [CrossRef]

70. Stanwick, P.A.; Stanwick, S.D. The determinants of corporate social performance: An empirical examination. *Am. Bus. Rev.* **1998**, *16*, 86.

71. Cai, Y.; Jo, H.; Pan, C. Vice or Virtue? The Impact of Corporate Social Responsibility on Executive Compensation. *J. Bus. Ethic.* **2011**, *104*, 159–173. [CrossRef]

72. Frye, M.B.; Nelling, E.; Webb, E. Executive Compensation in Socially Responsible Firms. *Corp. Gov. Int. Rev.* **2006**, *14*, 446–455. [CrossRef]

73. Miles, P.C.; Miles, G. Corporate social responsibility and executive compensation: Exploring the link. *Soc. Responsib. J.* **2013**, *9*, 76–90. [CrossRef]

74. Wood, D.J.; Jones, R.E. Stakeholder mismatching: A theoretical problem in empirical research on corporate social performance. *Int. J. Organ. Anal.* **1995**, *3*, 229–267. [CrossRef]

75. Pfeffer, J.; Langton, N. The Effect of Wage Dispersion on Satisfaction, Productivity, and Working Collaboratively: Evidence from College and University Faculty. *Adm. Sci. Q.* **1993**, *38*, 382. [CrossRef]

76. Collins, C.J.; Smith, K.G. Knowledge Exchange and Combination: The Role of Human Resource Practices in the Performance of High-Technology Firms. *Acad. Manag. J.* **2006**, *49*, 544–560. [CrossRef]

77. Lyon, T.; Lu, Y.; Shi, X.; Yin, Q. How do investors respond to Green Company Awards in China? *Ecol. Econ.* **2013**, *94*, 1–8. [CrossRef]

78. Xu, X.D.; Zeng, S.X.; Tam, C.M. Stock Market's Reaction to Disclosure of Environmental Violations: Evidence from China. *J. Bus. Ethic.* **2011**, *107*, 227–237. [CrossRef]

79. Huang, P.; Zhang, X.; Deng, X. Survey and analysis of public environmental awareness and performance in Ningbo, China: A case study on household electrical and electronic equipment. *J. Clean. Prod.* **2006**, *14*, 1635–1643. [CrossRef]

80. Qi, G.; Zeng, S.; Yin, H.; Lin, H. ISO and OHSAS certifications: How stakeholders affect corporate decisions on sustainability. *Manag. Dec.* **2013**, *51*, 1983–2005. [CrossRef]

81. Van Rooij, B.; Lo, C.W.H. Fragile convergence: Understanding variation in the enforcement of China's industrial pollution law. *Law Policy* **2010**, *32*, 14–37. [CrossRef]

82. Zeng, S.X.; Xu, X.D.; Yin, H.T.; Tam, C.M. Factors that Drive Chinese Listed Companies in Voluntary Disclosure of Environmental Information. *J. Bus. Ethic.* **2011**, *109*, 309–321. [CrossRef]

83. Green, J.R.; Stokey, N.L. A Comparison of Tournaments and Contracts. *J. Poli. Econ.* **1983**, *91*, 349–364. [CrossRef]

84. Baker, G.P.; Jensen, M.C.; Murphy, K.J. Compensation and incentives: Practice vs. theory. *J. Financ.* **1988**, *43*, 593–616. [CrossRef]

85. Prendergast, C. The Provision of Incentives in Firms. *J. Econ. Lit.* **1999**, *37*, 7–63. [CrossRef]

86. Becker, B.E.; Huselid, M.A. The Incentive Effects of Tournament Compensation Systems. *Adm. Sci. Q.* **1992**, *37*, 336. [CrossRef]

87. Connelly, B.L.; Tihanyi, L.; Crook, T.R.; Gangloff, K.A. Tournament theory: Thirty years of contests and competitions. *J. Manag.* **2014**, *40*, 16–47. [CrossRef]

88. Kini, O.; Williams, R. Tournament incentives, firm risk, and corporate policies. *J. Financ. Econ.* **2012**, *103*, 350–376. [CrossRef]

89. Kale, J.R.; Reis, E.; Venkateswaran, A. Rank-Order Tournaments and Incentive Alignment: The Effect on Firm Performance. *J. Financ.* **2009**, *64*, 1479–1512. [CrossRef]

90. Shen, C.H.-H.; Zhang, H. Tournament Incentives and Firm Innovation. *Rev. Financ.* **2017**, *22*, 1515–1548. [CrossRef]

91. Hvide, H.K. Tournament Rewards and Risk Taking. *J. Labor Econ.* **2002**, *20*, 877–898. [CrossRef]

92. Hvide, H.K.; Kristiansen, E.G. Risk taking in selection contests. *Games Econ. Behav.* **2003**, *42*, 172–179. [CrossRef]
93. Gaba, A.; Tsetlin, I.; Winkler, R.L. Modifying Variability and Correlations in Winner-Take-All Contests. *Oper. Res.* **2004**, *52*, 384–395. [CrossRef]
94. Gilpatric, S.M. Risk Taking in Contests and the Role of Carrots and Sticks. *Econ. Inq.* **2009**, *47*, 266–277. [CrossRef]
95. Goel, A.M.; Thakor, A.V. Overconfidence, CEO selection, and corporate governance. *J. Financ.* **2008**, *63*, 2737–2784. [CrossRef]
96. Musacchio, A.; Farias, A.M.; Lazzarini, S.G. *Reinventing State Capitalism*; Harvard University Press: Cambridge, MA, USA, 2014.
97. Peng, M.W.; Wang, D.Y.L.; Jiang, Y. An institution-based view of international business strategy: A focus on emerging economies. *J. Int. Bus. Stud.* **2008**, *39*, 920–936. [CrossRef]
98. Li, H.; Meng, L.; Wang, Q.; Zhou, L.-A. Political connections, financing and firm performance: Evidence from Chinese private firms. *J. Dev. Econ.* **2008**, *87*, 283–299. [CrossRef]
99. Zhou, K.Z.; Gao, G.Y.; Zhao, H. State Ownership and Firm Innovation in China: An Integrated View of Institutional and Efficiency Logics. *Adm. Sci. Q.* **2016**, *62*, 375–404. [CrossRef]
100. Zheng, H.; Zhang, Y. Do SOEs outperform private enterprises in CSR? Evidence from China. *Chin. Manag. Stud.* **2016**, *10*, 435–457. [CrossRef]
101. Wu, H.; Li, S.; Ying, S.X.; Chen, X. Politically connected CEOs, firm performance, and CEO pay. *J. Bus. Res.* **2018**, *91*, 169–180. [CrossRef]
102. Chen, V.Z.; Li, J.; Shapiro, D.M.; Zhang, X. Ownership structure and innovation: An emerging market perspective. *Asia Pac. J. Manag.* **2013**, *31*, 1–24. [CrossRef]
103. Li, J.; Tang, Y. CEO Hubris and Firm Risk Taking in China: The Moderating Role of Managerial Discretion. *Acad. Manag. J.* **2010**, *53*, 45–68. [CrossRef]
104. Farag, H.; Meng, Q.; Mallin, C. The social, environmental and ethical performance of Chinese companies: Evidence from the Shanghai Stock Exchange. *Int. Rev. Financ. Anal.* **2015**, *42*, 53–63. [CrossRef]
105. Usman, M.; Farooq, M.U.; Zhang, J.; Makki, M.A.M.; Khan, M.K. Female directors and the cost of debt: Does gender diversity in the boardroom matter to lenders? *Manag. Audit. J.* **2019**, *34*, 374–392. [CrossRef]
106. Huang, M.; Li, M.; Liao, Z. Do politically connected CEOs promote Chinese listed industrial firms' green innovation? The mediating role of external governance environments. *J. Clean. Prod.* **2021**, *278*, 123634. [CrossRef]
107. Xu, E.; Yang, H.; Quan, J.M.; Lu, Y. Organizational slack and corporate social performance: Empirical evidence from China's public firms. *Asia Pac. J. Manag.* **2014**, *32*, 181–198. [CrossRef]
108. Marquis, C.; Qian, C. Corporate Social Responsibility Reporting in China: Symbol or Substance? *Organ. Sci.* **2014**, *25*, 127–148. [CrossRef]
109. Shen, W.; Lin, C. Firm Profitability, State Ownership, and Top Management Turnover at the Listed Firms in China: A Behavioral Perspective. *Corp. Gov. Int. Rev.* **2009**, *17*, 443–456. [CrossRef]
110. Hung, M.Y.; Wong, T.; Zhang, T. Political considerations in the decision of Chinese SOEs to list in Hong Kong. *J. Account. Econ.* **2012**, *53*, 435–449. [CrossRef]
111. Li, D.; Lin, H.; Yang, Y.-W. Does the stakeholders—Corporate social responsibility (CSR) relationship exist in emerging countries? Evidence from China. *Soc. Responsib. J.* **2016**, *12*, 147–166. [CrossRef]
112. Li, D.; Huang, M.; Ren, S.; Chen, X.; Ning, L. Environmental Legitimacy, Green Innovation, and Corporate Carbon Disclosure: Evidence from CDP China 100. *J. Bus. Ethic.* **2018**, *150*, 1089–1104. [CrossRef]
113. Usman, M.; Javed, M.; Yin, J. Board internationalization and green innovation. *Econ. Lett.* **2020**, *197*, 109625. [CrossRef]
114. Haß, L.H.; Müller, M.A.; Vergauwe, S. Tournament incentives and corporate fraud. *J. Corp. Financ.* **2015**, *34*, 251–267. [CrossRef]
115. Vo, T.T.N.; Canil, J. CEO pay disparity: Efficient contracting or managerial power? *J. Corp. Financ.* **2019**, *54*, 168–190. [CrossRef]
116. Manner, M.H. The Impact of CEO Characteristics on Corporate Social Performance. *J. Bus. Ethic.* **2010**, *93*, 53–72. [CrossRef]
117. Liu, S.; Yan, M.-R. Corporate Sustainability and Green Innovation in an Emerging Economy—An Empirical Study in China. *Sustainability* **2018**, *10*, 3998. [CrossRef]
118. Khan, F.U.; Zhang, J.; Dong, N.; Usman, M.; Ullah, S.; Ali, S. Does privatization matter for corporate social responsibility? Evidence from China. *Eur. Bus. Rev.* **2021**, *11*, 497–515. [CrossRef]
119. Meyer, J.W.; Rowan, B. Institutionalized Organizations: Formal Structure as Myth and Ceremony. *Am. J. Sociol.* **1977**, *83*, 340–363. [CrossRef]
120. Conyon, M.J.; He, L. Executive compensation and corporate governance in China. *J. Corp. Financ.* **2011**, *17*, 1158–1175. [CrossRef]
121. Waldman, D.A.; Siegel, D. Defining the socially responsible leader. *Leadersh. Q.* **2008**, *19*, 117–131. [CrossRef]
122. Carpenter, M.A.; Sanders, W.G. Top management team compensation: The missing link between CEO pay and firm performance? *Strat. Manag. J.* **2002**, *23*, 367–375. [CrossRef]
123. Jamali, D.; Mirshak, R. Corporate Social Responsibility (CSR): Theory and Practice in a Developing Country Context. *J. Bus. Ethic.* **2006**, *72*, 243–262. [CrossRef]
124. Ioannou, I.; Serafeim, G. What drives corporate social performance? The role of nation-level institutions. *J. Int. Bus. Stud.* **2012**, *43*, 834–864. [CrossRef]
125. Matten, D.; Moon, J. "Implicit" and "explicit" CSR: A conceptual framework for a comparative understanding of corporate social responsibility. *Acad. Manag. Rev.* **2008**, *33*, 404–424. [CrossRef]

126. Cohen, M.; Gould, F.; Bentur, J. Bt rice: Practical steps to sustainable use. *Int. Rice Res. Notes* **2000**, *25*, 4–10.
127. Schmoch, U.; Rammer, C.; Legler, H. *National Systems of Innovation in Comparison: Structure and Performance Indicators for Knowledge Societies*; Springer Science & Business Media: Berlin/Heidelberg, Germany, 2006.

International Journal of
Environmental Research and Public Health

Article

Linking Green Human Resource Practices and Environmental Economics Performance: The Role of Green Economic Organizational Culture and Green Psychological Climate

Syed Mehmood Ali Shah [1], Yang Jiang [1,*], Hao Wu [1,*], Zahoor Ahmed [2,3], Irfan Ullah [2] and Tomiwa Sunday Adebayo [4,5]

[1]   Northeast Asian Research Center, Jilin University, Changchun 130012, China; mehmood.shah100@yahoo.com
[2]   School of Management and Economics, Beijing Institute of Technology, Beijing 100081, China; zahoorahmed83@yahoo.com (Z.A.); irfanullahkhan9214@gmail.com (I.U.)
[3]   Department of Business Administration, Faculty of Management Sciences, ILMA University, Karachi 75190, Pakistan
[4]   Department of Business Administration, Faculty of Economic and Administrative Science, Cyprus International University, Nicosia, Northern Cyprus, TR-10, Mersin 99080, Turkey; twaikline@gmail.com
[5]   Department of Finance & Accounting, AKFA University, 1st Deadlock, 10th Kukcha Darvoza Street, Tashkent 100012, Uzbekistan
*   Correspondence: jyer415@jlu.edu.cn (Y.J.); wuh@jlu.edu.cn (H.W.)

Citation: Shah, S.M.A.; Jiang, Y.; Wu, H.; Ahmed, Z.; Ullah, I.; Adebayo, T.S. Linking Green Human Resource Practices and Environmental Economics Performance: The Role of Green Economic Organizational Culture and Green Psychological Climate. *Int. J. Environ. Res. Public Health* **2021**, *18*, 10953. https://doi.org/10.3390/ijerph182010953

Academic Editors: Pasquale Avino, Massimiliano Errico, Aristide Giuliano and Hamid Salehi

Received: 28 August 2021
Accepted: 12 October 2021
Published: 18 October 2021

Publisher's Note: MDPI stays neutral with regard to jurisdictional claims in published maps and institutional affiliations.

Abstract: An eco-friendly environment with green strategies can help to achieve better environmental performance. However, literature on the relationship between green human resource management practices (GHRMP) and sustainable environmental efficiency (SEF) is limited. Moreover, there is limited knowledge about the factors that could mediate the relationship between GHRMP and SEF. Therefore, the present study examines the impact of green human resource management practices mediating through green psychological climate (GPC) and green organizational culture (GOC) for better environmental efficacy. For this purpose, the primary data on variables are collected by using structured assessment tools and analyzed through regression models. Unlike previous studies, this study adopts a mediation model and unfolds not only the role of green human resource practices in psychological climate and green organizational culture but also clarifies the mediating role of GPC and GOC in sustainable environmental efficiency. The findings unfolded that ecological factors such as green psychological climate, green organizational culture, and sustainable environmental efficiency are positively affected by green human resources management. In addition, green organizational culture and green psychological climate positively mediate the relationship between GHRMP and SEF. This study recommends adopting green human resource management strategies and increasing technical innovations to improve sustainability and economic performance.

Keywords: human resource management (HRM); green psychological climate (GPC); green organizational culture (GOC); environmental concerns; economic performance

1. Introduction

Human activities have increased global warming up to 1.0 degrees Celsius over pre-industrial levels. This global warming is leading to climate change, which in turn increases unexpected weather changes over time, such as devastating storms and droughts, heatwaves, and forest fires, ecological destruction, and environmental degradation. Thus, climate change is considered to be among the most challenging issues of the twenty-first century [1]. The significant impact of climate change on a local, regional, and international scale can be felt around the world, while the corporate sector is also at the forefront of all sustainability controversies [2]. Companies may also play an essential part in dealing with ecological problems [3] and today's magic word is "environmental consciousness," which has infiltrated every part of our lives and workplaces.

Late in life, our personal and professional habits started to affect the planet negatively, and we could not afford to ignore the implications. We need to change our living habits for better consequences. Undoubtedly, the corporate community participates in discussions on environmental issues and hence plays an essential role in resolving ecological threats. Green human resource practices is a concept that aids in creating a green workforce in a sustainable community that recognizes and values environmental stewardship.

As described by [4], green HRM is the application of HRM policies to encourage companies in utilizing their resources efficiently and promoting environmental stewardship, thus improving employee morale and happiness. Others describe green HRM as using HRM policies, principles, and practices to promote the productive usage of business resources while avoiding environmental harm caused by organizations [5]. Human resource management (HRM) is an essential administrative factor that deals with the most critical resource in organizations. Therefore, sustainability has become an important goal of HRM since environmental performance cannot be achieved without human efforts.

Moreover, [6] pointed out the degree of environmental commitment that determines the resulting set of environmental targets. Substantial work relevant to the manufacturing sector was performed in ecological improvement. A factor of 1.5 can be attained when pollution control is the strategic objective [5] while several scholars suggest a factor of four as a practical short-term eco-productivity objective [7]. Reduced environmental costs, above a multiplier of four, appear appropriate in terms of drastic changes coming from eco-innovation, and environmental load reductions up to a factor of 50 were listed for sustainability. The achievement of the fundamental global goal is heavily dependent on population growth, size estimations, supply, and demand dynamics [8]. In addition, we assert that the most critical part of development is green human capital management. In this context, analyzing the role of green practices for human resource management (HRM) in sustainable environmental management is the sole objective of this paper.

2. Literature Review and Hypotheses Development

An organization's human resources department plays an essential role in developing the longevity of an enterprise [9]. Green management can be considered as to how an enterprise handles the climate with different techniques [10,11]. This term covers green procurement, green placement, green preparation, green success assessments, green rewards, and compensation. This activity must be effectively maintained. It is essential to learn how these healthcare institutions manage these activities in their fields. For hour managers, the critical issue in organizations will be working on numerous practices which improve their employees' environmental performance and green behavior. A detailed analysis is needed to determine how green HRM activities are supposed and handled within these organizations in different practices. This specific report makes efforts to understand the connection between green human resource management strategies, environmental efficiency, and employees' green behavior.

Because of the growing global environmental deprivation confronting current and future generations, there is a stronger focus on environmental sustainability and sustainable development growth (SDG) over the last decade [12–14]. Indeed, the world was identified as the most critical market concern of the 1990s [15]. Since then, it has spawned the idea of "Go Green," which is a worldwide buzzword among academics, the financial and environmental, and the public because of disturbing climate change and global warming [10,16–19]. This can be seen in the case of industries and other financial institutions where environmental issues affect living beings, such as global warming, acid rain, air and water pollution, ozone layer depletion, and climate change, and these are addressed through voluntary codes of conduct, including the United Nations Environment Programme-Finance Initiative (UNEP-FI), the Equator Principles for Project Finance, and others [20].

Environmental issues are perceived to be a very significant selling aspect that has led to the green marketing idea, which focuses on designing marketing campaigns that fulfill consumers' environmental wishes. Like conventional marketing, greening is related to the concept of social marketing by identifying green practices and perceptions that could be used to develop green ideas and goods for encouraging the inclusion of fiscal, social, and environmental considerations in the implementation of values. Green initiatives may be considered a socially conscious policy that reaffirms an organization's dedication to its corporate social obligations and acts as a benchmark to improve the efficiency of an enterprise. For most sectors worldwide, this is often deemed as an obstacle to incorporate adequate and acceptable green policies essential for modern enterprises. It is necessary to remember the environmental commitment to companies that is becoming an important consideration for consumers to support their decisions.

Environmental performance evaluation (EPE) is a process developed by the International Standard Organization (ISO) to assist management in understanding and making decisions about an organization's environmental performance by data collection and analysis and obtaining evaluation-based information on environmental performance. This study aims to provide the readers with basic understandings of green HRM, multiple valuable works on green HRM by others, and an improved range of green activities that can be used to build a healthy green workplace. The following are several attempts to suggest such HR environmental policies: to study the impact of green HRM practices on organizations' environmental performance, mediating through the green climate and moderated through green organizational culture. Green HRM practices play a significant role in industry to boost environment-related issues. Organizations have enough chances of growth by being green and making a new responsive environment, which helps in substantial operational savings by decreasing their carbon impression.

By adopting green practices, firms can considerably reduce service costs by using energy-efficient, low-waste technologies. Green HRM can create a culture of concern for the well-being and health of our fellow employees.

Green HRM can improve the environment by recycling waste material, which makes the environment clean and sustainable. A sustainable environment enhances the air quality and ensures the health of the public. This study investigates the organization's green HRM practices and finds ways to implement them to maintain the environment and improve business performance.

By adopting green practices, the environmental performance becomes sustainable, which helps the organization to achieve competitive advantages. A sustainable environment also makes society green, which makes sure the health safety of an individual. Green HRM practices in an organization and society ensure the sustainability of the domain.

2.1. Green Human Resource Management and Environmental Performance

The concept of green management was initiated as a part of business strategy during the 1990s [21]. However, in the 2000s, it began to gain power. In essence, Wehrmeyer and Parker (1996) coined the term green HRM as the organized and intended integration of traditional HRM activities with the organization's environmental priorities [22]. Moreover, [23] stressed the dire need to create correlations between human resource practices and sustainable development. In this context, a modern concept of economic growth is discussed that aims to promote politically, ecologically, and socially sustainable development through trade and foreign policy, financial and fiscal policies, agricultural, and industrial policies [24]. The literature on the degree to which HRM is green is frequently analyzed on a spectrum of traditional HR practices: employment research, recruiting and placement, induction, preparation, assessment of results, and awards [24].

Designing and implementation of new roles and positions to concentrate exclusively on environmental aspects of organizations can be referred to as reen work design [25]. It includes combining various roles, duties, and obligations in each work relevant to environmental protection. In other terms, the addition of environmental factors into the work definition and, at the same time, the integration of green competence in job specification is an intriguing part of green job-research [25].

Many businesses have already begun to adopt sustainability practices for environmental protection and each work description has started to directly provide at least one ecological maintenance requirement and even ecological concern [24]. The green recruiting and selection process is based on environmentally friendly methods for recruitment such as internet resources and reduced paper use during the recruitment phase, and the use of green attitudes at selection [26]. The green skills of individuals are essential to environmental success [27]. Therefore, it is worth remembering individuals who value sustainability practices and adopt simple ecological activities such as recycling, carpooling, and energy saving. On the other side, even applicants who respect ethical commitments will be drawn to environmentally friendly and "green employer" organizations [22]

Furthermore, [25] notes that green induction entails familiarizing new hires with the organization's greening activities and empowering them to show green interpersonal citizenship. New workers must have a genuine understanding and strategy for their organizational environmental culture [28]. Organizations can take two approaches to this, i.e., general green induction and green work induction [29]. Under general green leadership, companies provide prospective members with basic knowledge on strategies and procedures on environmental sustainability. New hires are focused on sustainability projects unique to their work during specific eco induction. Today, all these tactics in companies prove their merit.

Environmental consciousness should be considered a precondition for achieving triumphant environmental success by corporate workers at all levels. Therefore, environmental education is essential to improve the mindset and actions of the organizational participants [29]. The focus was placed on providing such instruction for workers who would promote conservation and waste management practices. Training workers to generate green workspace research and energy conservation, execute rotations to prepare green management for the future, and improve green personal abilities may be regarded as valuable green training and growth strategies.

The green performance assessment is a performance assessment based on green standards that has a different aspect to the success in greening the performance feedback interview [30]. Such green activity will show its merit since once an action is assessed to evaluate a person, its perceived value increases, and efforts are made to adhere. Thus, using green measures in the performance assessment scheme will speed up employee adoption [17]. Moreover, [31] have welcomed the inclusion of environmental sustainability priorities in the performance assessment scheme, as it guarantees that workers' progress is regularly provided with reviews.

This involves the adaption of programs to reward the development of renewable skills, the usage of monetary and non-monetary environmental incentives (such as benefits, cash premiums, sabbaticals, gifts), and connectint green pay and reward systems to the green system activities [5]. It must be emphasized that green habits are rewarded in the workplace, and carbon footprints are reduced. This can be seen as a possible instrument for promoting organizational environmental practices.

Green empowerment: encouraging corporate employees to make environmental choices and encourage them to take accountability for their actions that lead to cost appreciation, a sense of identity, and improved ecological efficiency. Such green behavior strengthens employees' commitment to sustainability programs and their happiness after achieving their environmental targets.

People and their harmful activities have affected the world and the company [4,32]. Green human resources management functions can build an environmentally sustainable ecosystem for companies. The HR professionals and the cooperation of workers are responsible for protecting the environment [33]. Different activities in HRM may be used to render it environmentally friendly. So, according to Cherian and Jacob (2012), the organization will achieve good results if it integrates human resources activities with environmental policies. Furthermore, [31] indicated that if companies do not include their workers in ecological activity, they will affect their ecological success. However, [34] noted later that many companies now rely very actively on protecting their community by their workers.

Green recruiting and selection are regarded as the strongest HRM methods in presenting applicants with the green HRM plan. One of the challenges facing HR people these days is to recruit and attract skilled people. Many organizations are still struggling to maintain their climate. It might help us get candidates who take the climate and sustainability seriously. This would help companies hit their environmental sustainability goal. Moreover, [34] green training and advancement are often deemed an essential human resources strategy for organizational success in environmental performance. Environmental training is considered an essential method for developing human resources [33,35]. This type of training can create an employee ethos, such that green sustainable policies, including waste reduction and a pro-active attitude to the environment, are developed. According to the [31] training curriculum, such activities as environmental protection techniques, energy conservation, safety recycling, and waste management may be used.

Green HRM practices can encourage people to engage in environmentally friendly behavior [36]. The relationship between green HRM practices, such as green employment, green recruitment, green choice, green education, and green rewards, and the positive environmental performance is up to date by many scholars [37]. With the help of green practices, organizations can achieve sustainability and competitive advantages. The essential elements of green HRM are sustainability and environmental activities, but a lack of resources in previous studies [37].

Business organizations showcase themselves as natural conservatives in a range to pull incredibly savvy experts with excellent green information, adopting green practices and maintainability issues (Rawashdeh and Karim Al-Adwan, 2012). (Cherian and Jacob, 2012) pointed out that organizations focused on greening human resource purposes may be more efficient and generate positive outcomes. Conversely, companies not associated with employees who engage in green activities may be less efficient in ecological performance. Furthermore, [31] suggested that ecological enterprises should focus on fascinating and selecting applicants with knowledge of the environment.

According to the author of [38], the impact of HRM practices on business execution can be seen in two ways. The central approach is a methodology guide that considers the effect of a series of standard HRM activities on the success of an organization. The second strategy offers unique insight by exploring the impact of such HRM operations on the company's execution. Since environmental efficiency drives business success, businesses in different industries use strategic ecological improvement practices to achieve a competitive edge [39].

The economic and ecological success is highly dependent on its management policy [1]. Changes to the organization's environmental performance policy, facilities, and staff responses to it are often part of the procedure. A significant boundary state is a transition and organizations address environmental concerns such as energy and water conservation through employee behavior improvement, greenhouse gas pollution reduction, increased recycling efforts, and increased usage of public transportation.

Employee involvement in environmental management initiatives is essential, and they are most involved in partnering with companies concerned about environmental concerns [40]. Using acceptable HRM procedures at all levels of the organizational phase, a review of companies with ISO 14001 qualification and their findings shows that organizations with high worker motivation have improved environmental efficiency.

Some considerations are critical when implementing green HRM activities, such as selecting applicants with a solid experienced environmental understanding and providing staff with regular environmental training [30].

The efficiency of the organizational climate can be explained as activities that have a positive environmental effect. Organizations must carry out practical ecological programs to do this [41]. Both pieces of research indicated the diverse GHRM practices that have a favorable and essential effect on organizational environmental practices. This would also give the rivals a competitive edge. After the discussion mentioned below, hypotheses are developed:

H1: *GHRM has a positive and considerable impact on GPC.*

H2: *GOC is positively influenced by GHRM.*

2.1.1. Green Psychological Climate

The green climate was identified as the climate for businesses that achieve sustainability objectives by implementing environmentally-friendly policies [42]. A green psychological environment is thus a consciousness of individual green policies, processes, and practices that represent the organization's green values [5]. If the organization prepares a robust environmental policy, it signals the integrity of its employees at the core of the enterprise [33]. Through following green HRM policies, companies deliver messages to workers on their environmental concerns above and beyond merely economic incentives to include employees in green decisions and activities [31].

There is an association between the understanding of corporate policy and the productive green actions of employees through the green psychology of a corporate world (Norton et al., 2014). Employees cannot invest in the working climate as they are not directly liable for energy expenses and supplies [42]. It is, therefore, necessary to explain the green duty of the company. Regarding the proper conception and evaluation of jobs; adequate green award helps articulate green workplace responsibility, fosters employee understanding of green benefit, and encourages employee interest in green business. Green psychological climates mediate the connection between green HRM and environmental success in the current study.

H3: *GPC has a positive impact on SEF.*

H4: *GPC positively mediates between the GHRM and P and SEF.*

2.1.2. Green Organizational Culture

The culture of an organization is supposed as "green" if the employees of the organization think and act outside the profit-seeking purposes to maximize the optimistic impact of organizational operations while at the same time reducing destructive operational events on the natural environment [43]. Organizations with a green culture tend to measure and change several policies to solve problems associated with the environment. Such an organization takes steps to integrate policies for environmental development in the organizational mission and vision (Afum, Agyabeng-Mensah, and Owusu, 2020). Having a committed green culture applies pressure and prompts manufacturers to stay faithful to corporate values.

As stated by [41], an organization desires to increase environmental performance which emphasizes the top management and takes a sensible struggle to invest in other organizational members concerning ecological initiatives. Such organizations integrate green practices into their mission statements to place corporate members and grow personnel skillful at solving environmental issues to achieve better quality environmental performance [44].

One of the main reasons for accepting a green culture policy is to confirm that the idea of environmental sustainability fills the thinking of all organizational employees. As organizations accept a green culture based on a persuasive strategy, where all corporate employees are fully involved, environmental performance will likely improve. Therefore, the current study will investigate the environmental performance of green HRM practices by moderating the role of green organizational culture.

Previous research has shown that a green culture will benefit the workers of companies where they work, thereby influencing the association between green HRM and environmental change. To better explain the partnership, the present research examines green HRM activities utilizing ecological efficiency as an intervening element in the presence of a green organizational culture.

The sustainability program must be recognized as a company-wide goal that includes all aspects of business activity [45]. The corporate culture improves sustainability and gains competitive advantages. Specifically, organizational culture development embraces the sustainability issues that are a vital source of sustainable competitive advantage.

The definition of corporate culture also appeared in green literature. Green corporate culture is a collection of principles, icons, assumptions, and organizations, representing the commitment or willingness to be an environmentally sustainable organization. Organizational culture is characterized as "a set of common mental assumptions that direct the action and perception of an organization by identifying appropriate behavior for various circumstances [46]. Symbolism for environmental conservation and security inside a green society forms the attitudes and activities of association participants.

Many authors argued that green organizations must make drastic behavioral changes to address environmental problems. Greening aims to increase the quality of non-renewable and renewable resource use, minimize emissions, and modify organizations and procedures that perform operations in a reasonable way concerning the environment [47].

Businesses choose to follow a green culture approach as management respect and express concern in protecting the environment [48]. OGC is essential in helping businesses turn their renewable success approach into a green one. Management issues in industrial enterprises facing environmental constraints compromise two competing objectives. Through selecting the optimum level of green performance and competitive advantage, profits may be reduced and green performance to increase benefit [49]. Companies without green culture can need limited capital to invest in a green strategy. These services may be allocated to critical corporate needs rather than environmental regulations by its management.

H5: *GOC has a positive impact on SEF.*

H6: *GOC positively mediates between the GHRM and P and SEF.*

2.2. Model Development

Two hypotheses were used to describe this theoretical research model. In the first place, we used the ability-motivation-opportunity (AMO) theory. According to the AMO theory, capacity, motivation, and opportunity are the components in HRM activities that describe how HRM actions increase an organization's human resources by increasing their competence, resulting in better outcomes, less waste, and a sustainable green climate. In addition, our study further supports the principle of the SVF, which confirms how personal beliefs affect employee behavior. Moreover, [42] has observed that personal environmental thoughts play a significant role in the eco-friendly actions of workers. Therefore, the SVF theory supports the structure suggested in this research by providing an organization with an atmosphere that promotes the personal beliefs of its workers and thus improves its environmental efficiency. In our study, the dependent variable is sustainable environmental efficiency (SEF). Because different multinationals may integrate the ethical responsibility and institutional push for green behaviors in their business operations with the use of environmental management practices to enhance their environmental performance. For this purpose, better results can only be achieved by focusing on sustainable environmental efficiencies and these have their impact on management systems. GHRM progressively investigates human resources and their environmental efficacy. The investigation of the relationship between GHRM and SEF can better explain the mechanisms through which they impact environmental and economic performances [50]. In contrast, the green human resources management (GHRM), green physiological climate (GPC), and green organizational culture (GOC) are used as a mediator between GHRM and SEF. Figure 1 shows the research model for determining the mediating impacts.

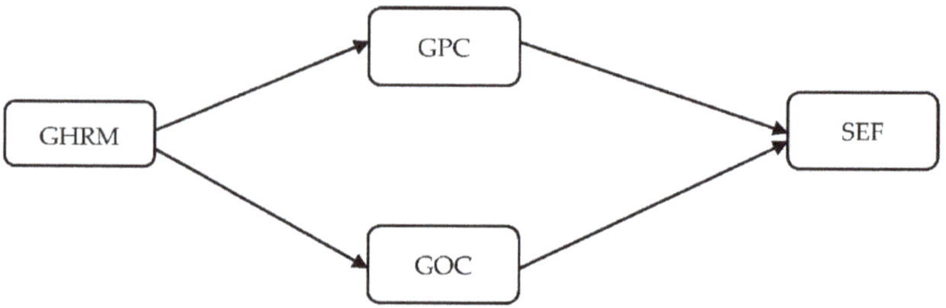

Figure 1. Conceptual model.

3. Methodology

3.1. Data Collection and Measurement

A quantitative study approach is chosen for the present study. Primary data is collected using questionnaire techniques and used for analysis by computing descriptive statistics and performing regression analysis to evaluate mediation and moderation effects within the variables. This study evaluated sustainability management strategies in human resources, environmental efficiency, and workers' green behavior. The research focuses on the highest middle and lower category of staff in various organizations. Data is obtained by stratified random sampling, in which a sample size of 480 workers was taken from the highest, medium, and lower levels. The questionnaire was prepared for research staff to assess green HRM policies, environmental efficiency, and green conduct. The population of the study has consisted of Chinese firms. In this analysis, primary parameters are retrieved that reflect the Chinese marketing background and marketing activities. The details began with the respondents' demographic profiles. The second section of the questionnaire consisted of ecological HRM activities, environmental and progressive behavior. Input on green HRM was provided on green preparation, green compensation and reimbursement, green performance assessment, and green promotion. Strongly agree = 5, agree = 4, Neutral = 3, Disagree = 2, Strongly disagreed = 1. All elements except demographics were assessed at the five-point scale. Table 1 describes the variables used in the study. Below Table 1 indicated variables with a description of items and the Likert Scale, which has been used for data collection.

Table 1. Variables with the description used under the present study.

Variable	Description	5-Point Likert Scale
GHRM	It provides a modern concept to promote sustainable development by following innovative policies on environmental aspects of organizations and encourage people to engage in green behavior.	Strongly agree = 5, agree = 4, Neutral = 3, Disagree = 2, Strongly disagreed = 1
GPC	Providing green climate by implementing environmentally-friendly policies of organizations to explains the green duty and mediates the link between green HRM and environmental success for both employees and companies.	Strongly agree = 5, agree = 4, Neutral = 3, Disagree = 2, Strongly disagreed = 1
GOC	Organizations with green culture evaluate the diverse policies to solve environment-related problems and improve the environmental performance for better development and achieve the required target and vision.	Strongly agree = 5, agree = 4, Neutral = 3, Disagree = 2, Strongly disagreed = 1
SEF	It pushes green behaviors in organizations with the implementation of environmental management policies for enhanced performance.	Strongly agree = 5, agree = 4, Neutral = 3, Disagree = 2, Strongly disagreed = 1

3.2. Data Analysis and Measurement of Model

SMART PLS 3.3 (SmartPLS GmbH, Schleswig-Holstein, Germany) is used for data analysis to employ the PLS algorithms with bootstrapping to 5000 substitute samples [51] Smart PLS is one of the leading software applications for partial least squares structural equation modeling (PLS-SEM). PLS is an emerging multivariate data analysis method, making it easy for researchers, academics, or even journal editors to let inaccurate applications of PLS-SEM go unnoticed. It was established by [52] and the software has gained acknowledgment since its introduction in 2005. It is munificently accessible to academics and researchers and has a gracious and responsive user interface and enhanced reporting aspects. Different tests were used for data validity and reliability such as CFA, rho_A, data reliability, AVE, and CR. Various items were used to improve the reliability and validity of data computed by the 5-point Likert scale, such as 1 for strongly disagree to 5 strongly agree. The mostly 5-point Likert scale is used in research because less than five and more than 7-Likert scales are less accurate [51]. Table 2 presenting the demographic values of the data, and according to Table 3, all the values of CA, rho_A, CR, AVE, and FL is greater than the standard, e.g., the standard value for CA, rho_A, and CR is equivalent or superior to 0.7 and 0.5 for AVE [52]. Table 4 shows the discriminant validity of the model.

To get the final solution, the confirmatory factor analysis was completed on the data, the root mean square error of approximation (RMSEA), the standardized root mean square residual (SRMR) [51], normed fit indexed (NFI) [53] the squared Euclidean distance (d_ULS), the geodesic distance (d_G) values, among others, were all contained by the expected range, indicating that there was no considerable common method variance (CMV) among the data assembled. The anticipated structural model was then anticipated to assess path evaluation and complete model fit. According to the evaluation, demonstration data is fit and satisfactory.

Table 2. Demographic information.

	Frequency	Percentage
Gender		
Male	254	53%
Female	226	47%
Age		
21–30 years old	198	41%
31–40 years old	185	39%
41–50 years old	97	20%
Education of Respondents		
Bachelors	240	50%
Masters	206	43%
PhD. Degree	34	7%
Type of Occupation		
Government Job	78	16%
Private Job	146	30%
Student	94	20%
Businessman	130	27%
Housewife	32	7%

Table 3. Factor analysis, data validity, and reliability.

Constructs	FL	CA	rho_A	CR	AVE
GHRMP		*0.984*	*0.984*	*0.985*	*0.814*
GHRMP1	0.811				
GHRMP10	0.926				
GHRMP11	0.856				
GHRMP12	0.923				
GHRMP13	0.898				
GHRMP14	0.945				
GHRMP15	0.954				
GHRMP2	0.864				
GHRMP3	0.905				
GHRMP4	0.885				
GHRMP5	0.887				
GHRMP6	0.956				
GHRMP7	0.907				
GHRMP8	0.886				
GHRMP9	0.917				
GOC		*0.917*	*0.925*	*0.936*	*0.710*
GOC1	0.847				
GOC2	0.829				
GOC3	0.763				
GOC4	0.820				
GOC5	0.936				
GOC6	0.852				
GPC		*0.941*	*0.950*	*0.955*	*0.810*
GPC1	0.875				
GPC2	0.932				
GPC3	0.907				
GPC4	0.850				
GPC5	0.933				
SEF		*0.967*	*0.968*	*0.974*	*0.884*
SEF1	0.951				
SEF2	0.927				
SEF3	0.944				
SEF4	0.924				
SEF5	0.955				

Standard method variance (CMV) is a generous systematic error that is frequently recognized as a feasible subject in standardized data analysis as a method of logical fallacy. Although we presented unidentified assessments on the respondents, the outcomes of this investigation may still be biased due to the circumstance that the predicted variables and predictor factors were derived from a particular survey.

Table 4. Discriminant validity.

Constructs	GHRMP	GOC	GPC	SEF
GHRMP	0.902			
GOC	0.823	0.843		
GPC	0.800	0.688	0.900	
SEF	0.994	0.825	0.802	0.940

4. Results

As per findings the major fit indicators are Chi-square = 7526.023, SRMR = 0.043, NFI = 0.727, d_ULS = 0.913, d_G = 8.141 and RMS theta = 0.156. The compositions and the pathways statement a substantial portion of the variation in the endogenous constructs postulated. Table 5 is showing the R square values, Like, GOC = 0.678, GPC = 0.640 and SEF = 0.785. It means that model is significant and well explained, like 68%, 64% and 78%, respectively.

Table 5. R Square Values.

	R Square
GOC	0.678
GPC	0.640
SEF	0.785

Figure 2 shows the outer loadings and path analysis values from H1, H2, H3, and H5. Table 6 also contains beta, SD, T Statistics, and *p* Values. These values show that H1, H2, H3, and H5 have a strong positive relationship. According to the conclusions, the effect of "GHRMP" on "GPC" and GOC is ($\beta = 0.800$, $p = 0.000$) and ($\beta = 0.823$, $p = 0.000$), respectively. The impact of "GPC" on "SEF" is ($\beta = 0.445$, $p = 0.000$), and "GOC has the impact on "SEF" is ($\beta = 0.520$, $p = 0.000$).

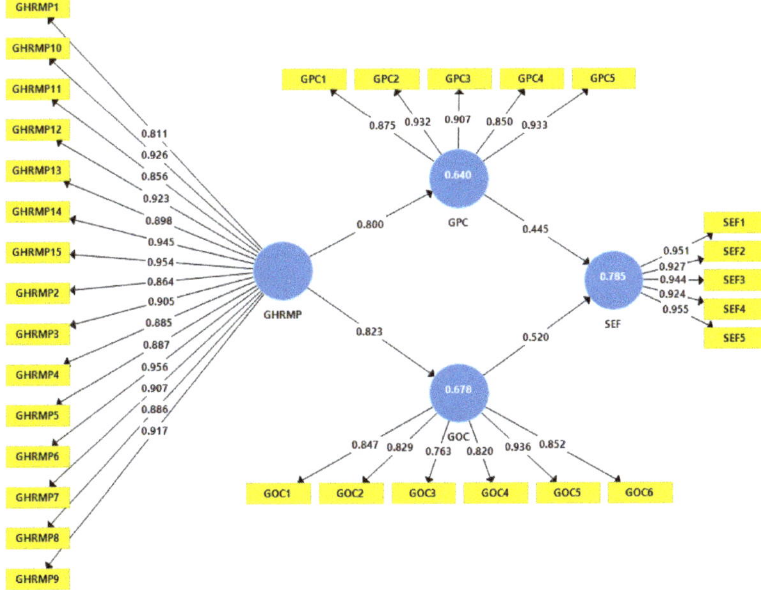

Figure 2. Factor loadings and direct path analysis.

Table 6. Direct path coefficient.

	Hypotheses	β	SD	T Statistics	*p* Values
H1	**GHRMP -> GPC**	0.800	0.029	27.812	0.000
H2	**GHRMP -> GOC**	0.823	0.025	32.823	0.000
H3	**GPC -> SEF**	0.445	0.066	6.689	0.000
H4	**GOC -> SEF**	0.520	0.063	8.261	0.000

Figure 3 and Table 7 are showing the mediation analysis of the model. According to results, "GPC" and "GOC" strongly mediates the between "GHRMP" and "SEF". E.g.,: ($\beta = 0.356$, $p = 0.000$) and ($\beta = 0.428$, $p = 0.000$) respectively.

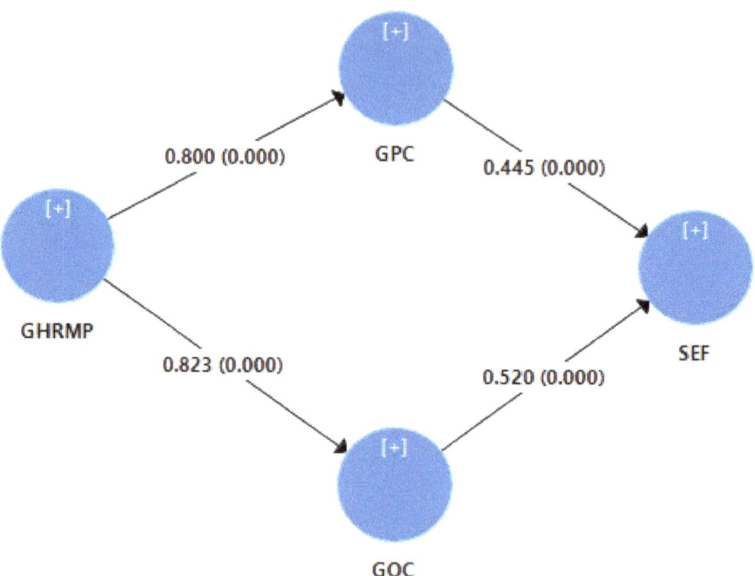

Figure 3. Mediator analysis.

Table 7. Mediator analysis.

	Hypotheses	β	SD	T Statistics	*p* Values
H5	**GHRMP -> GPC -> SEF**	0.356	0.063	5.616	0.000
H6	**GHRMP -> GOC -> SEF**	0.428	0.062	6.871	0.000

5. Discussion

This study has far-reaching implications for theory and practice. This GHRMP study adds to the persistent need for further research to bring together the various HRM outlets for achieving environmental sustainability. Environmental concerns are commonly regarded as a significant selling point. The goal of green marketing is to create marketing campaigns that satisfy consumers' ecological desires. Green ventures can be seen as an economically responsible approach that reaffirms a company's commitment to corporate social responsibility. This is often regarded as an impediment to the introduction and application of appropriate and suitable green policies in most industries worldwide. It is crucial to note that a company's environmental responsibility is getting increasingly important as customers become more influential in their purchasing decisions.

Prospective employees are more apt to see GHRM-enabled companies as the place to work for their dreams, and they are more likely to show long-term environmental productivity. These results demonstrate GHRM's potential to help organizations becoming more environmentally conscious. This knowledge would encourage physicians to effectively integrate GHRM into strategy, practice, and workforce training programs to attract qualified candidates. This has ramifications for organizational relationships, as organizations will emphasize the environmental friendliness of their hiring messages. Given that the report's respondents were final-year students, businesses will likely benefit from open dialogue about their environmental commitments and achievements at various stages of the campus recruitment and selection process. Eco-conscious organizations would gain a strategic edge in recruiting technical workers and, as a result, win the fight for talent in this highly dynamic business environment if they had to make open and meaningful disclosures of their green practices. To build faith and show genuine concern for the environment, it is recommended that organizations back up their claims and accomplishments with facts. Pro-environmental information in employment advertisements has been identified as attractive, prestigious, and deliberate information because organizations that advocate environmental conservation are likely to positively impact potential recruits in the hiring process [54].

The present study also affects organizations that are not known for their sustainability activities. As a result, those organizations will either be overlooked or considered by potential employees while considering possible job opportunities. This study contributes to the limited academic literature on sustainable HRM literature and sustainable environmental efficiency. Previous literature does not analyze the mediating role of GPC and GOC in the nexus between GHRMP and SEF. By cultivating a positive corporate citizen role, the organizations that practice GHRM influence prospective workers' choices to join and collaborate for them. Organizations known for their business citizenship are perceived as ideal places to work, and business citizenship increases their attractiveness. The increased appeal of these organizations has influenced people's desire to work with them. As a result, this research established organizational attractiveness as an important psychological mechanism for understanding GHRM's impact on prospective employees' attitudes and affective reactions to the business. The studies of [50,54] blend a positive sustainability message with an ecological green community and climate somewhat support the findings of this study. By explaining the boundary restrictions of the GHRM/SEP relationship, the study has implications for the organizational selection process, which considers the intimate environment and green culture of prospective employees. To attract high-quality applicants, the corporation must hire environmentally conscious individuals who will support the organization's GHRM efforts by proactively engaging in sustainability events and displaying green attitudes.

6. Conclusions

Human resource management (HRM) is an essential management factor that deals with one of the most valuable assets of a human resources organization. The whole HRM background is presently under consideration in the context of sustainability. By extending this statement, we argue that the essential part of sustainability is green human resource management. In this article, we concentrate specifically on green human resource management (GHRM), in which human resources management (HRM) manages an organization's ecosystem. The study of [44] describes green HRM as using HRM policies to facilitate efficient resource use within companies and the cause of the ecosystem that further enhances employee productivity and satisfaction. Others characterize green HRM as using HRM policies, perceptions, and practices to support the sustainable utilization of business capital and prevent harm from environmental issues among organizations [55].

Organizations must understand the value of green human resource activities and their effect on environmental sustainability to compete in today's business world. The primary goal of this paper was to investigate the impact of green human resource management activities on environmental performance. According to the association study findings, both environmental efficiency and green attitudes have a strong influence on green HRM activities. This study evaluated sustainability management strategies in human resources, environmental efficiency, and workers' green behavior. The research focuses on the highest middle and lower category of staff in various organizations. Data is obtained by stratified random sampling, in which a sample size of 480 workers was taken from the highest, medium, and lower levels. All elements except demographics were assessed at the five-point scale.

Furthermore, the research sought to include comprehensive mediation and moderation models to emphasize the interaction more clearly. Because of the regression study, the organizational environment is a mediating variable. In contrast, green corporate culture serves as a moderating variable, reducing the influence of the contingent and independent variables. The findings prove that GHRMP and SEF relationship is positively mediated by GOC and GPC.

6.1. Theoretical Implications

This paper highlights the emergence and dissemination of green HRM activities that positively support environmental performance. GHRM activities reflect a company's vital assets as they are a sustainable competitive advantage. GHRM researchers aim to use the company's resource-based perspective [56] to clarify the importance of GHRM activities in environmental efficiency and sustainability [57]. For example, [58]) argues that a green atmosphere or a green culture inside a company can be a source of sustainable, triumphant environmental success. Moreover, [59] argues that a GHRM framework can be a unique source of sustainable competitivity, particularly when its components have a high level of green experience both internally and externally. In short, most GHRM theories concentrate on the causal mechanism by which internal capital or processes such as green climate and culture may contribute to both corporate success, and environmental performance. Management fashion theory and institutional theory concentrate on the aspects of the outside world around organizations that affect organizations' activities and resulting outcomes. Management is a comparatively transitory common belief spread by fashion designers that contributes to rational management success through a specific management technique. Management modes, including aesthetic fashion, are typically defined by quick, gloomy swings in management.

6.2. Policy Recommendations

Due to the increasing amount of environmental concerns, organizations are under raising pressure to respond properly and for the purpose to implement sustainable business methods, and by following sustainable policies, the development of green human resource management (HRM) practices benefit both the organization and its stakeholders. It makes a strong vision to investigate how green HRM practices might improve environmental economic performance and its sustainable impact [60]. GHRM is an expansion of HRM and is a key player in developing policies, legislation and directing public awareness campaigns to educate employees about the importance of a green environment. GHRM policies and their practices are not limited to specific enterprises, it is for the whole area of any business and is responsible for green careers and green concerns in an environmentally friendly approach for better environmental and economically green performances [61]. Environmental protection through pollution deterrence, resource management, and lessened energy utilization provoke organizations to add green practices into their supply chain for green psychological climate, human resource strategies, and practices. It proposed the impact of green human resource management and green climate methods on effective market, social, financial, and environmental performance [62]. There is a dire need to uncover more effects of GHRM policies for employee satisfaction, economic performance, the link between GHRM policy practices and green management, reputation, quality, external and internal factors of organizations, employees interact with the organization, and most importantly, how these integrated GHRM policies are beneficial for better environmental performances [63]. Here, all the HRM activities, programs, and initiatives will be considered in terms of green psychological climate sustainability. An increment in these elements of green HRM results can help to achieve the organization's green strategic goals.

This research also explained how environmentally effective green HRM activities are required and how workers learn green behavior. These respondents should be used further about the studies' direction to make up a more comprehensive survey. It may also be seen in other markets, such as the automotive and utility sectors. The current study recommends that GPC and GOC can be strengthened to achieve more environmental benefits of GHRMP in organizations. The present results urge the researchers to re-consider GHRM policies for more clarifications behind the present needs of organizations for developing a green environment. The significance of originality in promoting competitiveness, as well as the urgency of resource preservation, embracing environmentally driven technologies look to be a natural choice for promoting sustainable development for organizations and better economic performances.

Author Contributions: Conceptualization, H.W. and Y.J.; methodology, S.M.A.S.; software, Z.A.; validation, T.S.A., I.U. and Z.A.; formal analysis, Y.J.; investigation, S.M.A.S.; resources, H.W.; data curation, I.U.; writing—original draft preparation, S.M.A.S.; writing—review and editing, Z.A.; visualization, Y.J.; supervision, H.W.; project administration, H.W.; funding acquisition, H.W. All authors have read and agreed to the published version of the manuscript.

Funding: This research is funded by the Major Project (grant numbers 16JJD790013) of the Key Research Base of Humanities and Social Sciences of China Ministry of Education.

Institutional Review Board Statement: Our study did not focus on human participants/animals. Following national legislation and institutional requirements, this study did not require ethical review and approval. That our institute does not require moral approval.

Informed Consent Statement: Not applicable.

Data Availability Statement: The datasets generated for this study are available on request to the corresponding author.

Conflicts of Interest: No conflicts of interest found.

References

1. Al-Tuwaijri, S.A.; Christensen, T.E.; Hughes, K. The relations among environmental disclosure, environmental performance, and economic performance: A simultaneous equations approach. *Account. Organ. Soc.* **2004**, *29*, 447–471. [CrossRef]
2. Moscardo, G. Connecting People with Experiences. *Res. Agenda Sustain. Tour* **2019**. Available online: https://www.elgaronline.com/view/edcoll/9781788117098/9781788117098.00013.xml (accessed on 27 September 2021).
3. Su, Z.-W.; Umar, M.; Kirikkaleli, D.; Adebayo, T.S. Role of political risk to achieve carbon neutrality: Evidence from Brazil. *J. Environ. Manag.* **2021**, *298*, 113463. [CrossRef] [PubMed]
4. Machado, A.M.; De Souza, W.M.; De Pádua, M.; Machado, A.R.D.S.R.; Figueiredo, L.T.M. Development of a One-Step SYBR Green I Real-Time RT-PCR Assay for the Detection and Quantitation of Araraquara and Rio Mamore Hantavirus. *Viruses* **2013**, *5*, 2272–2281. [CrossRef] [PubMed]
5. Dumont, J.; Shen, J.; Deng, X. Effects of Green HRM Practices on Employee Workplace Green Behavior: The Role of Psychological Green Climate and Employee Green Values. *Hum. Resour. Manag.* **2016**, *56*, 613–627. [CrossRef]
6. Thoresen, C.E. Spirituality and Health: Is There a Relationship? *J. Health Psychol.* **1999**, *4*, 291–300. [CrossRef] [PubMed]
7. von Weizsäcker, F.; Wieland, S.; Köck, J.; Offensperger, W.; Offensperger, S.; Moradpour, D.; Blum, H.E. Gene therapy for chronic viral hepatitis: Ribozymes, antisense oligonucleotides, and dominant negative mutants. *Hepatology* **1997**, *26*, 251–255. [CrossRef]
8. Rennings, K.; Wiggering, H. Steps towards indicators of sustainable development: Linking economic and ecological concepts. *Ecol. Econ.* **1997**, *20*, 25–36. [CrossRef]
9. Wirtenberg, J.; Harmon, J.; Russell, W.G.; Fairfield, K.D. HR's role in building a sustainable enterprise: Insights from some of the world's best companies. *Hum. Resour. Plan.* **2007**, *30*, 10–21.
10. Anwar, A.; Sinha, A.; Sharif, A.; Siddique, M.; Irshad, S.; Anwar, W.; Malik, S. The nexus between urbanization, renewable energy consumption, financial development, and CO_2 emissions: Evidence from selected Asian countries. *Environ. Dev. Sustain.* **2021**, 1–21. [CrossRef]

11. Phan, C.T.; Jain, V.; Purnomo, E.P.; Islam, M.; Mughal, N.; Guerrero, J.W.G.; Ullah, S. Controlling environmental pollution: Dynamic role of fiscal decentralization in CO_2 emission in Asian economies. *Environ. Sci. Pollut. Res.* **2021**, 1–10. [CrossRef]
12. Chien, F.; Anwar, A.; Hsu, C.-C.; Sharif, A.; Razzaq, A.; Sinha, A. The role of information and communication technology in encountering environmental degradation: Proposing an SDG framework for the BRICS countries. *Technol. Soc.* **2021**, *65*, 101587. [CrossRef]
13. Shan, S.; Ahmad, M.; Tan, Z.; Adebayo, T.S.; Li, R.Y.M.; Kirikkaleli, D. The role of energy prices and non-linear fiscal decentralization in limiting carbon emissions: Tracking environmental sustainability. *Energy* **2021**, *234*, 121243. [CrossRef]
14. Adebayo, T.S.; Kirikkaleli, D. Impact of renewable energy consumption, globalization, and technological innovation on environmental degradation in Japan: Application of wavelet tools. *Environ. Dev. Sustain.* **2021**, 1–26. [CrossRef]
15. Grove, R.; Grove, R.H. *Green Imperialism: Colonial Expansion, Tropical Island Edens and the Origins of Environmentalism, 1600–1860*; Cambridge University Press: Cambridge, UK, 1996; 560p.
16. Gao, C.; Ge, H.; Lu, Y.; Wang, W.; Zhang, Y. Decoupling of provincial energy-related CO_2 emissions from economic growth in China and its convergence from 1995 to 2017. *J. Clean. Prod.* **2021**, *297*, 126627. [CrossRef]
17. Samina, Q.S.; Hossain, M.N. *Current Position of Banks in the Practice of Green Banking in Bangladesh: An Analysis on Private Sector Commercial Banks in Bangladesh*; Social Science Research Network: Rochester, NY, USA, 2019. Available online: https://papers.ssrn.com/abstract=3308682 (accessed on 27 September 2021).
18. Kihombo, S.; Vaseer, A.I.; Ahmed, Z.; Chen, S.; Kirikkaleli, D.; Adebayo, T.S. Is there a tradeoff between financial globalization, economic growth, and environmental sustainability? An advanced panel analysis. *Environ. Sci. Pollut. Res.* **2021**, 1–11. [CrossRef]
19. Lin, X.; Zhao, Y.; Ahmad, M.; Ahmed, Z.; Rjoub, H.; Adebayo, T.S. Linking Innovative Human Capital, Economic Growth, and CO_2 Emissions: An Empirical Study Based on Chinese Provincial Panel Data. *Int. J. Environ. Res. Public Health* **2021**, *18*, 8503. [CrossRef]
20. World Bank. World Development Indicators. 2020. Available online: http://data.worldbank.org/country (accessed on 12 January 2021).
21. Mandip, G.; Ali, S.F.; Barkha, G.; Godulika, D.; Kamna, L. Emotional Intelligence as a Forecaster of Job Satisfaction amongst the Faculty of Professional Institutes of Central Indian City, Indore. *ISCA J. Manag. Sci.* **2012**, *1*, 7.
22. Jose Chiappetta Jabbour, C. How green are HRM practices, organizational culture, learning, and teamwork? A Brazilian study. *Ind. Commer. Train.* **2011**, *43*, 98–105. [CrossRef]
23. Ahmad, F.; Draz, M.U.; Ozturk, I.; Su, L.; Rauf, A. Looking for asymmetries and nonlinearities: The nexus between renewable energy and environmental degradation in the Northwestern provinces of China. *J. Clean. Prod.* **2020**, *266*, 121714. [CrossRef]
24. Shaikh, S.A.; Rabaiotti, J.R. Characteristic trade-offs in designing large-scale biometric-based identity management systems. *J. Netw. Comput. Appl.* **2010**, *33*, 342–351. [CrossRef]
25. Opatha, H.H.P. Green Human Resource Management a Simplified Introduction. 2013. Available online: http://dr.lib.sjp.ac.lk/handle/123456789/3734 (accessed on 27 September 2021).
26. Sinha, A.; Mishra, S.; Sharif, A.; Yarovaya, L. Does green financing help to improve environmental & social responsibility? Designing SDG framework through advanced quantile modelling. *J. Environ. Manag.* **2021**, *292*, 112751. [CrossRef]
27. Subramanian, N.; Abdulrahman, M.D.; Wu, L.; Nath, P. Green competence framework: Evidence from China. *Int. J. Hum. Resour. Manag.* **2015**, *27*, 151–172. [CrossRef]
28. Wagner, M.; Van Phu, N.; Wehrmeyer, W. The relationship between the environmental and economic performance of firms: An empirical analysis of the European paper industry. *Corp. Soc. Responsib. Environ. Manag.* **2002**, *9*, 133–146. [CrossRef]
29. Khan, A.; Majeed, S.; Sayeed, R. *Women Education in India and Economic Development Linkages: A Conceptual Study*; Social Science Research Network: Rochester, NY, USA, 2020. Available online: https://papers.ssrn.com/abstract=3631689 (accessed on 27 September 2021).
30. Opatha, H.H.P.; Arulrajah, A.A. Green Human Resource Management: Simplified General Reflections. *Int. Bus. Res.* **2014**, *7*, 101. [CrossRef]
31. Renwick, D.W.; Redman, T.; Maguire, S. Green Human Resource Management: A Review and Research Agenda. *Int. J. Manag. Rev.* **2012**, *15*, 1–14. [CrossRef]
32. Adebayo, T.S.; Odugbesan, J.A. Modeling CO_2 emissions in South Africa: Empirical evidence from ARDL based bounds and wavelet coherence techniques. *Environ. Sci. Pollut. Res.* **2020**, *28*, 9377–9389. [CrossRef]
33. Arulrajah, A.A.; Opatha, H.H.D.N.P.; Nawaratne, N.N.J. Employee green performance of job: A systematic attempt towards measurement. *Sri Lankan J. Hum. Resour. Manag.* **2016**, *6*, 37. [CrossRef]
34. Hadjri, M.I.; Perizade, B.; Zunaidah, Z.; Wk, W.F. Green Human Resource Management dan Kinerja Lingkungan: Studi Kasus pada Rumah Sakit di Kota Palembang. *Inovbiz J. Inov. Bisnis* **2020**, *8*, 182–192. [CrossRef]
35. Kihombo, S.; Ahmed, Z.; Chen, S.; Adebayo, T.S.; Kirikkaleli, D. Linking financial development, economic growth, and ecological footprint: What is the role of technological innovation? *Environ. Sci. Pollut. Res.* **2021**, 1–11. [CrossRef]
36. Cherian, J.; Jacob, J. A Study of Green HR Practices and Its Effective Implementation in the Organization: A Review. *Int. J. Bus. Manag.* **2012**, *7*, 25. [CrossRef]
37. Rawashdeh, A.M.; Al-Adwan, I. The impact of human resource management practices on corporate performance: Empirical study in Jordanian commercial banks. *Afr. J. Bus. Manag.* **2012**, *6*, 10591–10595.

38. Bowen, D.E.; Ostroff, C. Understanding HRM-Firm Performance Linkages: The Role of the "Strength" of the HRM System. *Acad. Manag. Rev.* **2004**, *29*, 203. [CrossRef]
39. Rodríguez-Antón, J.M.; del Mar Alonso-Almeida, M.; Celemín, M.S.; Rubio, L. Use of different sustainability management systems in the hospitality industry. The case of Spanish hotels. *J. Clean Prod.* **2012**, *22*, 76–84. [CrossRef]
40. Harvey, D.M.; Bosco, S.M.; Emanuele, G. The impact of "green-collar workers" on organizations. *Manag Res. Rev.* **2010**, *33*, 499–511. [CrossRef]
41. Jackson, S.E.; Seo, J. The greening of strategic HRM scholarship. *Organ. Manag. J.* **2010**, *7*, 278–290. [CrossRef]
42. Chou, C.-J. Hotels' environmental policies and employee personal environmental beliefs: Interactions and outcomes. *Tour. Manag.* **2014**, *40*, 436–446. [CrossRef]
43. Roscoe, S.; Subramanian, N.; Jabbour, C.J.C.; Chong, T. Green human resource management and the enablers of green organizational culture: Enhancing a firm's environmental performance for sustainable development. *Bus. Strategy Environ.* **2019**, *28*, 737–749. [CrossRef]
44. Dangelico, R.M. Improving Firm Environmental Performance and Reputation: The Role of Employee Green Teams. *Bus. Strat. Environ.* **2014**, *24*, 735–749. [CrossRef]
45. Oriade, A.; Osinaike, A.; Aduhene, K.; Wang, Y. Sustainability awareness, management practices and organizational culture in hotels: Evidence from developing countries. *Int. J. Hosp. Manag.* **2021**, *92*, 102699. [CrossRef]
46. Tahir, R.; Athar, M.R.; Faisal, F.; Shahani Nun, N.; Solangi, B. *Green Organizational Culture: A Review of Literature and Future Research Agenda*; Social Science Research Network: Rochester, NY, USA, 2019. Available online: https://papers.ssrn.com/abstract=3497251 (accessed on 27 September 2021).
47. Francis, C.; Elmore, R.; Ikerd, J.; Duffy, M. Greening of Agriculture. *J. Crop. Improv.* **2007**, *19*, 193–220. [CrossRef]
48. Klassen, R.D.; Vachon, S. Collaboration and evaluation in the supply chain: The impact on plant-level environmental investment. *Prod. Oper. Manag.* **2009**, *12*, 336–352. [CrossRef]
49. Russo, M.V.; Fouts, P.A. A resource-based perspective on corporate environmental performance and profitability. *Acad Manag. J.* **1997**, *40*, 534–559.
50. Longoni, A.; Luzzini, D.; Guerci, M. Deploying environmental management across functions: The relationship between green human resource management and green supply chain management. *J. Bus. Ethic.* **2016**, *151*, 1081–1095. [CrossRef]
51. Hair, J.F.; Risher, J.; Sarstedt, M.; Ringle, C.M. When to use and how to report the results of PLS-SEM. *Eur. Bus. Rev.* **2019**, *31*, 2–24. [CrossRef]
52. Ringle, C.; da Silva, D.; Bido, D. *Structural Equation Modeling with the SmartPLS*; Social Science Research Network: Rochester, NY, USA, 2015. Available online: https://papers.ssrn.com/abstract=2676422 (accessed on 27 September 2021).
53. Marsh, H.W.; Hocevar, D. Application of confirmatory factor analysis to the study of self-concept: First- and higher-order factor models and their invariance across groups. *Psychol. Bull.* **1985**, *97*, 562–582. [CrossRef]
54. Shetye, T.; Baker, B.A.; Thompson, L.F. Effects of Pro-Environmental Recruiting Messages: The Role of Organizational Reputation. *J. Bus. Psychol.* **2009**, *24*, 341–350.
55. Zoogah, D.B. The Dynamics of Green HRM Behaviors: A Cognitive Social Information Processing Approach. *Ger. J. Hum. Resour. Manag.* **2011**, *25*, 117–139. [CrossRef]
56. Barney, J. Firm Resources and Sustained Competitive Advantage. *J. Manag.* **1991**, *17*, 99–120. [CrossRef]
57. Wright, P.M.; Dunford, B.B.; Snell, S.A. Human resources and the resource-based view of the firm. *J. Manag.* **2001**, *27*, 701–721. [CrossRef]
58. Mcmahan, J. Self-Defense and the Problem of the Innocent Attacker. *Ethics* **1994**, *104*, 252–290. [CrossRef]
59. Gerhart, K.L.; White, R.G.; Cameron, R.D.; Russell, D.E. Estimating Fat Content of Caribou from Body Condition Scores. *J. Wildl. Manag.* **1996**, *60*, 713. [CrossRef]
60. Yusoff, Y.M.; Nejati, M.; Kee, D.M.H.; Amran, A. Linking Green Human Resource Management Practices to Environmental Performance in Hotel Industry. *Glob. Bus. Rev.* **2018**, *21*, 663–680. [CrossRef]
61. Malik, S.Y.; Hayat Mughal, Y.; Azam, T.; Cao, Y.; Wan, Z.; Zhu, H.; Thurasamy, R. Corporate Social Responsibility, Green Human Resources Management, and Sustainable Performance: Is Organizational Citizenship Behavior towards Environment the Missing Link? *Sustainability* **2021**, *13*, 1044. [CrossRef]
62. Acquah, I.S.K.; Agyabeng-Mensah, Y.; Afum, E. Examining the link among green human resource management practices, green supply chain management practices and performance. *Benchmarking Int. J.* **2021**, *28*, 267–290. [CrossRef]
63. Pham, N.T.; Hoang, H.T.; Phan, Q.P.T. Green human resource management: A comprehensive review and future research agenda. *Int. J. Manpow.* **2019**, *41*, 845–878. [CrossRef]

International Journal of
*Environmental Research
and Public Health*

Article

How to Promote Low-Carbon Economic Development? A Comprehensive Assessment of Carbon Tax Policy in China

Weijiang Liu [1,2,3], Yangyang Li [3,*], Tingting Liu [3], Min Liu [3] and Hai Wei [4]

[1] Center for Quantitative Economics, Jilin University, Changchun 130012, China; liuwj@jlu.edu.cn
[2] Northeast Revitalization and Development Research Institute, Jilin University, Changchun 130012, China
[3] Business School, Jilin University, Changchun 130012, China; 18326921663@163.com (T.L.);
minliu19@mails.jlu.edu.cn (M.L.)
[4] School of Cyber Security, Gansu University of Political Science and Law, Lanzhou 730070, China;
weihai94@163.com
* Correspondence: sunnyyyli@163.com

Abstract: Facing the increasingly severe environmental problems, the development of a green and sustainable low-carbon economy has become an international trend. In China, the core issue of low-carbon economic development is effectively resolving the contradiction between the exploitation and utilization of fossil energy and greenhouse gas emissions (mainly carbon emissions). Based on the SAM matrix, we established a static Computable General Equilibrium (CGE) model to simulate the impact of carbon tax policies on energy consumption, carbon emissions, and macroeconomics variables under 10, 20, and 30% emission reductions. Meanwhile, we analyze the impact of different carbon tax recycling mechanisms under the principle of tax neutrality. We find that the carbon tax effectively reduces carbon emissions, but it will negatively impact economic development and social welfare. A reasonable carbon tax recycling system based on the principle of tax neutrality can reduce the negative impact of carbon tax implementation. Among the four simulated scenarios of carbon tax cycle, the scenario of reducing residents' personal income tax is most conducive to realizing the "double dividend" of carbon tax.

Keywords: carbon tax; low-carbon economy; CO_2 emissions; double dividend; CGE model; tax neutrality; carbon tax recycling system

Citation: Liu, W.; Li, Y.; Liu, T.; Liu, M.; Wei, H. How to Promote Low-Carbon Economic Development? A Comprehensive Assessment of Carbon Tax Policy in China. *Int. J. Environ. Res. Public Health* **2021**, *18*, 10699. https://doi.org/10.3390/ijerph182010699

Academic Editors: Pasquale Avino, Massimiliano Errico, Aristide Giuliano and Hamid Salehi

Received: 14 September 2021
Accepted: 10 October 2021
Published: 12 October 2021

1. Introduction

The fundamental way for humanity to cope with climate change and realize the coordinated development of energy, environment, and economy is to implement and develop a low-carbon economy [1,2]. The global economic growth and the gradual increase in production activities have accelerated carbon emissions [3]. At the same time, the rapid growth of carbon emissions has seriously threatened the survival of human beings and the sustainable development of the social environment [4]. In recent years, more than 30 countries around the world have established the goal of achieving carbon neutrality around the middle of this century [5]. As the world's largest carbon dioxide emitter, China also pledged at the 2020 United Nations General Assembly to achieve carbon neutrality by 2060, and carbon dioxide emissions peaking before 2030. However, judging from the current situation, for countries in the process of high-speed industrialization, energy demand will show an upward trend for a long time. Therefore, controlling greenhouse gas emissions is still a key consideration in formulating environmental policies [6].

In order to promote the development of a low-carbon economy and reduce the impact of carbon emissions on the social environment, governments of various countries have begun to explore ways to reduce carbon emissions [7]. For example, implementing carbon taxes and carbon emission trading schemes are effective policy tools to reduce carbon emissions [8,9]. Among them, carbon tax policy has shown promising results in controlling carbon emissions [10–12]. The carbon tax is an environmental protection tax, which aims to reduce carbon emissions and alleviate global warming. Because the levy of a carbon tax will increase energy products' prices, the demand for energy products will decrease, reducing carbon emissions [13]. Secondly, the market mechanism will encourage clean energy to achieve sustainable energy development and the double emission reduction effect of the carbon tax [14]. However, the carbon tax will also bring some negative effects [15], such as increasing the tax burden of enterprises and residents, affecting their social welfare. In order to reduce these adverse effects and ensure the carbon tax system's implementation, reasonably using the incentive mechanism to design the carbon tax and integrating it into the existing tax system or adapting it to tax system reform is significant to improving the environment and strengthening the tax system's income redistribution [16,17]. Therefore, when formulating carbon tax policies, it is necessary to determine a reasonable carbon tax recovery mechanism based on the principle of tax neutrality.

The principle of tax neutrality refers to keeping taxes and expenditures unchanged, achieving budget neutrality, minimizing distortions to the market as much as possible, and ensuring the market's pure competitiveness. It does not aim to increase fiscal revenue, but returns most of the carbon tax revenue to taxpayers in the form of subsidies and compensation, thereby increasing the acceptability of the carbon tax system and reducing the impact of the carbon tax on social welfare. A carbon tax system that complies with the principle of tax neutrality will also reduce the "extra burden" of the carbon tax itself and the negative impact of the regressive nature of the carbon tax on taxpayers, especially low-income households. In the initial stage of the implementation of the carbon tax, to enhance the acceptability of carbon tax, the tax rate will usually be gradually increased from a lower level, so as to give play to the emission reduction function of the carbon tax system, which is in line with the principle of tax neutrality and the goal of sustainable economic development. The "tax neutrality principle" requires that taxpayers' tax burdens be balanced and stable at the macro level. Besides, China is implementing "structural tax cuts". Therefore, taxpayers' tax burdens in other areas should be reduced while carbon taxes are levied; it will realize the "double dividend" of improving enterprises' and residents' welfare and enhancing carbon emission reduction. The carbon tax recycling mechanism ensures that the overall income and expenditure remain unchanged and conform to tax neutrality. Meanwhile, it can reduce the impact of other distorting taxes and achieve tax burden shifting [18]. It also considers the environmental and economic goals of carbon tax collection, as making it more systematic can achieve the double dividend of reducing emissions and increasing social welfare. A carbon tax will only be accepted and supported by the public if it can achieve the goal of reducing carbon emissions. The double dividend effect of the carbon tax recycling mechanism further guarantees public support, making the implementation of the carbon tax more likely.

As a favorable tool for policy analysis, scholars have adopted different CGE models (such as static CGE, dynamic CGE, etc.) to study the effect of carbon tax policy implementation, confirming that the CGE model has significant advantages over other measurement tools in the field of policy simulation [19,20]. This paper conducts empirical analysis by establishing an environmentally Computable General Equilibrium (CGE) model for analyzing carbon tax policies. However, most previous studies have only analyzed the effects of the carbon tax and paid less attention to the synergy of multiple policies. In addition, these studies lack effective incentive policies to implement the carbon tax policy better. Additionally, they have seldom paid attention to the distortion of resource allocation caused by the additional burden of taxation. Differently, the contribution of this article is to uphold the principle of tax neutrality to design-related carbon tax simulations. In formulating the carbon tax revenue use mechanism, a carbon tax recycling system that reflects the principle of tax neutrality has been incorporated, and the revenue from the carbon tax will be used to reduce the tax burden of distorting taxes such as corporate income tax, so as to keep the overall tax burden from increasing. We contribute to incorporate the carbon tax, comprehensively analyzing carbon taxes' socio-economic impact on the country's low-carbon economy; meanwhile, we verify the existence of a double dividend. Finally, this article provides relevant policy recommendations for the government to design a reasonable carbon tax.

The study is structured as follows: Section 2 addresses the literature review. Section 3 outlines the methodology and data description. Section 4 presents the scenarios setting. The empirical results are presented in Section 5. Finally, Section 6 provides the discussion and conclusion.

2. Literature Review

2.1. Research on Influencing Factors of Carbon Emissions

Since the 21st century, China's economy has developed rapidly, and the energy provided by fossil fuels has greatly promoted the development of industries. With the massive exploitation and use of fossil energy, global air pollution, and the greenhouse effect have affected human survival and sustainable development [21,22]. In order to control carbon emissions effectively, it is essential to conduct in-depth research on the factors affecting its emissions, and it has become a hotspot in energy and environmental policy research.

The signing of the "Kyoto Protocol" in 1997 brought China's climate change research and governance into a new chapter [23–25]. Scholars around the world have conducted intense discussions on greenhouse gas emissions and governance. Among them, many scholars have demonstrated that the greenhouse effect has varying degrees of negative impact on different industries [26–28]. Pearce et al. [29] studied the impact of climate change on Employment in Australia and drew the conclusion that climate change may bring about employment growth and decline in different fields. Awan et al. [30] have suggested that firm environmental management capabilities at the production level have considerable value for addressing CO_2 emissions and minimizing environmental pollution. It requires a varied degree of competencies, in the product life cycle stage at the level of process and product to ensure and improve pollution performance. Research on climate governance relies on the decomposition and calculation of reliable carbon emissions influencing factors. Zhao et al. [31] studied the relationship between China's import and export and carbon emissions, confirming that the changing trend of low-carbon industry structure can effectively reduce carbon emissions growth.

In studying the influencing factors of carbon emissions, the decomposition identity of carbon emission factors proposed by Japanese scholar Kaya is widely used [32]. On this basis, Ang and Liu [33] proposed a new decomposition method—LMDII, which quantitatively decomposed the influencing factors of carbon emissions and confirmed that this method has a perfect effect on greenhouse impact emissions from various fields. For example, Liaskas et al. [34] use this method to analyze the influencing factors of industrial carbon dioxide emissions in E.U. countries. Zhang et al. [35] conduct a complete factor decomposition on China's energy carbon dioxide emissions and emission intensity during 1991–2006. In terms of application, scholars have done a lot of research. For example, Schipper et al. [36] applied the AWD (Adaptive-Weighting-Divisia) method to analyze the carbon dioxide emission trends of 13 countries. It is believed that energy intensity and energy consumption structure can explain most carbon emission intensity changes for most countries. Davis et al. [37] used the AWD method to analyze the reasons for the decline in energy intensity and carbon emissions in the United States from 1996 to 2000 and believe that energy structure adjustment is not the main reason, but weather changes are the main reason.

Based on the above analysis, it can be concluded that energy consumption is closely related to carbon dioxide emissions. Implementing measures to control energy consumption, such as a carbon tax, is the most effective means to reduce carbon dioxide emissions. Next, we will review the current major carbon emission mechanisms and choose the emission reduction mechanism that best suits China's national conditions.

2.2. Research on Carbon Emission Mechanism

Carbon taxes and carbon trading systems are carbon pricing policy tools commonly used by countries to promote carbon emission reduction, which will affect the cost of enterprises and individuals [38]. Currently, most countries have different emission reduction targets. The main reason is that countries are in different development stages, so the carbon emissions are also very different. As the world's largest carbon dioxide emitter, some regions in China have begun to implement the carbon trading system. In recent years, as countries have become increasingly aware of carbon taxes, it is necessary to assess whether carbon taxes effectively achieve emission reduction targets [39]. Next, we mainly conduct a comparative analysis of the two main carbon emission reduction systems, carbon tax and emission trading.

From the perspective of implementation costs, the carbon tax implementation cost and supervision cost are lower than the carbon trading system [40]. The carbon tax can be directly implemented as a tax item of environmental protection tax, reducing the cost of use [41]. The carbon trading system lacks coercive force in carbon emission reduction. In the carbon trading environment, the cost of carbon emission reduction depends on the uncontrollable market, which easily brings unreasonable emission reduction costs and greatly reduces the emission effect [42]. From the perspective of economic impact, under the principle of tax neutrality, the carbon tax has the characteristics of "double dividend" [43,44]. The collection of the carbon tax is not only conducive to achieving carbon emission reduction and improving the environment, but also guiding companies to make business decisions that are conducive to their own development and environmental protection based on the determined carbon tax price, making the economy more efficient [45]. However, the carbon emission price that relies on the market price mechanism is uncertain and cannot provide companies with an accurate assessment basis. For social welfare, Timilsina and Shrestha [46] used a static CGE model to find that the carbon tax has a smaller impact on the total social welfare compared with other emission reduction policy tools. When carbon tax revenue is used to reduce the existing indirect tax rate of non-energy products, the impact on the total social welfare is minimal.

2.3. Research on CGE Model of Carbon Tax

The carbon tax has become an emission reduction measure strongly recommended by economists and international organizations to reduce greenhouse emissions. Beginning in the 1990s, many scholars started to use the CGE model to simulate national carbon taxes [19,20], and then analyze the impact of carbon taxes on the energy system and economic variables such as GDP, welfare, investment, imports, and exports. At present, the CGE model has become one of the most mainstream tools for analyzing energy, environmental, and climate policies globally.

Garbaccio et al. [47] used a dynamic CGE model to simulate and analyze the impact of the carbon tax on China's economic development under the co-existence of a planned economy and a market economy. Shrestha and Marpaung [48] used the CGE model to study the impact of carbon taxes on the Indonesian power sector and proposed that the levy of carbon taxes can lead to a substantial increase in prices in the power sector and a significant decline in energy consumption. Wissema and Dellink [49] used the CGE model to analyze the impact of levying a carbon tax in Ireland on its economy and proposed that a carbon tax levying would significantly affect Irish consumption patterns. Allan et al. [50] used an energy-economy-environment CGE model to study Scottish carbon taxes' economic and environmental impacts. The study found that carbon taxes may achieve double dividends when carbon tax revenues are recycled through income taxes. Lin and Jia [51] analyzed the impact of carbon tax policies on the energy environment and economy by constructing CGE models of different carbon tax usage scenarios. The study found that the negative impact of the carbon tax on GDP is acceptable, and the higher the carbon tax rate, the greater the carbon dioxide emission reduction of the carbon tax policy. Fu et al. [52] developed a

CGE model to examine the social impact of carbon tax policy, finding that a carbon tax is a powerful driving force to reduce carbon emissions and promote the energy revolution.

In summary, domestic and foreign scholars have adopted the CGE model to conduct related research on the emission reduction effects of carbon tax policies, proving that the CGE model has significant advantages over other measurement tools in policy simulation.

This paper conducts empirical analysis by establishing an environmentally Computable General Equilibrium (CGE) model for analyzing carbon tax policies. Specific simulation research on the impact of carbon tax policies implemented to control carbon dioxide emissions on energy consumption, carbon dioxide emissions, and sectoral economic variables; analyze different carbon tax cycle policies' impacts on China's macroeconomic variables under the principle of tax neutrality, and explore how to realize double dividend. The above analysis can provide references for the implementation and policy formulation of China's carbon tax policy.

3. Methodology and Data

3.1. Description of the CGE Model

The CGE model adopted in this study has been widely used to simulate and analyze energy and environmental policies [53–55], systematically analyzing policy's impact under the general equilibrium framework. This model is based on Walras' general equilibrium theory, with producers, consumers, government, and foreign sectors as the basic economic units, and it can analyze the interaction of one or more variable disturbances on other variables and the entire economy [56–58]. Figure 1 portrays the detailed model skeleton.

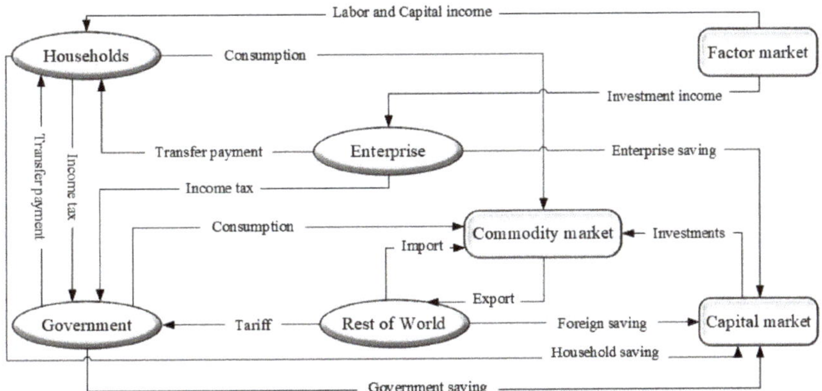

Figure 1. Skeleton of proposed CGE model.

The main structure of the production module is a five-layer nested structure. Given the importance of energy inputs to carbon emissions and the substitution effect among different energy sources, the energy inputs (coal, oil, and gas) are described by Constant Elasticity of Substitution (CES) functions and form an energy synthesis factor. The energy synthesis factor is in turn combined with capital through the CES function to form the energy capital factor, like many other studies [59,60]. Then, combined with labor through CES function to form the energy-capital-labor factor, that is, the value-added composites. This allows substitution between various input factors. The energy-capital-labor factor and non-energy intermediate inputs are combined through a Leontief function to form sectoral outputs. Most research on the CGE model separates oil and natural gas, but the separation method is quite rough and does not consider the difference in intermediate inputs. The main fossil energy is coal, so this article does not separate the two industries.

Referring to previous studies [61,62], the trade module functions are the constant elasticity transformation (CET) and Armington functions, assuming the distribution of domestically produced products and the domestic demand. The household consumption function generally adopts the Stone-Geary utility function, but its parameters are difficult to estimate. At the same time, this article is a static analysis. Therefore, the household consumption function adopts a simple linear function form. The closed rule adopts the neoclassical closure rule, assuming that labor and capital prices are endogenous. The labor and capital market achieve full employment. The social welfare function is measured by Hichsian equivalent variation. Specifically, it is based on the commodity price before implementing the policy and measures the household utility level change after implementing the policy.

Carbon dioxide emissions associated with fuel combustion and carbon taxes are introduced into the environmental module. The carbon tax is levied on carbon dioxide emissions related to fuel. By calculating the total carbon tax ratio on each fossil energy to the total fossil energy demand, the Ad valorem tax rate for fossil energy is obtained.

3.2. Data Sources

3.2.1. Source of Basic Data

The basic data of this CGE model is obtained from the 2017 social accounting matrix (SAM) table, which is based on China's 2017 input-output table and relevant data on customs, tax revenues, international payments, and capital flows. The SAM table includes nine sectors[1], three institutions (households, enterprises and government), and three production inputs (labor, capital and energy). The capital, government, and foreign inputs come from the 2018 China Statistical Yearbook (http://www.stats.gov.cn/tjsj/ndsj/, accessed on 6 October 2021) and the 2018 China Financial Yearbook (https://www.epsnet.com.cn/, accessed on 6 October 2021). Households savings are taken from the 2017 Flow of Funds Statement (http://www.stats.gov.cn/tjsj/ndsj/, accessed on 6 October 2021). The carbon dioxide emissions data come from the International Energy Statistics (https://www.eia.gov/, accessed on 6 October 2021) on the carbon dioxide emissions of China's three fossil energy sources.

3.2.2. Parameter Calibration

The model's parameters that need to be calibrated mainly include substitution elasticity coefficient, share parameter, and carbon dioxide emission coefficient. The substitution elasticity coefficients in the production and trade function are generally obtained through econometric methods or consulting relevant experts. The setting of alternative elastic parameters in this article mainly refers to previous literature [63]. The shared parameter is calibrated by the elasticity of substitution and the base year data of the variable. Using statistical data from the International Energy Statistics, the carbon dioxide emission coefficient is calculated based on the fossil energy sources' carbon dioxide emissions and the actual energy consumption.

4. Scenarios Setting

4.1. Carbon Tax Design

In the climate policy simulation of this model, carbon dioxide emissions can be directly calculated. That is why we use carbon dioxide emissions as the basis for calculating the carbon tax. The final investment and consumption demand of fossil energy accounts for a relatively small proportion of the total demand for fossil energy. Therefore, this article assumes that no carbon tax is levied on the final demand part, and only the intermediate input of fossil energy is taxed. The specific carbon tax design is as follows:

$$CTAX_i = tc \cdot \sum_j E_{i,j} \cdot \theta_i \tag{1}$$

$$CTAX_j = tc \cdot \sum_i E_{i,j} \cdot \theta_i \tag{2}$$

$$TCTAX = \sum_i CTAX_i \tag{3}$$

where, $CTAX_i$, $CTAX_j$, and $TCTAX$ are the carbon tax levied on the intermediate input of fossil energy i, carbon tax levied by sector j and the total carbon tax, respectively; tc represents the amount of carbon tax levied per ton of carbon dioxide emissions, that is, carbon tax; $E_{i,j}$ represents the energy input of fossil energy i in the sector j; θ_i is the unit energy carbon dioxide emission coefficient of energy i.

Based on the above calculations, we can convert the carbon tax rate of fossil energy into an ad valorem tax rate, that is, the carbon tax ratio on certain fossil energy to the value of the domestic demand for that fossil energy. The calculation formula in Equation (4).

$$tc_i = \frac{CTAX_i}{PQ_i \cdot QQ_i} \tag{4}$$

where, PQ_i and QQ_i, respectively, represent the demand and price of fossil energy i. As a result, the price of fossil energy demand will become $(1 + tc_i) \cdot PQ_i$. This will increase the cost of using fossil energy in the production function energy input, and government revenue will increase due to carbon tax.

4.2. Simulation Scenario Design

4.2.1. Carbon Emission and Energy Simulation Analysis

The ad valorem tax rates levied on different fossil energy sources are affected by the energy's carbon dioxide emission coefficient. The price of fossil energy and the production cost of enterprises will increase due to the levy of carbon taxes, affecting the consumption demand of different energy sources. Meanwhile, the ratio of fossil energy input to total input in different sectors is different. The production functions and the substitution elasticities of various production factors are inconsistent, affecting the fossil energy demand of different sectors. Therefore, we simulated and analyzed the impact of the carbon tax on various energy, sector, and macroeconomic variables under the emission reduction scenarios of 10, 20, and 30% reduction in total carbon dioxide emissions, respectively.

4.2.2. Carbon Tax Recycling Simulation Analysis

Levying a carbon tax can improve the environment, but it will have an adverse impact on economic development, residents' income, and social welfare. At the same time, it will increase corporate production costs and decrease corporate profits. For producers, it will increase investment in alternative energy sources, which will improve the efficiency of fossil energy utilization and the development of new energy. For consumers, rising energy prices will cause them to look for alternatives such as clean energy. If there is no good spending policy when the carbon tax is levied, its redistributive effect will be limited even if the policy is already very sound. The makers of carbon taxes should shift the tax burden to those directly responsible, so as to provide effective incentives to reduce carbon emissions.

The principle of tax neutrality requires that taxpayers' tax burdens are balanced and stable at the macro level. The carbon tax recycling mechanism ensures that the overall income and expenditure remain unchanged, in line with the principle of tax neutrality. Using carbon dioxide taxes to reduce the tax rates of existing taxes (such as income taxes, capital taxes, etc.) can reduce the impact of carbon taxes on related industries. At the same time, we will realize the double dividend of environmental taxes [26]. The first dividend is to improve the efficiency of fossil energy utilization, thereby reducing environmental pollution and improving the ecological environment. The second dividend reduces other tax rates or government transfer payments while imposing a carbon tax, increasing social welfare, and enhancing corporate competitiveness.

In summary, this study starts from the principle of tax neutrality, assuming three carbon tax recycling scenarios to compare the impacts on the social economy and verify the carbon tax double dividend. In the simulation analysis, Business as Usual (BaU) leverages a carbon tax under the scenario of reducing carbon dioxide emissions by 20%. The simulation scenario is to set up different carbon tax recycling methods. The specific settings for the simulating scenarios are designed, as listed in Table 1.

Table 1. Scenarios setting.

Design Exploration	Scenarios	Description
Carbon tax policy	Business as Usual (BaU)	A carbon tax is levied on the intermediate input of energy in the production sector, and no carbon tax is levied on the final demand sector.
Carbon tax cycle	Scenario 1	Based on the BaU scenario, reduce the resident income tax rate and maintain government revenue neutrality.
	Scenario 2	Based on the BaU scenario, reduce the enterprise income tax rate and maintain government revenue neutrality.
	Scenario 3	Based on the BaU scenario, reduce the enterprise indirect tax rate and maintain government revenue neutrality [1].

[1] Since the department's indirect tax rate is not equal, an equal percentage reduces its indirect tax rate.

5. Results

5.1. Energy and Carbon Emissions Simulation Results

5.1.1. Energy Effect

With the increase of carbon dioxide emission reduction, the carbon tax level gradually increases, as shown in Table 2. Meanwhile, the ad valorem carbon tax rate levied on fossil energy (coal, oil, and natural gas) gradually increases, with coal having the highest tax rate. When the emission reduction reaches 30%, coal's ad valorem tax rate will reach 41.95%. As for oil and natural gas, it is relatively low. In different emission reduction scenarios, fossil energy contributes differently to the total CO_2 emission reduction. We can see that the reduction task is mainly coal energy. This finding is consistent with other studies [51,52].

Table 2. The carbon tax and energy statistics.

Scenarios	Energy	10%	20%	30%
Carbon tax (yuan/ton)		30.50	71.35	128.20
Fossil energy tax rate	Coal	0.11	0.24	0.42
	O-G	0.03	0.06	0.11
Fossil energy reduction contribution	Coal	97.46%	96.87%	96.10%
	O-G	2.54%	3.13%	3.90%
Energy consumption change	Coal	−12.22%	−24.29%	−36.14%
	O-G	−1.26%	−3.10%	−5.78%
	Electricity	−0.44%	−0.92%	−1.46%

Note: O-G stands for oil and gas.

When carbon dioxide emission reductions decrease, the reduction ratio of energy consumption also increases. Among them, the reduction ratio of coal is the largest. When the emission reduction falls from 30 to 10%, the reduction ratio of coal consumption drops from 36.14 to 12.22%. However, the degree of electricity decline is not large, with a change of only 1.02%.

5.1.2. Sectoral Effects

Under different emission reduction scenarios, various sectors' demand for fossil energy (coal, oil, and natural gas) and electricity is depicted in Figure 2. The bars express the percentage change in sectors' energy consumption. The demand for coal in all sectors has fallen sharply. With the increase in emissions reductions, the extent of the decline has continued to increase. Taking the transportation industry as an example, when the carbon dioxide emission reduction is 10%, the demand for coal decreased by 12.86%; when the emission reduction is 30%, the sector's coal consumption reduces by 38.16%. The oil and natural gas consumption has fallen in most sectors, but the power sector has risen. When the emission reductions are 10, 20, and 30%, oil and gas consumption in the power sector has increased by 1.24, 2.21, and 2.66%, respectively. The electricity energy consumption in various sectors can also be seen in Figure 2. Overall, except for the electric power sector, the demand for electric energy in all sectors has declined. With the increase in emission reductions, the decline has increased. Taking the service industry as an example, when the emission reduction varies from 10 to 30%, the decline in electricity and energy consumption increases from 1.26 to 4.53%. When the emission reduction varies from 10 to 30% for the power sector, the power consumption will increase from 1.71 to 6.00%.

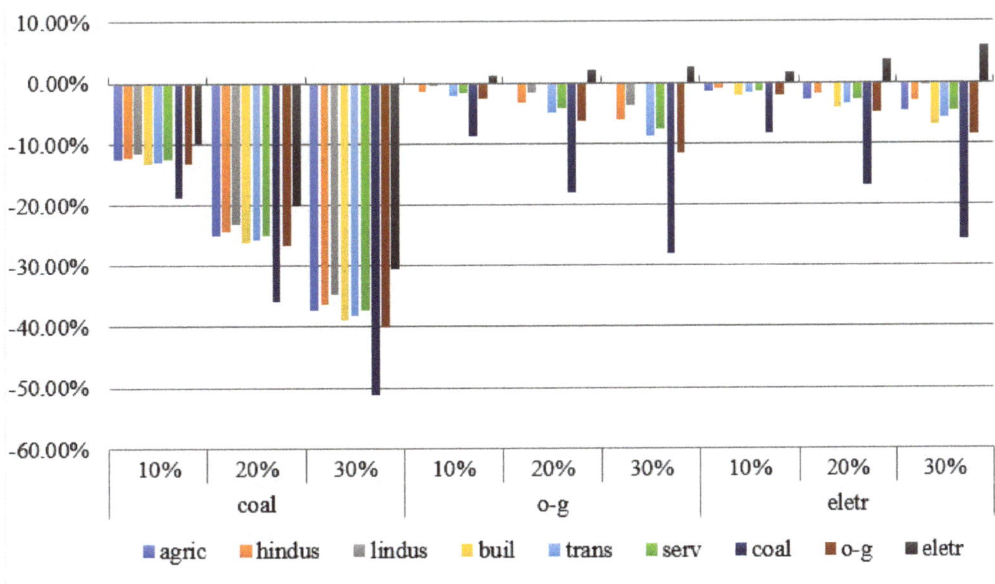

Figure 2. Consumption of each sector. The nine sectors and their classification are as follows: Agriculture (agric), Heavy industry (hindus), Light industry (lindus), Building industry (buil), Transportation industry (trans), Service industry (serv), Coal industry (coal), Oil and gas industry (o-g), Electricity industry (eletr).

The changes in carbon dioxide emissions and emission intensity of various sectors are shown in Figure 3. The bars express the percentage change in sectors' emissions and emissions intensity of carbon dioxide. It is obvious that the levy of a carbon tax reduces the carbon dioxide emissions of sectors. Among them, sectors with high demand for fossil energy, such as coal, have a significant reduction in carbon dioxide emissions, while oil and natural gas sectors have a relatively low reduction. The CO_2 emission intensity of each sector is obtained by comparing the total CO_2 emission of each sector with the sector's nominal GDP. Due to the levy of carbon taxes, the total CO_2 emissions of the sectors have decreased to varying degrees, but some sectors have increased for sector GDP. This leads to complex changes in the intensity of carbon dioxide emissions among sectors.

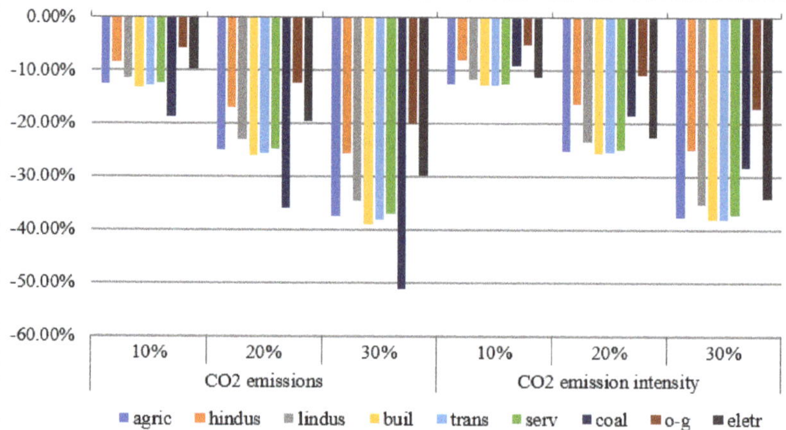

Figure 3. The sectors' emissions and emissions intensity of carbon dioxide.

5.1.3. Macroeconomic Variables Effects

The levy of carbon taxes lead to a decline in nominal GDP and real GDP, and with the continuous increase in emission reductions, the fall has gradually increased [41]. Table 3 shows the impact of carbon taxes on different macroeconomic variables. For households, the levy of a carbon tax leads to a rise in total income, and it increases with the emission reductions growth. The decline in households' demand leads to a rise in households' savings. With the increase of the carbon tax, social welfare has fallen more and more. The social welfare drops from -80.206 billion yuan to -295.336 billion yuan, when the emission reduction ranges from 10 to 30%. For enterprises, levying a carbon tax leads to a decline in income. Their savings have also fallen, and the decline is even greater with the increase in emission reductions. For the government, the main revenue comes from taxes. With the increase in carbon tax revenue, government revenue and savings have increased significantly. When carbon dioxide emissions are reduced by 10, 20, and 30%, government revenue will increase by 1.65, 3.43, and 5.4%, respectively. As emission reductions continue to increase, the reduction in carbon dioxide emission intensity gradually increases.

Table 3. Statistics of macroeconomic variables.

Scenarios	10%	20%	30%
Nominal GDP	-0.09%	-0.20%	-0.33%
Real GDP	-0.11%	-0.28%	-0.52%
Household income	0.03%	0.06%	0.08%
Household demand	-0.25%	-0.55%	-0.92%
Household saving	0.03%	0.06%	0.08%
Enterprise income	-0.20%	-0.43%	-0.70%
Enterprise saving	-0.20%	-0.42%	-0.70%
Government income	1.65%	3.43%	5.40%
Government saving	1.65%	3.43%	5.40%
CO_2 emission intensity	-9.91%	-19.84%	-29.77%
Social welfare	-802.06	-1767.84	-2953.36

5.2. Carbon Tax Recycle Simulation Results

5.2.1. The Impact on Institutions

The changes of various economic institutions in different simulation scenarios are depicted in Table 4. Regarding households' income, under the four scenarios, households' labor income remains unchanged, and the income of capital decreases. Households' capital income fell by 0.43, 0.54, 0.42, and 0.28% under the baseline scenario and the three simulated scenarios. Therefore, as Table 4 shows, the total income of residents under the four scenarios declines, and compared with the baseline scenario, the decline in the simulated scenario is greater. Households' demand level rose by 1.02% in scenario 1 and declined in other scenarios. Households' savings level fell the most in scenario 1, which was 0.08%. In scenarios 2, 3, the decline was 0.06 and 0.04%, respectively. It can be seen from the baseline scenario that the levy of a carbon tax caused the social welfare of residents to drop to −1767.84. Compared with the baseline scenario, households' social welfare in scenarios 1 imposes a carbon tax while reducing resident income tax and 3 imposes carbon tax while reducing corporate indirect tax under the carbon tax cycle improved to 3276.83 and −337.08.

Table 4. Statistics of macroeconomic variables of institutions.

Heading	Heading	Scenarios			
		BaU	Scenario 1	Scenario 2	Scenario 3
Households	Social welfare	−1767.84	3276.83	−2203.59	−337.08
	Labor income	0.00%	0.00%	0.00%	0.00%
	Capital income	−0.43%	−0.54%	−0.42%	−0.28%
	Total income	0.06%	−0.08%	−0.06%	−0.04%
	Demand	−0.55%	1.02%	−0.69%	−0.11%
	Savings	0.06%	−0.08%	−0.06%	−0.04%
Enterprises	Total income	−0.43%	−0.54%	−0.42%	−0.28%
	Savings	−0.42%	−0.54%	2.48%	−0.28%
Government	Total income	3.43%	0.00%	0.00%	0.00%
	Demand	3.18%	−0.20%	−0.26%	0.29%

For the enterprise, the income of the enterprise fell in all four scenarios. The decline rates of the baseline scenario and simulated scenarios 1, 2, and 3 are 0.43, 0.54, 0.42, and 0.28%, respectively. It is worth noting that under scenario 2, which is to impose a carbon tax while reducing corporate income tax, the income of enterprises has fallen, but due to the reduction of the corporate income tax rate, the level of corporate savings has been significantly increased. The corporate savings have increased significantly by 2.48%. In other scenarios, corporate savings have fallen. As for the government, the total revenue is fixed under these three simulation scenarios. Under scenarios 1 and 2, the government's demand drops by 0.2 and 0.26%, respectively. However, in scenario 3, government demand rises by 0.29%.

In general, scenario 1 imposes a carbon tax while reducing residents' income tax, which increases the level of residents' demand and social welfare. Scenario 2 reduces the corporate income tax rate, increasing corporate savings but resulting in a more significant decline in residents' consumption and income. Compared with the baseline scenario, the social welfare of residents has fallen even more. Scenario 3 reduces enterprises' indirect tax rates while increasing government consumption. Compared with the baseline scenario where only a carbon tax is levied, residents' social welfare has improved.

5.2.2. The Impact on the Economy

The changes in GDP are shown in Table 5. Both nominal GDP and real GDP declined in the four scenarios. Compared with the baseline scenario, nominal GDP in scenario 3 has the largest decrease, which is 0.86%, scenarios 1 and 2 fall by 0.25 and 0.17%, respectively. Scenario 1 has a larger reduction of real GDP than the BaU, which is 0.28%. For total investment, scenario 2 increases by 1.21%. Scenarios 1 and 3 reduce by 0.48% and 0.24%, respectively.

Table 5. Statistics of the economy and carbon emissions.

Heading	Scenarios			
	BaU	Scenario 1	Scenario 2	Scenario 3
Nominal GDP change	−0.20%	−0.25%	−0.17%	−0.86%
Real GDP change	−0.28%	−0.28%	−0.26%	−0.23%
Total investment change	−0.19%	−0.48%	1.21%	−0.24%
CO_2 emission intensity change	−19.84%	−19.80%	−19.86%	−19.30%
Carbon tax (yuan/ton)	71.35	71.88	73.44	76.97

5.2.3. The Impact on Carbon Emissions

As shown in Table 5, under the premise of reducing carbon dioxide emissions by 20%, the baseline scenario's carbon dioxide emission intensity has dropped by 19.84%. The changes in scenarios 1, 2, and 3 are almost the same as the BaU, with a decrease of 19.8, 19.86, and 19.30%, respectively, which can also achieve carbon emission reduction. Compared with the BaU, the carbon tax changes very little. The four scenarios' carbon tax prices are 71.35, 71.88, 73.44, and 76.97 yuan/ton.

6. Discussion and Conclusions

6.1. Discussion

In this article, using the 2017 input–output table and other official data sources, we constructed a static CGE model with nine sectors to simulate the impact of China's low-carbon economic policies on the social economy. The specific analysis includes the impact of carbon tax policies on energy consumption, carbon dioxide emissions, sectoral economic, and macroeconomic variables under different carbon dioxide emission reductions. Besides, we analyze the impact of carbon tax cycle mechanisms on macro-social economic variables under the principle of tax neutrality and test the "double dividend" theory.

When a carbon tax is imposed, the ad valorem tax rate levied on fossil energy gradually increases as the total emission reductions increase. Among them, the tax rate of coal is higher than that of oil and natural gas. The results are consistent with some previous studies [51,52]. Obviously, most of the contribution to emission reduction comes from coal, which is mainly due to the high carbon emission coefficient of coal and that coal occupies a dominant position in China's energy consumption structure. With the gradual increase in CO_2 emission reductions, energy consumption has gradually decreased, coal has the largest decline, while electricity has a smaller decline. According to Wang's research findings [9], carbon dioxide emissions in China's fossil energy consumption mainly come from coal, and the empirical results of this paper also confirm this conclusion. A carbon tax would have the biggest impact on coal than any other energy source. Therefore, companies will reduce the demand for coal to cut down production costs. As a secondary energy source, electricity consumes a large amount of fossil energy. Still, it is indirectly affected by collecting carbon taxes and does not directly emit CO_2, so it has little effect on electricity consumption. From the impact of carbon taxes on different sectors, it is found that the demand for coal in sectors has dropped significantly, and the decline has increased with the gradual increase in emissions reductions. In terms of oil and gas consumption, the consumption in sectors other than the power sector has declined. The main reason is that the mutual substitution of input factors between sectors and the increase in other input factors' prices for the power sector has led to a rise in oil and natural gas demand. Due to a carbon tax levy, sectors' emissions have declined when total CO_2 emission reductions gradually increase. Among them, sectors with high demand for fossil energy, such as coal, significantly reduce CO_2 emissions.

Various scenario simulations have an impact on macroeconomic variables. In the case of only levying a carbon tax, with the gradual increase in emission reductions, the decline in macroeconomic variables such as GDP, real GDP, social welfare, and total corporate income gradually increase. The total income of residents and the government has gradually increased. Furthermore, to reduce the negative impact of the carbon tax, we simulated carbon tax recycling measures such as reducing indirect corporate tax and household income tax while collecting carbon tax. While levying carbon taxes, reducing residents' income tax increases residents' demand, thus increasing residents' social welfare. Besides, imposing a carbon tax has improved the environment, thus realizing the "double dividend" of the carbon tax. While imposing carbon taxes, the corporate income tax rate is lowered. Because carbon taxes impact the production process, enterprises' capital price and total income are still falling. However, due to the reduction of corporate income tax rates, corporate savings have increased significantly. The decline in residents' capital income leads to a substantial decrease in residents' income, savings, and consumption, which cause a decrease in residents' social welfare and cannot realize the "double dividend" of the carbon tax. While levying carbon taxes, the corporate indirect tax rate is lowered. Since indirect taxes only occur in the distribution of domestic products, companies can pass the tax burden on to consumers, affecting domestic production product demand and demand prices. Therefore, corporate income and savings will decline; the decline in residents' capital income will also decrease. However, due to the reduction of indirect taxes, the demand price of products has fallen. Therefore, the consumption demand of residents has risen. At the same time, compared with just levying a carbon tax, residents' social welfare has increased.

6.2. Conclusions

The purpose of this study is to explore how the carbon tax policy that promotes the development of a low-carbon economy should be effectively implemented, so as to achieve green and sustainable development. Studies have found that carbon tax has a positive impact on reducing corporate carbon dioxide emissions, promoting the development of more energy-saving emission reduction technologies, and exploring more renewable resources, which is conducive to the realization of green and sustainable development goals. The research results also show that the levy of a carbon tax will adversely affect economic development, residents' income, and social welfare. However, introducing a suitable carbon tax recycling mechanism when formulating carbon tax policies can reduce the impact of the carbon tax on related industries. The implementation of a carbon tax recycling system based on the principle of tax neutrality, on the one hand, is conducive to achieving carbon emission reduction targets. On the other hand, it can improve the level of social welfare of residents, so as to achieve the double dividend of carbon tax.

Implementing a reasonable carbon tax recycling system (such as reducing the personal income tax of residents when formulating carbon tax policy) can reduce carbon emissions, promote economic growth, and achieve the "double dividend" effect. Currently, China is implementing a structural tax reduction policy. Whether the carbon tax can be implemented well is a challenge and an opportunity to further improve China's taxation system. The carbon tax system's design must conform to the overall direction of "tax reduction and burden reduction", by upholding the principle of tax neutrality, and gradually achieving the goal of green emission reduction.

Some limitations of this research will be addressed in future work. On the one hand, a dynamic CGE model will be established on the basis of the carbon tax static CGE model constructed in this article, so as to better reflect the long-term impact of the carbon tax policy. On the other hand, in future works, our research can be extended to analyze different market-based tools, such as tradable emission permits.

Author Contributions: W.L. designed and conceived this article; Y.L. contributed to set the model, run the program and write the most of this paper; T.L. and Y.L. processed the data and constructed the SAM table; T.L. and M.L. combed the literature, proofread the text of this article, and translated the article into English; M.L. and H.W. revised the charts of this article. All authors have read and agreed to the published version of the manuscript.

Funding: This study was financially supported by the research on "Factor Allocation and Industrial Upgrading Policies for Stable Economic Growth under the New Normal", Major Project of Key Research Bases of Humanities and Social Sciences, Ministry of Education, China (16JJD790015); Northeast Revitalization and Development Program of Jilin University, China," Research on Monitoring and Influencing Mechanism of Small and Micro Enterprise's Lifespan of Jilin Province". (No: 21dbzx05).

Institutional Review Board Statement: Not applicable.

Informed Consent Statement: Not applicable.

Data Availability Statement: Energy Information Administration (https://www.eia.gov/, accessed on 6 October 2021), China Financial Yearbook (https://www.epsnet.com.cn/, accessed on 6 October 2021), China Statistical Yeabook (http://www.stats.gov.cn/tjsj/ndsj/, accessed on 6 October 2021), and International Energy Statistics (https://www.eia.gov/, accessed on 6 October 2021).

Conflicts of Interest: The authors declare no conflict of interest.

References

1. Xiao, J.; Zhen, Z.; Tian, L.; Su, B.; Chen, H.; Zhu, A. Green Behavior towards Low-Carbon Society: Theory, Measurement and Action. *J. Clean. Prod.* **2020**, *278*, 123765. [CrossRef]
2. Bao, J.-Q.; Miao, Y.; Chen, F. Low Carbon Economy: Revolution in the Way of Human Economic Development. *China Ind. Econ.* **2008**, *4*, 017.
3. Ji, X.; Zhang, Y.; Mirza, N.; Umar, M.; Rizvi, S.K.A. The impact of carbon neutrality on the investment performance: Evidence from the equity mutual funds in BRICS. *J. Environ. Manag.* **2021**, *297*, 113228. [CrossRef] [PubMed]
4. Shi, Y.; Han, B.; Zafar, M.W.; Wei, Z. Uncovering the driving forces of carbon dioxide emissions in Chinese manufacturing industry: An intersectoral analysis. Environ. *Sci. Pollut. Res.* **2019**, *26*, 31434–31448. [CrossRef]
5. Bouscayrol, A.; Chevallier, L.; Cimetiere, X.; Clenet, S.; Lemaire-Semail, B. EPE'13 ECCE Europe, a carbon-neutral conference. *EPE J.* **2018**, *28*, 43–48. [CrossRef]
6. Usama, A.; Kraslawski, A.; Huiskonen, J. Governing interfirm relationships for social sustainability: The relationship between governance mechanisms, sustainable collaboration, and cultural intelligence. *Sustainability* **2018**, *10*, 4473.
7. Sun, C.; Ma, T.; Ouyang, X.; Wang, R. Does service trade globalization promote trade and low-carbon globalization? Evidence from 30 countries. *Emerg. Mark. Financ. Trade.* **2021**, *57*, 1455–1473. [CrossRef]
8. Ulucak, R.; Kassouri, Y. An assessment of the environmental sustainability corridor: Investigating the non-linear effects of environmental taxation on CO_2 emissions. *Sustain. Dev.* **2020**, *28*, 1010–1018. [CrossRef]
9. Wang, H.; Chen, Z.; Wu, X.; Nie, X. Can a carbon trading system promote the transformation of a low-carbon economy under the framework of the porter hypothesis? Empirical analysis based on the PSM-DID method. *Energy Policy* **2019**, *129*, 930–938. [CrossRef]
10. Bruvoll, A.; Larsen, B.M. Greenhouse gas emissions in Norway: Do carbon taxes work? *Energy Policy* **2004**, *32*, 493–505. [CrossRef]
11. Floros, N.; Vlachou, A. Energy demand and energy-related CO_2 emissions in Greek manufacturing: Assessing the impact of a carbon tax. *Energy Econ.* **2005**, *27*, 387–413. [CrossRef]
12. Jia, L.; Jinyu, B.; Yi, D.; Xiaohong, C.; Xiang, L. Impact of energy structure on carbon emission and economy of China in the scenario of carbon taxation. *Sci. Total Environ.* **2020**, *762*, 143093.
13. Metcalf, G.E. On the economics of a carbon tax for the United States. *Brook. Pap. Econ. Act.* **2019**, *2019*, 405–484. [CrossRef]
14. Burtraw, D.; Krupnick, A.; Palmer, K.; Paul, A.; Toman, M.; Bloyd, C. Ancillary benefits of reduced air pollution in the U.S. from moderate greenhouse gas mitigation policies in the electricity sector. *J. Environ. Econ. Manag.* **2003**, *45*, 650–673. [CrossRef]
15. Meng, S.; Siriwardana, M.; McNeill, J. The environmental and economic impact of the carbon tax in Australia. *Environ. Resour. Econ.* **2013**, *54*, 313–332. [CrossRef]
16. Tietenberg, T.H. Reflections—Carbon pricing in practice. *Rev. Environ. Econ. Policy* **2013**, *7*, 313–329. [CrossRef]
17. Wang, W.; You, X.; Liu, K.; Wu, Y.J.; You, D. Implementation of a Multi-Agent Carbon Emission Reduction Strategy under the Chinese Dual Governance System: An Evolutionary Game Theoretical Approach. *Int. J. Environ. Res. Public Health* **2020**, *17*, 8463. [CrossRef] [PubMed]
18. Yamazaki, A. Jobs and climate policy: Evidence from British Columbia's revenue-neutral carbon tax. *J. Environ. Econ. Manag.* **2017**, *83*, 197–216. [CrossRef]
19. Meng, X. Will Australian carbon tax affect the resources boom? Results from a CGE model. *Nat. Resour. Res.* **2012**, *21*, 495–507. [CrossRef]
20. Mohammed, T. A CGE analysis of the macroeconomic effects of carbon dioxide emission reduction on the Algerian economy. *Springer Proc. Bus. Econ.* **2017**, 1–20. [CrossRef]
21. Xu, X.; Xu, X.F.; Chen, Q.; Che, Y. The impact on regional "resource curse" by coal resource tax reform in China—A dynamic CGE appraisal. *Resour. Policy* **2015**, *45*, 277–289. [CrossRef]

22. Sofia, D.; Gioiella, F.; Lotrecchiano, N.; Giuliano, A. Mitigation strategies for reducing air pollution. *Environ. Sci. Pollut. Res.* **2020**, *27*, 19226–19235. [CrossRef]
23. Jakeman, G.; Hester, S.; Woffenden, K.; Fisher, B.S. Kyoto protocol: The first commitment period and beyond. *Aust. Commod.* **2002**, *9*, 176–197.
24. Kutlu, L. Greenhouse Gas Emission Efficiencies of World Countries. *Int. J. Environ. Res. Public Health* **2020**, *17*, 8771. [CrossRef] [PubMed]
25. Settimo, G.; Avino, P. The Dichotomy between Indoor Air Quality and Energy Efficiency in light of the Onset of the COVID-19 Pandemic. *Atmosphere* **2021**, *12*, 791. [CrossRef]
26. Pearce, D. The Role of Carbon Taxes in Adjusting to Global Warming. *The Economic Journal.* **1991**, *101*(407), 938. [CrossRef]
27. Chikaire, J.; Nwakwasi, R.N.; Anyoha, N.O.; Nnadi, F.N. Short note potential impacts of climate change on African agriculture. *Int. J. Nat. Appl. Sci.* **2010**, *6*, 493–496.
28. Hao, F.; van Brown, B.L. An analysis of environmental and economic impacts of fossil fuel production in the U.S. from 2001 to 2015. *Soc. Nat. Resour.* **2018**, *32*, 693–708. [CrossRef]
29. Pearce, A.; Stilwell, F. 'Green-collar' jobs: Employment impacts of climate change policies. *J. Aust. Political Econ.* **2008**, *62*, 120–138.
30. Awan, U.; Kraslawski, A.; Huiskonen, J. Progress from Blue to the Green World: Multilevel Governance for Pollution Prevention Planning and Sustainability. *Semant. Sch.* **2019**.
31. Zhao, Y.; Zhang, Z.; Wang, S.; Wang, S. CO_2 emissions embodied in China's foreign trade: An investigation from the perspective of global vertical specialization. *China World Econ.* **2014**, *22*, 102–120. [CrossRef]
32. Yoichi, K. *Impact of Carbon Dioxide Emission on GNP Growth: Interpretation of Proposed Scenarios*; Presentation to the Energy and Industry Subgroup, Response Strategies Working Group; IPCC: Paris, France, 1989.
33. Ang, B.W.; Liu, F.L. A new energy decomposition method: Perfect in decomposition and consistent in aggregation. *Energy* **2001**, *26*, 537–548. [CrossRef]
34. Liaskas, K.; Mavrotas, G.; Mandaraka, M.; Diakoulaki, D. Decomposition of industrial CO_2 emissions: The Case of European Union. *Energy Econ.* **2000**, *22*, 383–394. [CrossRef]
35. Zhang, M.; Mu, H.; Ning, Y. Accounting for energy-related CO_2 emission in China, 1991–2006. *Energy Policy* **2009**, *37*, 767–773. [CrossRef]
36. Schipper, L.; Murtishaw, S.; Khrushch, M.; Ting, M.; Karbuz, S.; Unander, F. Carbon emissions from manufacturing energy use in 13 IEA countries: Long-term trends through 1995. *Energy Policy* **2001**, *29*, 667–688. [CrossRef]
37. Davis, W.B.; Sanstad, A.H.; Koomey, J.G. Contributions of weather and fuel mix to recent declines in U.S. energy and carbon intensity. *Energy Econ.* **2003**, *25*, 375–396. [CrossRef]
38. Haites, E. Carbon taxes and greenhouse gas emissions trading systems: What have we learned? *Clim. Policy* **2018**, *18*, 955–966. [CrossRef]
39. Lee, C.F.; Lin, S.J.; Lewis, C.; Chang, Y.F. Effects of carbon taxes on different industries by fuzzy goal programming: A case study of the petrochemical-related industries, Taiwan. *Energy Policy* **2007**, *35*, 4051–4058. [CrossRef]
40. Newell, R.G.; Pizer, W.A. Regulating stock externalities under uncertainty. *J. Environ. Econ. Manag.* **2003**, *45*, 416–432. [CrossRef]
41. Avi-Yonah, R.S.; Uhlmann, D.M. Combating global climate change: Why a carbon tax is a better response to global warming than cap and trade. *Stanf. Environ. Law J.* **2009**, *28*, 3–50. [CrossRef]
42. Stram, B.N. A new strategic plan for a carbon tax. *Energy Policy* **2014**, *73*, 519–523. [CrossRef]
43. Orlov, A.; Grethe, H.; McDonald, S. Carbon taxation in Russia: Prospects for a double dividend and improved energy efficiency. *Energy Econ.* **2013**, *37*, 128–140. [CrossRef]
44. Dissou, Y.; Karnizova, L. Emissions cap or emissions tax? A multi-sector business cycle analysis. *J. Environ. Econ. Manag.* **2016**, *79*, 169–188. [CrossRef]
45. Wittneben, B.B.F. Exxon is right: Let us re-examine our choice for a cap-and-trade system over a carbon tax. *Energy Policy* **2009**, *37*, 2462–2464. [CrossRef]
46. Timilsina, G.R.; Shrestha, R.M. General equilibrium analysis of economic and environmental effects of carbon tax in a developing country: Case of Thailand. *Environ. Econ. Policy Stud.* **2002**, *5*, 179–211. [CrossRef]
47. Garbaccio, R.F.; Ho, M.S.; Jorgenson, D.W. Why Has the Energy-Output Ratio Fallen in China? *Energy J.* **1999**, *20*, 63–91. [CrossRef]
48. Shrestha, R.M.; Marpaung, C.O.P. Supply- and demand-side effects of carbon tax in the Indonesian power sector: An integrated resource planning analysis. *Energy Policy* **1999**, *27*, 185–194. [CrossRef]
49. Wissema, W.; Dellink, R. AGE analysis of the impact of a carbon energy tax on the Irish economy. *Ecol. Econ.* **2007**, *61*, 671–683. [CrossRef]
50. Allan, G.; Lecca, P.; McGregor, P.; Swales, K. The economic and environmental impact of a carbon tax for Scotland: A computable general equilibrium analysis. *Ecol. Econ.* **2014**, *100*, 40–50. [CrossRef]
51. Lin, B.; Jia, Z. The energy, environmental and economic impacts of carbon tax rate and taxation industry: A CGE based study in China. *Energy* **2018**, *159*, 558–568. [CrossRef]
52. Fu, Y.; Huang, G.; Liu, L.; Zhai, M. A factorial CGE model for analyzing the impacts of stepped carbon tax on Chinese economy and carbon emission. *Sci. Total Environ.* **2021**, *759*, 143512. [CrossRef]

53. Paroussos, L.; Fragkos, P.; Capros, P.; Fragkiadakis, K. Assessment of carbon leakage through the industry channel: The E.U. perspective. *Technol. Forecast. Soc. Chang.* **2015**, *90*, 204–219. [CrossRef]
54. Lu, Y.; Liu, Y.; Zhou, M. Rebound effect of improved energy efficiency for different energy types: A general equilibrium analysis for China. *Energy Econ.* **2017**, *62*, 248–256. [CrossRef]
55. Zhao, Y.; Li, H.; Xiao, Y.; Liu, Y.; Cao, Y.; Zhang, Z.; Wang, S.; Zhang, Y.; Ahmad, A. Scenario analysis of the carbon pricing policy in China's power sector through 2050: Based on an improved CGE model. *Ecol. Indic.* **2018**, *85*, 352–366. [CrossRef]
56. Pui, K.L.; Othman, J. Economics and environmental implicationsof fuel efficiency improvement in Malaysia: A computable general equilibriumapproach. *J. Clean. Prod.* **2017**, *156*, 459–469. [CrossRef]
57. Li, W.; Zhang, Y.; Lu, C. The impact on electric power industryunder the implementation of national carbon trading market in China: Adynamic CGE analysis. *J. Clean. Prod.* **2018**, *200*, 511–523. [CrossRef]
58. Tran, T.M.; Siriwardana, M.; Meng, S.; Nong, D. Impactof an emissions trading scheme on Australian households: A computable general equilibrium analysis. *J. Clean. Prod.* **2019**, *221*, 439–456. [CrossRef]
59. Lin, B.; Jia, Z. Economic, energy and environmental impact of coal-to-electricity policy in China: A dynamic recursive CGE study. *Sci. Total Environ.* **2019**, *698*, 134241. [CrossRef] [PubMed]
60. Li, N.; Zhang, X.; Shi, M.; Hewings, G.J.D. Does China's air pollution abatement policy matter? An assessment of the Beijing-Tianjin-Hebei region based on a multi-regional CGE model. *Energy Policy* **2019**, *127*, 213–227. [CrossRef]
61. Zhou, Y.; Fang, W.; Li, M.; Liu, W. Exploring the impacts of a low-carbon policyinstrument: A case of carbon tax on transportation in China. *Resour. Conserv. Recycl.* **2018**, *139*, 307–314. [CrossRef]
62. Zhao, Y.; Zhang, Y.; Wei, W. Analysis of the energy-environment-economysystem in China based on the dynamic CGE model. *J. Syst. Eng.* **2014**, *29*, 581–591.
63. Zhao, Y.; Zhang, Y.; Wei, W. Quantifying international oil price shocks on renewable energy development in China. *Appl. Econ.* **2020**, *53*, 329–344. [CrossRef]

International Journal of
Environmental Research and Public Health

MDPI

Article

An Analysis of the Impact of the Emissions Trading System on the Green Total Factor Productivity Based on the Spatial Difference-in-Differences Approach: The Case of China

Susheng Wang [1,2], Gang Chen [1,*] and Xue Han [1]

[1] School of Economics and Management, Harbin Institute of Technology Shenzhen, Shenzhen 518055, China; wangss@sustech.edu.cn (S.W.); hanxue0960@163.com (X.H.)

[2] Department of Finance, Southern University of Science and Technology, Shenzhen 518055, China

* Correspondence: 13b956005@stu.hit.edu.cn; Tel.:+86-158-1553-5026

Abstract: How to effectively identify the spatial effect of the emissions trading system(ETS) on urban green total factor productivity(GTFP) generated through the linkage of economic factors between cities is a necessary part of scientifically evaluating the effect of ETS policy in emerging-market countries. This study aims to examine the spatial effect, mechanism, and heterogeneity of the ETS on urban GTFP based on the panel data of 281 cities from 2004 to 2017 in China, applying spatial difference-in-differences(DID) Durbin model (SDID-SDM) with multidimensional fixed effect (FE). The results show that ETS significantly improves the GTFP of the pilot cities, produces a spatial spillover effect and the results are robust to the placebo test, propensity score matching SDID (PSM-SDID) test, and Carbon-ETS interference test. Further analysis shows that the policy effect is mainly driven by improving energy efficiency, promoting green innovation, and optimizing the industrial structure. In addition, we found that ETS performs better in regions with a high degree of marketization, strong environmental law enforcement, and a low proportion of coal consumption. In general, the identification method of this study can be used as a scientific reference for conducting similar research in other emerging countries.

Keywords: emissions trading system; green total factor productivity; spatial difference-in-difference; energy efficiency; green innovation; industry structure; spatial heterogeneity

Citation: Wang, S.; Chen, G.; Han, X. An Analysis of the Impact of the Emissions Trading System on the Green Total Factor Productivity Based on the Spatial Difference-in-Differences Approach: The Case of China. *Int. J. Environ. Res. Public Health* **2021**, *18*, 9040. https://doi.org/10.3390/ijerph18179040

Academic Editors: Pasquale Avino, Massimiliano Errico, Aristide Giuliano, Hamid Salehi and Elena Rada

Received: 26 June 2021
Accepted: 25 August 2021
Published: 27 August 2021

Publisher's Note: MDPI stays neutral with regard to jurisdictional claims in published maps and institutional affiliations.

1. Introduction

China is the world's largest developing country and an important member of emerging-market countries. In the past four decades, the gross domestic product (GDP) of China has maintained an average of more than 9%. Even under the negative impact of the 2008 international financial crisis, it has maintained a high growth rate of more than 6.5% (excluding the impact of price factors). However, such rapid economic growth data is accompanied by excessive energy consumption and serious environmental pollution problems [1–3]. In 2010, the former Ministry of Environmental Protection, the National Bureau of Statistics, and the former Ministry of Agriculture jointly issued the "First National Pollution Source Census Bulletin" (census in 2010). The communique data shows that the total emissions of major pollutants are 2.32 million tons of sulfur dioxide(SO_2), 30,289,600 tons of chemical oxygen demand, and 179.777 million tons of nitrogen oxides. To deal with the environmental challenges caused by pollutant emissions, many developed countries such as the European Union and the United States first launched the emissions trading system(ETS), such as the US SO_2 emission allowance trading project, nitrogen oxides (NO_X) trading project, etc. [4–6]. China formally approved 11 pilot provinces (cities) including Zhejiang, Jiangsu, Inner Mongolia, Hubei, and Hunan as national-level pilot units in 2007 and actively explored and implemented the paid use and transaction system of pollution rights. ETS is an important institutional arrangement aimed at using market mechanisms to reduce pollutant emissions, which can internalize corporate emission reduction costs through total pollutant control and quota trading. The policy goal is to establish a long-term mechanism for energy conservation and emission reduction [7,8]. Therefore, investigating the impact mechanism of the pilot emission trading system on green total factor productivity is crucial to how the government uses market-oriented environmental policy tools to deal with the dual challenges of green sustainable development and high-quality economic growth.

Cities are the main spatial carrier of environmental pollution control, and they are the key to achieving the goal of total pollutant emission, promoting the sound operation of the ETS, and then achieving the goal of green and high-quality development; the environmental regulation goals at the provincial level that are broken down to the enterprise execution level are also implemented at the city level. Due to the circulation connection of resources and factors between cities, the impact of the emission trading system within the city on the emission behavior of enterprises will inevitably produce a certain degree of spatial externality on the surrounding areas through the economic connection between cities [9–11]. Nevertheless, previous studies have seldom paid attention to the mechanism of the city-level ETS on the green total factor productivity(GTFP) of the city itself and surrounding areas, and the spatial effect of this market-oriented environmental policy cannot be ignored [12,13].

Theoretically, within the city, the influence of ETS on the territorial GTFP is mainly realized by influencing the decisionmaking of micro-enterprises. Specifically, the pilot policy transfers the cost of emission reduction directly to pollutant producers, who can make flexible choices among pollutant emission quota trading, emission reduction decisions (enhancing energy efficiency, launching green innovation activities, etc.), and location decisions to effectively respond to the policy pressure of the pilot policy [14,15]. In general, when the cost of pollutant emissions is lower than its revenue, these companies choose to purchase pollutant emission quotas; when the revenue from pollutant emissions cannot cover the cost, companies often make active emission reduction decisions such as optimizing resource allocation, improving energy efficiency, and investing in green technologies innovation, etc., for example, when the benefits of green innovation are higher than the cost of emission reduction, companies often choose green technology innovation to solve the pollution problem; and when the benefits of green innovation cannot cover local emission reduction costs, companies often choose to relocate which makes the green innovation benefits of the new location higher than the emission reduction costs. Producers' emission reduction decisions, green innovation decisions, and relocation decisions at the macro level drive capital and economic factors to gradually withdraw from polluting industries to clean industries, affecting energy efficiency, technological innovation and the adjustment of industrial structure, and then improving the overall GTFP of the city. Among cities, due to the policy pressure of ETS, some emission companies cannot cover emission reduction costs through green innovation and other emission reduction decisions. These companies will choose to withdraw from the local market and relocate. The relocation of these polluting companies can increase the overall GTFP level of the original location. On the other hand, because these companies have technical efficiency advantages relative to the new location, they passively improve the technical efficiency of the new location, thereby generating the spatial spillover effect of GTFP [16].

We use China's ETS to empirically test the impact mechanism of market-oriented environmental tools on GTFP, which has the following significance: firstly, China's rapid economic growth is accompanied by serious environmental pollution problems, such as excessive energy consumption and excessive pollutant emissions. These are typical problems that have occurred or are about to occur in developing countries and some emerging-market countries [17]. Secondly, the pilot areas approved by the Chinese government have different geographical locations in which the spatial heterogeneity of humanities, economics, and geography is significant. It is possible to comprehensively examine the possible potential spatial heterogeneity of the effects in areas with different economic development levels and different cultural and geographical characteristics [15]. In addition, the ETS is a typical market-oriented environmental policy tool, and the effect of the policy is sensitive to the degree of marketization in the pilot area and the intensity of environmental law enforcement, and it is a challenge to China, which has been criticized by the West for "low marketization and strong government intervention". Therefore, it has great practical value to examine the effect of the ETS in the context of government intervention to understand how environmental governance policies improve GTFP in a complex market environment.

In summary, this study is based on the panel data of 281 cities across the country from 2004 to 2017 and takes the spatial difference-in-differences(DID) Durbin model (SDID-SDM) as the benchmark model to empirically examine the impact mechanism of the ETS on GTFP. This study carried out a parallel trend test and found that there is no significant difference between the treatment group and the control group before the implementation of ETS, which means the treatment group and the control group meet the parallel trend assumption. The estimation results of the SDID-SDM model show that the GTFP in the pilot cities has increased significantly, and the ETS also has a positive spatial spillover effect on the GTFP in the surrounding areas of the pilot cities. In addition, to eliminate the potential influence of selectivity bias and possible confounding factors, a series of tests such as placebo test, propensity score matching SDID (PSM-SDID) test, and triple-difference test to exclude the effects of Carbon-ETS were further conducted, and the empirical results remained robust.

In addition, we also conducted empirical tests on the three potential mechanisms of the ETS on GTFP. The mechanism analysis results show that the ETS may achieve the policy effect of improving GTFP by improving energy efficiency, promoting green technology innovation, and optimizing industrial structure. Secondly, by grouping the samples according to the degree of marketization, environmental enforcement, and energy consumption endowment level, we thoroughly investigate the heterogeneity of policy effects. The estimation results show that the positive policy effect of the ETS on GTFP performs better in regions with a high degree of marketization, strong environmental law enforcement, and a low proportion of coal consumption.

The marginal contribution of this research to the existing literature is mainly in the following three aspects: first of all, this study enriches the empirical test of the spatial effect of ETS in improving GTFP. Previous studies were mainly based on the perspective of panel data [18–20], ignoring the circulation of resources and factors between cities, and the impact of ETS on corporate emissions behavior will inevitably affect the policy effect through the cross-regional economic connections, which forms the spatial externality of policy effects. It means that the application of panel data model estimation results may lead to model setting bias. This study applies the SDID-SDM approach to estimate the policy effect to avoid this model setting bias which may truly reflect the GTFP promotion effect of the ETS.

Secondly, this study enriches the empirical test of the effectiveness of the ETS in emerging-market countries and developing countries. Previous studies tend to focus on the evaluation of environmental policy effects in developed countries such as in Europe and the United States [15,21], and they easily ignored the effects of environmental policies in countries with poor economic and social development. On the one hand, underdeveloped areas themselves lack the motivation for active market-oriented environmental regulation. On the other hand, it is also related to the reality that underdeveloped regions will undertake the transfer of polluting industries due to the global industrial layout. Developed countries in Europe and the United States have a relatively developed market and economic system and are equipped to effectively implement market-oriented environmental policies. It is worthy of thorough analysis and research on whether emerging-market countries that are under market-oriented construction and developing countries with relatively underdeveloped economic and social development can effectively take market-oriented policy tools to achieve green and sustainable development. Our research results show that even in emerging-market countries and developing countries, market-oriented environmental policy, such as the ETS, can still exert policy effects and bring positive spatial economic effects on GTFP.

Finally, based on the benchmark model, this research further examines the impact mechanism of the ETS on urban GTFP through mechanism analysis and enriches the spatial dimension of the existing ETS mechanism analysis.

The study is organized as follows: Section 2 summarizes the relevant literature and policy implementation background of China's ETS. Section 3 introduces the empirical research design based on the SDID-SDM. Section 4 reports the estimation results of the benchmark model. Section 5 analyzes the influence mechanism of the policy effect, and Section 6 examines the heterogeneity of the policy effect. Section 7 details the different robustness tests carried out, and the last section gave the main conclusions and policy recommendations of this study on the impact of China's ETS on GTFP.

2. Literature

2.1. Green Innovation Effect of ETS

In 1990, Article 4 of the US "Clean Air Act" amendment proposed the "Acid Rain Plan", which was approved by Congress to use SO_2 emission allowance trading as a means of emission reduction. It was an early and successful environmental policy tool to achieve reductions through market-based emissions allowance trading [22,23]. Pollutant emission trading refers to the use of emission quota trading to reduce the emissions of major pollutants and reduce the negative impact on the environment. It aims to use market competition and price mechanisms to guide producers to emission reduction behaviors and to achieve the goal of total pollutant control. At the same time, it reduces the overall cost of pollution control in society and realizes green technological innovation. In terms of existing research on the policy effects of the emission trading system, scholars have focused on the following aspects: one is emission reduction effect, which is the core policy goal [24,25]; the other is the economic growth effect, which is the policy economic development goal [26,27]; the third is the green innovation effect which is the efficiency goal, and few scholars have conducted analysis and research on the relationship between the ETS and GTFP [28–30]. The first two aspects are the key research content of environmental economics in the past decades, and the green innovation effect of environmental regulations has been the research focus in recent years; of note, the research on the impact of GTFP is becoming an academic hotspot. This study summarizes previous studies on the impact of the ETS on green technology innovation and GTFP, and it shows that the academic community has not reached a consensus on the green innovation effect of the ETS, and the existing research conclusions mainly have three aspects, namely: promotion theory, Inhibition theory and (U-shaped and inverted U-shaped relationship, etc.) nonlinear relationship theory.

First of all, many scholars support the promotion theory based on the classic "Porter Hypothesis" [31,32], that is, environmental policies represented by the ETS have a positive effect on green technology innovation and improves GTFP [33,34]. Zhang L et al. (2019) based on the empirical analysis of the regulatory data of listed companies in seven pilot provinces and cities showed that China(CN)-ETS is significantly positively correlated with green innovation, while market competition has weakened this positive correlation [30]. However, some other scholars in the empirical research found that the classic "Porter hypothesis" conclusion did not appear, and environmental policies brought more negative externalities, that is, environmental policies represented by the ETS restrained the green technological innovation and hindered the green technological innovation process of enterprises [35,36]. In addition, some studies found that both the promotion theory and the suppression theory are valid. They are just the results of different effects shown in different stages of nonlinear policy effects. Therefore, the impact of environmental policies represented by the ETS on GTFP is a process of non-linear change with both promotion and suppression effects [20].

In summary, the research on the impact of ETS on GTFP is important, but the academic community has not yet reached a consensus. Moreover, the sample areas for this type of research are mostly developed countries such as regions in Europe and the United States, and they lack extensive attention to relatively underdeveloped economies such as emerging-market countries and developing countries. However, it is precisely these relatively underdeveloped economies that are facing more complex environmental issues. In recent years, scholars have begun to pay attention to the treatment effects of market-oriented environmental policy tools in emerging-market countries and developing countries [37–39], such as Oliveira et al. (2019), who applied the Economic Projection and Policy Analysis (EPPA6) model to assess ETS cooperation between Brazil and Europe. This showed that a domestic ETS reduces emissions and promotes technological substitution towards alternative energy for both participants [40]. Therefore, we use China's ETS as a research sample which is a typical representative of emerging-market countries and developing countries, to examine the identification of the policy effects of the ETS. It will help to investigate whether economies facing the dual pressure of economic growth and sustainable development can adopt market-oriented environmental policy tools to a green effect, and how the influence mechanism works on the green economy effect.

2.2. ETS in China

To reverse the extensive economic growth mode with high pollution, high energy consumption, high emissions, and low efficiency and to quickly realize the sustainable development of a green economy, the Chinese government has conducted a great deal of environmental policy exploration, especially about the issue of pollutant emissions [41,42]. In 1987, the Minhang District of Shanghai launched a water pollutant discharge transaction. In 1988, the former National Environmental Protection Agency approved 18 cities including Shanghai, Beijing, Tianjin, Shenyang, Xuzhou, and Changzhou as pilot areas developing water pollution discharge permits. In 2003, the former State Environmental Protection Administration cooperated with the US Environmental Protection Association to carry out training on total sulfur dioxide control and emissions trading across the country. In 2007, the Ministry of Finance together with the former Ministry of Environmental Protection and the National Development and Reform Commission successively approved 11 provinces (cities) including Tianjin, Hebei, Shanxi, Inner Mongolia, Jiangsu, Zhejiang, Henan, Hubei, Hunan, Chongqing and Shaanxi as national pilot units to carry out the pilot work of the ETS. In 2014, Qingdao City of Shandong Province was included in the pilot program (see Figure 1). The emission trading training in cooperation with the U.S. Environmental Protection Association makes China's ETS design similar to that of the United States, including total amount control, initial allocation, quota period, quota use, and penalties for violations, while the specific content is subject to adaptive adjustments by the local government which are significant regional differences, such as transaction methods (competitive transactions, negotiated transactions, public auctions, quota transfers, etc.) and transaction price systems (paid use prices, transaction benchmark prices, government repurchase prices).

Based on the statistical data officially released by China, the ETS in China has achieved its policy results. The first is the effect of pollutant emission reduction. The "Second National Pollution Source Census Bulletin" jointly issued by the three departments in 2020 (the census period is 2017) shows that compared with the data of the first national pollution source census, the emissions of pollutants in 2017 such as the sulfur dioxide, chemical oxygen demand, and nitrogen oxides decreased by 72%, 46%, and 34% respectively compared with 2007. The second is the scale of emissions trading. As of August 2018, the total amount of paid use fees for emission rights collected on the primary market was 11.77 billion yuan, and the cumulative transaction amount in the secondary market was 7.23 billion yuan. It shows that China's pilot policy for emissions trading has achieved goals in pollutant reduction and economic benefits, but official statistics have not disclosed the green innovation effect of the pilot policy, so it is necessary to conduct a more in-depth analysis to examine whether the pilot policy achieves the policy goal of improving GTFP and to deeply explore the internal mechanism of the pilot policy affecting GTFP and the heterogeneity of policy effects.

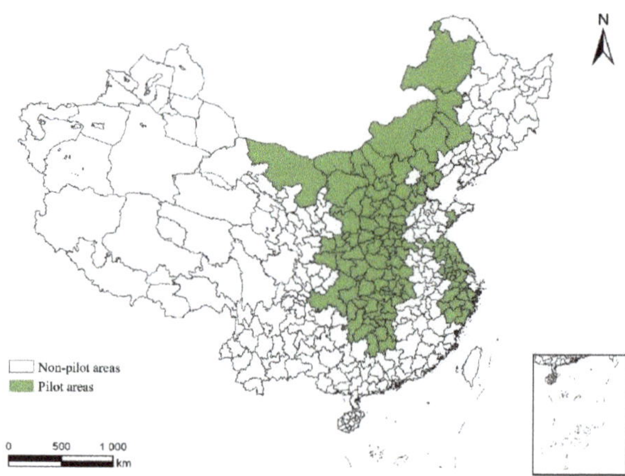

Figure 1. Distribution of ETS pilot areas in China.

3. Research Design

3.1. Samples and Data

In 2007, the Ministry of Finance, the former Ministry of Environmental Protection, and the National Development and Reform Commission successively approved 11 provinces (cities) including Tianjin, Hebei, Shanxi, Inner Mongolia, Jiangsu, Zhejiang, Henan, Hubei, Hunan, Chongqing and Shaanxi as national pilot units to carry out the pilot work of the ETS. In 2014, Qingdao City in Shandong Province was included in the pilot program. This study uses 2003–2017 as sample period, deleting the cities adjusted by administrative divisions during the sample period, and replacing them with newer administrative divisions, such as Chaohu City in Anhui Province and Longnan City in Gansu Province, also deleting areas with missing data such as Tibet et al. In addition, Hong Kong, Macao, and Taiwan, all with different statistical calibers, were deleted. Finally, the panel data of 281 cities from 2003 to 2017 were selected as the research data set for empirical testing. Economic data at the city level were deflated based on prices in 2003. The input-output data required for GTFP calculations including undesired output and control variable data are all from the "China Statistical Yearbook", "China Regional Economic Statistical Yearbook", and "China City Statistical Yearbook". The city patent authorization data comes from the China Research Data Service Platform (CNRDS) database.

3.2. Variables

3.2.1. Dependent Variable—GTFP

The measurement of GTFP is mainly based on the malmquist-luenberger(ML) index method of the slacks-based measure(SBM) directional distance function [43]. The input factors include labor (employed population), capital (capital stock), and energy (industrial electricity). The expected output is the actual GDP of the region after price deflation. The undesired output is industrial smoke and dust emissions (tons), wastewater emissions, sulfur dioxide emissions (tons), and PM 2.5.

3.2.2. Key Explanatory Variable—ETS Dummy

The key explanatory variable of this study is the multi-period combined dummy variable of the ETS, which is composed of the pilot group dummy variable and the pilot time group dummy variable. When a city belongs to the pilot group, the pilot group dummy variable is 1; otherwise, it is 0. The pilot cities include 12 provinces (cities), which are Tianjin, Hebei, Shanxi, Inner Mongolia, Jiangsu, Zhejiang, Henan, Hubei, Hunan,

Chongqing, Shaanxi, and Shandong Qingdao, that is, there are 109 cities in the treatment group, and the remaining 172 cities are the control group. In addition, since the approval time of the above 109 pilot cities is not uniform and belongs to the multi-phase DID situation, it is assumed that when city i is approved as a pilot in year t, the value of the city will be 1 for each subsequent year (because the months of approval for the pilot project are all at the end of the current year, so the second year of approval is used as the starting point of the pilot. For example, Qingdao City in Shandong Province was approved as a pilot in December 2014, so 2015 is the starting year of the Qingdao pilot.).

3.2.3. Control Variables

Based on existing literature [18,20], we controlled for a set of variables to capture the influence factors of the GTFP. The control variables mainly include economic development level (measured by per capita GDP), population size (measured by population density), energy consumption scale (measured by industrial electricity consumption), industrial structure (measured by industrial output value to GDP), and innovation level (measured by the number of invention patents).

3.3. Empirical Model

We applied SDID-SDM to study the impact of ETS on GTFP. The SDID-SDM model is setting as follows:

$$gtfp_{it} = \alpha_0 + \rho(\Omega' \otimes W')_{it} gtfp_{it} + \alpha_1 DID_{it} + \gamma_1(\Omega' \otimes W')_{it} DID_{it} + \gamma_2(\Omega' \otimes W')_{it} control_{it} + \alpha_2 control_{it} + city_i + year_t + \varepsilon_{it} \quad (1)$$

where the dependent variable $gtfp_{it}$ denotes the GTFP measured by SBM-ML in city i at year t. the independent variable DID_{it} denotes the dummy variable of ETS, which equals 1 if the city i at year t is approved as the pilots; otherwise, it equals 0. $control_{it}$ denotes control variables. $(\Omega' \otimes W')_{it}$ denotes the space-time weight matrix, while Ω' denotes the temporal weight matrix, and W' denotes the spatial weight matrix. α_0 denotes the constant, ρ denotes the coefficients of spatial lag of dependent variable, α_1 denotes the coefficients of the independent variable, α_2 denotes the coefficients of control variables, γ_1 denotes the coefficients of spatial lag of independent variable, and γ_2 denotes the coefficients of spatial lag of control variable. $city_i$ is urban fixed effects absorbing all unobserved city-specific, time-invariant factors that may influence the dependent variable. $year_t$ is the year fixed effect to control for the general macroeconomic factors affecting all cities. ε_{it} is a random error.

4. Empirical Analysis

4.1. Parallel Trend Test

The key identification hypothesis of the DID model is that non-pilot areas provide effective counterfactual changes for the policy treatment effects of pilot areas [44,45], that is, before the implementation of the ETS, urban GTFP maintained relatively stable changes, while there is a significant difference between the treatment group and the control group after the pilot implementation. To ensure the basic assumption is met, this study follows the parallel trend test method of multi-period DID, and performed SDID-SDM regression for the first three years and the last three years of the treatment period. The regression results show that before the implementation of the emission trading system, there was no systematic difference in the time trend between the pilot area and non-pilot area which means it satisfies the parallel trend assumption. It should be noted that there is a certain time lag effect in the emission trading system, as shown in Figure 2, three years after the treatment time.

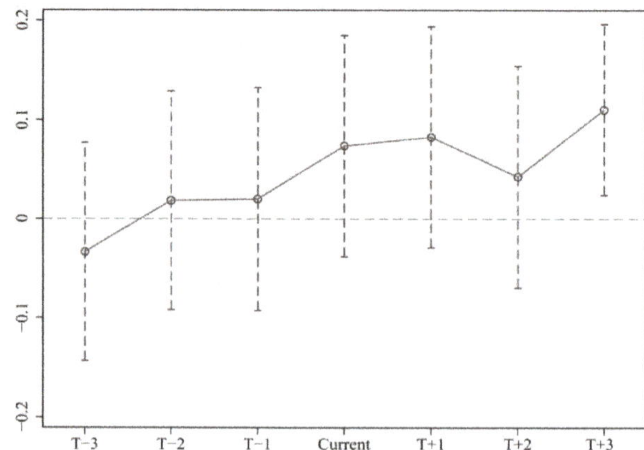

Figure 2. Parallel trend test.

4.2. Baseline Regression

As shown in Table 1, column (1) is the estimated result of the panel-DID model as a comparison to investigate the average treatment effect when there is no spatial dimension, and column (2) is the estimated result of Equation (1). The results show that after controlling for city-fixed effects, year-fixed effects, and control variables, without considering spatial effects, the ETS has a significant positive effect on GTFP in pilot cities. After considering the spatial effects in this study, the ETS still significantly improves the GTFP of the pilot cities, and the coefficient is significant at the 1% confidence level. In addition, the ETS also drives the growth of GTFP in the surrounding areas of the pilot cities, and the spatial lag coefficient is significant at the 5% confidence level. This finding is consistent with existing relevant research conclusions and theoretical hypotheses [18,20,46], that is, after the implementation of the ETS, the GTFP of the territorial cities and surrounding areas increased significantly.

Table 1. Baseline regression results.

Model	Panel-DID	SDID-SDM
Variables	**(1)**	**(2)**
DID	0.491 *** (11.57)	0.662 *** (2.93)
W × DID		2.190 ** (2.11)
Control	Y	Y
Year-fe	Y	Y
City-fe	Y	Y
Obs.	3934	3934
R^2	0.440	0.472

Note: DID is short for difference-in-differences, SDID-SDM is short for spatial difference-in-differences Durbin model. The parentheses are the *t*-values. *** and ** represent significant levels at 1% and 5%, respectively.

5. Mechanism Analysis

As discussed in Section 1, the ETS may affect GTFP changes through energy efficiency, green innovation, and industrial structure. We empirically tested these potential impact mechanisms respectively.

5.1. Impact of Energy Efficiency

When faced with the policy pressure of the emission trading system, producers in pilot cities often choose different emission strategies based on the relationship between emission quota expenditures and emission benefits and between emission reduction costs and emission reduction benefits. When the emission quota expenditure is higher than the emission income, producers will adopt corresponding energy-saving and emission-reduction measures under policy pressure, such as improving energy efficiency by reforming energy technology and changing energy consumption structure [47]. The improvement of energy efficiency can significantly improve the allocation efficiency of energy resources, thereby realizing the improvement of GTFP. In addition, there are still some companies that cannot cover costs even if they improve energy efficiency. These producers often choose green technology innovation or relocation. Relocation will not only improve the overall energy efficiency level of the original location but also relies on the energy efficiency advantages of the original location relative to the new location to indirectly improve the energy efficiency level of the new location, and then turns out a positive spatial spillover effect. We used the ratio of industrial electricity consumption to regional GDP, that is, the level of energy consumption per unit of GDP, to measure city-level energy efficiency and examine the mechanism through the following model:

$$
\begin{aligned}
gtfp_{it} &= \alpha_0 + \rho(\Omega' \otimes W')_{it} gtfp_{it} + \gamma_2(\Omega' \otimes W')_{it} control_{it} + \alpha_2 control_{it} + city_i + year_t + \varepsilon_{it} \\
&+ (\alpha_1 DID_{it} + \alpha_3 ee_{it} + \alpha_4 DID_{it} \times ee_{it}) + (\Omega' \otimes W')_{it}(\gamma_1 DID_{it} + \gamma_3 ee_{it} + \gamma_4 DID_{it} \times ee_{it})
\end{aligned}
\tag{2}
$$

Table 2 column (1) gives the estimated results of Equation (2). It is shown that the interaction coefficient of the emission trading pilot DID and energy efficiency is significantly positive at the 5% level, and the spatial lag coefficient of the interaction term is significantly positive at the 10% level, indicating that the emission trading system can improve the energy efficiency of territorial cities, and it can also positively promote GTFP in the surrounding areas of the pilot cities.

Table 2. Results of the impact mechanism analysis.

Model Variables	Energy Efficiency (1)	Green Innovation (2)	Industry Structure (3)
$DID \times ee$	0.036 ** (2.21)		
$W \times DID \times ee$	0.413 * (1.65)		
$DID \times gti$		0.280 * (1.83)	
$W \times DID \times gti$		3.725 * (1.91)	
$DID \times str$			−0.295 ** (−2.26)
$W \times DID \times str$			1.857 * (1.75)
Control	Y	Y	Y
Year-fe	Y	Y	Y
City-fe	Y	Y	Y
Obs.	3934	3934	3934
R^2	0.356	0.359	0.355

Note: DID is short for difference-in-differences. The parentheses are the *t*-values. ** and * represent significant levels at 5%, and 10%, respectively.

5.2. Impact of Green Innovation

The impact of the ETS on green technology innovation is mainly reflected in two aspects. On the one hand, the classic Porter hypothesis, strict and flexible environmental regulations can provide incentives for clean technology innovation ([48]). The ETS will not only stimulate quota income companies to obtain higher quota sales profits through innovation but also stimulate emission companies that purchase quotas to reduce pollution costs and increase profits through green innovation. These two ways work together to enhance the overall green technological innovation of the city. On the other hand, for those companies that are unwilling to innovate or have low innovation gains, they will relocate the companies from the pressure area through relocation decisions, thereby improving the green efficiency in the region, and the companies that move away will also increase the technological efficiency of the new location. Therefore, it produces the spatial spillover effect of green innovation, which we call the Porter spatial effect. We use the ratio of the number of green patents to the total number of patent grants to measure the level of green innovation at the city level and examine the above mechanism through the following model:

$$
\begin{aligned}
gtfp_{it} = {} & \alpha_0 + \rho(\Omega' \otimes W')_{it}gtfp_{it} + \gamma_2(\Omega' \otimes W')_{it}control_{it} + \alpha_2 control_{it} + city_i + year_t + \varepsilon_{it} \\
& + (\alpha_1 DID_{it} + \alpha_3 gti_{it} + \alpha_4 DID_{it} \times gti_{it}) + (\Omega' \otimes W')_{it}(\gamma_1 DID_{it} + \gamma_3 gti_{it} + \gamma_4 DID_{it} \times gti_{it})
\end{aligned}
\tag{3}
$$

Table 2 column (2) gives the estimated results of Equation (3). The interaction coefficient of the emission trading pilot DID and green innovation is significantly positive at the 10% level, and the spatial lag coefficient of the interaction term is significantly positive at the 10% level, indicating that the emission trading system can improve the green innovation of territorial cities, and it can also positively promote GTFP in the surrounding areas of the pilot cities which verifies the Porter spatial effect of the ETS.

5.3. Impact of Industry Structure

The ETS can also affect the macro-industrial structure of cities [49] and affect GTFP through industrial structure adjustments. In the pilot areas, due to severe environmental protection policy pressures, operating costs of pollutant producers have significantly increased, such as the cost of pollutant emission quotas, pollutant reduction cost, monitoring and verification cost of pollutant emissions, etc., which directly or indirectly affect the resource allocation decision of the enterprise which may gradually fade out of polluting industries; this will reduce the proportion of the secondary industry [15,50]. In addition, the total emission limit and the market-oriented quota trading mechanism will promote the gradual transfer of pollution capital from the pollution industry to the clean industry or other industries, promote the green innovation of the pollution industry, and further accelerate the shrinkage of the traditional pollution industry. Therefore, the ETS can theoretically affect GTFP by affecting the adjustment of the industrial structure. We use the proportion of the secondary industry to measure the city-level industrial structure, and the estimation model is set as follows:

$$
\begin{aligned}
gtfp_{it} = {} & \alpha_0 + \rho(\Omega' \otimes W')_{it}gtfp_{it} + \gamma_2(\Omega' \otimes W')_{it}control_{it} + \alpha_2 control_{it} + city_i + year_t + \varepsilon_{it} \\
& + (\alpha_1 DID_{it} + \alpha_3 str_{it} + \alpha_4 DID_{it} \times str_{it}) + (\Omega' \otimes W')_{it}(\gamma_1 DID_{it} + \gamma_3 str_{it} + \gamma_4 DID_{it} \times str_{it})
\end{aligned}
\tag{4}
$$

Table 2 column (3) gives the estimated results of Equation (4). The interaction coefficient between the DID and the industrial structure is significantly negative at the 5% level, and the spatial lag coefficient of the interaction term is significantly positive at the 10% level, indicating that ETS can increase urban GTFP by reducing the proportion of the secondary industry in pilot cities, and it can also increase the GTFP in the surrounding areas by increasing the proportion of the secondary industry in the surrounding cities to produce a positive spatial spillover effect.

6. Heterogeneity Analysis

The effective implementation of market-based environmental governance policies depends on the pilot areas with a high level of marketization. When the pilot areas have a low level of marketization, a market-based environmental policy, such as ETS, will be greatly restricted and then unable to handle the internalization of pollution emissions cost, and the quota revenue cannot cover the cost of reduction well, which makes the policy effect greatly discounted [18]. Therefore, the treatment effect of the ETS will show spatial heterogeneity due to the different degrees of regional marketization. Secondly, the effective implementation of market-oriented environmental policies is also closely related to the implementation of regional environmental enforcement [51,52]. Strong environmental enforcement by local governments will improve environmental governance efficiency. As known, the environmental enforcement of environmental policies in developing countries and some emerging-market countries is weak, and previous studies have shown that the effectiveness of environmental policy is greatly reduced in areas where environmental enforcement is low (poor supervision). Regulatory power rent seeking is an important cause of environmental economic corruption that leads to weak environmental enforcement [53]. In addition, the region's energy consumption endowment (mainly the proportion of coal consumption) is also an important source of the spatial heterogeneity of environmental policy effects [54].

According to the research of Li R, Ramanathan R (2018, [55]), Hou B et al. (2020, [18]), we adopt the provincial marketization data disclosed by Fan Gang et al. (2011, [56]), according to the median of the total marketization index divides the sample cities into two groups, namely, regions with a higher degree of marketization and regions with a lower degree of marketization. Secondly, different from Hou B et al. (2020, [18]), this study used the ratio of the number of environmental administrative punishment cases to the total energy consumption in the provincial environmental law enforcement data disclosed in the "China Environmental Statistics Yearbook" to measure the intensity of regional environmental enforcement. This measurement can avoid differences in environmental enforcement caused by differences in energy consumption between provinces. Cities are also grouped according to the median of the number of cases and divided into two groups: strong law enforcement and weak law enforcement. Similarly, we used provincial energy structure data to measure urban energy consumption endowment, which is measured by the proportion of provincial coal consumption in energy consumption based on the data from the "China Energy Statistics Yearbook", to group it by the median, and classify it into high coal consumption group and low coal consumption group. It should be noted that the above three data sources are all province-level data. It is reasonable to use province-level data to assign value to cities for which spatial heterogeneity of China's environmental enforcement and marketization is mainly driven by provincial differences because differences in culture, economic development, and energy consumption endowments within the province are generally significantly smaller than inter-provincial differences. The groupings in the heterogeneity analysis all adopted the data before the pilot period which may somehow avoid possible selectivity deviations, and in this study, we adopted the data from 2007.

Table 3 shows the estimated results of the heterogeneity analysis of policy effects. Columns (1)–(2) are the estimated results of different marketization levels. Coefficients in column (1) are significant, and column (2) is not significant, indicating that the average treatment effect of the pilot policies in regions with a high degree of marketization on GTFP is still significant, and the ETS in regions with a low degree of marketization have no significant impact on GTFP. Secondly, the coefficients of the DID and W*DID in column (1) are higher than those of column (2), indicating that the pilot program in regions with a higher degree of marketization performs better. This heterogeneous result is relatively easy to understand. The ETS is a typical market-oriented environmental policy, and the degree of marketization of the pilot city will directly affect the effectiveness of the pilot policy on market. The higher the marketization, the clearer the pilot policy signals will be communicated, and the easier it will be to influence enterprises' pollution discharge decisions, green innovation decisions, and relocation decisions through policy pressures, resulting in better effects on GTFP.

Table 3. Heterogeneity analysis results.

Model	Marketization Level		Environmental Enforcement		Energy Consumption Endowment	
	(1)	(2)	(3)	(4)	(5)	(6)
Variables	High	Low	Strong	Weak	Heavy	Light
DID	0.758 ***	0.014	0.680	0.433 **	0.611 ***	11.394 ***
	(3.08)	(0.03)	(1.06)	(2.29)	(2.74)	(5.25)
W × DID	3.189 ***	2.496	−6.246 ***	6.064 *	2.732 **	−6.434 ***
	(3.76)	(1.20)	(−2.77)	(1.958)	(2.24)	(−2.59)
Control	Y	Y	Y	Y	Y	Y
Year-fe	Y	Y	Y	Y	Y	Y
City-fe	Y	Y	Y	Y	Y	Y
Obs.	2268	1666	2240	1694	2450	1484
R^2	0.434	0.501	0.486	0.519	0.412	0.536

Note: DID is short for difference-in-differences. The parentheses are the *t*-values. ***, ** and * represent significant levels at 1%, 5%, and 10%, respectively.

Columns (3)–(4) are the estimated results of different environmental enforcement. The coefficient of the DID in column (3) is not significant, the coefficient of the W × DID is significant, and the estimation results in column (4) are significant. It shows that the ETS in different environmental enforcement groups still significantly affects GTFP, and there are differences between groups in treatment effects. It shows that the average treatment effect of the regional pilot policies with strong environmental enforcement on GTFP is mainly achieved through negative spatial externalities, while the impact of the regional pilot policies with weak enforcement on GTFP is consistent with the benchmark model. Secondly, the coefficient of the W × DID in column (3) is negative, which may indicate that areas with strong environmental law enforcement will increase the burden of environmental PR costs for companies in the pilot city. Compared with areas with weak law enforcement, more companies will consider relocation decisions, and these relocation companies only relocate due to the strong environmental enforcement rather than the pressure from ETS, which makes the new location have a green productivity advantage over the old location. Therefore, this relocation decision will increase the GTFP of the old location while reducing that of the new location. The objective reason is that excessively strict environmental enforcement has distorted the market-oriented allocation process of the ETS.

Columns (5)–(6) are the estimation results of different energy consumption endowments. The coefficients of columns (5)–(6) are all significant. The coefficient of DID in column (5) is significantly lower than column (6), the W*DID coefficient in column (5) is positive, and the W*DID coefficient in column (6) is negative, which means there are significant differences between the group. The DID coefficient in column (5) is significantly lower than that in column (6). It indicates that the effect of the ETS in areas with a higher proportion of coal consumption is significantly lower than that in areas with a lower proportion of coal consumption, which means that pilot cities in regions with the lower coal consumption are more inclined to use clean technologies (non-fossil energy) for production, so producers' emission reduction decisions are more inclined to improve energy efficiency, carry out green innovation activities, etc., which then more significantly improve productivity. Moreover, these tendencies to clean technology indicate that pilot cities in regions with a low proportion of coal consumption have higher innovation gains. It will attract companies with higher green productivity in surrounding cities to move to pilot cities, which will cause negative spatial externality, so the coefficient of the W*DID in column (6) is negative. The effects of pilot policies in areas with a higher proportion of coal consumption are different. Producers in high coal consumption areas tend to make emission reduction decisions in improving energy efficiency while purchasing emissions quota. The green innovation income of these producers is relatively weaker than the first two, and the effect is consistent with the benchmark model, that is, it positively promotes GTFP in territorial cities and has a positive spatial spillover effect on surrounding cities.

Above all, we believe that the treatment effect of the ETS on GTFP shows significant heterogeneous characteristics under different marketization levels, environmental enforcement efforts, and energy consumption endowments, and the promoting effects perform better in regions with high marketization, strong environmental enforcement, and low coal consumption.

7. Robustness Test

7.1. Placebo Test

To further eliminate the influence of other unknown factors on the selection of pilot cities, this study conducted 999 samplings in all 281 cities and randomly selected 108 cities as the virtual treatment group for each sampling (the original number of treatment groups was 108), and the remaining 173 cities were used as a randomized control group. We estimated the placebo test by adopting the SDID-SDM approach, and the results are as shown in Figure 3. Figure 3a,b are, respectively, the distribution diagrams of the coefficients of DID and WDID in the random sampling estimation results. It can be found that the t value of most sampling estimation coefficients changes within a small range, and the significance fails (that is, the pass rate is still low even under the 10% confidence level), indicating that the ETS did not show a significant treatment effect in the random sampling simulation. It can be considered that the conclusion of the treatment effect identified by the benchmark model estimation passed the placebo test.

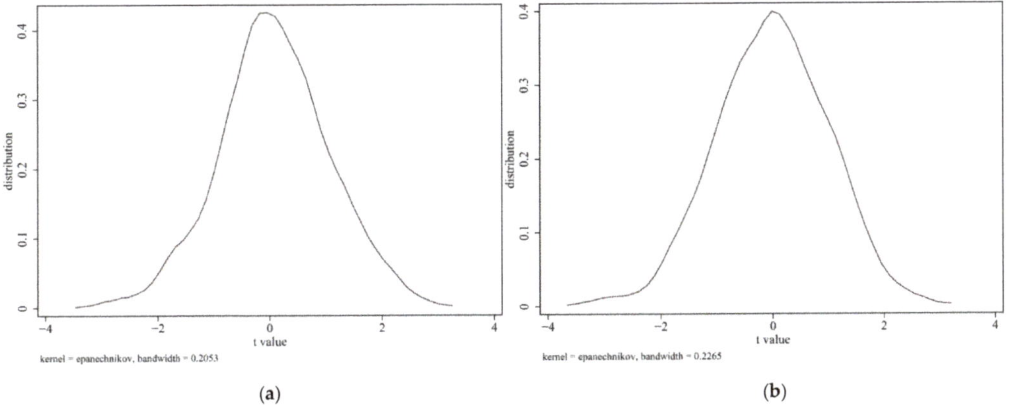

Figure 3. Placebo test. (**a**) shows the coefficient's t distribution of *DID* calculated in the placebo test; (**b**) shows the coefficient t distribution of $W \times DID$ calculated in the placebo test.

7.2. PSM-SDID

Although the ETS has a significant promotion and positive spatial spillover effect on GTFP, the result may be caused by potential selectivity bias [49]. Therefore, we used the propensity score matching (PSM) method to solve the problem of selective bias that may exist in the grouping by identifying and matching to form a new treatment group and control group and then continued to use the SDID-SDM model to identify and evaluate the treatment effect. The results of the estimation of the PSM-SDID method are shown in Table 4 (1). This study found that the ETS still significantly improves the GTFP of the territorial cities and has a positive spatial spillover effect on the GTFP of the surrounding cities, indicating that the research conclusions of the benchmark model are robust.

Table 4. Estimation results in PSM-SDID and effect with Carbon ETS.

Model Variables	PSM-SDID (1)	Carbon ETS (2)
DID	0.640 *** (2.84)	
W × *DID*	3.110 ** (2.56)	
DDD		0.195 * (1.87)
W × *DDD*		43.441 *** (6.63)
Control	Y	Y
Year-fe	Y	Y
City-fe	Y	Y
Obs.	3201	3934
R^2	0.150	0.271

Note: DID is short for difference-in-differences, PSM-SDID is short for propensity score matching spatial difference-in-differences, ETS is short for emissions trading system, DDD is short for difference-in- difference-in-difference. The parentheses are the t-values. ***, ** and * represent significant levels at 1%, 5%, and 10%, respectively.

7.3. Carbon ETS

Many existing policy studies on the carbon ETS have used the ETS as a confusing factor. These studies control the possible confusion caused by the ETS by constructing dummy variables of the ETS [18]. Therefore, in contrast to these studies, this study constructed a combined dummy variable for the carbon ETS, that is, during the treatment period, the value of the dummy variable of carbon emission trading pilot is 1but otherwise 0. Then we bought it into the benchmark SDID-SDM model, and adopted the difference-in difference-in-differences(DDD) method to identify and estimate the model. The results are shown in column (2) of Table 4. After controlling the confusing effects of the carbon ETS, the ETS still significantly increases the GTFP of the territorial cities and has a positive spatial spillover effect on the GTFP of the surrounding cities, indicating that the results of the policy effects are robust.

8. Discussion and Conclusions

8.1. Discussion

To accelerate the transition from extensive economic growth to green and high-quality development, the Chinese government has carried out more environmental policy explorations. Among them, the market-oriented environmental policy tool ETS has achieved remarkable green economic benefits. Adopting the SDID-SDM model with multidimensional FE, we examined the spatial policy impact of ETS on urban GTFP.

Benchmark regression results showed that, after controlling for individual fixed effects, time fixed effects, and the influence of control variables, the ETS policy implemented by China has effectively increased the GTFP of territorial cities, which is consistent with the existing research conclusions [18,20]. In addition, as mentioned above, the effect of ETS policy will accompany the commodity trade between cities and the circulation of resource elements to produce spatial correlation, which is ignored or less mentioned in previous studies [20,30]. Ignoring the impact of inter-city spatial interactions on policy effects will bias the identification results of ETS policy effects. In this regard, we effectively identified the spatial effects of ETS policies by applying the SDID-SDM model.

Subsequently, we also discussed in detail how ETS influences urban GTFP through energy efficiency, green technology innovation, and industrial structure and the identification of spatial effects under the corresponding mechanism. This verifies the important role of urban spatial connections in energy efficiency improvement, green innovation, and industrial structure adjustment [57]. The possible explanation behind this is that we believe it is related to the location decision making of pollutant companies. The research on heterogeneous enterprise location selection theory shows that enterprise location selection will directly or indirectly affect regional efficiency, regional innovation, regional industrial structure, and regional spatial structure [58], and the policy pressure of ETS will drive some pollutant companies to relocate. The spatial redistribution process of a large number of pollutant companies will promote ETS policy effects to show more spatial effects through macro-level energy efficiency, green technology innovation, and industrial structure.

In addition, the heterogeneity analysis results show that the level of marketization, environmental law enforcement, and regional energy consumption endowments are important sources of heterogeneity in the effectiveness of ETS policies. Firstly, a higher level of market-oriented reform will bring about a more obvious effect of improving territorial GTFP and the spatial spillover effect of surrounding GTFP. The policy effect of lower regions is not significant. This result on the one hand confirms that China's market reform has achieved certain results and refutes the doubts of some foreign scholars [20,59]. It also shows the spatial heterogeneity of China's market-oriented reforms. In the future, China needs to continue to explore market-oriented reforms in areas with low levels of marketization and further guide the market to play an important role in the allocation of resource elements. At the same time, it also means that China should extend the successful experience of emissions trading pilots to more regions; expand other possible market-oriented environmental policy tools, including intertemporal emissions rights, cross-regional trade, and green financial instrument design, etc.; accelerate the improvement of the regional green policy system; actively integrate into the international environmental governance system; learn from the advanced environmental governance experience of developed countries; and enhance the right to speak in international environmental governance.

Secondly, areas with strong environmental law enforcement have a negative siphon effect, and areas with low environmental enforcement have a positive spillover. This result confirms that the difference in supervision will have opposite spatial effects. When the supervision is strong, the companies that choose to actively respond to the strategy of polluting companies will benefit more, which will form a trend of gathering in regions with strong supervision, and those pollutant companies that respond negatively will spread to the surrounding area and then cause a negative siphon effect. On the contrary, in regions with loose supervision, the profits of pollutant companies that adopt active strategies have dropped significantly. More companies choose to purchase pollutant emission rights for negative response strategies such as pollutant emission. In turn, innovative enterprises move abroad, and there is a positive spatial spillover effect. In this regard, we propose the following suggestions: one is to strengthen the supervision of local government environmental law enforcement departments [60], take a zero tolerance stance for environmental corruption, increase the rent-seeking cost of local government officials' environmental supervision powers and avoid the harmful impact of environmental economic corruption on the burden of enterprises; the other is to increase the transparency of the environmental law enforcement process, make the law enforcement process open and transparent, and disclose information on law enforcement documents.

Finally, the heterogeneity brought about by the endowment of urban energy consumption shows that coal consumption has a certain path dependence. To break this path, we must start with the energy consumption structure, use compulsory or semi-compulsory policy tools to guide areas with a high proportion of coal consumption to carry out energy efficiency improvement and green innovation activities and guide the use of clean technology.

8.2. Conclusions

In general, our research provides empirical support for the significant spatial effects of ETS's impact on urban GTFP and shows that the spatial correlation between cities has significant spatial effects in terms of policy effect, influence mechanism, heterogeneity, etc., which makes up for the neglect of the existing ETS policy effect identification research on the spatial dimension. In addition, the identification method of this study can be used as a scientific reference for conducting similar research in other emerging countries.

Although this research provided some valuable findings and enlightenment for the government's decision making and research in the field of emission reduction and green growth, it inevitably has certain limitations. First of all, our research only takes China's ETS as a case. No cases of other emerging-market countries were introduced, and there is a lack of more extensive identification verification. Secondly, there is a lack of finer-grained analysis on the heterogeneity of policy spatial effects. For example, the introduction of spatial measurement technologies such as multiscale geographically weighted regression(MGWR) and geographical and temporal weighted regression (GTWR) allows finer-grained space correlation effects, which will provide help for the precise identification of policy effects. Finally, we did not conduct a more in-depth analysis of areas with weak environmental law enforcement and did not examine in detail the possible role and influence mechanism of corruption in the heterogeneity of spatial effects.

Author Contributions: Conceptualization, G.C., S.W. and X.H.; methodology, G.C.; software, G.C.; validation, G.C. and X.H.; formal analysis, X.H.; data curation, G.C.; writing—original draft preparation, G.C.; writing—review and editing, G.C. and S.W.; visualization, G.C. and X.H.; supervision, S.W.; project administration, G.C. and S.W.; funding acquisition, G.C. All authors have read and agreed to the published version of the manuscript.

Funding: This research was funded by the Ministry of Education of Humanities and Social Science Project (No. 17YJCZH063), Shenzhen Philosophy and Social Sciences Planning 2020 Project (No. SZ2020B012).

Institutional Review Board Statement: Not applicable.

Informed Consent Statement: Not applicable.

Data Availability Statement: Data available on request due to restrictions privacy. The data presented in this study are available on request from the corresponding author. The data are not publicly available due to privacy.

Conflicts of Interest: The authors declare no conflict of interest.

References

1. Vennemo, H.; Aunan, K.; Lindhjem, H.; Seip, H.M. Environmental Pollution in China: Status and Trends. *Rev. Environ. Econ. Policy* **2009**, *3*, 209–230. [CrossRef]
2. Lu, Z.N.; Chen, H.; Hao, Y.; Wang, J.; Song, X.; Mok, T.M. The dynamic relationship between environmental pollution, economic development and public health: Evidence from China. *J. Clean. Prod.* **2017**, *166*, 134–147. [CrossRef]
3. Chen, S. Environmental pollution emissions, regional productivity growth and ecological economic development in China. *China Econ. Rev.* **2015**, *35*, 171–182. [CrossRef]
4. Solomon, B.D. Global CO_2 emissions trading: Early lessons from the US acid rain program. *Clim. Chang.* **1995**, *30*, 75–96. [CrossRef]
5. Chestnut, L.G.; Mills, D.M. A fresh look at the benefits and costs of the US acid rain program. *J. Environ. Manag.* **2005**, *77*, 252–266. [CrossRef]
6. Barreca, A.I.; Neidell, M.; Sanders, N. Long-run pollution exposure and mortality: Evidence from the Acid Rain Program. *J. Public Econ.* **2021**, *200*, 104440. [CrossRef]
7. Zhang, D.; Karplus, V.J.; Cassisa, C.; Zhang, X. Emissions trading in China: Progress and prospects. *Energy Policy* **2014**, *75*, 9–16. [CrossRef]
8. Yan, Y.; Zhang, X.; Zhang, J.; Li, K. Emissions trading system (ETS) implementation and its collaborative governance effects on air pollution: The China story. *Energy Policy* **2020**, *138*, 111282. [CrossRef]
9. Zhang, C. Using multivariate analyses and GIS to identify pollutants and their spatial patterns in urban soils in Galway, Ireland. *Environ. Pollut.* **2006**, *142*, 501–511. [CrossRef]
10. Cheng, Z. The spatial correlation and interaction between manufacturing agglomeration and environmental pollution. *Ecol. Indic.* **2016**, *61*, 1024–1032. [CrossRef]
11. Liu, K.; Lin, B. Research on influencing factors of environmental pollution in China: A spatial econometric analysis. *J. Clean. Prod.* **2018**, *206*, 356–364. [CrossRef]

12. Wang, J.; Wang, K.; Shi, X.; Wei, Y.-M. Spatial heterogeneity and driving forces of environmental productivity growth in China: Would it help to switch pollutant discharge fees to environmental taxes? *J. Clean. Prod.* **2019**, *223*, 36–44. [CrossRef]

13. Yu, D.J.; Li, J. Evaluating the employment effect of China's carbon emission trading policy: Based on the perspective of spatial spillover. *J. Clean. Prod.* **2021**, *292*, 126052. [CrossRef]

14. Yang, Z.; Fan, M.; Shao, S.; Yang, L. Does carbon intensity constraint policy improve industrial green production performance in China? A quasi-DID analysis. *Energy Econ.* **2017**, *68*, 271–282. [CrossRef]

15. Hu, Y.; Ren, S.; Wang, Y.; Chen, X. Can carbon emission trading scheme achieve energy conservation and emission reduction? Evidence from the industrial sector in China. *Energy Econ.* **2020**, *85*, 104590. [CrossRef]

16. Li, J.; Du, Y.X. Spatial effect of environmental regulation on green innovation efficiency: Evidence from prefectural-level cities in China. *J. Clean. Prod.* **2021**, *286*, 125032. [CrossRef]

17. Cai, X.; Lu, Y.; Wu, M.; Yu, L. Does environmental regulation drive away inbound foreign direct investment? Evidence from a quasi-natural experiment in China. *J. Dev. Econ.* **2016**, *123*, 73–85. [CrossRef]

18. Hou, B.; Wang, B.; Du, M.; Zhang, N. Does the SO2 emissions trading scheme encourage green total factor productivity? An empirical assessment on China's cities. *Environ. Sci. Pollut. Res.* **2019**, *27*, 6375–6388. [CrossRef]

19. Yao, S.; Yu, X.; Yan, S.; Wen, S. Heterogeneous emission trading schemes and green innovation. *Energy Policy* **2021**, *155*, 112367. [CrossRef]

20. Shi, D.; Li, S.L. Emissions Trading System and Energy Use Efficiency—Measurements and Empirical Evidence for Cities at and above the Prefecture Level. *China Ind. Econ.* **2020**, *138*, 5–23.

21. Todorović, J.Đ.; Đorđević, M.; Ristić, M. Environmental taxes as the instrument of environmental policy in developing countries. Нови Економист **2018**, *12*, 45–52. [CrossRef]

22. Joskow, P.L.; Schmalensee, R. The political economy of market-based environmental policy: The US acid rain program. *J. Law Econ.* **1998**, *41*, 37–84. [CrossRef]

23. Chan, H.R.; Chupp, B.A.; Cropper, M.L.; Muller, N.Z. The impact of trading on the costs and benefits of the Acid Rain Program. *J. Environ. Econ. Manag.* **2018**, *88*, 180–209. [CrossRef]

24. Zhang, Y.; Li, S.; Luo, T.; Gao, J. The effect of emission trading policy on carbon emission reduction: Evidence from an integrated study of pilot regions in China. *J. Clean. Prod.* **2020**, *265*, 121843. [CrossRef]

25. Tang, K.; Liu, Y.; Zhou, D.; Qiu, Y. Urban carbon emission intensity under emission trading system in a developing economy: Evidence from 273 Chinese cities. *Environ. Sci. Pollut. Res.* **2020**, *28*, 5168–5179. [CrossRef]

26. Vespermann, J.; Wald, A. Much Ado about Nothing?—An analysis of economic impacts and ecologic effects of the EU-emission trading scheme in the aviation industry. *Transp. Res. Part A Policy Pract.* **2011**, *45*, 1066–1076. [CrossRef]

27. Zhang, W.; Li, J.; Li, G.; Guo, S. Emission reduction effect and carbon market efficiency of carbon emissions trading policy in China. *Energy* **2020**, *196*, 117117. [CrossRef]

28. Rogge, K.S.; Schneider, M.; Hoffmann, V.H. The innovation impact of the EU Emission Trading System—Findings of company case studies in the German power sector. *Ecol. Econ.* **2011**, *70*, 513–523. [CrossRef]

29. Lyu, X.; Shi, A.; Wang, X. Research on the impact of carbon emission trading system on low-carbon technology innovation. *Carbon Manag.* **2020**, *11*, 183–193. [CrossRef]

30. Zhang, L.; Cao, C.; Tang, F.; He, J.; Li, D. Does China's emissions trading system foster corporate green innovation? Evidence from regulating listed companies. *Technol. Anal. Strat. Manag.* **2018**, *31*, 199–212. [CrossRef]

31. Porter, M.E. America's Green Strategy. *Sci. Am.* **1991**, *264*, 168. [CrossRef]

32. Ambec, S.; Barla, P. A theoretical foundation of the Porter hypothesis. *Econ. Lett.* **2002**, *75*, 355–360. [CrossRef]

33. Lv, M.; Bai, M. Evaluation of China's carbon emission trading policy from corporate innovation. *Financ. Res. Lett.* **2021**, *39*, 101565. [CrossRef]

34. Tang, H.-L.; Liu, J.-M.; Mao, J.; Wu, J.-G. The effects of emission trading system on corporate innovation and productivity-empirical evidence from China's SO2 emission trading system. *Environ. Sci. Pollut. Res.* **2020**, *27*, 21604–21620. [CrossRef] [PubMed]

35. Feng, C.; Shi, B.; Kang, R. Does Environmental Policy Reduce Enterprise Innovation?—Evidence from China. *Sustainability* **2017**, *9*, 872. [CrossRef]

36. Yi, M.; Fang, X.; Wen, L.; Guang, F.; Zhang, Y. The Heterogeneous Effects of Different Environmental Policy Instruments on Green Technology Innovation. *Int. J. Environ. Res. Public Health* **2019**, *16*, 4660. [CrossRef] [PubMed]

37. Philibert, C. How could emissions trading benefit developing countries. *Energy Policy* **2000**, *28*, 947–956. [CrossRef]

38. Sahoo, N.R.; Mohapatra, P.K.J.; Sahoo, B.K.; Mahanty, B. Rationality of energy efficiency improvement targets under the PAT scheme in India–A case of thermal power plants. *Energy Econ.* **2017**, *66*, 279–289. [CrossRef]

39. Ba, F.; Thiers, P.R.; Liu, Y. The evolution of China's emission trading mechanisms: From international offset market to domestic Emission Trading Scheme. *Environ. Plan. C Politi-Space* **2018**, *36*, 1214–1233. [CrossRef]

40. Oliveira, T.D.; Gurgel, A.C.; Tonry, S. International market mechanisms under the Paris Agreement: A cooperation between Brazil and Europe. *Energy Policy* **2019**, *129*, 397–409. [CrossRef]

41. Muldavin, J. The paradoxes of environmental policy and resource management in reform-era China. *Econ. Geogr.* **2000**, *76*, 244–271. [CrossRef]

42. Wu, J.; Deng, Y.; Huang, J.; Morck, R.; Yeung, B. *Incentives and Outcomes: China's Environmental Policy*; No. w18754; National Bureau of Economic Research: Cambridge, MA, USA, 2013.
43. Li, Y.; Chen, Y. Development of an SBM-ML model for the measurement of green total factor productivity: The case of pearl river delta urban agglomeration. *Renew. Sustain. Energy Rev.* **2021**, *145*, 111131. [CrossRef]
44. Qiao, Z.; Li, Z. Do foreign institutional investors enhance firm innovation in China? *Appl. Econ. Lett.* **2018**, *26*, 1125–1128. [CrossRef]
45. Marcus, M.; Sant'Anna, P.H.C. The role of parallel trends in event study settings: An application to environmental economics. *J. Assoc. Environ. Resour. Econ.* **2021**, *8*, 235–275.
46. Li, X.; Shu, Y.; Jin, X. Environmental regulation, carbon emissions and green total factor productivity: A case study of China. *Environ. Dev. Sustain.* **2021**, 1–21. [CrossRef]
47. MacGill, I.; Outhred, H.; Nolles, K. National Emissions Trading for Australia: Key Design Issues and Complementary Policies for Promoting Energy Efficiency, Infrastructure Investment and Innovation. *Australas. J. Environ. Manag.* **2004**, *11*, 78–87. [CrossRef]
48. Wang, Y.; Sun, X.; Guo, X. Environmental regulation and green productivity growth: Empirical evidence on the Porter Hy-pothesis from OECD industrial sectors. *Energy Policy* **2019**, *132*, 611–619. [CrossRef]
49. Zang, J.; Wan, L.; Li, Z.; Wang, C.; Wang, S. Does emission trading scheme have spillover effect on industrial structure upgrading? Evidence from the EU based on a PSM-DID approach. *Environ. Sci. Pollut. Res.* **2020**, *27*, 12345–12357. [CrossRef] [PubMed]
50. Albrizio, S.; Kozluk, T.; Zipperer, V. Environmental policies and productivity growth: Evidence across industries and firms. *J. Environ. Econ. Manag.* **2017**, *81*, 209–226. [CrossRef]
51. Shinkuma, T.; Sugeta, H. Tax versus emissions trading scheme in the long run. *J. Environ. Econ. Manag.* **2016**, *75*, 12–24. [CrossRef]
52. Jacobsen, L.B.; Nielsen, M.; Nielsen, R. Gains of integrating sector-wise pollution regulation: The case of nitrogen in Danish crop production and aquaculture. *Ecol. Econ.* **2016**, *129*, 172–181. [CrossRef]
53. Nemesio, I.V. Strengthening environmental rule of law: Enforcement, combatting corruption, and encouraging citizen suits. *Geo. Int'l Envtl. L. Rev.* **2014**, *27*, 321.
54. Liu, C.; Ma, C.; Xie, R. Structural, Innovation and Efficiency Effects of Environmental Regulation: Evidence from China's Carbon Emissions Trading Pilot. *Environ. Resour. Econ.* **2020**, *75*, 741–768. [CrossRef]
55. Li, R.; Ramanathan, R. Exploring the relationships between different types of environmental regulations and environmental performance: Evidence from China. *J. Clean. Prod.* **2018**, *196*, 1329–1340. [CrossRef]
56. Fan, G.; Wang, X.; Zhu, H. *NERI Index of Marketization of China's Provinces 2011 Report*; Economic Science Press: Beijing, China, 2011; pp. 273–283.
57. Song, M.; Chen, Y.; An, Q. Spatial econometric analysis of factors influencing regional energy efficiency in China. *Environ. Sci. Pollut. Res.* **2018**, *25*, 13745–13759. [CrossRef] [PubMed]
58. Tang, K.; Qiu, Y.; Zhou, D. Does command-and-control regulation promote green innovation performance? Evidence from China's industrial enterprises. *Sci. Total Environ.* **2020**, *712*, 136362. [CrossRef]
59. Allen, F.; Qian, J.; Qian, M. Law, finance, and economic growth in China. *Financ. Econ.* **2005**, *77*, 57–116. [CrossRef]
60. Zhang, B.; Chen, X.; Guo, H. Does central supervision enhance local environmental enforcement? Quasi-experimental evidence from China. *J. Public Econ.* **2018**, *164*, 70–90. [CrossRef]

International Journal of *Environmental Research and Public Health*

Linking Innovative Human Capital, Economic Growth, and CO_2 Emissions: An Empirical Study Based on Chinese Provincial Panel Data

Xi Lin [1,2,*], Yongle Zhao [1], Mahmood Ahmad [3], Zahoor Ahmed [4], Husam Rjoub [5] and Tomiwa Sunday Adebayo [6,*]

[1] Business School, Hohai University, Nanjing 211100, China; zhao_yongle@163.com
[2] Business School, Guilin University of Technology, Guilin 541004, China
[3] Business School, Shandong University of Technology, Zibo 255000, China; mahmood@sdut.edu.cn
[4] Department of Economics, Faculty of Economics and Administrative Sciences, Cyprus International University, Mersin 10, Haspolat 99040, Turkey; zahoorahmed83@yahoo.com
[5] Department of Accounting and Finance, Faculty of Economics and Administrative Sciences, Cyprus International University, Mersin 10, Haspolat 99040, Turkey; hrjoub@ciu.edu.tr
[6] Department of Business Administration, Faculty of Economics and Administrative Science, Cyprus International University, Nicosia, Northern Cyprus, Mersin TR-10, Turkey
* Correspondence: linxi@glut.edu.cn (X.L.); twaikline@gmail.com (T.S.A.)

Citation: Lin, X.; Zhao, Y.; Ahmad, M.; Ahmed, Z.; Rjoub, H.; Adebayo, T.S. Linking Innovative Human Capital, Economic Growth, and CO_2 Emissions: An Empirical Study Based on Chinese Provincial Panel Data. *Int. J. Environ. Res. Public Health* **2021**, *18*, 8503. https://doi.org/10.3390/ijerph18168503

Academic Editors: Pasquale Avino, Massimiliano Errico, Hamid Salehi and Aristide Giuliano

Received: 6 July 2021
Accepted: 9 August 2021
Published: 11 August 2021

Abstract: To study the economic and environmental effects of human capital, previous studies measure human capital based on education; however, this approach has many shortcomings because not all educated people are innovative human capital. Hence, this study introduces the concept of innovative human capital by developing a new index that measures human capital based on the number of patents every one million R&D staff full-time equivalent. After this, this paper studies the impact of innovative human capital on CO_2 emissions in China. The provincial panel data of 30 Chinese provinces from 2003 to 2017 is analyzed using the fixed effect, ordinary least squares, and system generalized method of moments (SYS-GMM). The analysis revealed that innovative human capital alleviates environmental deterioration in China. The findings unfold the existence of the environmental Kuznets curve (EKC) considering innovative human capital in the model. It implies that Chinese economic development will eventually support environmental sustainability if China continues to develop its innovative human capital. Among the control variables, economic structure, population density, and energy intensity stimulate environmental degradation by increasing CO_2 emissions. However, FDI has a negative relationship with CO_2 emissions. Lastly, the study proposes comprehensive policies to increase innovative human capital for environmental sustainability.

Keywords: innovative human capital; CO_2 emission; Chinese provinces; economic growth

1. Introduction

Economic growth is humans' eternal objective and the focus of economists across the world. However, with the development of a global economy since the 1960s, environmental problems have become serious and concerned various international organizations and countries. Thus, in 1980, the United Nations Environment Program (UNEP) appealed for sustainable development with two other global organizations. The Earth Summit was held in Rio and published the very famous declaration "Rio Declaration on Environment and Development" in 1992. Afterward, the Kyoto protocol and Paris agreement were signed under the United Nations (UN) framework convention on climate change [1]. China actively joined the two agreements and has made efforts to pursue green growth since then. The idea of green growth has originated from the Asia and Pacific Region. Green growth provides the idea of harmonizing growth and ecological sustainability while upgrading the eco-efficiency of growth and increasing the synergies between the economy and the environment [2]. At present, the Chinese economy is transferring to a higher-quality, more balanced, and greener growth path. Chinese central government has taken many measures to reduce the environmental pressures caused by economic growth. China has made some progress in the fields, such as energy and carbon intensities, per capita carbon emission intensities and environmentally related taxes, etc. However, there are lots of problems waiting to be solved in the largest CO_2 emitting country.

According to the 2011 OECD Ministerial Council Meeting, innovation sustained by a sturdy intellectual property rights system supports economic growth, conserves the environment, and creates employment. In this regard, China has invested so much in scientific and technological innovation for decades and ranked second for global gross expenditure on R&D (GERD). According to the statistics, the environment-related patents have increased 15 times for the period 2000–2012 in the context of China. This increase is much more than the increase in the innovation of the OECD countries. However, the inventions' quality and protection still need to be improved. In addition, the share of GERD in GDP and the share of research personnel in the overall employment are also less than many OECD nations [3]. China is the number one country in terms of total CO_2 emissions in the world; meanwhile, per capita carbon emission has also kept increasing [4]. As clearly shown in Figure 1, total carbon emissions increased from 138.47 metric tons to 328.86 metric tons over the period of 2003–2017 in China. In 2007, the Intergovernmental Panel on Climate Change (IPCC) stated that the CO_2 emissions emitted by humankind contribute to 90% of global warming. The current technology is unlikely to decrease carbon emission compared to other pollutants [5]. It is noteworthy that more effective technologies still depend on relentless innovation by Research and Development (R&D) personnel. R&D personnel are defined as "specific human capital whose knowledge and skills are less transferable and have a narrower scope of applicability" [6]. This specific human capital can also be called as innovative human capital, and it is a unique resource to promote productivity and technological progress. Thus, innovative human capital plays a vital role in economic development [7]. More precisely, innovative human capital can be defined as a kind of heterogeneous capital that always possesses cutting-edge knowledge and skills in a specific professional field, continuously carries out innovative activities, and obtains innovative output, so that the marginal income can continue to increase.

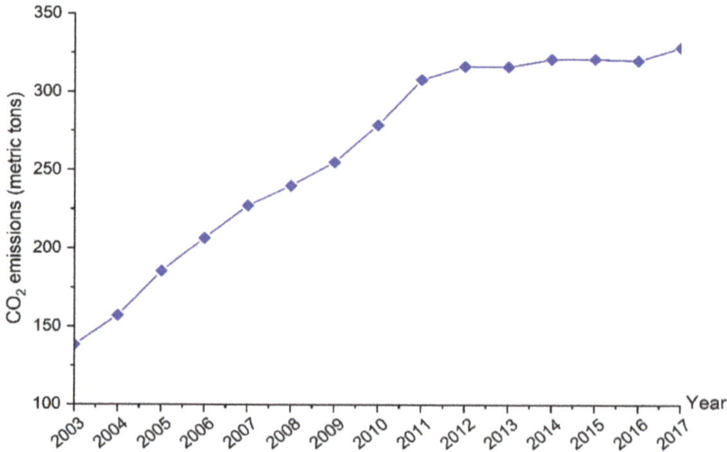

Figure 1. Annual trend of CO_2 emissions in China over 2003–2017. Data source: China energy statistical yearbook (https://data.cnki.net/Yearbook/ accessed on 16 June 2021).

Although innovative human capital has the strongest innovative ability to create clean production technologies compared to the other kinds of human capital, few scholars paid attention to the relationship between innovative human capital, green growth, and environmental sustainability. At present, many researchers are still mainly discussing the relationship between the quantity of human capital, economic growth, and environmental quality. They usually use educational attainment to measure the quantity of human capital and neglect the quality difference of human capital. However, the quality of human capital may be more relevant to environmental sustainability and green growth. The quality aspects of human capital may have greater potential in explaining growth [8].

The measurement of human capital, including innovative human capital has always been a controversial issue in academic circles. In the past, many scholars used people with tertiary education to represent innovative human capital [9–12]. Although this measuring method is widely used, it may cause the following several problems: first, this method omits the qualitative difference among the people with tertiary education because the people's educational attainment only reflects the differences in the amount of human capital [13]. Second, college students are usually included in the number of people receiving higher education, which amplifies the stock of innovative human capital to a large extent. Third, not all the people who have received higher education are innovative human capital, and those who are engaged in scientific and technological work are the main part of innovative human capital. Thus, it is very necessary and meaningful to compute the innovative human capital by addressing all these weaknesses and probe the influence of innovative human capital on CO_2 emissions in China. It is a vital point for China to strictly control CO_2 emissions and effectively execute green development. To avoid the above problems, this paper followed the footsteps of Romer [14] and mainly considered science and technology staff when measuring the innovative human capital.

Hence, this new proxy for innovative human capital will more likely capture the ability of innovative human capital to promote innovation which can play a mounting role in increasing energy efficiency and output, decreasing environmental degradation. In addition, this will reduce the chances of unreliable results and misleading policies that may arise due to the use of inappropriate measures of human capital. For example, using the education-based human capital, the recent work of Ahmed et al. [15] found that human capital is positively linked with emissions in Latin American countries. We believe that such surprising outcomes can arise due to the weaknesses of previous proxies of human capital. Additionally, education-based measures are mainly based on the view that education promotes awareness that leads to environmental sustainability; however, our measure directly promotes innovation and technological advancement, which are more relevant for environmental sustainability.

Against this backdrop, the authors intend to explore the relationship between innovative human capital and CO_2 emissions in China using an appropriate measure of innovative human capital. The article uses 15-year Chinese provincial panel data to examine the impact of innovative human capital on emissions in an EKC framework. The study extends the literature and measures innovative human capital by developing an index based on the number of patents every one million R&D staff full-time equivalent. The study has both theoretical meaning and practical applications. Concerning theoretical contribution, the paper will enrich and strengthen the present literature on human capital and environment nexus. It is conducive for us to better understand the impact mechanism of human capital for green growth. On the practical meaning, the research is aiming at providing some advice for the Chinese government to formulate more targeted strategies towards environmental sustainability and green growth.

Additionally, this work is equally important for other developed and developing countries because it presents and discusses a new measure of innovative human capital. In recent literature, studies linking innovative human capital with different economic and environmental indicators are growing both in China and other countries. Thus, the methodology used in the study can be followed for measuring innovative human capital in other countries and regions. This will be helpful to apprehend the role of innovative human capital in economic development and environmental sustainability and formulate appropriate policies to achieve environmental and economic objectives.

The remainder of this article is arranged as follows: Part 2 discusses theoretical background and literature review. Part 3 presents the model and variables. Part 4 presents the empirical strategy. Part 5 is about the interpretation and discussion of results, and the last part (Part 6) presents the conclusion and policy directions.

2. Literature Review and Hypothesis Development

2.1. Innovative Human Capital and CO_2 Emissions Nexus

Analyzing the impact of human capital on environmental quality has recently gained some attention from scholars. An increasing body of literature has focused on explaining the environmental impact of human capital by using the traditional education-based proxy [12,16,17]. Scholars have explored the relationship between human capital and environmental issues using both the panel and the single country-level data. From a panel perspective, Yao et al. [18,19] used the data of 20 OECD economies and found that educated people prefer clean energy consumption over dirty energy. They uncover that advanced human capital measured by tertiary education has a negative influence on CO_2 emissions i.e., an extra year of tertiary schooling is connected with a reduction in emissions between 50.1% and 65.8%.

Alvarado et al. [16] also found that economic development cannot reduce energy consumption from fossil sources, but human capital does decrease the non-renewable energy in 27 OECD countries from 1980 to 2015. Pablo-Romero and Sánchez-Braza [20] empirically studied the relationships between energy, physical, and human capital in a larger region including OECD, NAFTA, BRIC, East European, East Asian, and EU15 countries. They found that there are substitutability relationships between human capital and energy utilization. Khan [21] argued that the influence of economic development on emissions is dependent upon the human capital level, after a certain level of human capital, CO_2 emissions will be reduced, and environmental awareness and friendly technologies will be promoted. Hao et al. [22] specially investigated the effects of human capital on CO_2 emissions for G7 countries for the period 1991–2017 and discovered that human capital can reduce CO_2 emissions. Khan et al. [23] also concluded that human capital can enhance renewable energy consumption of G7 countries by analyzing the data from 1995 to 2017. In addition, Khan et al. [24] found that improvement in human capital can intensify the negative relationship between CO_2 emissions and fiscal decentralization in G7 countries. The positive role of human capital in promoting renewable energy consumption was also found in some studies for African countries [25,26].

Proceeding to the research on individual countries, Mahmood et al. [27] used the EKC model to empirically explore the effects of economic growth and renewable energy on CO_2 emissions by adding human capital for Pakistan and discovered that the human capital mitigates CO_2 emissions. Bano et al. [28] explored the human capital and emissions nexus in Pakistan and revealed that the human capital alleviates emissions without hindering economic growth. Ahmed et al. [4] investigated the effect of natural resources, human capital, and urbanization on environmental quality in China and found that human capital mitigates environmental deterioration and has a moderating effect in promoting sustainable urbanization. Xin and Lyu [29] proved that the EKC model can explain the nexus between technological innovation and pollution in China's major cities, and human capital strengthens the effect of technological innovation on pollution. Wu and Tian [30] also confirmed the EKC hypothesis at the provincial level in China and pointed out that intermediate-level education is positively correlated with environmental pollution, but higher-level education has the opposite impact. However, the finding of Huang et al. [31] is partly different from Wu and Tian [30]. They separated human capital into four categories, including primary, knowledgeable, skilled, and institutional, and discovered that these four types of human capital have negative impacts on carbon emissions intensity among the eastern, central, and western regions in China [29]. Table 1 presents the summary of literature on human capital and environmental degradation.

Table 1. Summary of literature on human capital and environmental quality nexus.

Reference	Country	Period	Human Capital Proxy	Results
Danish et al. [17]	Pakistan	1971–2014	Human capital index (HCI)	HC mitigates ecological footprint
Yao et al. [19]	OEDC countries	1870–2014	Total number of average schooling years	HC has a favorable impact on the environment
Khan et al. [21]	122 countries	1980–2014	Primary, secondary, tertiary, and average years of schooling	HC improves the quality of the environment
Hao et al. [22]	G-7	1991–2017	Human capital index (HCI)	HC improves environmental quality
Mahmood et al. [27]	Pakistan	1980–2014	Human capital index (HCI)	HC mitigates emissions
Wang and Wu [7]	China and India	2013–2018	Stock of technological innovative professionals (10 thousand)	Improves air quality (reduces PM2.5)
Khan et al. [24]	OECD countries	1990–2018	Returns to education by incorporating information from the labor market	HC decreases emissions
Bano et al. [28]	Pakistan	1971–2014	Number of students enrolled in secondary school for general education per capita, number of students enrolled in secondary school for vocational education per capita, total number of students enrolled in secondary school as a percentage of gross enrollment, and human capital index (HCI)	HC reduces emissions
Huang et al. [31]	China	1998–2017	Heterogeneous human capital based on skilled and interprovincial migration rate.	HC decreases carbon intensity
Zafar et al. [32]	United States	1970–2015	Human capital index (HCI)	HC decreases pollution
Ahmed and Wang Ahmed et al. [33]	India	1971–2014	Human capital index (HCI)	HC improves environmental quality
Ahmed et al. [34]	G-7	1971–2014	Human capital index (HCI)	HC improves environmental quality
Ahmed et al. [35]	15 LCA countries	1995–2017	Human capital index (HCI)	HC increases emissions

Note. HC—human capital; LCA—Latin American and Caribbean countries; HCI refers to human capital index based on the average years of schooling and projected rate of return.

Synthesizing the above shreds of evidence, we hypothesize that innovative human capital can promote environmental sustainability through relevant research and development activities. Hence, we formulate the following hypothesis:

H1. *Innovative human capital poses a positive impact on environmental quality.*

2.2. Economic Growth and CO$_2$ Emissions Nexus

The relationship between economic growth and environmental degradation has been a hot issue over the last two decades. In 1991, Grossman and Krueger [36] first used the concept of the EKC to study the impact of economic growth on environmental quality and found an inverted U-shaped relationship between per capita GDP and pollution. Since then, a bulk of studies devoted to scrutinizing the association between GDP and environmental quality based on EKC, but conclusions are different. For instance, Narayan and Narayan [37] investigated the impact of income on environmental quality in 43 countries. Their results confirmed the existence of the EKC as environmental degradation decreases with the increase in income over time. Similar findings are also reported by numerous scholars, for instance, Pata [38] for Turkey, Song et al. [39] for the United States and China, Danish et al. [40] for BRICS countries, Narayan et al. [41] for the panel of 181 countries. Yang et al. [42] also revealed the presence of EKC in OECD countries, suggesting that economic growth causes environmental degradation at an early level of development but improves environmental quality after reaching a certain level.

On the contrary, some researchers believed that the EKC is invalid. For instance, Akbostanci et al. [43] tested the EKC hypothesis using the data of Turkey over 1992–2001. Their results do not support the EKC for both panel and times series data. Likewise, using the spatial econometric approach, Wang and Ye [44] investigated the impact of income on environmental quality in China. Their results unveiled that income monotonously increases CO_2 emissions, indicating that environmental degradation will not decline with the increase in income. In addition, some studies did not fully unfold the EKC for their entire sample. Al-Mulali et al. [45] found that EKC is valid for upper-middle and high-income countries but not in low and lower-middle income countries. Similarly, Guangyue and Deyong [46] tested the EKC for China's regional carbon emissions. Their results indicated that EKC is valid for the central and eastern regions but not for the western region. Therefore, we formulate the following hypothesis:

H2. *There is an inverted U-shaped relationship between economic growth and CO_2 emissions.*

This detailed literature review represents that several studies evaluated the human capital and CO_2 emissions nexus, but all these studies used the human capital based on education. We have already pointed out the limitations of such a measurement in Section 1. Hence, this study uses a better measurement of human capital and examines its effects on CO_2 emissions using the EKC framework.

3. Theoretical Background and Model Construction

3.1. Theoretical Background

The concept of innovative human capital originated from some scholars who thought about the heterogeneity of human capital. They proposed that human capital can be divided into two categories in the form of marginal return: if the kind of human capital increases marginal return during some special period, it will be known as idiosyncratic human capital; conversely, it will be known as coessential human capital [47–49]. Based on this opinion, Yao [50] first raised the concept of innovative human capital that he thought can move the production boundary outward. Why does innovative human capital own such an ability to increase production efficiency? The reason is that innovative human capital can innovate knowledge, technology, and information [51]. Subsequently, some scholars took further research on innovative human capital and classified it into many different kinds [52,53]. From the theory of marginal return, the connotation of innovative human capital and idiosyncratic human capital are the same.

Innovative human capital can make an impact on the economy from micro, industry, and also on macro levels. At the micro level, the enterprise system can affect employees' innovative behaviors from the three aspects, including incentive mechanism, transaction costs, and uncertainty of innovation [54]. More importantly, innovative human capital is the critical determinant of firm performance differential [55]. Enterprises' competitive advantage is significantly and positively influenced by innovative human capital [56]. Additionally, from the viewpoint of innovation propensity, for small firms, innovative human capital can be more desirable and valuable [57]. University graduates are usually regarded as innovative human capital by researchers around the world. Xia Pan et al. [58] used the number of university graduates to measure higher education and found the total number of university graduates reduce innovation at the provincial level, but the graduates from the reputed higher education institutions increase firm-level innovation at the provincial level in China. Peng [59] undertook deeper empirical research in which he discovered that graduates with master's degrees can bring radical innovation, and graduates with bachelor's degrees can push progressive innovation for firms.

At the industry level, Xia Pan et al. [58] further pointed out that elite university graduates are positively connected with the innovation of privately owned enterprises and insignificantly connected with the innovation of the state- and foreign-owned enterprises. However, in high-tech industries, this positive connection was more prominent. Zhu and Li [60] concretely conducted an empirical study of the technological innovative human capital's impact on the manufacturing industry competitiveness based on the data of Guangdong province in China and found that the technological innovative human capital plays a key role in the innovation of the industry. Besides, some scholars researched the impact of innovative human capital on different Chinese industries like the pharmaceutical manufacturing, e-commerce, and high-tech industries, and they all found that the innovative human capital has a positive impact on these industries to gain competitive advantage [61–63].

At the macro level, there is a new trend that innovative human capital (IHC) measured by the number of university graduates has gradually increased the contribution, and the other kinds of human capital have gradually decreased the contribution to Chinese economic growth [9]. The effect of IHC on China's economic growth is several times more than that of basic human capital [64]. Innovative human capital plays a more and more decisive role in production [65]. Although the Chinese economy has been mainly facilitated by investment, economic growth is increasingly dependent on innovative human capital [66]. Innovative human capital mainly promotes technological progress through technological innovation and is essential to realize sustained and rapid economic development [67]. However, from the perspective of regional comparison, China's innovative human capital has obvious regional differences [10]. With the continuous deterioration of the environment, it is obviously not enough to focus only on the impact of IHC on economic growth, but it is also essential to explore the role of innovative human capital in environmental protection.

Theoretically, economic growth can affect carbon emissions through three main channels, i.e., scale, composition, and technique [68]. The scale effect implies that more production requires extra material during the early stage of development, which generates more waste and pollution. The composition effect indicates that the level of pollution and raw material pattern used in the manufacturing process depends on the country's economic sectors. For instance, there will be less pollution in the countries possessing a larger services sector. Therefore, structural transformation along with economic development affects environmental quality. The third channel is the technique effect which implies that more advanced and green technologies create less material-intensive goods and less pollution at a high level of economic growth [69].

3.2. Model Construction

The EKC framework proposed by Grossman and Krueger [68,70] has been widely used by scholars in China and other countries and has been verified to varying degrees. Hence, following previous studies [33,71], this paper decides to build the following econometric model based on the EKC framework.

$$y = \alpha_0 + \beta_1 x + \beta_2 x^2 + \mu \tag{1}$$

In Equation (1), y measures environmental impact, x refers to GDP per capita; α_0 denotes the constant, β_1 and β_2 are the parameters to be estimated relating to y; finally, μ denotes the interference term. As the study uses China's provincial-level panel data and meanwhile considers the factor of innovative human capital influencing CO_2 emissions, the Equation (1) can be transformed into the following form:

$$CO_{2_{it}} = \alpha_0 + \alpha_1 GDP_{it} + \alpha_2 GDP_{it}^2 + \beta_1 IHC + \mu_{it} \tag{2}$$

In Equation (2), subscripts i and t, respectively, denote province and year. GDP denotes GDP per capita. IHC means innovative human capital and β_1 is its parameter.

For alleviating omitted variable bias, we sequentially added several control variables, which are possibly connected with variation in CO_2 emissions. Therefore, Equation (2) can be represented as follows:

$$CO_{2_{it}} = \alpha_0 + \alpha_1 GDP_{it} + \alpha_2 GDP_{it}^2 + \beta_1 IHC_{it} + \sum_{j=1}^{n} \gamma_j M_{jit} + \mu_{it} \tag{3}$$

In Equation (3), subscripts i and t, respectively, denote province and year. M refers to the control variables. In the selection of control variables, after referring to the existing literature, we decided to control four variables, including population density, economic structure, energy intensity, and FDI. Therefore, Equation (3) can be transformed as follows:

$$CO_{2_{it}} = \sum_{j=1}^{p} \alpha_j CO2_{i,t-j} + \beta_1 GDP_{it} + \beta_2 GDP_{it}^2 + \beta_3 IHC_{it} + \beta_4 POP_{it} + \beta_5 STRU_{it} + \beta_6 ENER_{it}$$
$$+ \beta_7 FDI_{it} + v_i + \varepsilon_{it} \tag{4}$$

3.3. Variables and Data Sources

In the above equation, CO_2 is the dependent variable. The independent variables are GDP and IHC. Besides, there are four control variables, including POP, STRU, ENER, and FDI. Table 2 provides more detail on these variables.

Table 2. Variable's measurement, symbol, and data source.

Variables	Symbol	Measurement	Data Source
CO_2 emissions	CO_2	Total annual emissions based on different fossil fuel usage (measured in terms of metric tons)	CESY
Economic growth	GDP	GDP per capita (constant 2003 prices in RMB)	CPSY
Innovative human capital	IHC	The number of patents every one million R&D staff full-time equivalent.	CPSY and CSTY
Population density	POP	Population per square km of land	CPSY
Economic structure	STRU	The proportion of annual industrial added value in GDP	CPSY and CSTY
Energy intensity	ENER	Energy consumption (ton of standard coal equivalent) divided by GDP	CESY
Foreign direct investment	FDI	FDI's inflows percentage of GDP	CPSY

Note: CESY—China energy statistical yearbook (https://data.cnki.net/Yearbook/ accessed on 16 June 2021); CPSY—Chinese provincial statistical yearbook (http://www.stats.gov.cn/tjsj/ndsj/ accessed on 16 June 2021); CSTY—China science and technology statistical yearbook (http://www.stats.gov.cn/ztjc/ztsj/kjndsj/ accessed on 16 June 2021); CESY—China environmental statistical yearbook (https://data.cnki.net/yearbook/Single/ accessed on 16 June 2021).

In Table 1, the data on three independent variables including real GDP per capita, population density, and energy intensity are directly extracted from both Chinese provincial statistical yearbooks and China environmental statistical Yearbooks. The data on the other three independent variables (IHC, economic structure, and FDI) is calculated based on the relative data originating from both Chinese provincial statistical yearbooks and China science and technology statistical yearbooks. The paper's study period is from 2003 to 2017, so we use the price of 2003 to denote real GDP per capita. The starting period of the research (2003) is selected based on CO_2 emissions data (some fossil fuel emissions data is unavailable before 2003) and 2017 is linked with the data availability of innovative human capital. For calculating the quantity of CO_2 emissions, the paper refers to method 1 of IPCC [72] and takes the following formula:

$$CE_{it} = \sum_{n=1}^{n} EF_n Activity_{itn} \tag{5}$$

In Equation (5), CE_{it} indicates the CO_2 emissions of the i-th province in the T-year. EF_n denotes the emission factor of the n-th source. $Activity_{itn}$ means the consumption of the n-th source of the i-th province in the T-year. For all provinces in China, the main fossil energy consumption is oil consumption (including gasoline, kerosene, diesel, and fuel oil), coal consumption (raw coal and coke), natural gas consumption, and cement production. Therefore, the paper will calculate the total annual CO_2 emissions of the above eight emission sources for 30 provinces (except Tibet) during the period under analysis. The data of the total consumption of the sources can be collected from China energy statistical yearbooks. About the emission factors of different energy sources, the paper refers to "the guidelines for the compilation of provincial greenhouse gas inventories" published by the Chinese national development and reform commission in 2011. The emission factors of the eight energies are shown in Table 3.

Table 3. Carbon emission factors.

Serial Number	Emission Source Name	Emission Factor
1	Gasoline	2.9251
2	Kerosene	3.0179
3	Diesel	3.0959
4	Fuel oil	3.1705
5	Raw coal	1.9003
6	Coke	2.8604
7	Natural gas	2.1622
8	Cement production	0.538

Keeping in view the limitations of education-based proxies of human capital discussed in Section 1, this paper follows Romer [14] and measures human capital based on science and technology staff. In all kinds of Chinese yearbooks, there is no complete data of science and technology or R&D staff from 2003 to 2017, but the complete data of R&D staff full-time equivalent were provided for the same period. Therefore, the paper designs an index named "innovative efficiency" to denote the innovative human capital. The index's formula (developed by the author) is represented as follows:

$$\text{Innovative efficiency}_{it} = \frac{\text{Total number of patent}_{it}}{\text{Total number of R\&D staff's full} - \text{time equivalent}_{it}} \times 10000 \quad (6)$$

In this Equation (6), the innovative efficiency represents innovative human capital. The total number of patents$_{it}$ means the amount of the authorized patents of the i-th province in the t-th year. The total number of R&D staff full-time equivalent$_{it}$ indicates the amount of the total working hours divided by average annual working hours in full-time R&D jobs of the i-th province in the t-th year. The data for calculating innovative efficiency was extracted from both Chinese provincial statistical yearbooks and China science and technology statistical yearbooks. Thus, all variables used in this article and their data sources were discussed.

4. Econometric Strategy

This study relied on panel data estimation techniques to empirically analyze the impact of economic growth, innovative human capital, population density, economic structure, energy intensity, and foreign direct investment on carbon emissions. Ahmed et al. [15] suggested that panel data estimation models have several advantages over time series data, such as it provides robust results and counters the issue of heterogeneity, multicollinearity, and endogeneity. Thus, following the studies of Al-Mulali et al. [45] and Dogan et al. [73], we employed panel ordinary least squares estimator (POLS) and fixed-effect method for the baseline model. Fixed effect regression accounts for unobserved time-invariant among individual characteristics, and that may lead to biased results.

Further, this study employs the system generalized method of moment (SYS-GMM) developed by Blundell and Bond [74]. The panel ordinary least square and fixed effect model are criticized for the inefficient results due to unobserved correlation with the lags of regressors [75]. To overcome this issue, Arellano and Bond [76] proposed a GMM method to estimate a dynamic panel model that eliminates the countries' specific heterogeneity by using the first difference of dependent variable. However, in the simulation studies, Blundell and Bond [74] demonstrated that the first difference GMM has poor precision; when the autoregressive parameter is relatively large, and the time-series observations are small, it may lead to large finite sample bias. To counter this issue, Blundell and Bond [74] developed the system GMM, which relies on the lagged difference of a response variable as an instrument for the equations at level and lagged as instruments for the dependent variable equation at the first difference. Following studies of Muhammad [77] and Ibrahim and Ajide [78], a standard SYS-GMM estimator can be stated as follows:

$$
\begin{aligned}
CO2_{it} = {} & \alpha + \beta_1(CO2_{it-1} - CO2_{it-2\varphi}) + \beta_2(GDP_{it} - GDP_{it-2\varphi}) + \beta_3(GDP_{it}^2 - GDP_{it-2\varphi}^2) \\
& + \beta_4(IHC_{it} - IHC_{it-2\varphi}) + \beta_5(POP_{it} - POP_{it-2\varphi}) + \beta_6(STRU_{it} - STRU_{it-2\varphi}) \\
& + \beta_7(ENER_{it} - ENER_{it-2\varphi}) + \beta_8(FDI_{it} - FDI_{it-2\varphi}) + \eta_t + \mu_i + \varepsilon_{it}
\end{aligned} \tag{7}
$$

where i exhibits 30 cross-sectional provinces of China, t is the time from 2003 to 2017. φ represents auto-regression coefficient which is based on a year lag assumed sufficient to control for past information, η_t and μ_i are time-specific and country-specific effects, respectively, while ε denotes error term.

5. Results and Discussion

Descriptive statistics given in Table 4 show those carbon emissions in Chinses provinces are less volatile than economic growth. Economic growth increased from 3603 (per capita constant 2003) to 99783 during 2003–2017. The innovative human capital has a minimum value of 264.25 and a maximum value of 6776.56, indicating a significant upward trend. On average, the population density is 436.75, with a maximum value of 3826.0. Additionally, structural changes in Chinese provinces are varying between 11.84 and 59.240. Moreover, the average foreign direct investment is 3%, and it shows a deviation value of 2.338.

Table 4. Descriptive statistics.

Variable	Obs	Mean	Std. Dev.	Min	Max
CO_2	450	261.529	181.965	15.600	842.200
GDP	450	25884.9	18029.369	3603	99783.0
IHC	450	2143.165	1402.827	264.25	6776.56
POP	450	436.756	633.98	7.000	3826.0
STRU	450	39.195	8.403	11.84	59.240
ENER	450	1.172	0.740	0.250	6.740
FDI	450	3.022	2.338	0.040	15.330

Table 5 shows the outcome of the pairwise correlation matrix. It reveals a positive correlation between GDP and CO_2. Innovative human capital and population density also show a positive correlation with emissions. The correlation between energy intensity and carbon emissions is negative. On the contrary, structural change and foreign direct investment also depict a positive correlation towards carbon emissions.

Table 5. Pairwise correlations.

	CO_2	GDP	IHC	POP	STRU	ENER	FDI
CO_2	1.000						
GDP	0.389 *	1.000					
GDP^2	0.380 *	0.999 *	1.000				
IHC	0.369 *	0.651 *	0.645 *	1.000			
POP	0.285 *	0.435 *	0.445 *	0.320 *	1.000		
STRU	0.546 *	0.015	0.006	−0.089	0.009	1.000	
ENER	−0.261 *	−0.740 *	−0.739 *	−0.694 *	−0.573 *	0.162 *	1.000
FDI	0.018	0.198 *	0.204 *	0.072	0.569 *	0.037	−0.254 *

* $p < 0.1$.

Table 6 represents the outcome of panel OLS and fixed effects results. The findings specify that economic growth poses a positive effect on CO_2 emissions in China. The coefficient is significant in both models, indicating that a 1% increase in GDP will increase CO_2 emissions by 2.479% and 2.376%. Meanwhile, the coefficient of GDP square is negative, which implies that the EKC hypothesis exists in China's provinces. These findings corroborate with the results of Ahmad et al. [79] for emerging countries, Pata [38] for Turkey, and Ahmed et al. [80] for Japan. Thus, our result supports hypothesis 2 (H2).

Table 6. Regression results.

	OLS (1)		Fixed Effect (2)	
Variables	**Coefficient**	**Std. Error**	**Coefficient**	**Std. Error**
GDP	2.479 **	1.129	2.376 ***	0.359
GDP^2	−0.115 **	0.057	−0.062 ***	0.019
IHC	−0.358 ***	0.055	−0.052 ***	0.018
POP	0.187 ***	0.047	0.257 **	0.110
STRU	0.049 ***	0.003	0.193 ***	0.053
ENER	0.200 **	0.094	0.710 ***	0.065
FDI	−0.178 ***	0.042	−0.034 **	0.014
Constant	−13.448 **	5.603	−12.414 ***	1.73
Observation	450	-	450	-
R^2	0.525	-	0.788	-
Hausman test (Prob)	-	-	22.40 (0.002)	-
Province FE	-	-	Yes	-
Year FE	-	-	Yes	-

*** $p < 0.01$, ** $p < 0.05$.

Interestingly, the coefficient of innovative human capital is significant and negative at a 1% level, signifying a mitigating effect of IHC on environmental degradation. The result portrays that innovative human capital in China is actively working for the betterment of environmental quality. Thus, our results support hypothesis 1 (H1). This finding contradicts the conclusion of Ahmed et al. [15] who disclosed that education-based human capital negatively affects the environmental quality in Latin American nations because educational attainment in these countries leads to more economic activities and more energy consumption. This also opposes the outcomes of Balaguer and Cantavella [81] who concluded that human capital in Australia increased CO_2 emissions for the majority of the years from 1950 to 2014. This result also contradicts the claim of Hassan et al. [71] that human capital in Pakistan does not influence environmental quality. Conversely, our finding is in line with some of the studies that illustrated a positive association between environmental quality and education-based human capital, for instance, Ahmed and Wang [33] for India, Ahmed et al. [34] for G7 nations, and Zafar et al. [32] for the United States.

Our outcome is unique and different from previous studies since instead of using the traditional education-based human capital which decreases or increases environmental degradation in previous studies, we employed innovative human capital based on R&D staff which is more relevant because it is directly linked with innovation. Thus, innovative human capital can improve environmental quality in different ways, such as increasing environmental-related technological innovation and sustainable usage of natural resources [69]. Therefore, innovative human capital can be used as a valuable tool to cope with environmental challenges. Previous studies on human capital and environment nexus illustrate different findings possibly because human capital in those studies is based on education and covers the quantity dimensions rather than quality dimensions. Education and associated awareness may pose a very small influence on environmental quality compared to the innovation and technological advancement that are traits of our innovative human capital. Indeed, innovation and technological advancement can influence technological efficiency and environmental quality [1,82]. Thus, this finding highlights the need to use an appropriate measure of innovative human capital to estimate its role in economic and environmental quality for achieving economic and environment-related goals.

The results further indicate that the population density exerts a positive effect on carbon emissions. China ranks as one of the most populous countries globally, with a population estimated at 1.4 billion as of 2017. The population density was 138.5 per square kilometer in 2003 and 472 in 2017, which shows a 244% increase. Thus, greater population density stimulates human activities, which exert significant pressure on natural resources and the environment. Our findings are consistent with the results of Rahman and Alam [83] for Bangladesh and Ohlan [84] for India but against the results of Meng and Han [85].

The coefficients of structural change and energy intensity are 0.193% and 0.710%. These are statistically significant at a 1% level, indicating that if structural change and energy intensity increase (decrease) by 1%, then carbon emissions will increase (decrease) by 0.193% and 0.710%. The positive effect of structural change on carbon emission shows that structural change in China is not environmentally friendly. More precisely, the structural change deteriorates environmental quality. Energy intensity also damages environmental quality in China. The coefficient of FDI is negative and statistically significant. The negative effect of foreign direct investment on carbon emissions opposes the notion of the pollution haven hypothesis. These results support the view that foreign direct inflows promote environmental sustainability in China. It implies that China has strictly regulated its FDI and foreign investors do not transfer dirty technology to China.

Our findings of structural change are in line with the work of Ahmad et al. [79] but contrary to the results of Marsiglio et al. [86], and Ali et al. [87], who found that structural change improves environmental quality. The difference in results is because the structural change in China has increased industrialization that mainly relies on pollutant fossil fuels that degrade environmental quality. The outcome of energy intensity supports the previous work of Amin and Dogan [86] and Zhang and Zhou [87]. The result of FDI matches the outcome of Saud et al. [88] in the context of belt and road nations but contradicts the outcome of Shahbaz et al. [89], who found the deteriorating effect of FDI in France.

The outcomes of system GMM shown in Table 7 indicate that the results are consistent with the OLS and fixed effect estimates. The results depict that the coefficient for GDP is positive while the square of GDP is negative, which validates the EKC hypothesis in China. The innovative human capital has a negative and significant value, and the effects of structural change, population density, and energy intensity on carbon emissions are significant and positive. By employing the system GMM method the long-run coefficients of GDP, GDP square, innovation human capital, population density, structural change, energy intensity, and foreign direct investment are 0.543, −0.027, −0.027, 0.028, 0.174, 0.044, and −0.018, respectively.

Table 7. System GMM results.

Variables	Coefficient	Stand Error	T-Value	*p*-Value
$L.CO_2$	0.896 ***	0.049	18.38	0.000
GDP	0.543 ***	0.131	4.14	0.000
GDP^2	−0.027 ***	0.007	−4.12	0.000
IHC	−0.027 **	0.010	−2.62	0.014
POP	0.028 **	0.012	2.29	0.029
STRU	0.174 *	0.086	2.01	0.054
ENER	0.044 ***	0.015	2.87	0.008
FDI	−0.018 ***	0.005	−3.39	0.002
Constant	−3.307	7.590	−0.44	0.666
Sargan	121.76	-	-	-
P(Sargan)	0.100	-	-	-
AR(1)	0.534	-	-	-
AR(2)	0.013	-	-	-
Year fixed effect	Yes	-	-	-
ID fixed effect	Yes	-	-	-
Observations	415	-	-	-

Sargan test specifies the over-identification test for the restriction in system GMM estimation. The AR (1) and AR (2) Arellano–Bond test represents the first and second-order autocorrelation in the first difference. *** $p < 0.01$, ** $p < 0.05$, * $p < 0.1$.

6. Conclusions and Policy Suggestions

This paper inspected the role of innovative human capital (IHC) in CO_2 emissions using a provincial dataset from 30 Chinese provinces from 2003 to 2017 and measuring innovative human capital based on the number of patents every one million R&D staff full-time equivalent. The results of fixed effect, OLS, and SYS-GMM methods revealed that innovative human capital decreases emissions in China and helps to form the EKC between CO_2 emissions and GDP. In addition to this, economic structure, FDI, and energy intensity increase emissions, while population density lessens emission levels. The finding of the study reveals that China's economic development will not be detrimental to environmental quality if IHC could be continuously developed.

Based on the above empirical results and conclusions, the paper proposes the following policies to realize carbon neutrality and to finally embark on the road of green development:

China needs to intensify efforts to foster innovative human capital. The paper's empirical result shows that innovative human capital can bring a positive influence on reducing CO_2 emissions. However, compared with the United States, Britain, Germany, and other Western developed countries, the quantity or quality of China's innovative human capital is still at a comparatively low level. Therefore, from the national level, the Chinese central government needs to do some strategic planning to nurture innovative human capital. In this regard, launching different training programs for human capital, and focusing on increasing R&D staff by offering some subsidies and benefits on R&D in different sectors of the economy can help to foster innovative human capital. Moreover, policies should be designed to develop collaboration between universities and industries and research funding for universities should be increased to stimulate innovation. The collaboration of industries with the universities will enable them to reap the benefits of innovation and develop advanced technologies, which in turn will increase energy efficiency and reduce emissions.

The independent variable STRU denotes the proportion of annual industrial added value in GDP. The coefficient of STRU is positive indicating that it adds to emissions. Hence, the Chinese government should optimize the economic structure of China. There are two paths to optimize it. One is limiting or stopping the development of the heavy-polluting industry and promoting the application of environmentally friendly production technologies among all industries. The other is continually increasing the proportion of the modern service industry in the economic structure. The independent variable energy intensity increases emissions. The main problem is that the consumption percentage of fossil energy is still relatively high in China. The Chinese government must take strong measures to decrease the consumption of fossil energy, at the same time, efforts are needed to increase the use of renewable energy. Population density also drives emissions. China is the country that has the largest population in the world, and its average population density is also comparably high. Fortunately, in recent years, China's population growth has been strictly controlled, and the growth rate has reduced. The Chinese government needs to initiate centralized city development to reduce the adverse effects of population density. Additionally, the focus should be on continuously improving public transportation and discouraging private vehicle ownership through different policies and taxes.

The coefficient of FDI in this study is negative. China is the largest developing country in the world. FDI is very important to China, especially when China has just implemented the policy of opening up. Through FDI, the Chinese government can make use of foreign investment to increase the research and development of green production technology, and this can also help to foster innovative human capital. Although China is now the second-largest economy in the world, its GDP per capita ranking is still relatively low. FDI is still very useful for China's social and economic development, so China should keep collaborating well with developed countries and fully use its FDI to boost economic progress and environmental sustainability. In this regard, policies should be launched to ease foreign investment, and unnecessary barriers should be eliminated. The Chinese government should devise policies to direct FDI flow to the areas with relatively low innovative human capital. This will also promote the research and development of environmentally friendly production technology. Thus, China can embark on the road of green development and realize the strategic goals of reducing CO_2 emissions as soon as possible.

This research explores a new topic and therefore, an extensive research gap exists that future studies can address. The innovative human capital based on R&D staff is directly linked with innovation and technological advancement. Thus, this measure is more appropriate to capture the impacts of IHC on the environment. However, this measure ignores the research output of university students (Master and Doctoral level). Therefore, future studies may use proxies including R&D staff and educational attainment for innovative human capital and compare the results of both methods. In addition, this paper used FDI inflows and ignored the potential reverse technology spillover. Future studies can expand the research on FDI in China and other countries by considering the reverse technology spillover effect. Researchers can examine the role of innovative human capital in economic growth for different nations using the new measure presented in this paper. Additionally, the impact of IHC on different environmental indicators can be investigated.

Author Contributions: Conceptualization, design of the work, writing—original manuscript, results interpretation, revision, X.L.; Supervision, comments, review, guidance, project administration, Y.Z.; Software, writing—original manuscript, empirical analysis M.A.; Writing—original manuscript, writing—reviewing and editing, Z.A.; Reviewing and editing, validation, H.R.; Revising the manuscript, editing, correcting mistakes T.S.A. All authors have read and agreed to the published version of the manuscript.

Funding: This research received no external funding.

Institutional Review Board Statement: Not applicable.

Informed Consent Statement: Not applicable.

Data Availability Statement: Data is readily available at request.

Conflicts of Interest: The authors declare no conflict of interest.

References

1. Kihombo, S.; Saud, S.; Ahmed, Z.; Chen, S. The effects of research and development and financial development on CO_2 emissions: Evidence from selected WAME economies. *Environ. Sci. Pollut. Res.* **2021**, *2021*, 1–11. [CrossRef]
2. Danish; Ulucak, R. How do environmental technologies affect green growth? Evidence from BRICS economies. *Sci. Total Environ.* **2020**, *712*, 136504. [CrossRef]
3. Linster, M.; Yang, C. China's Progress Towards Green Growth: An International Perspective. *OECD Green Growth Pap.* **2018**, *5*, 1–41. [CrossRef]
4. Ahmed, Z.; Asghar, M.M.; Malik, M.N.; Nawaz, K. Moving towards a sustainable environment: The dynamic linkage between natural resources, human capital, urbanization, economic growth, and ecological footprint in China. *Resour. Policy* **2020**, *67*, 101677. [CrossRef]
5. Song, M.; Zhu, S.; Wang, J.; Zhao, J. Share green growth: Regional evaluation of green output performance in China. *Int. J. Prod. Econ.* **2020**, *219*, 152–163. [CrossRef]
6. Becker, G.S. *Human Capital*; The University of Chicago Press: Chicago, IL, USA; London, UK, 1993. Available online: https://press.uchicago.edu/ucp/books/book/chicago/H/bo3684031.html (accessed on 19 May 2021).
7. Wang, F.; Wu, M. Does air pollution affect the accumulation of technological innovative human capital? Empirical evidence from China and India. *J. Clean. Prod.* **2020**, *285*, 124818. [CrossRef]
8. Islam, R.; Ang, J.; Madsen, J. Quality-adjusted human capital and productivity growth. *Econ. Inq.* **2014**, *52*, 757–777. [CrossRef]
9. Hu, Y.; Liu, Z. Effects of Different Types of Human Capitals on Economic Growth. *Popul. Econ.* **2004**, *2*, 55–58.
10. Huang, J.; Xie, L.; Zhong, M. An empirical study on the relationship between innovative human capital and economic growth in China. *Sci. Technol. Prog. Policy* **2009**, *26*, 1–4.
11. Bai, Y. Research on technical efficiency of innovative human capital in China. *Commer. Res.* **2016**, *2016*, 156–163. [CrossRef]
12. Liu, Z.; Li, H.; Hu, Y.; Li, C. Human Capital Structure Upgrading and Economic Growth: A Reconsideration of Disparities among China's Eastern, Central and Western Regions. *Econ. Res. J.* **2018**, *53*, 50–63.
13. Zhao, L.; Jia, B.; Hu, M. The Impact of Human Capital on Carbon Emission Density in Chinese Provincial Based on the Spatial Econometric Analysis. *Popul. Dev.* **2014**, *20*, 2–10. [CrossRef]
14. Romer, P.M. Increasing Returns and Long-Run Growth. *J. Polit. Econ.* **1986**, *94*, 1002–1037. [CrossRef]
15. Ahmed, Z.; Nathaniel, S.P.; Shahbaz, M. The criticality of information and communication technology and human capital in environmental sustainability: Evidence from Latin American and Caribbean countries. *J. Clean. Prod.* **2021**, *286*, 125529. [CrossRef]
16. Alvarado, R.; Deng, Q.; Tillaguango, B.; Méndez, P.; Bravo, D.; Chamba, J.; Alvarado-Lopez, M.; Ahmad, M. Do economic development and human capital decrease non-renewable energy consumption? Evidence for OECD countries. *Energy* **2021**, *215*, 119147. [CrossRef]
17. Danish; Hassan, S.T.; Baloch, M.A.; Mahmood, N.; Zhang, J.W. Linking economic growth and ecological footprint through human capital and biocapacity. *Sustain. Cities Soc.* **2019**, *47*, 101516. [CrossRef]
18. Yao, Y.; Ivanovski, K.; Inekwe, J.; Smyth, R. Human capital and energy consumption: Evidence from OECD countries. *Energy Econ.* **2019**, *84*, 104534. [CrossRef]
19. Yao, Y.; Ivanovski, K.; Inekwe, J.; Smyth, R. Human capital and CO_2 emissions in the long run. *Energy Econ.* **2020**, *91*, 104907. [CrossRef]
20. del Pablo-Romero, M.P.; Sánchez-Braza, A. Productive energy use and economic growth: Energy, physical and human capital relationships. *Energy Econ.* **2015**, *49*, 420–429. [CrossRef]
21. Khan, M. CO_2 emissions and sustainable economic development: New evidence on the role of human capital. *Sustain. Dev.* **2020**, *28*, 1279–1288. [CrossRef]
22. Hao, L.-N.; Umar, M.; Khan, Z.; Ali, W. Green growth and low carbon emission in G7 countries: How critical the network of environmental taxes, renewable energy and human capital is? *Sci. Total Environ.* **2021**, *752*, 141853. [CrossRef] [PubMed]
23. Khan, Z.; Malik, M.Y.; Latif, K.; Jiao, Z. Heterogeneous effect of eco-innovation and human capital on renewable & non-renewable energy consumption: Disaggregate analysis for G-7 countries. *Energy* **2020**, *209*, 118405. [CrossRef]
24. Khan, Z.; Ali, S.; Dong, K.; Li, R.Y.M. How does fiscal decentralization affect CO_2 emissions? The roles of institutions and human capital. *Energy Econ.* **2021**, *94*, 105060. [CrossRef]
25. Mehrara, M.; Rezaei, S.; Razi, D.H. Determinants of Renewable Energy Consumption among ECO Countries; Based on Bayesian Model Averaging and Weighted-Average Least Square. *Int. Lett. Soc. Humanist. Sci.* **2015**, *54*, 96–109. [CrossRef]
26. Akintande, O.J.; Olubusoye, O.E.; Adenikinju, A.F.; Olanrewaju, B.T. Modeling the determinants of renewable energy consumption: Evidence from the five most populous nations in Africa. *Energy* **2020**, *206*, 117992. [CrossRef]
27. Mahmood, N.; Wang, Z.; Hassan, S.T. Renewable energy, economic growth, human capital, and CO_2 emission: An empirical analysis. *Environ. Sci. Pollut. Res.* **2019**, *26*, 20619–20630. [CrossRef] [PubMed]
28. Bano, S.; Zhao, Y.; Ahmad, A.; Wang, S.; Liu, Y. Identifying the impacts of human capital on carbon emissions in Pakistan. *J. Clean. Prod.* **2018**, *183*, 1082–1092. [CrossRef]
29. Xin, X.; Lyu, L. Spatial Differentiation and Mechanism of Technological Innovation Affecting Environmental Pollution in Major Chinese Cities. *Sci. Geogr. Sin.* **2021**, *41*, 129–139.

30. Wu, Y.; Tian, B. The extension of regional Environmental Kuznets Curve and its determinants: An empirical research based on spatial econometrics model. *Geogr. Res.* **2012**, *4*, 51–64.
31. Huang, C.; Zhang, X.; Liu, K. Effects of human capital structural evolution on carbon emissions intensity in China: A dual perspective of spatial heterogeneity and nonlinear linkages. *Renew. Sustain. Energy Rev.* **2021**, *135*, 110258. [CrossRef]
32. Zafar, M.W.; Zaidi, S.A.H.; Khan, N.R.; Mirza, F.M.; Hou, F.; Kirmani, S.A.A. The impact of natural resources, human capital, and foreign direct investment on the ecological footprint: The case of the United States. *Resour. Policy* **2019**, *63*, 101428. [CrossRef]
33. Ahmed, Z.; Wang, Z. Investigating the impact of human capital on the ecological footprint in India: An empirical analysis. *Environ. Sci. Pollut. Res.* **2019**, *26*, 26782–26796. [CrossRef]
34. Ahmed, Z.; Zafar, M.W.; Ali, S.; Danish. Linking urbanization, human capital, and the ecological footprint in G7 countries: An empirical analysis. *Sustain. Cities Soc.* **2020**, *55*, 102064. [CrossRef]
35. Grossman, G.M.; Krueger, A.B. *Environmental Impacts of a North American Free Trade Agreement*; National Bureau of Economic Research: Cambridge, MA, USA, 1991. [CrossRef]
36. Narayan, P.K.; Narayan, S. Carbon dioxide emissions and economic growth: Panel data evidence from developing countries. *Energy Policy* **2010**, *38*, 661–666. [CrossRef]
37. Pata, U.K. Renewable energy consumption, urbanization, financial development, income and CO_2 emissions in Turkey: Testing EKC hypothesis with structural breaks. *J. Clean. Prod.* **2018**, *187*, 770–779. [CrossRef]
38. Song, Y.; Zhang, M.; Zhou, M. Study on the decoupling relationship between CO_2 emissions and economic development based on two-dimensional decoupling theory: A case between China and the United States. *Ecol. Indic.* **2019**, *102*, 230–236. [CrossRef]
39. Baloch, M.A.; Mahmood, N.; Zhang, J.W. Effect of natural resources, renewable energy and economic development on CO_2 emissions in BRICS countries. *Sci. Total Environ.* **2019**, *678*, 632–638. [CrossRef]
40. Narayan, P.K.; Saboori, B.; Soleymani, A. Economic growth and carbon emissions. *Econ. Model.* **2016**, *53*, 388–397. [CrossRef]
41. Yang, X.; Li, N.; Mu, H.; Pang, J.; Zhao, H.; Ahmad, M. Study on the long-term impact of economic globalization and population aging on CO_2 emissions in OECD countries. *Sci. Total Environ.* **2021**, *787*, 147625. [CrossRef] [PubMed]
42. Akbostanci, E.; Türüt-Aşik, S.; Tunç, G.I. The relationship between income and environment in Turkey: Is there an environmental Kuznets curve? *Energy Policy* **2009**, *37*, 861–867. [CrossRef]
43. Wang, Z.; Ye, X. Re-examining environmental Kuznets curve for China's city-level carbon dioxide (CO_2) emissions. *Spat. Stat.* **2017**, *21*, 377–389. [CrossRef]
44. Al-Mulali, U.; Weng-Wai, C.; Sheau-Ting, L.; Mohammed, A.H. Investigating the environmental Kuznets curve (EKC) hypothesis by utilizing the ecological footprint as an indicator of environmental degradation. *Ecol. Indic.* **2015**, *48*, 315–323. [CrossRef]
45. Guangyue, X.; Deyong, S. An Empirical Study on the Environmental Kuznets Curve for China's Carbon Emissions: Based on Provincial Panel Data. *Chin. J. Popul. Resour. Environ.* **2011**, *9*, 66–76. [CrossRef]
46. Ding, D.; Liu, Z. From human capital to heterogeneous human capital. *Product. Res.* **1999**, *3*, 7–9.
47. Ding, D. Modern enterprise: A contract of heterogeneous human capital and homogeneous human capital. *Real Only* **2001**, 45–50. Available online: https://kns.cnki.net/kcms/detail/detail.aspx?dbcode=CJFD&dbname=CJFD2001&filename=WASI200106008&v=%25mmd2BJsVe2Vb1K3BUthF6pLipJiOwUV4c2NLTCxAq86VnLqyT4xDhgM26AketpS%25mmd2FlOwa.08-05-2021. (accessed on 19 May 2021).
48. Yang, H. Discussion about heterogeneous human capital. *Thinking* **2008**, 37–41. Available online: https://kns.cnki.net/kcms/detail/detail.aspx?dbcode=CJFD&dbname=CJFD2008&filename=SXZX200802008&v=TVQP5%25mmd2BK17x87kMNyg8hbZeVfvDtZzhQ2KxsbdPfBULED%25mmd2ByzMJwZN5nw2uzGC3%25mmd2FzU.09-05-2021. (accessed on 19 May 2021).
49. Yao, S. Discussion about innovative human capital. *Financ. Econ.* **2001**, 10–14.
50. Li, H.; Xi, Y. The Innovative Human Capital and Its Motivation. *J. Southwest Jiaotong Univ. Social Sci.* **2002**, *3*, 47–51.
51. Xi, J. Thinking of innovative human capital. *J. Cap. Norm. Univ. Sci.* **2005**, 67–70. Available online: https://kns.cnki.net/kcms/detail/detail.aspx?dbcode=CJFD&dbname=CJFD2005&filename=SDSD200505012&v=EIuoCvKtyPewTOVPLZ9TqjnBCfjIYd%25mmd2FOoon9Eqq7zRqlE05%25mmd2FyAM7dgFOpb8HZlEo.10-05-2021. (accessed on 19 May 2021).
52. Kong, X. Research on the classification of innovative human capital. *Sci. Technol. Manag. Res.* **2009**, *29*, 328–330.
53. Yao, S. Innovative human capital, institutions and performance of enterprises. *Contemp. Financ. Econ.* **2001**, *80*, 3–7.
54. Chen, L.; Yao, S. Creative Human Capital and Its Integration: The Determinant of Firm Performance Differentia. *Soc. Sci. Yunnan* **2004**, *1*, 45–49.
55. Gao, S.; Xu, L.; Ma, S.; Wang, X. Innovative Human Capital and Firm Competitive Advantages—The Moderating Effect of Members Deployment and Team Spirit in R&D Team. *Sci. Technol. Manag. Res.* **2016**, *36*, 180–186.
56. McGuirk, H.; Lenihan, H.; Hart, M. Measuring the impact of innovative human capital on small firms' propensity to innovate. *Res. Policy* **2015**, *44*, 965–976. [CrossRef]
57. Pan, X.; Gao, Y.; Guo, D.; Cheng, W. Does Higher Education Promote Firm Innovation in China? *Sustainability* **2020**, *12*, 7326. [CrossRef]
58. Peng, W. The Influence of Heterogeneity Innovation Human Capital on Enterprise Value Chain: An Empirical Study Based on the Listed Companies in China's Manufacturing Industry. *Financ. Sci.* **2019**, 120–132.

59. Zhu, Q.; Li, H. A Study of the Impact of Innovation-minded Human Resources on the Industrial Competitiveness-based on data of Guandong. *Contemp. Financ. Econ.* **2007**, 75–80. Available online: https://kns.cnki.net/kcms/detail/detail.aspx?dbcode=CJFD&dbname=CJFD2007&filename=DDCJ200707016&v=k1FfAfRnWEKn4Oy8zi9XJXPdlTP2qVXKoDW36DVAanhdyHuk60J9x5Na91t25k8D.11-05-2021. (accessed on 19 May 2021).

60. Gao, S.; Xu, L.; Wang, Y.; Zhao, S. Analysis on the effect of innovative human capital on the performance improvement of pharmaceutical manufacturing industry in Hebei Province. *Mod. Manag.* **2016**, *36*, 79–81. [CrossRef]

61. Liu, L. Research on how innovative human capital promotes the development of network commercial economy. *J. Commer. Econ.* **2016**, 117–118.

62. Zhang, H. Innovative human capital and high-tech enterprise development. *PriceTheory Pract.* **2003**, 59–60.

63. Jing, Y.; Liu, X. Research on the relationship between innovative human capital and economic growth in China (1990–2010). *Seeker* **2013**, 218–221.

64. Zhang, G.; Chen, C.; Cao, Y.; Xie, L. An Empirical Study of the Impact of Creative Human Capital on Economic Growth. *Sci. Technol. Prog. Policy* **2010**, *27*, 137–141.

65. Xie, L.; Huang, J. An analysis of innovative human capital, total factor productivity and economic growth. *Sci. Technol. Prog. Policy* **2009**, *26*, 153–157.

66. Liu, Z.; Zhang, W. Theoretical and Empirical Study on the Creative Human Capital and Technological Progress. *Sci. Technol. Prog. Policy* **2010**, *27*, 138–142.

67. Ahmad, M.; Jiang, P.; Murshed, M.; Shehzad, K.; Akram, R.; Cui, L.; Khan, Z. Modelling the dynamic linkages between eco-innovation, urbanization, economic growth and ecological footprints for G7 countries: Does financial globalization matter? *Sustain. Cities Soc.* **2021**, *70*, 102881. [CrossRef]

68. Grossman, G.M.; Krueger, A.B. Economic Growth and the Environment. *Q. J. Econ.* **1995**, *110*, 353–377. [CrossRef]

69. Hassan, S.T.; Xia, E.; Khan, N.H.; Shah, S.M.A. Economic growth, natural resources, and ecological footprints: Evidence from Pakistan. *Environ. Sci. Pollut. Res.* **2019**, *26*, 2929–2938. [CrossRef]

70. *IPCC 2019 Refinement to the 2006 IPCC Guidelines for National Greenhouse Gas Inventory*; Intergovernmental Panel on Climate Change: Hayama, Japan, 2019.

71. Dogan, E.; Aslan, A. Exploring the relationship among CO_2 emissions, real GDP, energy consumption and tourism in the EU and candidate countries: Evidence from panel models robust to heterogeneity and cross-sectional dependence. *Renew. Sustain. Energy Rev.* **2017**, *77*, 239–245. [CrossRef]

72. Blundell, R.; Bond, S. Initial conditions and moment restrictions in dynamic panel data models. *J. Econom.* **1998**, *87*, 115–143. [CrossRef]

73. Judson, R.A.; Owen, A.L. Estimating dynamic panel data models: A guide for macroeconomists. *Econ. Lett.* **1999**, *65*, 9–15. [CrossRef]

74. Arellano, M.; Bond, S. Some tests of specification for panel data:monte carlo evidence and an application to employment equations. *Rev. Econ. Stud.* **1991**, *58*, 277–297. [CrossRef]

75. Muhammad, B. Energy consumption, CO_2 emissions and economic growth in developed, emerging and Middle East and North Africa countries. *Energy* **2019**, *179*, 232–245. [CrossRef]

76. Ibrahim, R.L.; Ajide, K.B. Trade Facilitation, Institutional Quality, and Sustainable Environment: Renewed Evidence from Sub-Saharan African Countries. *J. Afr. Bus.* **2020**, *2020*, 1–23. [CrossRef]

77. Ahmad, M.; Ahmed, Z.; Majeed, A.; Huang, B. An environmental impact assessment of economic complexity and energy consumption: Does institutional quality make a difference? *Environ. Impact Assess. Rev.* **2021**, *89*, 106603. [CrossRef]

78. Ahmed, Z.; Zhang, B.; Cary, M. Linking economic globalization, economic growth, financial development, and ecological footprint: Evidence from symmetric and asymmetric ARDL. *Ecol. Indic.* **2021**, *121*, 107060. [CrossRef]

79. Balaguer, J.; Cantavella, M. The role of education in the Environmental Kuznets Curve. Evidence from Australian data. *Energy Econ.* **2018**, *70*, 289–296. [CrossRef]

80. Kihombo, S.; Ahmed, Z.; Chen, S.; Adebayo, T.S.; Kirikkaleli, D. Linking financial development, economic growth, and ecological footprint: What is the role of technological innovation? *Environ. Sci. Pollut. Res.* **2021**, *2021*, 1–11. [CrossRef]

81. Rahman, M.M.; Alam, K. Clean energy, population density, urbanization and environmental pollution nexus: Evidence from Bangladesh. *Renew. Energy* **2021**, *172*, 1063–1072. [CrossRef]

82. Ohlan, R. The impact of population density, energy consumption, economic growth and trade openness on CO_2 emissions in India. *Nat. Hazards* **2015**, *79*, 1409–1428. [CrossRef]

83. Meng, X.; Han, J. Roads, economy, population density, and CO_2: A city-scaled causality analysis. *Resour. Conserv. Recycl.* **2018**, *128*, 508–515. [CrossRef]

84. Marsiglio, S.; Ansuategi, A.; Gallastegui, M.C. The Environmental Kuznets Curve and the Structural Change Hypothesis. *Environ. Resour. Econ.* **2016**, *63*, 265–288. [CrossRef]

85. Ali, W.; Abdullah, A.; Azam, M. The dynamic relationship between structural change and CO2 emissions in Malaysia: A cointegrating approach. *Environ. Sci. Pollut. Res.* **2017**, *24*, 12723–12739. [CrossRef]

86. Amin, A.; Dogan, E. The role of economic policy uncertainty in the energy-environment nexus for China: Evidence from the novel dynamic simulations method. *J. Environ. Manag.* **2021**, *292*, 112865. [CrossRef]

87. Zhang, C.; Zhou, X. Does foreign direct investment lead to lower CO$_2$ emissions? Evidence from a regional analysis in China. *Renew. Sustain. Energy Rev.* **2016**, *58*, 943–951. [CrossRef]
88. Saud, S.; Chen, S.; Danish; Haseeb, A. Impact of financial development and economic growth on environmental quality: An empirical analysis from Belt and Road Initiative (BRI) countries. *Environ. Sci. Pollut. Res.* **2019**, *26*, 2253–2269. [CrossRef] [PubMed]
89. Shahbaz, M.; Nasir, M.A.; Roubaud, D. Environmental degradation in France: The effects of FDI, financial development, and energy innovations. *Energy Econ.* **2018**, *74*, 843–857. [CrossRef]

International Journal of
*Environmental Research
and Public Health*

MDPI

Article

Crop Production and Agricultural Carbon Emissions: Relationship Diagnosis and Decomposition Analysis

Jianli Sui and Wenqiang Lv *

School of Business, Jilin University, No. 2699 Qianjin Street, Changchun 130012, China; jlsui@163.com
* Correspondence: lvwq20@mails.jlu.edu.cn

Abstract: Modern agriculture contributes significantly to greenhouse gas emissions, and agriculture has become the second biggest source of carbon emissions in China. In this context, it is necessary for China to study the nexus of agricultural economic growth and carbon emissions. Taking Jilin province as an example, this paper applied the environmental Kuznets curve (EKC) hypothesis and a decoupling analysis to examine the relationship between crop production and agricultural carbon emissions during 2000–2018, and it further provided a decomposition analysis of the changes in agricultural carbon emissions using the log mean Divisia index (LMDI) method. The results were as follows: (1) Based on the results of CO_2 EKC estimation, an N-shaped EKC was found; in particular, the upward trend in agricultural carbon emissions has not changed recently. (2) According to the results of the decoupling analysis, expansive coupling occurred for 9 years, which was followed by weak decoupling for 5 years, and strong decoupling and strong coupling occurred for 2 years each. There was no stable evolutionary path from coupling to decoupling, and this has remained true recently. (3) We used the LMDI method to decompose the driving factors of agricultural carbon emissions into four factors: the agricultural carbon emission intensity effect, structure effect, economic effect, and labor force effect. From a policymaking perspective, we integrated the results of both the EKC and the decoupling analysis and conducted a detailed decomposition analysis, focusing on several key time points. Agricultural economic growth was found to have played a significant role on many occasions in the increase in agricultural carbon emissions, while agricultural carbon emission intensity was important to the decline in agricultural carbon emissions. Specifically, the four factors' driving direction in the context of agricultural carbon emissions was not stable. We also found that the change in agricultural carbon emissions was affected more by economic policy than by environmental policy. Finally, we put forward policy suggestions for low-carbon agricultural development in Jilin province.

Keywords: crop production; agricultural carbon emissions; EKC; decoupling; LMDI

Citation: Sui, J.; Lv, W. Crop Production and Agricultural Carbon Emissions: Relationship Diagnosis and Decomposition Analysis. *Int. J. Environ. Res. Public Health* **2021**, *18*, 8219. https://doi.org/10.3390/ijerph18158219

Academic Editors: Pasquale Avino, Massimiliano Errico, Aristide Giuliano and Hamid Salehi

Received: 29 June 2021
Accepted: 30 July 2021
Published: 3 August 2021

1. Introduction

Modern agricultural activities play an important role in greenhouse gas emissions due to their high material input, energy consumption, and pollutant discharge levels [1,2]. Globally, ensuring food security and coping with climate change caused by greenhouse gas emissions are the most common challenges today. In the past two decades, greenhouse gas emissions from agricultural activities have accounted for 10–14% of total global greenhouse gas emissions (in CO_2 equivalents) [3]. According to the 2019 Intergovernmental Panel on Climate Change (IPCC) special report on climate change and land [4], greenhouse gas emissions derived from agricultural activities reached 108–191 hundred million tons of CO_2 equivalents during 2007–2016, and they accounted for 21–37% of global greenhouse gas emissions. Being an agriculture-centered country, China's sown area has decreased from 117 million hectares to 116 million hectares over the last four decades, while the grain yield has increased from 321 million tons to 664 million tons, which is an increase of 107%.

Along with the grain yield per unit area increasing, the amount of chemical fertilizer used increased from 12.69 million tons in 1980 to 52.04 million tons in 2019, and agriculture was the second biggest source of carbon emissions [5]. In this regard, it is necessary for China to study the nexus of agricultural economic growth and carbon emissions.

Considering the ever-increasing speed in crop production, the relationship between agricultural activities and environmental degradation has been intensively studied in recent years, focusing on such areas as the environmental stress on the ecosystem caused by highly specialized crop production [6] and excessive fertilizer application due to the increasing numbers of part-time and aging farmers involved in the process of rural transformation [7–10]. Meanwhile, different methods have been applied in agri-environmental impact assessment. As is well known, the environmental Kuznets curve (EKC) describes an inverted-U relationship between environmental degradation and income [11]. Up to now, a large body of theoretical and empirical literature has dealt with EKC studies [12–15]. EKC studies have tested most of the primary pollutants and relevant indicators, including SO_2, NOx, carbon emissions, ecological footprint, etc. [16–18], and both emissions per capita and total emissions [19]; furthermore, these studies have been developed from the earliest simple quadratic functions of the levels of income to multivariate and multi-mediation models, intending to indicate underlying factors, such as globalization, trade openness, human capital, etc. [20–23]. In particular, many studies have tested the existence of the EKC hypothesis by considering CO_2 emissions with time-series or panel data; some results support the existence of an inverted U-shape [24,25], some support an N-shape [19], some support an inverted N-shape [21], and some support a linear shape without a turning point [26,27].

In scientific discussions on economic growth versus environmental degradation, EKC deals with the concept of decoupling [28,29]. The Organization for Economic Co-operation and Development (OECD) has described the synchronous changes taking place within economic growth and pollution emissions as different degrees of decoupling/coupling states [30], and the idea of decoupling environmental bads from economic goods has garnered policymakers' attention, with many scholars assessing the progress in the environmental-degradation–economic-growth relationship. In addition to the decoupling index proposed by the OECD [30], Tapio's elasticity coefficient has commonly been used and extended as a decoupling index that overcomes the shortcoming of selecting a base period [31]. In terms of empirical studies dealing with decoupling analysis in China, Tian et al. [32] used Tapio's model to analyze the relationship between agricultural activities and carbon emissions at the national level, and we found that weak decoupling and strong decoupling occurred most during the years 2001–2010, while Yang et al. [33] and Chen et al. [34] conducted further in-depth case studies of China's main grain-producing areas.

In policy terms, decomposition analysis is increasingly used in environmental policymaking, while the log mean Divisia index (LMDI) method is recommended for general use in the study of CO_2 emission changes as it meets all constraints, such as complete decomposition, consistency in aggregation, and satisfying the factor reversal test [35–39]. In China, many scholars have used the LMDI method to analyze the drivers of carbon emissions related to crop production on a national or regional basis. For instance, Li et al. [40] pointed out that agricultural economic growth in China had a strong driving effect on carbon emissions, while agricultural production efficiency, agricultural structure, and labor force had certain inhibiting effects on carbon emissions during 1993–2008; Li et al. [2] quantified the contributions of factors influencing agricultural carbon emissions in China from 1991 to 2015 and found that the driving factors varied significantly by region. In fact, in terms of empirical findings, Wei et al. [41] indicated that the state of the agricultural economy was the decisive factor in the increase in crop production carbon footprint, while agricultural investment, urbanization, and technological progress were important factors for reducing the crop production carbon footprint in Guangdong province. Zhao et al. [42] confirmed that the agricultural economic level and industrial structure were the main drivers promoting the increase in agricultural carbon emissions, while agricultural production efficiency and agricultural labor force had inhibitory effects on the increase in carbon emissions in Hunan province. As for Heilongjiang province, Chen et al. [34] showed that agricultural output value was the key driving force of agricultural carbon emission increases, while agricultural production efficiency was the primary driving force for agricultural carbon emission reduction.

The Chinese government has pledged to peak carbon emissions by 2030 and achieve carbon neutrality by 2060. In principle, measures to reduce carbon emissions in manufacturing include technological innovation, environmental regulation, and the market mechanism of carbon trading; however, this is more difficult for the agricultural sector due to its multiple carbon sources, strong randomness and dispersed process, and the limited effects of its market mechanisms. Li et al. [2] indicated that northeast China was the largest contributor to agricultural carbon emissions in the country, and economic factors were the main driving forces of the increase in carbon emissions. Jilin province is a major agricultural province, whose corn yield per unit area ranked first in China for many consecutive years, and it has made great contributions to ensuring national food security. Nevertheless, the agri-environment has been deteriorating daily, and a question in the context of this high yield arises as to the relationship between crop production and agricultural carbon emissions, which seriously affects sustainable agricultural development [43,44]. This paper takes Jilin province as the study area and tests the relationship between crop production and agricultural carbon emissions using EKC and decoupling analysis; then, it decomposes the driving factors of agricultural carbon emissions during 2000–2018, and finally, it puts forward suggestions for agricultural carbon emission mitigation.

The structure of this article is organized as follows. Section 2 explains the methodology and data. Section 3 presents the empirical results, using EKC and decoupling analysis to examine the relationship between crop production and agricultural carbon emissions, and then further decomposes the influencing factors of agricultural carbon emissions. Section 4 contains the discussion and policy implications. Section 5 concludes the study.

2. Methods and Data

2.1. CO_2 EKC Model

Based on the analysis framework of EKC proposed by Friedl and Getzner [19], Zhang et al. [45], and Lv et al. [46], considering the simulation accuracy of the CO_2 EKC model, this paper first describes the general functional form as a cubic polynomial and builds a cubic curve equation on the basis of two indicators: crop production (expressed as agricultural output value) and agricultural carbon emissions. In view of the target of carbon emission reduction in agriculture, the total values of agricultural carbon emissions and agricultural output value are chosen instead of the per capita index. Then, the existence of a cubic curve-fitting equation is examined according to stationarity and cointegration tests; if the cubic term is not statistically significant, it will be eliminated. Subsequently, the form of the quadratic polynomial will be re-estimated, etc. The CO_2 EKC may be U-shaped, inverted-U-shaped, N-shaped, inverted-N-shaped, or another shape, depending on the coefficients of the explanatory variable, and the CO_2 EKC equation is quantified as follows:

$$C_t = \beta_0 + \beta_1 G_t + \beta_2 G_t^2 + \beta_3 G_t^3 + \varepsilon_t \tag{1}$$

where C_t denotes the agricultural carbon emissions in year t; G_t represents the agricultural output value in year t; β_i is the coefficient of the explanatory variable, and ε_t illustrates the random error term. According to Ekins [47] and Friedl and Getzner [19], if $\beta_1 > 0$, $\beta_2 < 0$, and $\beta_3 > 0$, an N-shaped relationship between crop production and agricultural carbon emissions can arise; if $\beta_1 > 0$, $\beta_2 < 0$, and $\beta_3 = 0$, the inverted-U EKC hypothesis stands, and so on.

2.2. Decoupling Index

In this paper, the decoupling index (*DI*) is constructed following Tapio's elasticity coefficient [31], and it is computed using the following expression:

$$DI_t = \frac{\Delta C}{\Delta G} = \frac{(C_t - C_0)/C_0}{(G_t - G_0)/G_0} \qquad (2)$$

where DI_t is defined to explore the relationship between crop production and agricultural carbon emissions at two time points, t and 0 represent the last period and the base period, respectively, and the time interval is one year in this study; C_t and C_0 indicate agricultural carbon emissions in the last period and the base period, respectively; ΔC represents the change rate in agricultural carbon emissions from the last period to the base period; G_t and G_0 indicate the agricultural output value in the last period and the base period, respectively; and ΔG represents the change rate in agricultural output value from the last period to the base period.

According to the different decoupling elasticity values, six degrees of decoupling/coupling states were divided, including recessive decoupling, weak decoupling, strong decoupling, expansive coupling, weak coupling, and strong coupling. The specific grading standards and decoupling elasticity values are shown in Table 1.

Table 1. Degrees of decoupling agricultural carbon emissions from crop production.

Decoupling Degree	Change Rate in Agricultural Carbon Emissions (ΔC)	Change Rate in Agricultural Output Value (ΔG)	DI	Connotation
Expansive coupling	>0	>0	$DI > 1$	Agricultural economic growth occurs at the cost of accelerated agricultural carbon emissions.
Strong coupling	>0	<0	$DI < 0$	The worst state when the agricultural economy is in recession, while agricultural carbon emissions increase.
Weak coupling	<0	<0	$1 > DI > 0$	Agricultural carbon emission reduction rate is slower than agricultural economic recession.
Weak decoupling	>0	>0	$1 > DI > 0$	Agricultural carbon emission growth rate is slower than agricultural economic growth rate.
Strong decoupling	<0	>0	$DI < 0$	The ideal state in which agricultural economy grows, while agricultural carbon emissions decrease.
Recessive decoupling	<0	<0	$DI > 1$	Agricultural carbon emission reduction rate is faster than agricultural economic recession.

2.3. LMDI Method

Based on the research of Li et al. [2], Wu and Zhang [48], and Li et al. [22], an extended Kaya identity for agricultural carbon emissions in Jilin province can be expressed as Equation (3):

$$C = \frac{C}{G} \times \frac{G}{TG} \times \frac{TG}{AL} \times AL = CI \times SI \times EI \times AL \qquad (3)$$

where C represents agricultural emissions; G represents the agricultural output value; TG represents the gross output value of agriculture, forestry, animal husbandry, and fishery; AL represents the agricultural labor force scale; CI is defined as agricultural carbon emission intensity and is represented as the agricultural carbon emissions per unit of agricultural output value. This last is also an agricultural production efficiency factor—a decline in CI means agricultural production efficiency increases in a specific region, which directly or indirectly reveals a change in the relationship between crop production and agricultural carbon emissions; SI refers to the agricultural structure, which reflects the share of the agricultural output value in the gross agricultural output value; and EI reflects the agricultural economic development level, considering the agricultural labor force. Table 2 shows the meanings of all the symbols in Equation (3).

Table 2. Definition of variables in the LMDI method.

Variable	Symbol	Definition	Unit
Agricultural carbon emissions	C	Carbon emissions derived from crop production	tons
Agricultural output value	G	Value added of agriculture	100 million yuan
Gross agricultural output value	TG	Total value added of agriculture, forestry, animal husbandry, and fishery	100 million yuan
Agricultural carbon emission intensity	CI	Carbon emissions per unit of agricultural output value	tons/yuan
Agricultural structure effect	SI	The share of the agricultural output value in the gross agricultural output value	%
Agricultural labor force	AL	Rural population engaged in agricultural activities	person
Agricultural economic effect	EI	Gross agricultural output value divided by agricultural labor force	yuan per capita

We used the LMDI method to decompose the drivers of agricultural carbon emissions in Jilin province into four factors: agricultural carbon emission intensity effect (ΔCI), agricultural structure effect (ΔSI), agricultural economic effect (ΔEI), and agricultural labor force effect (ΔAL). ΔCI reflects the change in agricultural carbon emissions intensity, ΔSI reflects the change in the share of the agricultural output value out of the gross agricultural output value, ΔEI reflects the change in agricultural economic growth caused by the agricultural labor force, and ΔAL reflects the change in the scale of the agricultural labor force.

Based on plus decomposition, the total effects of agricultural CO_2 emissions are expressed as Equation (4):

$$\Delta C_{tot} = C_t - C_0 = \Delta CI + \Delta SI + \Delta EI + \Delta AL. \tag{4}$$

Each effect in Equation (4) is expressed as follows:

$$\Delta CI = \sum \frac{C_t - C_0}{\ln C_t - \ln C_0} \cdot \ln \left(\frac{CI_t}{CI_0}\right) \tag{5}$$

$$\Delta SI = \sum \frac{C_t - C_0}{\ln C_t - \ln C_0} \cdot \ln \left(\frac{SI_t}{SI_0}\right) \tag{6}$$

$$\Delta EI = \sum \frac{C_t - C_0}{\ln C_t - \ln C_0} \cdot \ln \left(\frac{EI_t}{EI_0}\right) \tag{7}$$

$$\Delta AL = \sum \frac{C_t - C_0}{\ln C_t - \ln C_0} \cdot \ln\left(\frac{AL_t}{AL_0}\right). \tag{8}$$

2.4. Data Sources and Processing

In this study, we used time series data from 2000 to 2018 to model the EKC, build a decoupling index, and analyze influence factors using the LMDI method.

The data for quantifying agricultural carbon emissions and the following decomposition analysis were collected from the Jilin Province Statistical Yearbook (2001–2019) and the China Rural Statistical Yearbook (2001–2019). Both the agricultural output value and the gross agricultural output value were deflated to represent a constant 2000 price.

The formula for calculating agricultural carbon emissions is as follows:

$$C = \sum_{i=1}^{6} E_i \times f_i \tag{9}$$

where C denotes agricultural carbon emissions derived from crop production activities; i represents the type of agricultural carbon sources, including 6 types of agricultural carbon sources (chemical fertilizers, pesticides, plastic films, agricultural machinery, agricultural plowing, and agricultural irrigation); E_i represents the amount of agricultural carbon source i; f_i is the i-type carbon emission coefficient of agricultural carbon sources (derived from Li et al. [2], Tian et al. [32], and Guo et al. [39]), and C, CH_4, and NO_2 are converted into CO_2 equivalents with reference to the IPCC [49].

Table 3 reports the descriptive statistical results of the main variables in this paper.

Table 3. Summary statistics of main variables.

Variable	Observations	Mean	Standard Deviation	Minimum	Maximum
Agricultural carbon emissions (10^4 tons)	19	488.93	119.02	308.96	683.04
Agricultural output value (10^6 yuan)	19	61,499.98	18,592.87	32,027.00	92,833.79
Gross agriculture output value (10^6 yuan)	19	12,000.00	33,873.04	60,940.00	17,100.00
Agricultural labor force (10^4 persons)	19	516.13	24.21	478.24	573.90
Agricultural economic effect (yuan per capita)	19	23,213.92	6665.63	11,791.80	35,696.27
Agricultural carbon emission intensity (tons/10^4 yuan)	19	0.82	0.16	0.65	1.15
Agricultural structure (%)	19	0.51	0.02	0.48	0.54

3. Results

3.1. Estimating a CO_2 EKC

Based on time series data from 2000 to 2018, we estimated the CO_2 emissions derived from crop production in Jilin province (Figure 1). Agricultural CO_2 emissions reached 3.09 million tons in the first year (2000); after a transitory upward trend, there was a slight decline in 2003, where the annual decline rate slid to −15.3% owing to the fact that farmers' willingness to grow food hit a low. The Chinese issued a series of policies of agricultural tax reduction and exemption in 2004, and then, the farmers recovered their confidence in agriculture and increased inputs in crop production, leading to agricultural CO_2 emissions increasing with a growth rate of 36.2% and reaching a phase peak of 5.91 million tons in 2007. Subsequently, there was a clear break in the increase in agricultural CO_2 emissions due to the global financial crisis in 2008, which then plunged to a new low (annual decline rate of −31.9%), which was even lower than in 2004. Under the stimulus of beneficial farming policies in 2009, agricultural CO_2 emissions climbed steadily and reached 6.83 million tons in 2018. The Chinese government put forward a policy of building a resource-conserving and environmentally friendly society in 2012, under its influence, the growth rates of agricultural CO_2 emissions declined in a fluctuating pattern during 2012–2018, while agricultural CO_2 emissions still showed a rising trend.

Figure 2 shows a scatter plot of agricultural CO_2 emissions and agricultural output value at a constant 2000-level price in Jilin province. An N-shaped relationship between crop production and agricultural carbon emissions can be seen. There were three obvious turning points in agricultural CO_2 emission levels during 2000–2018 in Jilin province, including a slight decline in 2003 (3.13 million tons of agricultural CO_2 emissions and 425.27 hundred million yuan of value added in agriculture), one phase peak in 2007 (5.91 million tons of agricultural CO_2 emissions and 516.51 hundred million yuan of value added in agriculture), and then a low in 2008 (4.02 million tons of agricultural CO_2 emissions and 594.51 hundred million yuan of value added in agriculture). This pattern can be explained by the agricultural policies and macroeconomic factors. Agricultural CO_2 emissions bottomed out in 2003 and rose after a series of agricultural policies issued in 2004, until being pushed back down again due to the global financial crisis in 2008; they increased sharply thereafter from 2009 to 2018, which was accompanied by the improving agricultural economy. From Figure 2 alone, we cannot infer the future relationship between agricultural CO_2 emissions and agricultural economic growth in Jilin province, that is, there is no indication of when another turning point will occur in the upcoming period.

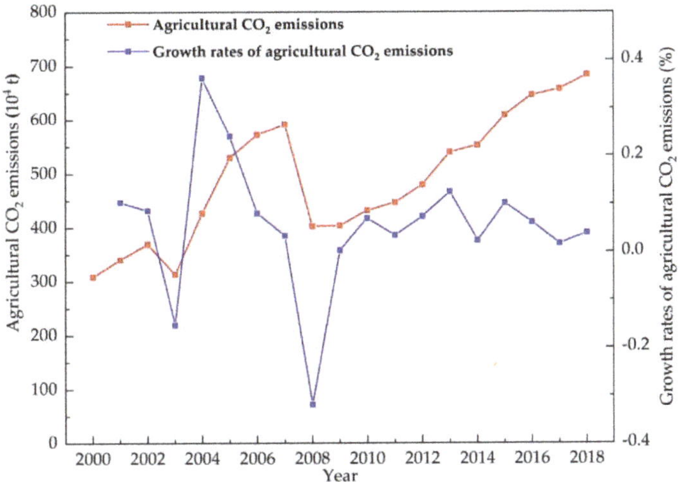

Figure 1. Agricultural CO_2 emissions and growth rates in Jilin province during 2000–2018.

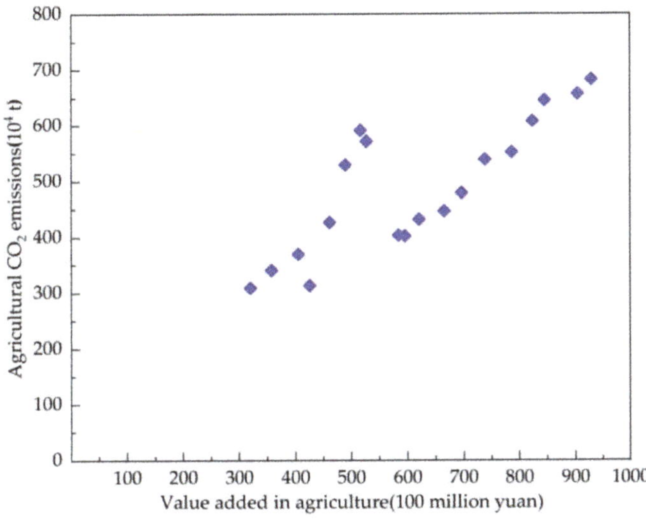

Figure 2. Scatter plot of agricultural CO_2 emissions and value added in agriculture in Jilin province during 2000–2018.

Importantly, the scatter plot presents only the correlation between crop production and agricultural CO_2 emissions during 2000–2018—a causal relationship cannot be identified without undertaking a statistical test, so an additional theoretical analysis is required.

The next analysis framework in this study, following Friedl and Getzner [19] and Zhang et al. [45], was as follows: (1) testing for stationarity of both dependent and independent variables in the time series; (2) examining whether both variables (agricultural CO_2 emissions and agricultural output value) are cointegrated; (3) further testing the most suitable functional form for depicting the development of agricultural CO_2 emissions in Jilin province.

Firstly, we conducted a unit root test. In order to overcome the defects of the small samples and prevent sequence spurious regression, we applied the ADF test to examine the stationarity of the dependent variable (CO_2 emissions, expressed as C) and the independent variable (agricultural output value, expressed as G) for 2000–2018. Table 4 shows the results of the unit root test, showing that both variables chosen in this paper have a significance level of 5% and are stationary series, which can be further tested to determine the long-term equilibrium relationship.

Table 4. Unit root test results.

Variable	Test Type	ADF Test	Critical Values at Significance Level			*p*-Value	Test Results
	(c, t, q)	Statistics	1%	5%	10%		
lnC	(c,t,3)	−3.833	−4.728	−3.760	−3.325	0.044	Stationary
lnG	(c,t,0)	−3.879	−4.572	−3.691	−3.287	0.036	Stationary

Note: In the test type (c, t, q), c, t, and q represent the constant, time trend, and lag order, respectively. The lag order is obtained based on the SIC criterion.

Secondly, as a preliminary step for testing the EKC hypothesis, we conducted a Granger causality test to examine the correlation between the selected variables. Table 5 shows the results of the Granger causality test; the agricultural output value is the Granger cause of agricultural CO_2 emissions at a significance level of 1% (not vice versa).

Table 5. Granger causality test results.

Null Hypothesis	F Statistics	*p*-Value	Test Results
lnG is not the Granger reason for lnC	14.655	0.003	Reject
lnC is not the Granger reason for lnG	0.835	0.550	Do not reject

Finally, we constructed an empirical EKC model for Jilin's agricultural CO_2 emissions. Based on the above test results, the regression model of agricultural CO_2 emissions and agricultural output value adopted the cubic functional form shown in Formula (10) and Table 6.

$$lnC = -501.44 + 245.34lnG - 38.87(lnG)^2 + 2.05(lnG)^3 \tag{10}$$

Table 6. Estimation of agricultural CO_2 EKC for Jilin province.

Explanatory Variables	Coefficient of Explanatory Variables
Constant	−501.44 *
lnG	245.34 *
$(lnG)^2$	−38.87 *
$(lnG)^3$	2.05 *
Adjusted R^2	0.71
F statistic	60.99

Note: * is at the significance level of 10%.

According to the analysis framework of CO2 EKC (in Section 2.1), along with the research of Ekins [47] and Friedl and Getzner [19] and the statistical quality of the estimated CO2 EKC (Table 6), we found that the coefficient of the explanatory variable lnG, β_1, is 245.34 > 0; the coefficient of the explanatory variable $(lnG)^2$, β_2, is −38.87 < 0; and the coefficient of the explanatory variable $(lnG)^3$, β_3, is 2.05 > 0, which validates the cubic functional form. In addition, the statistical quality of the estimation and the scatter plot (Figure 2) confirm one another, and an N-shaped relationship between crop production and agricultural carbon emissions results—that is, increasing agricultural CO2 emissions in the beginning, a decline in agricultural CO2 emissions in the middle, and an upward trend in agricultural CO2 emissions at the end. In this model, no possible turning point toward a decline in agricultural CO2 emissions appears after 2010, which shows the challenges faced by Jilin province to reduce agricultural carbon emissions.

3.2. Decoupling Analysis

As regards the criteria for decoupling/coupling degrees (Table 1), the results of the decoupling of agricultural carbon emissions from crop production are shown in Table 7.

Table 7. Decoupling states in agricultural carbon emissions in Jilin province during 2000–2018.

Year	Change Rate in Agricultural Carbon Emissions (ΔC)	Change Rate in Agricultural Output Value (ΔG)	DI	Decoupling States
2000–2001	0.103	0.116	0.886	Weak decoupling
2001–2002	0.086	0.133	0.643	Weak decoupling
2002–2003	−0.153	0.050	−3.075	Strong decoupling
2003–2004	0.362	0.084	4.300	Expansive coupling
2004–2005	0.241	0.062	3.909	Expansive coupling
2005–2006	0.080	0.077	1.040	Expansive coupling
2006–2007	0.034	−0.020	−1.664	Strong coupling
2007–2008	−0.319	0.152	−2.105	Strong decoupling
2008–2009	0.003	−0.019	−0.155	Strong coupling
2009–2010	0.070	0.063	1.109	Expansive coupling
2010–2011	0.034	0.073	0.473	Weak decoupling
2011–2012	0.074	0.046	1.597	Expansive coupling
2012–2013	0.125	0.060	2.078	Expansive coupling
2013–2014	0.024	0.066	0.360	Weak decoupling
2014–2015	0.102	0.047	2.143	Expansive coupling
2015–2016	0.062	0.026	2.408	Expansive coupling
2016–2017	0.017	0.070	0.247	Weak decoupling
2017–2018	0.039	0.028	1.432	Expansive coupling

There were four types of decoupling/coupling states between crop production and agricultural CO2 emissions during 2000–2018: expansive coupling state occurred for 9 years, followed by a weak decoupling state, which occurred for 5 years, and strong decoupling and strong coupling occurred for 2 years each.

Generally, strong decoupling means a positive change rate in agricultural output value, a negative change rate in agricultural CO2 emissions, and negative DI. It indicates that the development model of high inputs and high emissions in exchange for rapid agricultural economic growth is gradually shifting to a development model of low inputs and low emissions, and that the pressure on the rural ecological environment has been alleviated. Table 7 shows that the trend of strong decoupling only occurred in 2003 and 2008, wherein agricultural CO2 emissions decreased by −15.3% and −31.9%, respectively, while the agricultural output value increased by 5% and 15.2%, respectively. To some extent, special events contributed to the strong decoupling taking place at these two time points. For instance, farmers' willingness to grow food had been in decline since 1998 and hit a low in 2003, which decreased inputs into agricultural production and agricultural CO2 emissions, resulting in the strong decoupling state in 2003; additionally, in 2008, a similar decoupling state appeared in Jilin province, which was affected by the global financial crisis.

Strong coupling indicates the worst situation, with a negative change rate for agricultural output value, a positive change rate for agricultural CO_2 emissions, and negative DI. In terms of the occurrence of strong coupling in Jilin province from 2000 to 2018, the agricultural output value declined severely due to severe drought and the aftermath of the global financial crisis in 2007 and 2009, respectively, while change rates in agricultural CO_2 emissions were positive, which gave rise to strong coupling states in these two years.

Weak decoupling indicates a state with positive change rates in both agricultural output value and carbon emissions, and a value of DI ranging from 0 to 1. Table 7 shows that weak decoupling occurred for 5 years, although not consistently in the time series, as it occurred in the years 2001, 2002, 2011, 2014, and 2017. This indicates that agricultural carbon emission growth is somewhat restrained by the execution of various existing policies and measures; however, the absolute carbon emission reduction is smaller in this period than the agricultural economic growth, so agricultural carbon emissions are still rising, and carbon emission reduction measures should be further implemented.

According to the criteria of the expansive coupling state (Table 1), the change rates of both agricultural output value and agricultural carbon emissions in this period are positive, and thus, agricultural carbon emissions rise rapidly. As the most common outcome for Jilin province, the expansive coupling state accounted for 50% of the whole study period. For example, in 2018, the change rate of agricultural carbon emissions was 0.039, while that of agricultural output value was 0.028, and the DI was 1.432, which indicates agricultural economic growth occurred at the cost of accelerated agricultural carbon emissions in 2018.

Figures 3 and 4 show the variation in the values of decoupling elasticity. The DI clearly presents a fluctuating state, along with the different variation characteristics of both agricultural carbon emissions and agricultural output value. Notably, decoupling occurred only at specific time points, i.e., before the implementation of policies to strengthen agriculture and benefit farmers in 2003, or the global financial crisis in 2008. During 2009–2018, an expansive coupling state appeared for 6 years, inlaid and alternating with weak decoupling states.

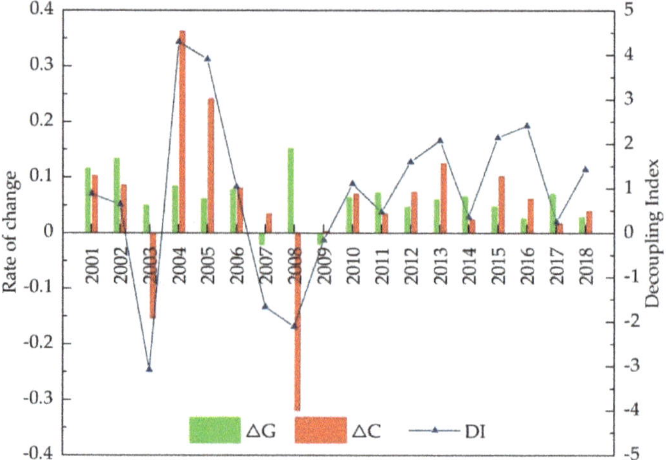

Figure 3. The change in the decoupling index of agricultural carbon emissions.

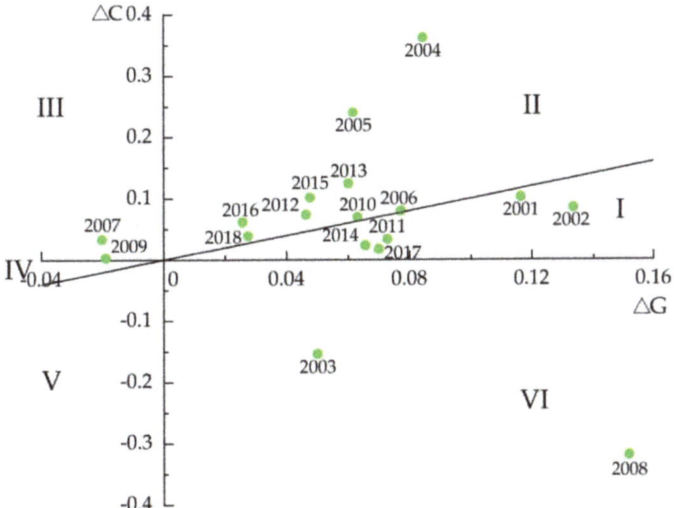

Figure 4. Decoupling distribution of agriculture. Note: I, II, III, IV, V, and VI represent weak decoupling, expansive coupling, strong coupling, weak coupling, recessive decoupling, and strong decoupling, respectively.

In light of the above, there was no stable decoupling between crop production and agricultural CO_2 emissions in Jilin province during 2000–2018, and it is common that carbon emissions increase when the agricultural economic value grows. Expansive coupling has appeared frequently in recent years, which indicates that the target of decoupling agricultural CO_2 emissions from crop production remains elusive in the coming years.

3.3. Results of LMDI Decomposition

According to the definition of CI in this paper, agricultural carbon emission intensity is calculated, and Figure 5 shows changes in agricultural carbon emission intensity in Jilin province during 2000–2018.

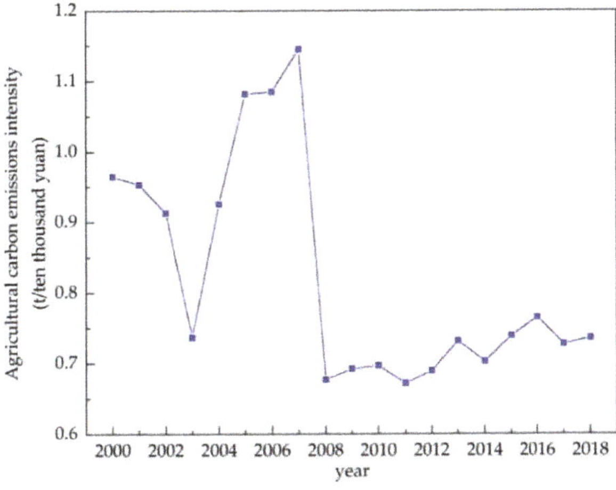

Figure 5. Agricultural carbon emission intensity in Jilin province during 2000–2018.

Based on the LMDI model and Equations (4)–(8), the decomposition results of carbon emission changes in agriculture in Jilin during 2000–2018 are illustrated in Table 8, and Figure 6 shows the contribution of four factors to agricultural carbon emissions. The total change in agricultural carbon emissions was 3.74 million tons between 2000 and 2018. Generally, the positive driving factors of agricultural carbon emissions included agricultural economic growth and agricultural structure, with contributions of 5.14 million tons and 0.35 million tons, respectively; agricultural economic growth played an especially significant role in the increase in agricultural carbon emissions in Jilin province, with a contribution of 93.56%. These decomposition results are consistent with those of Li et al. [40] and Guo et al. [39]. Agricultural carbon emission intensity and agricultural labor force were negative driving factors of agricultural carbon emissions, with cumulative contributions of –1.18 million tons and –0.58 million tons, respectively; notably, agricultural carbon emission intensity had the most important inhibitory impact on the change in agricultural carbon emissions, with a contribution rate of 66.99%, and agricultural labor force was also a negative factor that cannot be ignored, with a contribution rate of 33.01%.

From the perspective of policymaking, we focused on several key time points in our detailed decomposition analysis aimed at reducing agricultural carbon emissions. According to the above results of the CO_2 EKC model and decoupling analysis, turning points arose in the years 2003 and 2008, with consequential downward trends in agricultural carbon emissions, and strong decoupling states appeared; meanwhile, the only two negative decomposition effects (-56.6 million tons and -188.7 million tons, respectively) appeared. In 2003, except for the agricultural economic effect, which was a positive driving factor of agricultural carbon emissions, agricultural carbon emission intensity, agricultural structure, and agricultural labor force were all negative driving factors, especially agricultural carbon emission intensity, the contribution rate of which was 89.36%, reflecting its significant inhibitory effect.

Table 8. Decomposition results of agriculture carbon emission changes in Jilin province (10^4 t).

Year	Agricultural Carbon Emission Intensity Effect (ΔCI)	Agricultural Structure Effect (ΔSI)	Agricultural Economic Effect (ΔEI)	Agricultural Labor Force Effect (ΔAL)	Total Effects (ΔC_{tot})
2000–2001	−3.89	−9.81	47.04	−1.57	31.78
2001–2002	−15.22	0.59	47.45	−3.61	29.21
2002–2003	−73.17	−4.26	25.27	−4.45	−56.62
2003–2004	83.81	1.79	32.18	−4.26	113.51
2004–2005	74.25	−24.24	47.55	5.15	102.71
2005–2006	1.58	32.16	11.26	−2.53	42.47
2006–2007	32.00	−19.04	−8.51	15.60	19.35
2007–2008	−258.01	22.39	68.73	−21.82	−188.72
2008–2009	9.00	−28.54	−0.01	20.75	1.20
2009–2010	2.69	10.74	8.36	6.42	28.20
2010−2011	−15.98	8.74	−17.06	39.13	14.83
2011–2012	12.03	−5.59	40.30	−13.84	32.89
2012–2013	30.17	12.08	22.75	−5.14	59.86
2013–2014	−21.97	12.66	39.97	−17.92	12.74
2014–2015	29.23	2.44	33.60	−9.20	56.07
2015–2016	21.65	−3.73	40.45	−20.88	37.50
2016–2017	−32.88	23.33	42.38	−21.65	11.19
2017–2018	7.71	3.68	32.72	−18.19	25.92
2000–2018	−117.72	35.40	514.42	−58.01	374.09

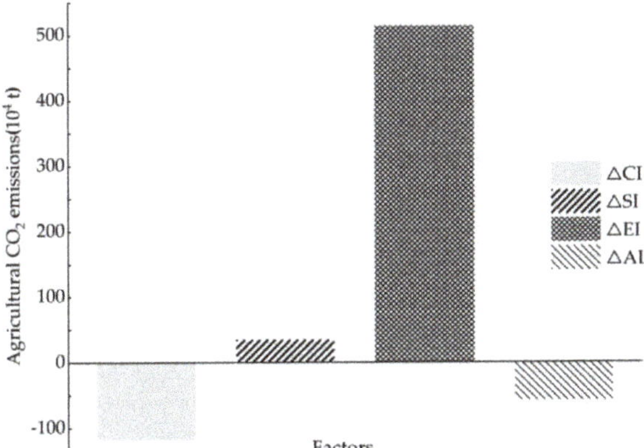

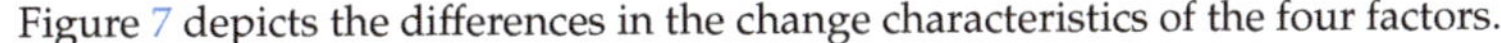

Figure 6. Total contribution of four factors to agricultural CO_2 emissions in Jilin province.

Figure 7 depicts the differences in the change characteristics of the four factors.

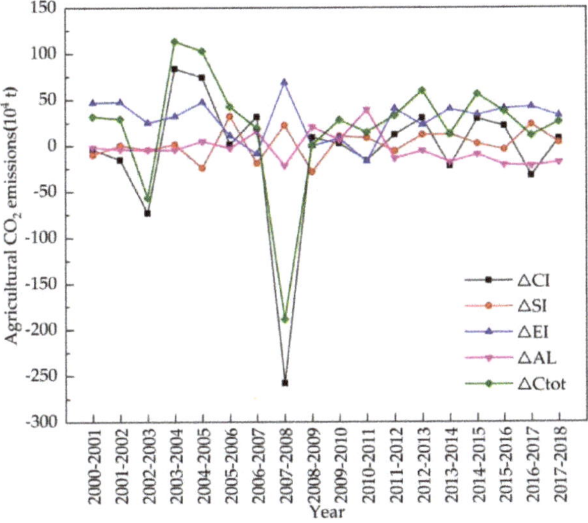

Figure 7. Characteristics of four factors' contribution to the variation in agricultural CO_2 emissions in Jilin province during 2000–2018.

In 2008, affected by the global financial crisis, on the one hand, agricultural product prices dramatically declined, and farmers' engagement in agriculture and their inputs into crop production thus greatly decreased. This led to agricultural carbon emission intensity and agricultural labor force both reaching a low, which resulted in agricultural carbon emissions dropping to −258 million tons and −21.8 million tons, respectively, and the contribution rate of agricultural carbon emission intensity reached 92.2%. On the other hand, considering the sharp decrease in agricultural labor force, the agricultural economic effect—namely, the gross agricultural output value divided by agricultural labor force—reached a peak, although it was far less strong than that of agricultural carbon emission intensity.

Another two important points appeared in 2007 and 2009, with a clear peak in 2007 in the N-shaped CO_2 EKC, and the beginning of the increase in agricultural carbon emissions after the global financial crisis in 2009; these were the only strong coupling states. Further decomposition analysis indicated that both agricultural carbon emission intensity and agricultural labor force acted as positive driving factors of agricultural carbon emissions, while agricultural structure and agricultural economic effect (a decrease in crop production due to drought or financial crisis) acted as inhibitory driving factors. Contrary to the decoupling states, the increases in both the population working in agriculture and the inputs into agricultural production counteracted the inhibitory effects of agricultural structure and the agricultural economic factor, which brought about the increase in agricultural carbon emissions.

Apart from the above four specific time points, most of the study period showed either an upward trend in agricultural carbon emissions, as seen from the N-shaped EKC, or expansive coupling alternating with weak decoupling according to the results of the decoupling analysis, with various decompositions and combinations. In terms of the varied trends after 2012 when industrial overcapacity was exacerbated, the agricultural economic effect still acted as the major driving factor of increase in agricultural carbon emissions; agricultural structure acted mostly as a positive driving factor; the agricultural labor force became an inhibitory factor, reducing agricultural carbon emissions with the promulgation of agricultural policy and market-oriented reforms in the process of rural transformation; the effect of agricultural carbon emission intensity presented no stable inhibitory trend, and it is important for Jilin province to further pursue its key role in agricultural carbon emission reductions.

4. Discussion and Policy Implications

(1) From a policy perspective, according to the requirements for a low-carbon economy, when the economy grows, carbon emissions should increase with either a steady or a negative growth trend. However, it is difficult to achieve this goal in the process of transformation in China. What we can do is to guide the transformation to a low-carbon economy in order to achieve a growth rate of carbon emission intensity that is relatively slower than the economic growth rate. Therefore, it is necessary to evaluate the economy–environment relationship with scientific methods and further conduct decomposition analyses of the influencing factors so as to provide policy suggestions.

Both the EKC hypothesis and decoupling analysis describe the dynamic relationship between economic development and environmental pollution, which have both internal connections and obvious differences. The EKC hypothesis expounds the nonlinear relationship of environmental pollution with the level of economic development, while a decoupling analysis reveals whether the relationship between economic development and environmental pressure changes synchronously. The EKC describes the long-term relationship between CO_2 emissions and economic growth, but it does not capture the short-term change in a particular year or phase. In contrast, a decoupling analysis measures the extent to which CO_2 emissions decouple from economic growth in the short term, and its indicators provide real-time data but do not identify long-term trends—i.e., the elastic index of decoupling describes the change rates of economic growth and pollution emissions but ignores the absolute changes in both [50]. In light of the work of Vehmas et al. [28], not only the shape of the EKC but also the decoupling results can stem from economic growth, environmental policy, or some other factors. In fact, the agricultural economic development level, industrial structure, rural labor force scale, and crop production inputs (including pesticides, chemical fertilizer, etc.) all have an influence on agricultural carbon emissions, but the EKC hypothesis and decoupling analysis only explain the nonlinear relationship between economic development and environmental pollution, and the effects of economic growth on environmental pollution cannot be expounded by means of the above methods. With this in mind, the latter method is the most important for policymaking.

In view of this, this paper first employed the EKC hypothesis and decoupling analysis to examine the relationship between agricultural carbon emissions and crop production in

Jilin province, a major grain-producing area, and then analyzed the influencing factors via LMDI decomposition, so as to provide a scientific basis for subsequent policymaking.

(2) The major findings in this paper indicate some special relationships among the results of the CO_2 EKC estimation, the decoupling analysis, and the decomposition analysis for Jilin province. Firstly, in terms of time points, strong decoupling coincided with declines in agricultural carbon emissions and the negative total decomposition effects of LMDI in 2003 and 2008 (as shown in Figure 2, Tables 7 and 8); additionally, strong coupling was partially connected with the peak in the increase in agricultural carbon emissions (such as in 2007, as shown in Figure 2 and Table 7). Secondly, on the whole, the agricultural economic effect played a significant role in the increase in agricultural carbon emissions in Jilin province, while agricultural carbon emission intensity had the most important inhibitory impact on agricultural carbon emissions (Figure 6), similarly to existing conclusions [32,39]. We also found that the positive or negative effects (or driving directions) of the four factors in agricultural carbon emissions were not stable, and macroeconomic changes and the implementation of related policies can be considered to have had an impact on agricultural carbon emissions in Jilin province during 2000–2018. Third, in terms of the policy effect, we found that agricultural carbon emissions were affected more by economic than environmental policy, based on integrating various time points at which economic policies or environmental policies were issued, such as the years 2004, 2008, 2012, and so on (Figures 1 and 7). The empirical results show that agricultural policies or macroeconomic changes affected farmers' willingness to grow food and increase production inputs, thus having a direct or indirect influence on the change in agricultural carbon emissions. Therefore, policy implications can be inferred from both agricultural and environmental policy.

(3) In terms of policy implications, there is no doubt that Jilin province will pursue agricultural economic growth, given that it is one of China's main grain-producing areas; however, it is limited by the pattern of the agricultural economic growth. The existing agricultural carbon emission reduction policies have not been effectively implemented in practice, and the efficiency of agricultural carbon emission reduction needs to be improved. On the one hand, the rapid improvement in rural living standards is an indisputable fact, which will have a reducing effect on total agricultural carbon emissions. On the other hand, agricultural carbon emission intensity, agricultural technology progress, and urbanization each play a role in reducing agricultural carbon emissions by developing energy-saving technology, improving agricultural productivity, adjusting the structure of urban and rural regions, etc.; however, at present, China's agricultural low-carbon technology is in the initial stage, and control over the overall change in agricultural carbon emissions needs to be further improved.

We should actively promote clean energy in rural areas and improve low-carbon agricultural technologies. Clean energy includes both natural and biomass energy. Advanced equipment and technologies producing low-emission clean energy should be actively introduced, and farmers should be encouraged to use clean energy to replace traditional high-emission energy.

We should increase the investment in science and technology and improve the efficiency of agricultural production through agricultural technology. High-yield and efficient agricultural production technology will provide important support for the development of modern agriculture, as well as being a reliable basic means to achieve the goal of low-carbon agriculture. Government departments should provide special funds for the development of low-carbon agricultural technologies and improve the approach to technology research [48].

We should establish a set of low-carbon agricultural ecological compensation technology systems, increase compensation intensity, and encourage farmers to participate actively in fallow and no-till, so as to reduce carbon sources and increase carbon sinks.

5. Conclusions

The following are the conclusions of the paper:

(1) Based on the results of the CO_2 EKC estimation, a long-term N-shaped EKC was found, which reflects that Jilin province is facing a dilemma between agricultural economic growth and agricultural carbon emissions, and the upward trend in agricultural carbon emissions has not changed with the development of the agricultural economy.

(2) In the short term, according to the results of the decoupling analysis, weak decoupling, strong decoupling, expansive coupling, and strong coupling occurred in alteration. Among them, expansive coupling occurred for 9 years in total, followed by weak decoupling, which occurred for 5 years, and then strong decoupling and strong coupling occurred for 2 years each. Strong decoupling occurred in 2003 and 2008, which was related to the macroeconomics and policies at these time points; strong coupling appeared due to a severe drought in 2007 and due to the aftermath of the global financial crisis in 2009. There was no stable evolutionary path from coupling to decoupling during the years 2000–2018, which currently remains true.

(3) Based on previous research, we used the LMDI method to decompose the driving factors of agricultural carbon emissions in Jilin province into four factors: agricultural carbon emission intensity effect, agricultural structure effect, agricultural economic effect, and agricultural labor force effect. From a policy-making perspective, we integrated the results of both the EKC and the decoupling analysis, and we conducted a detailed decomposition analysis, focusing on several key time points.

Overall, agricultural economic growth played a significant role in the increase in agricultural carbon emissions, while agricultural carbon emission intensity was the main factor behind the decline in agricultural carbon emissions in Jilin province, especially in the years 2003 and 2008, when turning points toward a downward trend in agricultural carbon emissions and strong decoupling states appeared. Another two important time points in the N-shaped CO_2 EKC included a clear peak that appeared in 2007 and the start of the increase in agricultural carbon emissions after the global financial crisis in 2009; these were the only two strong coupling states. Different from strong decoupling, both agricultural carbon emission intensity and agricultural labor force acted as positive driving factors of agricultural carbon emissions, while agricultural structure and agricultural economic growth acted as inhibitory driving factors of agricultural carbon emissions.

Most of the study period showed an upward trend for agricultural carbon emissions, as seen from the N-shaped EKC, and expansive coupling states alternated with weak decoupling states, especially after 2010, according to the results of the decoupling analysis, with various decompositions and combinations of driving factors. The efforts toward agricultural carbon emission reduction in Jilin province are clearly still ineffective.

Author Contributions: J.S. designed and conceived this article; W.L. processed the data and wrote the paper. Both authors have read and agreed to the published version of the manuscript.

Funding: This research was funded by the National Natural Science Foundation of China (grant number No.71573104); Young Academic Leaders Training Program of Jilin University (grant number No.2019FRLX10); and Interdisciplinary Integration Innovation Cultivation Project of Jilin University (grant number No.JLUXKJC2020301).

Institutional Review Board Statement: Not applicable.

Informed Consent Statement: Not applicable.

Data Availability Statement: China Rural Statistical Yearbook (https://data.cnki.net/trade/yearbook/single/n2019120190?z=z009), and Jilin Statistical Yearbook (http://tjj.jl.gov.cn/tjsj/tjnj/2019/ml/indexe.htm).

Conflicts of Interest: The authors declare no conflict of interest.

References

1. Xu, B.; Lin, B. Can Expanding Natural Gas Consumption Reduce China's CO_2 Emissions? *Energy Econ.* **2019**, *81*, 393–407. [CrossRef]
2. Li, N.; Wei, C.; Zhang, H.; Cai, C.; Song, M.; Miao, J. Drivers of the National and Regional Crop Production-Derived Greenhouse Gas Emissions in China. *J. Clean. Prod.* **2020**, *257*, 120503. [CrossRef]
3. FAOSTAT. Statistics Database. 2020. Available online: http://faostat.fao.org (accessed on 30 May 2021).
4. Mbow, C.; Rosenzweig, C.; Barioni, L.G.; Benton, T.G.; Herrero, M.; Krishnapillai, M.; Liwenga, E.; Pradhan, P.; Rivera-Ferre, M.G.; Sapkota, T.; et al. 2019: Food Security. In *Climate Change and Land: An IPCC Special Report on Climate Change, Desertification, Land Degradation, Sustainable Land Management, Food Security, and Greenhouse Gas Fluxes in Terrestrial Ecosystems*; Shukla, P.R., Skea, J., Calvo Buendia, E., Masson-Delmotte, V., Pörtner, H.-O., Roberts, D.C., Zhai, P., Slade, R., Connors, S., van Diemen, R., et al., Eds.; IPCC: Geneva, Switzerland, 2019.
5. National Bureau of Statistics of the People's Republic of China. Statistics Database. 2020. Available online: http://www.stats.gov.cn (accessed on 20 May 2021).
6. Silvia, C.; Roberto, E. Is There A Long-Term Relationship between Agricultural GHG Emissions and Productivity Growth? A Dynamic Panel Data Approach. *Environ. Resour. Econ.* **2014**, *58*, 273–302.
7. Huang, X.; Xu, X.; Wang, Q.; Zhang, L.; Gao, X.; Chen, L. Assessment of Agricultural Carbon Emissions and Their Spatiotemporal Changes in China, 1997–2016. *Int. J. Environ. Res. Public Health* **2019**, *16*, 3105. [CrossRef] [PubMed]
8. Ge, D.Z.; Long, H.L.; Zhang, Y.N.; Tu, S.S. Pattern and Coupling Relationship between Grain Yield and Agricultural Labor Changes at County Level in China. *Acta Geogr. Sin.* **2017**, *72*, 1063–1077.
9. Zhang, Y.; Long, H.; Li, Y.; Ge, D.; Tu, S. How Does Off-farm Work Affect Chemical Fertilizer Application? Evidence from China's Mountainous and Plain Areas. *Land Use Policy* **2020**, *99*, 104848. [CrossRef]
10. Ma, L.; Long, H.L.; Zhang, Y.N.; Tu, S.S. Spatio-Temporal Coupling Relationship between Agricultural Labor Changes and Agricultural Economic Development at County Level in China and its Implications for Rural Revitalization. *Acta Geogr. Sin.* **2018**, *73*, 2364–2377.
11. Grossman, G.M.; Krueger, A.B. Economic Growth and the Environment. *Q. J. Econ.* **1995**, *110*, 353–377. [CrossRef]
12. Apergis, N.; Christou, C.; Gupta, R. Are there Environmental Kuznets Curves for US State-level CO_2 Emissions? *Renew. Sustain. Energy Rev.* **2017**, *69*, 551–558. [CrossRef]
13. Badeeb, R.A.; Lean, H.H.; Shahbaz, M. Are too Many Natural Resources to Blame for the Shape of the Environmental Kuznets Curve in Resource-based Economies? *Resour. Policy* **2020**, *68*, 101694. [CrossRef]
14. Yang, X.; Lou, F.; Sun, M.; Wang, R.; Wang, Y. Study of the Relationship between Greenhouse Gas Emissions and the Economic Growth of Russia based on the Environmental Kuznets Curve. *Appl. Energy* **2017**, *193*, 162–173. [CrossRef]
15. Churchill, S.A.; Inekwe, J.; Ivanovski, K.; Smyth, R. The Environmental Kuznets Curve across Australian States and Territories. *Energy Econ.* **2020**, *90*, 104869. [CrossRef]
16. Ahmad, N.; Du, L.; Lu, J.; Wang, J.; Li, H.; Muhammad, Z.H. Modelling the CO_2 Emissions and Economic Growth in Croatia: Is There Any Environmental Kuznets Curve? *Energy* **2017**, *123*, 164–172. [CrossRef]
17. Dong, K.; Sun, R.; Jiang, H.; Zeng, X. CO_2 Emissions, Economic Growth, and the Environmental Kuznets Curve in China: What Roles Can Nuclear Energy and Renewable Energy Play? *J. Clean. Prod.* **2018**, *196*, 51–63. [CrossRef]
18. Sarkodie, S.A.; Adams, S.; Owusu, P.A.; Leirvik, T.; Ozturk, I. Mitigating Degradation and Emissions in China: The Role of Environmental Sustainability, Human Capital and Renewable Energy. *Sci. Total Environ.* **2020**, *719*, 137530. [CrossRef] [PubMed]
19. Friedl, B.; Getzner, M. Determinants of CO_2 Emissions in A Small Open Economy. *Ecol. Econ.* **2003**, *45*, 133–148. [CrossRef]
20. Pata, U.K. Environmental Kuznets Curve and Trade Openness in Turkey: Bootstrap ARDL Approach with A Structural Break. *Environ. Sci. Pollut. Res.* **2019**, *26*, 20264–20276. [CrossRef] [PubMed]
21. Kang, Y.Q.; Zhao, T.; Yang, Y.Y. Environmental Kuznets Curve for CO_2 Emissions in China: A Spatial Panel Data Approach. *Ecol. Indic.* **2016**, *63*, 231–239. [CrossRef]
22. Li, Z.; Song, Y.; Zhou, A.; Liu, J.; Pang, J.; Zhang, M. Study on the Pollution Emission Efficiency of China's Provincial Regions: The Perspective of Environmental Kuznets Curve. *J. Clean. Prod.* **2020**, *263*, 121497. [CrossRef]
23. Ugur, K.P.; Abdullah, E.C. Investigating the EKC hypothesis with Renewable Energy Consumption, Human Capital, Globalization and Trade Openness for China: Evidence from Augmented ARDL Approach with A Structural Break. *Energy* **2021**, *216*, 119220.
24. Ahmed, K.; Long, W. Environmental Kuznets Curve and Pakistan: An Empirical Analysis. *Proc. Econ. Financ.* **2012**, *1*, 4–13. [CrossRef]
25. Apergis, N.; Ozturk, I. Testing Environmental Kuznets Curve Hypothesis in Asian Countries. *Ecol. Indic.* **2015**, *52*, 16–22. [CrossRef]
26. Azomahou, T.; Laisney, F.; Van, P.N. Economic Development and CO_2 Emissions: A Nonparametric Panel Approach. *J. Public Econ.* **2006**, *90*, 1347–1363. [CrossRef]
27. Farhani, S.; Ozturk, I. Causal Relationship between CO_2 Emissions, Real GDP, Energy Consumption, Financial Development, Trade Openness, and Urbanizationin Tunisia. *Environ. Sci. Pollut. Res.* **2015**, *22*, 15663–15676. [CrossRef] [PubMed]
28. Vehmas, J.V.; Luukkanen, J.; Kaivo-oja, J. Linking Analyses and Environmental Kuznets Curves for Aggregated Material Flows in the EU. *J. Clean. Prod.* **2007**, *15*, 1662–1673. [CrossRef]

29. Fischer, K.M.; Swilling, M. *Decoupling Natural Resource Use and Environmental Impacts from Economic Growth*; United Nations Environment Programme: Nairobi, Kenya, 2011; pp. 1–174.
30. Organization for Economic Co-operation and Development (OECD). Indicators to Measure Decoupling of Environmental Pressure from Economic Growth. SG/SD(2002)1/FINAL. Available online: http://www.oecd.org/indicators-modelling-outlooks/1933638.pdf (accessed on 6 June 2021).
31. Tapio, P. Towards A Theory of Decoupling: Degrees of Decoupling in the EU and the Case of Road Traffic in Finland between 1970 and 2001. *Transp. Policy* **2005**, *12*, 137–151. [CrossRef]
32. Tian, Y.; Zhang, J.; Li, B. Research on China's Agricultural Carbon Emissions: Calculation, Spatial-Temporal Comparison and Decoupling Effect. *Resour. Sci.* **2012**, *34*, 2097–2105.
33. Yang, S.J.; Li, Y.B.; Yan, S.G. An Empirical Analysis of the Decoupling Relationship between Agricultural Carbon Emission and Economic Growth in Jilin Province. *IOP Conf. Ser. Mater. Sci. Eng.* **2018**, *392*, 062101.
34. Chen, H.; Wang, H.; Qin, S. Analysis of Decoupling Effect and Driving Factors of Agricultural Carbon Emission: A Case Study of Heilongjiang Province. *Sci. Technol. Manag. Res.* **2019**, *17*, 247–252.
35. Ang, B.W. Decomposition Analysis for Policy–Making in Energy: Which is the Preferred Method? *Energy Policy* **2004**, *32*, 1131–1139. [CrossRef]
36. Liu, B.; Zhang, X.; Yang, L. Decoupling Efforts of Regional Industrial Development on CO_2 emissions in China based on LMDI Analysis. *China Popul. Resour. Environ.* **2018**, *28*, 78–86.
37. Kim, S. LMDI Decomposition Analysis of Energy Consumption in the Korean Manufacturing Sector. *Sustainability* **2017**, *9*, 202. [CrossRef]
38. Zhao, Y.; Li, H.; Zhang, Z.; Zhang, Y.; Wang, S.; Liu, Y. Decomposition and scenario analysis of CO_2, emissions in China's power industry: Based on LMDI method. *Nat. Hazards* **2017**, *86*, 1–24. [CrossRef]
39. Guo, H.; Fan, B.; Pan, C. Study on Mechanisms Underlying Changes in Agricultural Carbon Emissions: A Case in Jilin Province, China, 1998–2018. *Int. J. Environ. Res. Public Health* **2021**, *18*, 919. [CrossRef]
40. Li, B.; Zhang, J.; Li, H. Research on Spatial-Temporal Characteristics and Affecting Factors Decomposition of Agricultural Carbon Emission in China. *China Popul. Resour. Environ.* **2011**, *21*, 80–86.
41. Wei, Z.; Qin, Q.; Kuang, Y.; Huang, N. Investigating Low-Carbon Crop Production in Guangdong Province, China (1993–2013): A Decoupling and Decomposition Analysis. *J. Clean. Prod.* **2017**, *146*, 63–70.
42. Zhao, X.; Song, L.; Tan, S. Study on Influential Factors of Agricultural Carbon Emission in Hunan Province based on LMDI Model. *Environ. Sci. Techno.* **2018**, *41*, 177–183.
43. Xu, Q.; Li, Y.; Yang, S. Measurement and Decomposition of Carbon Emission by the Process of Agricultural Modernization in Jilin Province. *J. Chin. Agric Mechan.* **2018**, *39*, 103–109.
44. Sun, J.; Zhao, K.; Niu, Y. Evaluation and Difference Analysis of Social and Economic Development Level of Three Major Grain-producing Areas–Based on the Perspective of Benefit Compensation for Major Grain-producing Areas. *Res. Agric. Mod.* **2017**, *38*, 581–588.
45. Zhang, H.; Jiang, Q.; Lv, J. Economic Growth and Food Safety: FKC Hypothesis Test and Policy Implications. *Econ. Res.* **2019**, *11*, 180–194.
46. Lv, D.; Wang, R.; Zhang, Y. Sustainability Assessment Based on Integrating EKC with Decoupling: Empirical Evidence from China. *Sustainability* **2021**, *13*, 655. [CrossRef]
47. Ekins, P. The Kuznets Curve for the Environment and Economic Growth: Examining the Evidence. *Environ. Plan. A* **1997**, *29*, 805–830. [CrossRef]
48. Wu, X.; Zhang, J. Agricultural Carbon Emissions at Provincial Scale in China: Growth Effect and Decoupling Effect. *J. Agrotech. Econo.* **2017**, *5*, 27–36.
49. Intergovernmental Panel on Climate Change (IPCC). *2006 IPCC Guidelines for National Greenhouse Gas Inventories Volume 4: Agriculture, Forestry and Other Land Use*; Eggleston, S.L., Buendia, K., Miwa, T.N., Tanabe, K., Eds.; Prepared by the National Greenhouse Gas Inventories Programme; Institute for Global Environmental Strategies: Hayama, Japan, 2006.
50. Xia, Y.; Zhong, M. Relationship between EKC Hypothesis and the Decoupling of Environmental Pollution from Economic Development: Based on Decoupling Partition of China Prefecture-Level Cities. *China Popul. Resour. Environ.* **2016**, *26*, 8–16.

International Journal of
*Environmental Research
and Public Health*

Article

Linking Economic Growth, Urbanization, and Environmental Degradation in China: What Is the Role of Hydroelectricity Consumption?

Tomiwa Sunday Adebayo [1,*], Mary Oluwatoyin Agboola [2], Husam Rjoub [3], Ibrahim Adeshola [4], Ephraim Bonah Agyekum [5] and Nallapaneni Manoj Kumar [6,*]

[1] Department of Business Administration, Faculty of Economics and Administrative Science, Cyprus International University, Nicosia, Northern Cyprus, TR-10 Mersin, Turkey
[2] College of Business, Dar Al Uloom University, 1 Mizan st. Al Falah, Riyadh 13314, Saudi Arabia; maryagboola@dau.edu.sa
[3] Department of Accounting and Finance, Faculty of Economics and Administrative Sciences, Cyprus International University, Mersin 10 99040, Turkey; hrjoub@ciu.edu.tr
[4] Department of Management Information Systems, School of Applied Sciences, Cyprus International University, Northern Cyprus, Via Mersin 10, Turkey; deshyengr@live.com
[5] Department of Nuclear and Renewable Energy, Ural Federal University Named after the First President of Russia Boris Yeltsin, 19 Mira Street, 620002 Ekaterinburg, Russia; agyekumephraim@yahoo.com
[6] School of Energy and Environment, City University of Hong Kong, Kowloon, Hong Kong, China
* Correspondence: twaikline@gmail.com (T.S.A.); mnallapan2-c@my.cityu.edu.hk (N.M.K.)

Citation: Adebayo, T.S.; Agboola, M.O.; Rjoub, H.; Adeshola, I.; Agyekum, E.B.; Kumar, N.M. Linking Economic Growth, Urbanization, and Environmental Degradation in China: What Is the Role of Hydroelectricity Consumption? *Int. J. Environ. Res. Public Health* **2021**, *18*, 6975. https://doi.org/10.3390/ijerph18136975

Academic Editors: Pasquale Avino, Massimiliano Errico, Aristide Giuliano and Hamid Salehi

Received: 31 May 2021
Accepted: 28 June 2021
Published: 29 June 2021

Publisher's Note: MDPI stays neutral with regard to jurisdictional claims in published maps and institutional affiliations.

Abstract: Achieving environmental sustainability has become a global initiative whilst addressing climate change and its effects. Thus, this research re-assessed the EKC hypothesis in China and considered the effect of hydroelectricity use and urbanization, utilizing data from 1985 to 2019. The autoregressive distributed lag (ARDL) bounds testing method was utilized to assess long-run cointegration, which is reinforced by a structural break. The outcome of the ARDL bounds test confirmed cointegration among the series. Furthermore, the ARDL revealed that both economic growth and urbanization trigger environmental degradation while hydroelectricity improves the quality of the environment. The outcome of the ARDL also validated the EKC hypothesis for China. In addition, the study employed the novel gradual shift causality test to capture causal linkage among the series. The advantage of the gradual shift causality test is that it can capture gradual or smooth shifts and does not necessitate previous information of the number, form of structural break(s), or dates. The outcomes of the causality test revealed causal connections among the series of interest.

Keywords: CO_2 emissions; hydroelectricity consumption; economic growth; urbanization; China

1. Introduction

Climate change has made sustainable development the primary policy aim, with the objective of reducing greenhouse gas (GHG) emissions [1,2]. China, the globe's second-biggest economy, is confronted with several environmental challenges. For instance, China's CO_2 emissions (CO_2) increased by 2.2 percent in 2018, accounting for 27.8 percent of total CO_2 emissions worldwide, and its average growth rate from 2007 to 2017 was 2.5 percent per year [3]. However, China has committed to mitigating CO_2 per unit GDP by 60–65 percent by 2030, relative to 2005 levels [4]. That is, China's economic expansion must be accompanied by minimal environmental dangers. It is widely known that China has made substantial economic progress over the last three decades, with China's GDP accounting for 14.08 percent of the world's GDP [5]. Since the 1990s, China's GDP growth rates have been regarded as among the fastest on the globe. The rise in living standards in China pushed the majority of the people to migrate to urban regions (which are more energy-intensive than rural regions) in search of a higher quality of life and more work possibilities, causing the urban population to skyrocket.

In China, for example, the urban population was 16 percent of the overall population in 1960. By 2019, the urban population accounted for 60% of China's overall population. As a result of this economic expansion, China's energy consumption accounted for 24 percent of the worldwide energy consumption and 34 percent of the world's energy use growth in 2018 [3]. Furthermore, between 2014 and 2016, the industrial and construction sectors absorbed the majority of China's energy, accounting for 70% of overall energy consumption [6]. China's growing energy usage has resulted in much higher levels of air pollution. China produces more GHGs emissions than does the rest of the industrialized world combined. China released 27% of the globe's GHGs emissions in 2019. This significant upsurge in CO_2 is attributable to an upsurge in fossil fuel utilization for energy generation, which accounted for about 86 percent of total primary energy use in China in 2020 [7]. As air pollution levels rise in China, the government is making more efforts to construct initiatives that encourage renewable energy [8]. For example, China has enacted many measures to facilitate green energy, including the policy processes of 2006 and 2009, as well as subsidy programs in 2003 and 2010, all of which were aimed at boosting the utilization of renewable energy in the nation [9].

In China, hydroelectricity is the most prominent renewable energy source. In general, other renewable energy sources accounted for just 5% of the whole energy mix in 2020, whereas hydroelectricity accounted for 8% of the overall energy mix in China [7]. Along with China's expanding economic activity, the consumption of hydroelectricity, specifically, and electricity, in general, is growing. Furthermore, China is among the top ten hydroelectricity producers in the world. This renewable energy source has the potential to help China reduce CO_2 emissions. During the period between 1980 and 2021, the significance of hydroelectricity in China rose as output levels nearly doubled. As a result, an increase in the levels of consumption and production of this form of energy might have had a significant impact on the reduction of environmental pollution. Hydropower can create energy without releasing greenhouse gases into the atmosphere. It can, nevertheless, result in environmental and societal risks, including degraded wildlife habitat, deteriorated water quality, impeded fish movement, and reduced recreational advantages on rivers. The destruction of forest, wildlife habitat, agricultural land, and beautiful regions occurs when land is flooded for a hydropower reservoir. The Three Gorges Dam in China, for example, required the relocation of whole villages to create space for reservoirs. Thus, though hydropower aids in mitigating CO_2, it also destroys ecosystems. Therefore, striking a balance is essential.

Notwithstanding the well-established studies, scholars have paid little attention to the influence of hydroelectricity on CO_2 emissions in China. Prior research looked at many pollution factors, such as energy usage and GDP [10–14], urban population [11,15–20], financial development [21–23], and trade openness [24–28]. Furthermore, because CO_2 emissions account for more than 76 percent of GHGs emissions, the majority of researchers utilized CO_2 as an indication of degradation of the environment [19,25,28,29]. The environmental Kuznets curve (EKC) theory has gained popularity amongst researchers as a significant instrument for environmental policy. According to the theory, in the early stages of a nation's economic development, a rise in GDP growth would produce more environmental damage until it gets to a point when the connection between GDP growth and environmental damage is negative. The extant research in China does not sufficiently address numerous facets of the EKC hypothesis.

Previous research has generally relied on utilization energy (e.g., coal) as a metric for energy usage. The hydroelectricity usage role in the EKC model has not been investigated significantly. Considering the renewable energy aggregate without taking into account the complex nature of its elements can obscure the varying effects of different types of utilization of energy and lead to incorrect policy conclusions for each element, particularly for hydroelectricity, which is distinct from other types of energy and has dissimilar effects on emissions in China. Furthermore, structural break problems have not been fully accounted for in the China model. Overlooking the potential of a structural break in the study might impair the ability to reject the null hypothesis of non-stationarity [30]. Moreover, the single break was incorporated into the ARDL approach to capture its effect on CO_2 emissions. Additionally, by enhancing the Toda–Yamamoto technique with a Fourier approximation, a novel causality approach is provided. This technique may capture gradual or smooth shifts and does not necessitate previous information of the number, form of structural break(s), or dates. We intend to add to the growing research by investigating the effects of hydroelectricity usage on environmental pollution in China.

The next segment presents the literature review in Section 2 and empirical methodology in Section 3. Section 4 presents findings and discussion. Section 5 concludes the empirical analysis.

2. Literature Review

This section of the research is divided into two distinct parts. The theoretical framework is centered on the EKC hypothesis and the empirical review discusses in detail the studies conducted by prior researchers.

2.1. Theoretical Framework

The framework is based on the EKC hypothesis, which has been frequently used in empirical research to investigate the impact of socioeconomic variables on environmental pollution. The scale, composite, and techniques effects of an economic boom can all have an influence on the degradation of the environment. According to the scale effect, economic growth first leads to environmental pollution since it demands more resources and energy, culminating in more waste and pollution [31]. The level of contamination and materials used in the manufacturing process, on the other hand, are influenced by the sectoral structure of a country. As a consequence, the composition effect predicts that a nations' structural move from the industrial to the service sector would reduce the negative environmental effects of economic development. Lastly, the technique effect demonstrates that when a nation's wealth grows, it embraces new and advanced technologies that increase output whilst reducing pollution [32]. Renewables such as hydroelectricity consumption are ways of mitigating environmental degradation. As such, it is expected to minimize the level of environmental degradation. Lastly, urbanization triggers economic growth that if not sustainable triggers degradation of the environment [33,34]. The number of people, their activities, and the increasing demands on resources all have an influence on the physical environment as a result of urbanization. Urbanization has significant health implications, owing to pollution and overcrowding.

In terms of the predicted signs of the variables' coefficients, it is commonly assumed that increased production leads to environmental degradation through increasing utilization of resources and energy. Due to unsustainable developmental patterns, the continual increase in output in developing countries such as the MINT economies poses a serious danger to the environment. In emerging countries such as China, environmental degradation worsens as income rises. In addition, GDP is anticipated to lead to increased energy use. Countries that consume a lot of non-renewable energy have higher economic growth, which puts more pressure on energy demand. Fossil fuels provide a significant amount of the energy required for growing economic activity. Increasing the use of fossil fuels increases CO_2 emissions. As a result of this assumption, we anticipate that China's GDP will have a beneficial influence on CO_2 emissions. $\left(\theta_1 = \frac{\delta CO_2}{\delta GDP} > 0 \right)$. Economic development, on the other hand, enhances environmental quality once it reaches a certain point. As a result of the above logic, GDP^2 is anticipated to decrease CO_2 emissions $\left(\theta_2 = \frac{\delta CO_2}{\delta GDP^2} < 0 \right)$. Urbanization is anticipated to trigger economic growth, which also leads to an upsurge in degradation of the environment. Thus, urbanization is anticipated to trigger CO_2 emissions

$\left(\theta_3 = \frac{\delta CO_2}{\delta URB} > 0 \right)$. Furthermore, hydroelectricity consumption, which is part of renewables, is predicted to enhance the quality of the environment. Thus, an upsurge in hydroelectricity consumption is anticipated to mitigate CO_2 emissions $\left(\theta_4 = \frac{\delta CO_2}{\delta HYDRO} < 0 \right)$.

2.2. Empirical Review

Over the years, several studies have been examined to assess the dynamic association between CO_2 emissions and hydroelectricity use, economic growth, and urbanization. Nonetheless, their findings are mixed, which is due to the timeframe of the studies, techniques employed, and characteristics of the country or countries of investigation. For instance, the study of Adebayo et al. [18] on the CO_2–GDP–EC association in Indonesia utilizing data between 1965 and 2018 showed that both GDP and utilization of energy trigger CO_2 emissions. Furthermore, the investigators applied a causality test and their outcome uncovered a one-way causal linkage from GDP and utilization of energy to CO_2. Likewise, in Malaysia, Zhang et al. [17] assessed the urbanization, GDP, and financial development effect on CO_2 by utilizing data from between 1970 and 2018 and applying wavelet and causality tests. Their empirical outcomes revealed that both urbanization and GDP caused degradation of the environment, while financial development mitigated CO_2. The study of Awosusi et al. [29] in South Korea on the association between CO_2 and globalization, utilization of energy, and energy use using data from 1965 to 2019 showed that globalization and GDP triggers CO_2. In China, Soylu et al. [13] assessed the CO_2–GDP–REC association using the novel wavelet coherence and causality approaches. The outcomes from their study showed that renewable energy use mitigates CO_2 while GDP increases CO_2. The causality test outcomes also showed that both REC and GDP can predict CO_2. Furthermore, the study validates the existence of EKC. The study of Orhan et al. [10] on the connection between CO_2 and trade openness, urbanization, and GDP in India showed that both urbanization and GDP harm the quality of the environment, while no significant link was fond between CO_2 and trade openness. Using China as a case study and applying the wavelet tools, Kirikkaleli [35] assessed the CO_2–urbanization connection between 1960 and 2018. The outcomes from the wavelet tool showed positive co-movement between urbanization and CO_2. The study of Umar et al. [36] on the connection between natural resources, CO_2, and GDP in China used data from 1980 to 2018. Their outcomes uncovered that GDP and natural resources increased the CO_2 level in China during the period of study. Solarin et al. [30] assessed the connection between CO_2, GDP, and hydroelectricity usage in India and China, utilizing data stretching from 1960 and 2014. The empirical research utilized VECM and granger causality tests to capture the connection, and their outcomes showed that hydroelectricity use mitigates CO_2, while GDP increases CO_2. Furthermore, the study confirmed the EKC hypothesis. Solarin and Ozturk [37] assessed the dynamics between GDP, CO_2, and hydroelectricity use in selected Latin America countries from 1970–2012. The investigators used panel causality tests and their outcome showed that GDP and hydroelectricity use triggers CO_2 emissions. The study of Apergis et al. [38] on the hydroelectricity consumption and economic growth nexus showed that hydroelectricity consumption triggers economic expansion. The study of Gyamfi et al. [39] on the CO_2–hydroelectricity use association showed that an increase in GDP triggers CO_2, while hydroelectricity use mitigates CO_2.

3. Empirical Methodology

According to the traditional EKC hypothesis, deterioration of the environment is proportional to GDP and GDP squared. Over the years, several studies (e.g., [18,30]) have incorporated utilization of energy as a CO_2 emissions determinant. In line with these prior studies, the present research formulated the model as follow:

$$CO_{2t} = \beta_1 + \beta_2 GDP_t + \beta_3 GDP_t^2 + \beta_4 URB_t + \beta_5 HYDRO_t + \beta_6 BD_t + \varepsilon_t \qquad (1)$$

In Equation (1), CO_2 stand for emissions per capita, GDP represents economic growth, which is measured as GDP per capita (USD Constant 2020), GDP^2 is economic growth squared, HYDRO stands for hydroelectricity consumption, which is measured as hydro (% electricity), and URB denotes urbanization, which is measured as urban population. The time trend is denoted by subscript t and BD denotes the dummy variable. The dummy variable was introduced to capture the structural break in CO_2. The structural shift was selected based on the structural break of CO_2 emissions at the level revealed by the Zivot and Andrew unit root test. The structural break was in the 2002. The shift in this period was probably due to the pattern of economic transformation that arose during that period in the country. The break date for CO_2 was 2002 in this empirical analysis. The data utilized in this study stretched from 1985 to 2019 (34 observations). The hydroelectricity consumption and CO_2 data were obtained from the British petroleum database, while urbanization and GDP data were gathered from the database of World Bank. In line with the studies of [13,30,40], we added GDP into the model. Moreover, following the studies of [17,29], we incorporated the URB into the model. Lastly, we introduced HYDRO into the model following the studies of [11,30,41,42]. The cointegration test may be performed using the F-test on the lagged levels of the variables [43]. The null and alternative hypotheses are no cointegration and there is cointegration among the variables. The research utilized the ARDL bounds test to capture the long-run association between CO_2 and the regressors. The ARDL bounds test is presented in Equation (2).

$$
\begin{aligned}
\Delta CO_{2t} &= \beta_1 + \sum_{i=1}^{k} \beta_2 \Delta CO_{2t-i} + \sum_{i=1}^{k} \beta_3 \Delta GDP_{t-i} + \sum_{i=1}^{k} \beta_4 \Delta GDP^2_{t-i} \\
&+ \sum_{i=1}^{k} \beta_5 \Delta URB_{t-i} + \sum_{i=1}^{k} \beta_6 \Delta HYDRO_{t-i} + \beta_7 CO_{2t-i} + \beta_8 GDP_{t-i} \\
&+ \beta_9 GDP^2_{t-i} + \beta_{10} URB_{t-i} + \beta_{11} HYDRO_{t-i} + \beta_{12} BD_t + \varepsilon_t
\end{aligned}
\tag{2}
$$

In Equation (2), the difference operator is depicted by Δ. All of the variables have already been specified. The study assesses the short-run coefficients after assessing the long-run connection between the series and determining the coefficients in the long-run:

$$
\begin{aligned}
\Delta CO_{2t} &= \beta_1 + \sum_{i=1}^{k} \beta_2 \Delta CO_{2t-i} + \sum_{i=1}^{k} \beta_3 \Delta GDP_{t-i} + \sum_{i=1}^{k} \beta_4 \Delta GDP^2_{t-i} \\
&+ \sum_{i=1}^{k} \beta_5 \Delta URB_{t-i} + \sum_{i=1}^{k} \beta_6 \Delta HYDRO_{t-i} + \beta_7 \Delta BD_t + \beta_8 ECM_{t-1} \\
&+ \varepsilon_t
\end{aligned}
\tag{3}
$$

The speed of adjustment coefficient is represented by β_8. The ECM represents the error correction term that measures the speed of adjustment of our model to the long-run equilibrium. The coefficient of ECM must be negative and statically significant.

The present research employed the gradual shift causality test to catch the causal linkage among the series of investigations. We utilized the Fourier Toda–Yamamoto causality test developed by [44] to capture structural shifts in Granger causality analysis—including gradual and smooth shifts termed the "gradual-shift causality test". The gradual shift equations are as follows:

$$
y_t = \sigma(t) + \beta_1 y_{t-1} + \ldots + \beta_{p+dmax} y_{t-(p+d)} + \varepsilon_t
\tag{4}
$$

where, y_t stand for CO_2, GDP, GDP^2, URB, and HYDRO; σ stands for intercept; β stands for matrices coefficient; the error term is illustrated by ε; and the time function is depicted by t. To catch the structural shift, the expansion of Fourier was incorporated and depicted as follow:

$$
\sigma(t) = \sigma_0 + \sum_{k=1}^{n} \gamma_{1k} \sin\left(\frac{2\pi kt}{T}\right) + \sum_{k=1}^{n} \gamma_{2k} \cos\left(\frac{2\pi kt}{T}\right)
\tag{5}
$$

where frequency approximation is depicted by k. The components of the single frequency are depicted in Equation (6):

$$\sigma(t) = \sigma_0 + \gamma_1 \sin\left(\frac{2\pi kt}{T}\right) + \gamma_2 \cos\left(\frac{2\pi kt}{T}\right) \tag{6}$$

where γ_{2k} and γ_{1k} are utilized in measuring displacement and frequency amplitude, respectively, and n stands for the frequency. By substituting Equation (6) into (4), the structural change is thus considered and defines the FY causality with cumulative frequencies (CF).

$$y_t = \sigma_0 + \gamma_1 \sin\left(\frac{2\pi kt}{T}\right) + \gamma_2 \cos\left(\frac{2\pi kt}{T}\right) + \beta_1 y_{t-1} + \ldots + \beta_{p+d} y_{t-(p+d)} + \varepsilon_t \tag{7}$$

Here, the null hypothesis is tested by utilizing the Wald statistic.

Recent research in the Granger causality literature has focused on critical values of bootstrapping to boost the strength of the test statistic in small sample sizes while also being robust in the data's unit root and cointegration features [45,46]. The specification issue in Equation (5) is to determine how many Fourier frequency and lag lengths there are. Using information criteria such as Schwarz or Akaikec to find p is a common method. This method may also be used to solve Equation (5), although with some changes because it also needs the frequency determination.

4. Results and Discussion

4.1. Findings

We commenced the analyses by presenting a brief description of the series. The present study utilized the Rader chart (Figure 1) to present the description of the variable. The GDP has the highest mean value, which is followed by URB, GDP, HYDRO, and CO_2. The value of the skewness shows that all the series of investigations are moderately skewed since their values are less than 1. Furthermore, URB and GDP are negatively skewed while CO_2 and GDP^2 are positively skewed. Moreover, the value of the kurtosis showed that all the series comply with normal distribution since their values are less than 3 (platykurtic). The value of the Jarque–Bera for all the series showed that all the series conform to normality.

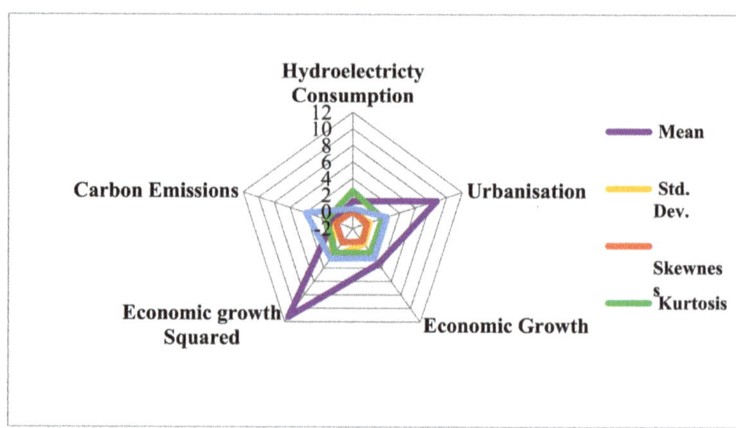

Figure 1. Descriptive statistics.

Furthermore, it was crucial to catch the series order of integration before commencing additional analysis. Based on this, the present study utilized the Zivot and Andrews (ZA) stationarity tests initiated by [47]. Though there are several conventional unit root tests (e.g., ADF, PP, KPSS, and EGLS) used in empirical analysis, this present research did not utilize these tests because they would yield misleading outcomes if there was proof of a structural change in the series [48,49]. The outcomes of the ZA are depicted in Table 1 and they showed that all the series are nonstationary at a level. Nonetheless, after the series' first difference is taken, all the series are stationary at first difference. We progressed to the cointegration test after confirming that the series is of order (1). The bounds test outcomes are shown in Table 2. The F-statistic (6.76) is greater than the lower and upper critical values at 1%, 5%, and 10% significance level. This implies that the null hypothesis of no cointegration is rejected, suggesting proof of long-run association between CO_2, GDP, GDP^2, URB, and HYDRO.

Table 1. ZA (intercept and trend).

	Level		First Difference	
	t-Statistic	Break Date	*t*-Statistic	Break Date
CO_2	−3.368	1996	−5.700 *	2002
URB	−1.752	2004	−6.028 *	2001
GDP	−4.217	2009	−5.293 **	2005
GDP^2	−4.192	1995	−5.804 **	2008
HYDRO	−4.032	2012	−5.835 *	2012

Note: * and ** stand for 1% and 5% significance level. BD denote break-date.

Table 2. Bound test.

F-Statistics		6.76 *			
Break Date		2002			
Cointegration		Yes			
10%		5%		1%	
F-statistics CV	2.204	3.320	2.615	3.891	3.572	5.112

Note: * represents a 1% level of significance.

After determining the presence of a long-run connection in the series, the next phase was to assess the effects of GDP, GDP^2, URB, and HYDRO on CO_2 and authenticate the EKC's existence. Table 3 shows the short- and long-term effects of GDP, GDP^2, URB, and HYDRO on CO_2. The outcomes from the long-run ARDL approach show that (i) GDP impacts CO_2 positively; (ii) URB impacts CO_2 emissions positively; (iii) there is a negative linkage between HYDRO and CO_2; (iv) GDP^2 impacts CO_2 emissions negatively, which confirms the EKC hypothesis; and (v) there is no significant linkage between BD and CO_2 emissions. Additionally, the outcomes of the short-run are comparable to the long-run outcomes except for the BD, which is the dummy variable, that has a positive impact on CO_2 emissions. The outcome of the ECM (−0.62) is negative and statistically significant, which confirms that errors from the previous periods can be corrected by the subsequent periods. The speed of adjustment was found to permit convergence among the variables in the long run with a negative and significant coefficient of the error correction model (ECM). The finding that the ECT was 0.62 indicates that there is cointegration among the variables, and this shows the ability of the model to experience a speed of adjustment at the rate of 62 percent to ascertain alignment to long-run equilibrium on CO_2 emissions due to the impact of the explanatory variables (urbanization, hydroelectricity consumption, and economic growth). Moreover, the present study conducted several diagnostic tests, which are depicted in Table 4. The outcomes showed that there are no heteroscedasticity or serial correlation in the model. In addition, there was no misspecification revealed by the RESET test. The outcomes of the CUSUM and CUSUM of Sq in Figure 2a,b uncovered that the model was stable at a 5% level of significance. Figure 3 shows the graphical outcomes of the ARDL long-run outcomes.

Table 3. ARDL long- and short-run outcomes.

Regressors	Long-Run Outcomes			Short-Run Outcomes		
	Coefficient	T-Statistics	*p*-Value	Coefficient	T-Statistics	*p*-Value
GDP	10.176 **	2.2101	0.0411	6.7884 *	3.8516	0.0013
GDP^2	−1.5234 ***	−2.0772	0.0533	−1.0392 *	−3.6959	0.0018
URB	4.9196 *	3.1177	0.0063	4.792 *	6.0073	0.0000
HYDRO	−0.1240 ***	−1.8521	0.0815	−0.2359 *	−4.3852	0.0005
BD	0.0158	1.4209	0.1734	0.2039 ***	1.8096	0.0904
ECM_{t-1}				−0.625*	−8.5033	0.0000
Diagnostic Tests						
R^2	0.99					
Adj R^2	0.98					
$\chi2$ ARCH	0.933 (0.555)					
$\chi2$ RESET	0.761 (0.458)					
$\chi2$ Normality	0.515 (0.772)					
$\chi2$ LM	2.041 (0.174)					

Note: 1%, 5%, and 10% levels of significance are illustrated by *, **, and ***, respectively.

Table 4. Gradual shift causality outcomes.

Dependent Variable	CO_2	GDP	GDP^2	URB	HYDRO
CO_2	1	20.070 *	18.622 *	7.4364	14.310 **
GDP	16.563 **	1	15.904 **	11.040	22.277 *
GDP^2	2.8363	16.224 **	1	14.945 **	23.781 *
URB	59.597 *	13.566 ***	13.550 **	1	38.204 *
HYDRO	21.584 *	23.651 *	17.812 **	8.176028	1

Note: 1%, 5%, and 10% levels of significance are illustrated by *, **, and ***, respectively.

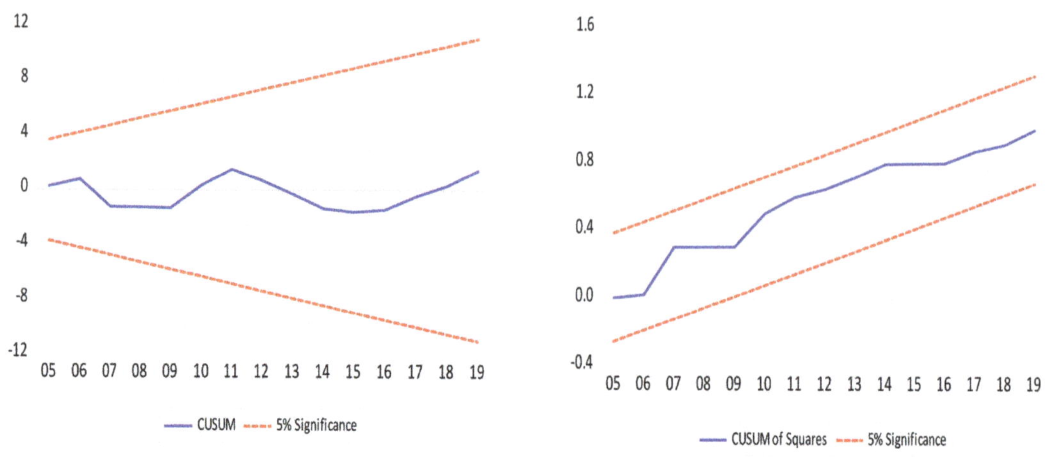

Figure 2. Stability test. (**a**): CUSUM, (**b**): CUSUM of Square.

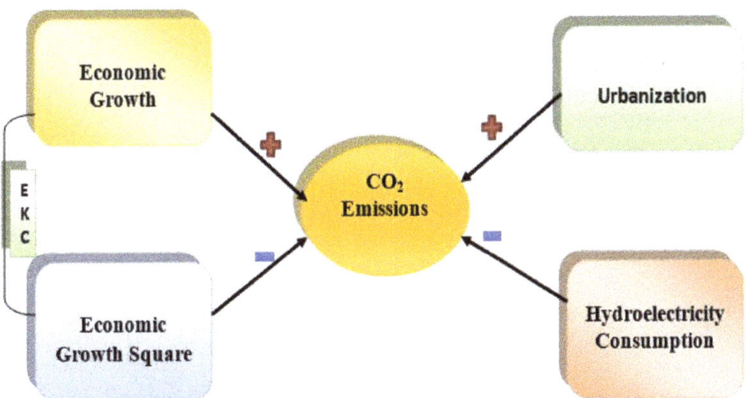

Figure 3. Graphical outcomes of the ARDL long-run estimation.

After the connection between CO_2 and GDP, GDP^2, URB, and HYDRO was established; the present study assessed the causal effect of GDP, GDP^2, URB, and HYDRO on CO_2 in China between 1985 and 2019. In doing so, our research introduced a gradual shift causality test that took into account structural changes. Conventional methods that look for abrupt changes are ineffective in identifying structural changes that develop gradually. To that end, we incorporated a Fourier approximation inside Toda and Yamamoto's (1995) Granger causality procedure. The Fourier approximation can capture smoothing or gradual changes without requiring past information of the forms of breaks, dates, or numbers [44,50]. The outcomes of the gradual shift causality test are depicted in Table 4. The outcomes showed bidirectional causal linkage between HYDRO and GDP; GDP and GDP^2, CO_2, and GDP; HYDRO and GDP^2; and lastly HYDRO and CO_2. Furthermore, there was proof of unidirectional causal linkage from CO_2 to GDP; from URB to GDP; from CO_2 to GDP; and from URB to HYDRO. Figure 4 presents the graphical outcomes of the gradual shift causality test.

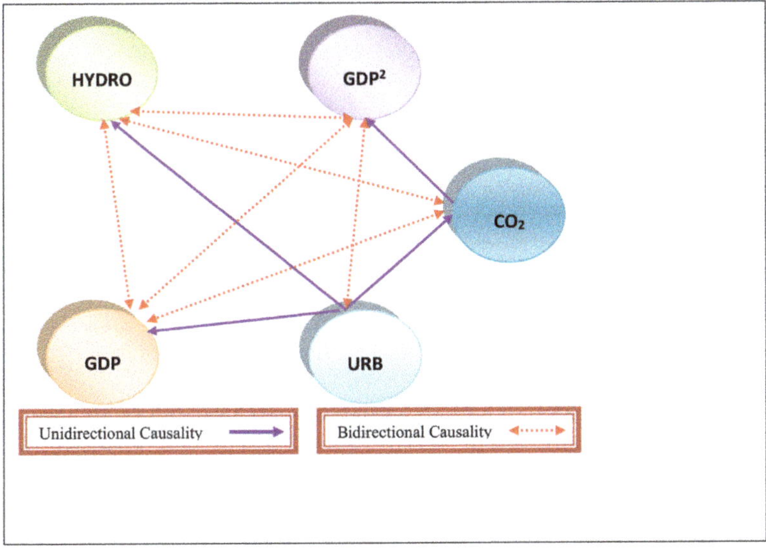

Figure 4. Graphical causality outcomes.

4.2. Discussions

The outcomes from the study support the EKC hypothesis in China. Furthermore, the causal connection between income growth and emissions illustrates that GDP (GDP squared) and CO_2 influence each other. Since there is an indication that causation flows from GDP to CO_2, this lends support to the validation of the EKC hypothesis in China. This outcome validates the studies of [51] for different regions, Murshed et al. [52] for South Asia nations, Pata and Caglar [53] for China, and Saint Akadırı et al. [54] for BRIC nations. Nonetheless, this study contradicts the studies of Ozturk and Al-mulali [55] for Cambodia, Al-mulali et al. [56] for Vietnam, and Govindaraju and Tang [57] for India and China who were unable to validate the EKC hypothesis.

The GDP–CO_2 positive association is braced by the fact that numerous industries in China consume fossil fuel. In 2020, China's total primary energy consumption constitutes coal (58%), petroleum and other liquids (20%), natural gas (8%), hydroelectricity (8%), other renewable (5%), and nuclear (2%) [7]. The fact that both GDP and CO_2 have been growing concurrently throughout the years adds to the evidence supporting a positive association between income and emissions. In addition, the rising rate of emissions can be reduced by increasing the usage of hydroelectricity and, maybe, focusing on less-energy-intensive industries. This implies that an upsurge in hydroelectricity aids in mitigating environmental degradation in China. This outcome is consistent with the studies of Solarin et al. [30], Kivyiro and Arminen [42], Gyamfi et al. [11], and Ridzuan et al. [41] who established a negative linkage between hydroelectricity and environmental deterioration.

The positive association between urbanization and CO_2 signifies that an upsurge in the urban population triggered degradation of the environment in China. This finding implies that China's urbanization contributed to an increase in energy demand fueled by fossil fuel resources, which in turn increased CO_2 emissions. This is not a surprising discovery given that CO_2 emissions originate from these networks (transportation, electrical appliances demand, and construction of commercial and residential buildings). In the early phases of urbanization, societies often seek flexibility and comfort by relying on personal vehicles rather than focusing on a sustainable environment. This outcome complies with the studies of Zhang et al. [17] for Malaysia, Sakiru et al. [30] for India and China, Awosusi et al. [29] for South Korea, and Kirikkaleli and Kalmaz [12], Olanrewaju et al. [21], and Rjoub et al. [58] for Turkey who found positive urbanization and CO_2 interconnection. The break in 2002 is related to China's major economic shift. China has been making strides toward a more open economy and boosting foreign trade with a number of nations. Many Chinese households' quality of life has improved, and the country has experienced promising economic growth [30]. The majority of the movement in the early 2000s is related to the pattern of economic transition that emerged in the nation during that time period. The regulatory and administrative reform of rural–urban migration policies, the tax system, foreign trade, international trade, and the financial system removed various legally binding constraints on economic expansion.

The outcomes from the gradual shift causality showed feedback causality between HYDRO and CO_2. Likewise, there is proof of bidirectional causal linkage between GDP and HYDRO, suggesting that HYDRO and CO_2 can predict each other. This outcome complies with the studies of Govindaraju and Tang [57] for India and China, Tiwari et al. [59], and Wang et al. [60]; however, it contradicts the outcome of Zhang and Cheng [61]. The findings of this study come as no surprise, as more usage of hydroelectricity may be related to lower use of fossil fuels. Hydroelectricity usage constitutes a minor portion of the energy mix, and there is still space for enhancement. In China, hydroelectricity accounted for 8% of the energy mix in 2020 [7]. The government of China has implemented programs and initiatives to promote the nation's usage of hydroelectricity. The Three Gorges Dam hydropower plant, the world's largest hydroelectricity system, is situated in China. It officially opened in 2003, and construction was finished in 2012 [6]. In 2019, China's investments in green energy stood at USD 83.4 billion, which represents 23% of the worldwide renewable energy investment. China is now becoming the largest global market for green energy [62].

The fact that China's metropolitan areas hold the economic key to the nation's success makes these regression outcomes unsurprising. Cities such as Beijing, Hangzhou, Guangzhou, Shanghai, and Shenzhen are China's commercial hubs. Since 2001, the urban populace in China has expanded by more than 200 million, accounting for more than 80% of the country's GDP [63]. As a result, the importance of urbanization cannot be overstated. This outcome is not unique for China, since cities and towns worldwide are the engines of economic success—over 80% of economic activities worldwide are produced in metropolitan areas.

5. Conclusions

The current research re-assesses the EKC in China and takes into consideration the role of hydroelectricity consumption and urbanization utilizing data stretching from 1985 to 2019. The research utilized a series of econometric techniques such as unit root, ARDL approach with a structural break, and gradual shift causality tests to assess these associations. The outcomes of the bounds test showed a long-run association between the variables of investigation. Moreover, we utilized the ARDL approach to catch long-run linkage between CO_2 emissions and HYDRO, GDP, GDP^2, and URB. The outcomes affirmed the EKC presence in China. Moreover, HYDRO exerts a negative effect on CO_2 with feedback causality between HYDRO and CO_2. With growing hydroelectricity in the energy mix, the usage of fossil fuels, which accounts for the majority of CO_2 emissions, is anticipated to drop. The policy relevance of these outcomes is that increased hydroelectricity use is anticipated to reduce emissions. Furthermore, there is a feedback causal linkage between utilization of hydroelectricity and GDP growth. This suggests that utilization of hydroelectricity has stimulated economic growth, with economic activity also positively influencing hydroelectricity consumption. This illustrates that energy plays a significant role in boosting economic growth, so that reducing energy consumption indiscriminately may have a negative impact on the nations' economic growth.

In China, incremental demand for energy, such as hydroelectricity consumption, has been growing as a result of continuing economic expansion, rising income levels, and increased availability of products and services. As a result, regulations that decrease the usage of hydroelectricity will have a negative impact on China's economic growth. Any hydroelectricity deficit will also hinder economic progress. Furthermore, a decrease in output will have a negative impact on the hydroelectricity demand. Shock to one of the series of interest will be felt in the others, and the feedback flow will keep the chain going. As a result, China would benefit from expansionary hydroelectricity plans. All of the remaining series have causal interconnections with urbanization. The one-way causation from urbanization to emission implies that urbanization is equally to blame for emissions. Furthermore, the unidirectional causal linkage from urbanization to income implies that urbanization is a determinant of economic expansion in China. As such, urbanization is a tool for economic progress. Policymakers who want to boost economic development in the long term by encouraging urbanization are likely to succeed.

Although the empirical outcome from the present study showed that hydroelectricity consumption impacts CO_2 emissions negatively, which aids in mitigating environmental degradation, it's negative environmental and societal risks, including degraded wildlife habitat, deteriorated water quality, impeded fish movement, and reduced recreational advantages on rivers, cannot be neglected. Therefore, future studies should investigate the influence of other renewables (e.g., solar energy, wind energy, geothermal energy, and biomass energy) on CO_2 emissions.

Furthermore, CO_2 emissions are viewed as a proxy for environmental damage, but they are not the sole proxy for the degradation of the environment. In the case of China, more studies should be conducted to study this linkage by integrating other determinants of environmental degradation.

Author Contributions: Conceptualization, T.S.A., M.O.A.; methodology, T.S.A., M.O.A., H.R., I.A.; software, T.S.A., M.O.A., H.R., I.A.; validation, T.S.A., M.O.A., H.R., I.A., E.B.A., N.M.K.; formal analysis, T.S.A., M.O.A., H.R., I.A., E.B.A., N.M.K.; investigation, T.S.A., M.O.A., H.R., I.A., E.B.A., N.M.K.; resources, T.S.A., M.O.A., H.R., I.A., E.B.A., N.M.K.; data curation, T.S.A., M.O.A., H.R., I.A., E.B.A., N.M.K.; writing—original draft preparation, T.S.A.; writing—review and editing, T.S.A., M.O.A., H.R., I.A., E.B.A., N.M.K.; visualization, T.S.A., M.O.A., H.R., I.A.; funding acquisition, N.M.K. All authors have read and agreed to the published version of the manuscript.

Funding: This research received no external funding.

Institutional Review Board Statement: Not applicable.

Informed Consent Statement: Not applicable.

Data Availability Statement: Data used for the study are available in the text.

Conflicts of Interest: The authors declare no conflict of interest.

References

1. Agyekum, E.; Ali, E.; Kumar, N. Clean Energies for Ghana—An Empirical Study on the Level of Social Acceptance of Renewable Energy Development and Utilization. *Sustainability* **2021**, *13*, 3114. [CrossRef]
2. Agyekum, E.B. Techno-economic comparative analysis of solar photovoltaic power systems with and without storage systems in three different climatic regions, Ghana. *Sustain. Energy Technol. Assess.* **2021**, *43*, 100906. [CrossRef]
3. BP Statistical Review—2019. Available online: https://www.bp.com/content/dam/bp/business-sites/en/global/corporate/pdfs/energy-economics/statistical-review/bp-stats-review-2019-china-insights.pdf (accessed on 30 May 2021).
4. Umar, M.; Ji, X.; Kirikkaleli, D.; Xu, Q. COP21 Roadmap: Do innovation, financial development, and transportation infrastructure matter for environmental sustainability in China? *J. Environ. Manag.* **2020**, *271*, 111026. [CrossRef]
5. Global Economy. China Percent of World GDP—Data, Chart. TheGlobalEconomyCom. 2021. Available online: https://www.theglobaleconomy.com/china/gdp_share/ (accessed on 31 May 2021).
6. EIA. International—U.S. In *Energy Information Administration (EIA) 2015.*. Available online: https://www.eia.gov/international/overview/world?Fips=ch (accessed on 30 May 2021).
7. BP. Statistical Review of World Energy 2020. 2021. Available online: https://www.bp.com/content/dam/bp/business-sites/en/global/corporate/pdfs/energy-economics/statistical-review/bp-stats-review-2020-full-report.pdf (accessed on 30 May 2021).
8. Agyekum, E.; Adebayo, T.; Bekun, F.; Kumar, N.; Panjwani, M. Effect of Two Different Heat Transfer Fluids on the Performance of Solar Tower CSP by Comparing Recompression Supercritical CO_2 and Rankine Power Cycles, China. *Energies* **2021**, *14*, 3426. [CrossRef]
9. Shokri, A.; Heo, E. Energy Policies to promote Renewable Energy Technologies; Learning from Asian Countries Experiences. In Proceedings of the Conference International Association for Energy Economics, Austin, TX, USA, 4–7 November 2012; pp. 1–10.
10. Orhan, A.; Adebayo, T.; Genç, S.; Kirikkaleli, D. Investigating the Linkage between Economic Growth and Environmental Sustainability in India: Do Agriculture and Trade Openness Matter? *Sustainability* **2021**, *13*, 4753. [CrossRef]
11. Gyamfi, B.A.; Bein, M.A.; Ozturk, I.; Bekun, F.V. The Moderating Role of Employment in an Environmental Kuznets Curve Framework Revisited in G7 Countries. *Indones. J. Sustain. Account. Manag.* **2020**, *4*, 241–248. [CrossRef]
12. Kirikkaleli, D.; Kalmaz, D.B. Testing the moderating role of urbanization on the environmental Kuznets curve: Empirical evidence from an emerging market. *Environ. Sci. Pollut. Res.* **2020**, *27*, 38169–38180. [CrossRef]
13. Soylu, Ö.; Adebayo, T.; Kirikkaleli, D. The Imperativeness of Environmental Quality in China Amidst Renewable Energy Consumption and Trade Openness. *Sustainability* **2021**, *13*, 5054. [CrossRef]
14. Tufail, M.; Song, L.; Adebayo, T.S.; Kirikkaleli, D.; Khan, S. Do fiscal decentralization and natural resources rent curb carbon emissions? Evidence from developed countries. *Environ. Sci. Pollut. Res.* **2021**, 1–12. [CrossRef]
15. He, X.; Adebayo, T.S.; Kirikkaleli, D.; Umar, M. Consumption-based carbon emissions in Mexico: An analysis using the dual adjustment approach. *Sustain. Prod. Consum.* **2021**, *27*, 947–957. [CrossRef]
16. Kalmaz, D.B.; Kirikkaleli, D. Modeling CO_2 emissions in an emerging market: Empirical finding from ARDL-based bounds and wavelet coherence approaches. *Environ. Sci. Pollut. Res.* **2019**, *26*, 5210–5220. [CrossRef] [PubMed]
17. Zhang, L.; Li, Z.; Kirikkaleli, D.; Adebayo, T.S.; Adeshola, I.; Akinsola, G.D. Modeling CO_2 emissions in Malaysia: An application of Maki cointegration and wavelet coherence tests. *Environ. Sci. Pollut. Res.* **2021**, *28*, 26030–26044. [CrossRef]
18. Adebayo, T.S.; Akinsola, G.D.; Kirikkaleli, D.; Bekun, F.V.; Umarbeyli, S.; Osemeahon, O.S. Economic performance of Indonesia amidst CO_2 emissions and agriculture: A time series analysis. *Environ. Sci. Pollut. Res.* **2021**, 1–15. [CrossRef]
19. Koengkan, M.; Fuinhas, J.A.; Santiago, R. The relationship between CO_2 emissions, renewable and non-renewable energy consumption, economic growth, and urbanisation in the Southern Common Market. *J. Environ. Econ. Policy* **2020**, *9*, 383–401. [CrossRef]

20. Khoshnevis Yazdi, S.; Golestani Dariani, A. CO 2 emissions, urbanisation and economic growth: Evidence from Asian countries. *Econ. Res. Ekon. Istraživanja* **2019**, *32*, 510–530. [CrossRef]
21. Adebayo, T.S.; Akinsola, G.D.; Odugbesan, J.A.; Olanrewaju, V.O. Determinants of Environmental Degradation in Thailand: Empirical Evidence from ARDL and Wavelet Coherence Approaches. *Pollution* **2021**, *7*, 181–196.
22. Jian, J.; Fan, X.; He, P.; Xiong, H.; Shen, H. The Effects of Energy Consumption, Economic Growth and Financial Development on CO_2 Emissions in China: A VECM Approach. *Sustainability* **2019**, *11*, 4850. [CrossRef]
23. Charfeddine, L.; Kahia, M. Impact of renewable energy consumption and financial development on CO_2 emissions and economic growth in the MENA region: A panel vector autoregressive (PVAR) analysis. *Renew. Energy* **2019**, *139*, 198–213. [CrossRef]
24. Shahbaz, M.; Tiwari, A.; Nasir, M. The effects of financial development, economic growth, coal consumption and trade openness on CO_2 emissions in South Africa. *Energy Policy* **2013**, *61*, 1452–1459. [CrossRef]
25. Mutascu, M. A time-frequency analysis of trade openness and CO_2 emissions in France. *Energy Policy* **2018**, *115*, 443–455. [CrossRef]
26. Adebayo, T.S.; Kirikkaleli, D.; Adeshola, I.; Oluwajana, D.; Akinsola, G.D.; Osemeahon, O.S. Coal Consumption and Environmental Sustainability in South Africa: The role of Financial Development and Globalization. *Int. J. Renew. Energy Dev.* **2021**, *10*, 527–536. [CrossRef]
27. Lv, Z.; Xu, T. Trade openness, urbanization and CO_2 emissions: Dynamic panel data analysis of middle-income countries. *J. Int. Trade Econ. Dev.* **2018**, *28*, 317–330. [CrossRef]
28. Dauda, L.; Long, X.; Mensah, C.N.; Salman, M.; Boamah, K.B.; Ampon-Wireko, S.; Dogbe, C.S.K. Innovation, trade openness and CO_2 emissions in selected countries in Africa. *J. Clean. Prod.* **2021**, *281*, 125143. [CrossRef]
29. Adebayo, T.S.; Awosusi, A.A.; Kirikkaleli, D.; Akinsola, G.D.; Mwamba, M.N. Can CO_2 emissiovns and energy consumption determine the economic performance of South Korea? A time series analysis. *Environ. Sci. Pollut. Res.* **2021**, 1–16.
30. Solarin, S.A.; Al-Mulali, U.; Ozturk, I. Validating the environmental Kuznets curve hypothesis in India and China: The role of hydroelectricity consumption. *Renew. Sustain. Energy Rev.* **2017**, *80*, 1578–1587. [CrossRef]
31. Dogan, E.; Inglesi-Lotz, R. The impact of economic structure to the environmental Kuznets curve (EKC) hypothesis: Evidence from European countries. *Environ. Sci. Pollut. Res.* **2020**, *27*, 12717–12724. [CrossRef]
32. Adebayo, T.S.; Kirikkaleli, D. Impact of renewable energy consumption, globalization, and technological innovation on environmental degradation in Japan: Application of wavelet tools. *Environ. Dev. Sustain.* **2021**. [CrossRef]
33. Adedoyin, F.F.; Alola, A.A.; Bekun, F.V. An assessment of environmental sustainability corridor: The role of economic expansion and research and development in EU countries. *Sci. Total Environ.* **2020**, *713*, 136726. [CrossRef]
34. Shahbaz, M.; Balsalobre-Lorente, D.; Sinha, A. Foreign direct Investment–CO_2 emissions nexus in Middle East and North African countries: Importance of biomass energy consumption. *J. Clean. Prod.* **2019**, *217*, 603–614. [CrossRef]
35. Kirikkaleli, D. New insights into an old issue: Exploring the nexus between economic growth and CO_2 emissions in China. *Environ. Sci. Pollut. Res.* **2020**, *27*, 1–10. [CrossRef]
36. Umar, M.; Ji, X.; Kirikkaleli, D.; Alola, A.A. The imperativeness of environmental quality in the United States transportation sector amidst biomass-fossil energy consumption and growth. *J. Clean. Prod.* **2021**, *285*, 124863. [CrossRef]
37. Solarin, S.A.; Ozturk, I. On the causal dynamics between hydroelectricity consumption and economic growth in Latin America countries. *Renew. Sustain. Energy Rev.* **2015**, *52*, 1857–1868. [CrossRef]
38. Apergis, N.; Chang, T.; Gupta, R.; Ziramba, E. Hydroelectricity consumption and economic growth nexus: Evidence from a panel of ten largest hydroelectricity consumers. *Renew. Sustain. Energy Rev.* **2016**, *62*, 318–325. [CrossRef]
39. Gyamfi, B.A.; Bein, M.A.; Bekun, F.V. Investigating the nexus between hydroelectricity energy, renewable energy, nonrenewable energy consumption on output: Evidence from E7 countries. *Environ. Sci. Pollut. Res.* **2020**, *27*, 25327–25339. [CrossRef]
40. Adebayo, T.S.; Udemba, E.N.; Ahmed, Z.; Kirikkaleli, D. Determinants of consumption-based carbon emissions in Chile: An application of non-linear ARDL. *Environ. Sci. Pollut. Res.* **2021**, 1–15. [CrossRef]
41. Ridzuan, A.R.; Albani, A.; Latiff, A.R.A.; Razak, M.I.M.; Murshidi, M.H. The Impact of Energy Consumption Based on Fossil Fuel and Hydroelectricity Generation towards Pollution in Malaysia, Indonesia And Thailand. *Int. J. Energy Econ. Policy* **2020**, *10*, 215–227. [CrossRef]
42. Kivyiro, P.; Arminen, H. Carbon dioxide emissions, energy consumption, economic growth, and foreign direct investment: Causality analysis for Sub-Saharan Africa. *Energy* **2014**, *74*, 595–606. [CrossRef]
43. Pesaran, M.H.; Shin, Y.; Smithc, R.J. Bounds testing approaches to the analysis of level relationships. *J. Appl. Econ.* **2001**, *16*, 289–326. [CrossRef]
44. Nazlioglu, S.; Gormus, N.A.; Soytas, U. Oil prices and real estate investment trusts (REITs): Gradual-shift causality and volatility transmission analysis. *Energy Econ.* **2016**, *60*, 168–175. [CrossRef]
45. Hatemi-J, A.; Irandoust, M. On the Causality Between Exchange Rates and Stock Prices: A Note. *Bull. Econ. Res.* **2002**, *54*, 197–203. [CrossRef]
46. Balcilar, M.; Ozdemir, Z.A.; Arslanturk, Y. Economic growth and energy consumption causal nexus viewed through a bootstrap rolling window. *Energy Econ.* **2010**, *32*, 1398–1410. [CrossRef]
47. Zivot, E.; Andrews, D.W.K. Further Evidence on the Great Crash, the Oil-Price Shock, and the Unit-Root Hypothesis. *J. Bus. Econ. Stat.* **2002**, *20*, 25–44. [CrossRef]

48. Adebayo, T.S.; Kalmaz, D.B. Ongoing Debate Between Foreign Aid and Economic Growth in Nigeria: A Wavelet Analysis. *Soc. Sci. Q.* **2020**, *101*, 2032–2051. [CrossRef]
49. Wang, K.-H.; Liu, L.; Adebayo, T.S.; Lobon, O.-R.; Claudia, M.N. Fiscal decentralization, political stability and resources curse hypothesis: A case of fiscal decentralized economies. *Resour. Policy* **2021**, *72*, 102071. [CrossRef]
50. Enders, W.; Lee, J. A Unit Root Test Using a Fourier Series to Approximate Smooth Breaks*. *Oxf. Bull. Econ. Stat.* **2011**, *74*, 574–599. [CrossRef]
51. Bibi, F.; Jamil, M. Testing environment Kuznets curve (EKC) hypothesis in different regions. *Environ. Sci. Pollut. Res.* **2021**, *28*, 13581–13594. [CrossRef]
52. Murshed, M.; Ali, S.R.; Banerjee, S. Consumption of liquefied petroleum gas and the EKC hypothesis in South Asia: Evidence from cross-sectionally dependent heterogeneous panel data with structural breaks. *Energy Ecol. Environ.* **2021**, *6*, 353–377. [CrossRef]
53. Pata, U.K.; Aydin, M. Testing the EKC hypothesis for the top six hydropower energy-consuming countries: Evidence from Fourier Bootstrap ARDL procedure. *J. Clean. Prod.* **2020**, *264*, 121699. [CrossRef]
54. Akadırı, S.S.; Alola, A.A.; Usman, O. Energy mix outlook and the EKC hypothesis in BRICS countries: A perspective of economic freedom vs. economic growth. *Environ. Sci. Pollut. Res.* **2021**, *28*, 8922–8926. [CrossRef]
55. Al-Mulali, U.; Ozturk, I.; Lean, H.H. The influence of economic growth, urbanization, trade openness, financial development, and renewable energy on pollution in Europe. *Nat. Hazards* **2015**, *79*, 621–644. [CrossRef]
56. Al-Mulali, U.; Saboori, B.; Ozturk, I. Investigating the environmental Kuznets curve hypothesis in Vietnam. *Energy Policy* **2015**, *76*, 123–131. [CrossRef]
57. Govindaraju, V.G.R.C.; Tang, C.F. The dynamic links between CO$_2$ emissions, economic growth and coal consumption in China and India. *Appl. Energy* **2013**, *104*, 310–318. [CrossRef]
58. Rjoub, H.; Odugbesan, J.A.; Adebayo, T.S.; Wong, W.-K. Sustainability of the Moderating Role of Financial Development in the Determinants of Environmental Degradation: Evidence from Turkey. *Sustainability* **2021**, *13*, 1844. [CrossRef]
59. Tiwari, A.K.; Shahbaz, M.; Hye, Q.M.A. The environmental Kuznets curve and the role of coal consumption in India: Cointegration and causality analysis in an open economy. *Renew. Sustain. Energy Rev.* **2013**, *18*, 519–527. [CrossRef]
60. Wang, S.; Zhou, D.; Zhou, P.; Wang, Q. CO$_2$ emissions, energy consumption and economic growth in China: A panel data analysis. *Energy Policy* **2011**, *39*, 4870–4875. [CrossRef]
61. Zhang, X.-P.; Cheng, X.-M. Energy consumption, carbon emissions, and economic growth in China. *Ecol. Econ.* **2009**, *68*, 2706–2712. [CrossRef]
62. CSIS. How Is China's Energy Footprint Changing? | ChinaPower Project n.d. Available online: https://chinapower.csis.org/energy-footprint/ (accessed on 30 May 2021).
63. World Bank. World Development Indicators. *Washington 2015.* Available online: https://datacatalog.worldbank.org/ (accessed on 30 May 2021).

International Journal of
Environmental Research and Public Health

Article

Do Environmental Stringency Policies and Human Development Reduce CO$_2$ Emissions? Evidence from G7 and BRICS Economies

Funda Hatice Sezgin [1,*], Yilmaz Bayar [2], Laura Herta [3] and Marius Dan Gavriletea [4]

[1] Department of Industrial Engineering, Istanbul University-Cerrahpaşa, 34320 Istanbul, Turkey
[2] Department of Economics, Bandirma Onyedi Eylul University, 10200 Bandirma, Turkey; yilmazbayar@yahoo.com
[3] Department of International Relations and German Studies, Faculty of European Studies, Babeş-Bolyai University, 400084 Cluj-Napoca, Romania; laura.herta@ubbcluj.ro
[4] Department of Business, Faculty of Business, Babeş-Bolyai University, 400084 Cluj-Napoca, Romania; marius.gavriletea@ubbcluj.ro
* Correspondence: hfundasezgin@yahoo.com

Abstract: This study explores the impact of environmental policies and human development on the CO$_2$ emissions for the period of 1995–2015 in the Group of Seven and BRICS economies in the long run through panel cointegration and causality tests. The causality analysis revealed a bilateral causality between environmental stringency policies and CO$_2$ emissions for Germany, Japan, the United Kingdom, and the United States of America, and a unilateral causality from CO$_2$ emissions to the environmental stringency policies for Canada, China, and France. On the other hand, the analysis showed a bilateral causality between human development and CO$_2$ emissions for Germany, Japan, the United Kingdom, and the United States of America, and unilateral causality from CO$_2$ emissions to human development in Brazil, Canada, China, and France. Furthermore, the cointegration analysis indicated that both environmental stringency policies and human development had a decreasing impact on the CO$_2$ emissions.

Keywords: environmental stringency policies; human development; CO$_2$ emissions; panel cointegration and causality analyses

Citation: Sezgin, F.H.; Bayar, Y.; Herta, L.; Gavriletea, M.D. Do Environmental Stringency Policies and Human Development Reduce CO$_2$ Emissions? Evidence from G7 and BRICS Economies. *Int. J. Environ. Res. Public Health* **2021**, *18*, 6727. https://doi.org/10.3390/ijerph18136727

Academic Editors: Pasquale Avino, Massimiliano Errico, Aristide Giuliano and Hamid Salehi

Received: 22 May 2021
Accepted: 21 June 2021
Published: 22 June 2021

Publisher's Note: MDPI stays neutral with regard to jurisdictional claims in published maps and institutional affiliations.

1. Introduction

The environmental problems, together with food security, global health, education, gender equality, poverty, are some of the most pressing global issues in the world.

Global warming, pollution, ozone and natural resources depletion, habitat and biodiversity loss, deforestation, human overpopulation, and waste disposal have been identified as the main global environmental issues [1,2] that need to be solved on a global scale. All countries are facing environmental challenges and need to ensure environmental protection, but there are significant differences from country to country related to strategies adopted to address all these challenges.

Researchers found a major gap between developed and developing countries in the ways in which environmental issues are perceived and understood, and the adoption of concrete strategies and actions for addressing environmental problems [3–5]. The problem of environmental degradation and protection must be viewed in a more complex and dynamic way: most developed countries have adopted environmental policies designed to protect the environment, but at the same time these countries had, and still have, the highest carbon dioxide emissions per capita [6]. In these circumstances, differentiated responsibilities and obligations must be imposed for countries based on their level of development.

New patterns of development based on environmental protection must be found, and developed nations must lead in finding proper regulations, standards, policies, technologies that facilitate and encourage the transition to environmentally sustainable economies and societies. Developing and least developed countries must be helped in their transition process to sustainable development, since they are confronted with issues such as the dominance of primary sectors, a significant percent of informal employment [7], a lack of new technologies, and a lack of knowledge that makes the transition more difficult compared to developed countries.

In this context, various environmental policies, including legal and market-based solutions such as property rights, environmental standards, and environmental taxes, have been developed to combat environmental degradation.

Over the last few decades, environmental policies have gained prominence in both government's strategic agendas and non-state actors' agendas, with such endeavors as the 2015 Paris Agreement marking ambitious steps in the struggle against greenhouse gas emissions. The substantial effort to reduce such emissions and to limit global warming implies complex and often controversial economic, social and even political stakes and has left a strong imprint on doctrine. The field of international relations has also been impacted by the tensions between leaders that have been opposed to harsher measures associated with environmental protection, or even denied climate change and the science behind it altogether, and those who have developed and implemented strategies, plans, etc. to respond to different requirements set out by national and even supranational regulatory frameworks.

In 2005, the G8 leaders agreed on a set of common goals and principles designed to address climate change, clean energy and sustainable development in Gleneagles, UK. The result was the Gleneagles Plan of Action, according to which, the G8 countries pledged to encourage and support the development of more efficient and lower-emitting vehicles [8]. This goal was consistent with previous commitments and joint efforts of G8, as declared and included in communiqués following summits and high level meetings: undertaking "domestically the steps necessary to reduce significantly greenhouse gas emissions", as pledged in 1998 in Birmingham; promoting "increasing global participation of developing countries in limiting greenhouse gas emissions", as decided in 1999 in Cologne; collaborating with international institutions in order "to encourage and facilitate investment in the development and use of sustainable energy", as agreed upon in 2000 in Okinawa [9].

The European Union (EU) has a long-term commitment to combat climate change, reflected by a range of indicators that have so far been included in two multiannual agendas, i.e., the Lisbon Strategy and Europe 2020, focusing mainly on the reduction in greenhouse gas emissions compared to the year of reference, 1990, energy efficiency and the share of renewable energy in total consumption [10]. The capacity of the EU to accomplish its climate change goals, albeit hindered (or perhaps unexpectedly aided) by a series of crises—from financial crises to the coronavirus outbreak—has been proved by the progress that has been made in the use of two main indicators provided by Europe 2020. In 2019, the EU's overall greenhouse gas emissions were registered at 24% below 1990 levels, i.e., exceeding the target set for 2020, amid the reduction in emissions generated by the transportation sector during the coronavirus pandemic. Renewable energy accounted for 19.7% of all energy consumed in the EU in 2019, while energy efficiency remains insufficiently tackled [11]. The EU has decided that 30% of all expenditure from the 2021–2027 multiannual financial framework—amounting to 1.08 billion euros—should be spent on climate-related goals, the most consistent financial allocation in this regard so far [12].

The research goal of this study is to evaluate the impact of environmental policies and human development on the CO_2 emissions for the period of 1995–2015 in a sample of G7 and BRICS economies, the top global CO_2 emitters. The UNDP Human Development Report released in 1990 stipulated that: "The basic objective of development is to create an enabling environment for people to enjoy long, healthy and creative lives." [13]. Since the

Human Development Index is considered the most used proxy for human development in the related literature [14], we have decided to use this indicator for our analyses.

The human development index is a statistical tool developed by the United Nations in 1990 [13] that assesses countries' social and economic development based on three key dimensions that have been used in our study to capture the broader view of human development.

We focused our research on this particular sample of countries since developed and major emerging economies are considered the largest contributors to CO_2 emissions [15]. Over the last few decades, countries such as China, the United States, and India significantly increased their CO_2 emissions [16] and we find this trend around the world. Not only the level of change CO_2 in the atmosphere but also the rate at which emissions are changing are the main concerns [15] that force us to adopt and implement urgent and innovative measures to reduce emissions. Ecosystems are capable of adapting to different changes in the environment, but increased emissions from the last few decades does not offer too much time to adapt.

Urgent measures and actions are still needed on a global scale, but also, studies that investigate their impact on CO_2 emissions are necessary. Since environmental policy stringency and environmental taxes are considered the main policy instruments for fighting against environmental degradation [17] and there is a lack of studies on this issue, we focused our research on the relationship between environmental stringency policies and CO_2 emissions. Additionally, we considered it appropriate to focus attention on human development based on the fact that usually, countries' level of development influence the demand for products and services that in turn will influence the CO2 emissions. It could be possible for developed countries to invest more in efficient technologies and products so they can decouple output from greenhouse gas emission, but still, more research is necessary to investigate this trend. In this context, we intend to make a contribution to the existing literature in two ways: firstly, the study will be one of the first studies exploring the interaction among environmental policy, human development, and CO_2 emissions for G7 and BRICS economies, and secondly, we will use econometric tests with cross-sectional dependence that will lead us to obtain relatively more robust results. In this context, the relevant literature is summarized in the forthcoming section, and then the dataset and method are briefly explained in Section 3. The empirical analysis is conducted in the fourth part of the research, and the paper ends with the Conclusions section.

2. Literature Review

The effectiveness of environmental policy on CO_2 emissions has been explored by a limited number of scholars and the general consensus was that environmental policy stringency index developed by OECD [18] can act as a good proxy of environmental policy. Botta and Kózluk [19] claimed that the Environmental Policy Stringency indicators should be considered "a first tangible effort to measure environmental policy stringency internationally over a relatively long-time horizon", even though they represent a simplification of the multifaceted and multidimensional approach of environmental policies. However, as explained by Wolde-Rufael and Mulat-Weldemeskel [17], despite the fact that environmental tax and environmental policy stringency have become pivotal policy instruments against environmental degradation, there is still a research gap with respect to their combined effectiveness in mitigating emissions especially for emerging economies.

Using data from 1990 to 2012, Ahmed and Ahmed [20] estimated CO_2 emissions in China until 2022 and found that stringent environmental policies can contribute to emissions reduction. Based on a recent analysis of 20 OECD countries, Ahmed [21] determined that environmental regulations promote green innovations in the countries examined. Moreover, stringent environmental policies supplemented by environmentally friendly innovations can act as a catalyst for sustainable development. The study of Wang et al. [22], conducted on a panel of 23 OECD countries during 1990–2015, pointed out that environmental policy stringency has a negative impact on CO_2, NOx, and SOx emissions, and a weak impact on PM2.5 emissions and PM2.5 exposure. The results can be explained by the fact that we have multiple sources of PM2.5 emissions and environmental policies are difficult to be implemented; also, the process of environmental policy stringency composite index construction does not focus attention on PM2.5 limits or restrictions. The empirical results of this study confirm the role of environmental policy stringency and also noted several shortcomings.

Wolde-Rufael and Mulat-Weldemeskel [23] explored the effect of environmental policy stringency on CO_2 emissions in Brazil, Russia, India, Indonesia, China, Turkey and South Africa over the 1993–2014 period through panel pooled mean group autoregressive distributive lag estimator and discovered an inverted U-shaped interaction between CO_2 emissions and environmental policy stringency. Same authors [17] analyzed the effect of environmental policy stringency and environmental tax on CO_2 emissions in seven emerging countries over the 1994–2015 duration through an augmented mean group estimator and discovered a U-shaped interaction between environmental policy stringency and CO_2 emissions and unilateral causality from the environmental policy stringency index and total environmental tax to CO_2 emissions. Furthermore, they found that total environmental tax and energy taxes negatively affected the CO_2 emissions, but no significant causality was discovered among energy and CO_2 taxes and CO_2 emissions.

Many studies emphasize the impact of environmental and energy policy on important economic outcomes, such as innovation, productivity and energy efficiency. Albrizio et al. [24] noted that stringency has exhibited an increasing trend in OECD countries over the last two decades. However, stricter environmental policies do not have a significant effect on aggregate productivity, as they have only short-term impacts. The most technologically advanced industries and firms have met with a small increase in productivity, as they are more likely to adapt, while the productivity of the least productive firms has dropped. The erection of barriers to entry and competition, as well as the preoccupation for the economic effects of environmental policies, vary markedly across the countries analyzed, but this variation is found not to be connected to the stringency of environmental policies. Therefore, in order to yield positive economic and environmental results, stringent environmental policies should be accompanied by as few barriers to entry and competition as possible.

The Porter hypothesis suggests that more stringent environmental policies can foster innovation and productivity [25]. Using a group of OECD countries, Albrizio et al. [26] analyzed the impact of changes in environmental policy stringency on productivity growth at industry and firm level and revealed different results based on the level of countries' technological development. They pointed out that in the most technologically advanced countries, a more rigid environmental policy leads to a short-term increase in industry level productivity, and also, they suggested that the most productive firms have been confronted with a temporary increase in productivity, whilst the less productive ones experienced a productivity growth decline. The study conducted by de Vries and Withagen [27] focused on the relation between environmental policy stringency and sulfur dioxide abatement, used as a proxy for innovation. Three different models of environmental stringency had been analyzed, and only one of them revealed a positive impact of environmental stringency on innovation. In the field of agriculture, Kara et al. [28] analyzed how federal environmental regulations influence agricultural production in the US, and found that stringent environmental regulations may increase the likelihood of adopting certain conservation practices.

Bieth [29] analyzed the impact of economic growth and human development on CO_2 emissions in six ASEAN economies over the 2007-208 duration through regression analysis and revealed a significant impact of economic growth and human development on CO_2 emissions.

The relationship between economic development and CO_2 emissions has been analyzed in the context of Environmental Kuznets Curve validity, and reached different findings depending on the country and the method employed in the analyses. However, the studies have generally employed real GDP per capita for economic development [30–33]. Our research attempts to fill the gap that still exists related to the use of more complex and reliable measures of economic development by using the human development index to proxy the economic development.

3. Data and Method

In our study, the impact of stringency policies and human development on CO_2 emissions was analyzed in G7 and BRICS economies over the 1995–2015 period through panel cointegration and causality analyses.

In the econometric model, CO_2 emissions were represented by CO_2 emissions in terms of metric tons per capita and environmental policy was represented by the environmental policy stringency index provided by the OECD, ranging from 0 (not stringent) to 6 (highest stringency degree) [18] and the stringency shows the degree of environmental policies putting the price on polluting or environmentally harmful behavior (see Botta and Koźluk [19] for detailed information about index construction.). Lastly, human development was proxied by the human development index of UNCTAD [34]. The index is calculated as a geometric mean of normalized indices of life expectancy, education, and gross national income [34]. The logarithmic forms of the series were used in the analyses and the data sources, and their symbols are displayed in Table 1.

Table 1. Data description.

Variables	Description	Definition	Data Sources
CO	CO_2 emissions (metric tons per capita)	"The total amount of carbon dioxide emitted by the country as a consequence of all relevant human (production and consumption) activities, divided by the population of the country" [6].	World Bank [6]
EPS	Environmental policy stringency index	"A country-specific and internationally-comparable measure of the stringency of environmental policy. Stringency is defined as the degree to which environmental policies put an explicit or implicit price on polluting or environmentally harmful behaviour" [18].	OECD [18]
HDI	Human development index	A statistical tool developed by the United Nations in 1990 that assesses countries' social and economic development based on three key dimensions: a long and healthy life, access to education, and a decent standard of living. Life expectancy index, education index and gross national income (GNI) are taken into consideration to calculate HDI [13].	UNCTAD [35]

Source: own processing.

The sample of the study consists of G7 nations (Canada, France, Germany, Italy, Japan, the United Kingdom, and the United States) and BRICS economies (Brazil, China, India, Russian Federation, and South Africa). All the series were annual, and the study period was from 1995–2015, because the environmental policy stringency index ended in 2015. The Stata 14.0, Gauss 10.0 and Eviews 10.0 were employed in the econometric analyses and the dataset summary characteristics are shown in Table 2. The mean of CO, EPS, and HDI series were, respectively, 8.63, 1.55, and 0.78 in the sample, but especially CO_2 emissions showed considerable variations among the countries. Descriptive statistics for variables included in our study, classified by country, is presented in Appendix A.

Table 2. Summary statistics of the dataset.

Characteristic	CO	EPS	HDI
Mean	8.634120	1.554643	0.788385
Median	8.768835	1.300000	0.851000
Maximum	20.17875	3.850000	0.938000
Minimum	0.841937	0.330000	0.461000
Std. Dev.	5.031014	1.042403	0.125613
Skewness	0.460112	0.629946	−0.813656
Kurtosis	2.678022	2.073179	2.446227

Source: own processing.

In the econometric part of the research, the tests of cross-sectional dependency and homogeneity were firstly conducted to decide which tests to employ for unit root, cointegration and causality analyses. Then, the Pesaran [36] CIPS unit root test, Westerlund and Edgerton [37] LM bootstrap panel cointegration test and Konya [37] bootstrap panel Granger causality test were conducted considering the existence of cross-sectional dependency and heterogeneity.

The Westerlund and Edgerton [37] LM bootstrap panel cointegration test considers the cross-sectional dependency among the series and yields effective results for small sample sizes and it also allows autocorrelation and heteroscedasticity in cointegrating equation. On the other hand, the Konya [38] bootstrap panel Granger causality test takes notice of both cross-sectional dependency and heterogeneity. The test is relied on Seemingly Unrelated Regressions (SUR) estimation, which yields more efficient results in the case of cross-sectional dependency among the series. The causality direction is investigated by Wald tests with bootstrap critical values. Furthermore, the test does not dictate any pretests [38].

4. Empirical Analysis

In the applied part of the research, first the cross-sectional dependence among the series was checked by employing tests of Pesaran et al. [39] $LM_{adj.}$, the Pesaran [40] LM CD, and the Breusch and Pagan [41] LM, and the findings are displayed in Table 3. The null hypothesis of cross-sectional independency was declined at 1% significance level and in turn, cross-sectional dependency among the series was reached.

Table 3. Results of cross-sectional dependency tests.

Test	Test Statistic	Probability Value
LM $_{adj}$	36.902	0.000
LM CD	34.771	0.000
LM	45.786	0.005

Note: H0: There is cross-sectional independency; H1: there is cross-sectional dependence. Source: own processing.

The homogeneity of the cointegration coefficients was checked through delta tilde tests of Pesaran and Yamagata [42], and the findings are displayed in Table 4. The null hypothesis of homogeneity was declined at 1% significance level and the cointegration coefficients were found to be heterogeneous.

Table 4. Results of homogeneity tests.

Test	Test Statistic	Probability Value
Delta tilde	27.413	0.000
Adjusted delta tilde	29.502	0.000

Note: H0: Slope coefficients are homogeneous; H1: slope coefficients are heterogeneous. Source: own processing.

The unit root existence at the series were analyzed by the Pesaran [36] CIPS unit root test, and the findings are displayed in Table 5, and three series were found to be I(1).

Table 5. Pesaran (2007) CIPS unit root test.

Variables	Level		First Differences	
	Constant	Constant + Trend	Constant	Constant + Trend
CO	−1.173	−1.215	−7.662 *	−8.035 *
EPS	−1.564	−1.739	−9.716 *	−9.994 *
HDI	−1.209	−1.296	−8.270 *	−8.619 *

Source: own processing. Note: * it is significant at 1% significance level.

The cointegrating relationship among environmental stringency policies, human development and CO_2 emissions were examined through the Westerlund and Edgerton [37] LM bootstrap cointegration test, and the findings are displayed in Table 6. Furthermore, the critical values were provided with 10,000 simulations and lag and lead values were taken as 2. Asymptotic probability values were derived from standard normal distribution. Therefore, bootstrap P values are considered in the case of cross-sectional existence. The null hypothesis suggesting the existence of significant cointegration relationship among the series were accepted, because the bootstrap p value was found to be higher than 10%.

Table 6. Pesaran (2007) CIPS unit root test.

LM_N^+	Constant			Constant and Trend		
	Test Statistic	Asymptotic p Value	Bootstrap p Value	Test Statistic	Asymptotic p Value	Bootstrap p Value
	8.361	0.344	0.398	9.557	0.369	0.412

Source: own processing.

The cointegrating coefficients were estimated through FMOLS (Fully Modified Ordinary Least Squares), and the findings are reported in Table 7. The results revealed that both EPS and HDI had a significant decreasing effect on the CO_2 emissions, but the decreasing impact of HDI on the CO_2 emissions was relatively higher when compared with the impact of EPS. The EPS and HDI had the largest decreasing impact on the CO_2 emissions in the United States of America, but EPS and HDI, relatively, had the least decreasing impact on the CO_2 emissions in India and China.

Table 7. Cointegration coefficients.

Countries	LnEPS	LnHDI
Brazil	−0.073 *	−0.147 *
Canada	−0.113 *	−0.165 *
China	−0.075 *	−0.107 *
France	−0.094 *	−0.135 *
Germany	−0.116 *	0.128 *
India	−0.054 *	−0.113 *
Italy	−0.103 *	−0.124 *
Japan	−0.119 *	−0.169 *
Russia	−0.101 *	−0.120 *
South Africa	−0.106 *	−0.119 *
United Kingdom	−0.121 *	−0.171 *
United States of America	−0.123 *	−0.175
Panel	−0.103 *	−0.142 *

Source: own processing. Note: * it is significant at 5% significance level.

The causality among environmental stringency policies, human development, and CO_2 emissions was checked through the Kónya [38] bootstrap panel Granger causality test given the presence of heterogeneity and cross-sectional dependency, and the findings are reported in Tables 8 and 9. First, the causality between CO_2 emissions (CO) and environmental stringency policies (EPS) was checked, and the findings are reported in Table 8. The findings revealed a bilateral causality between CO and EPS for Germany, Japan, the United Kingdom, and the United States of America and a unilateral causality from CO to the EPS for Canada, China, and France.

Table 8. Bootstrap Granger causality test between lnCO and lnEPS.

Countries	InCO Does Not Granger Cause InEPS				InEPS Does Not Granger Cause InCO			
	Wald Statistics	Bootstrap Critical Value			Wald Statistics	Bootstrap Critical Values		
		10%	5%	1%		10%	5%	1%
Brazil	37.45	55.24	58.11	61.70	27.45	44.67	46.02	49.36
Canada	78.13 ***	48.14	53.89	55.04	36.19	51.22	54.78	56.09
China	66.59 ***	54.19	58.21	60.88	31.27	48.44	50.19	51.38
France	64.23 ***	46.33	48.47	49.05	40.32	45.73	46.88	48.56
Germany	73.56 ***	50.13	54.66	57.29	69.44 **	68.15	71.36	73.07
India	31.86	48.16	51.45	54.14	39.86	53.86	54.21	56.99
Italy	43.58	52.41	56.78	59.39	37.07	45.38	46.22	47.03
Japan	79.21 ***	55.37	58.19	60.77	75.15 **	73.49	74.99	76.17
Russia	32.17	46.89	47.22	49.25	29.56	40.75	41.58	43.56
South Africa	44.12	61.23	64.89	65.57	31.47	44.86	45.73	48.19
United Kingdom	68.33 **	65.34	69.14	70.88	69.26 **	65.37	68.43	70.83
United States of America	73.89 ***	64.37	66.31	68.11	63.87 **	64.48	66.39	67.85

Source: own processing. Note: *** and ** respectively, indicate that it is significant at 1%, 5%, and 10% significance levels.

Table 9. Bootstrap Granger causality test between lnCO and lnHDI.

Countries	InCO Does Not Granger Cause InHDI				InHDI Does Not Granger Cause InCO			
	Wald Statistics	Bootstrap Critical Value (%)			Wald Statistics	Bootstrap Critical Value		
		10%	5%	1%		10%	5%	1%
Brazil	93.67 ***	67.89	71.45	78.23	46.34	56.78	61.13	64.87
Canada	92.49 ***	56.21	63.55	70.88	24.16	66.28	73.11	75.9
China	79.31 ***	60.47	66.23	69.16	19.85	34.27	40.19	42.52
France	55.73 *	47.24	56.89	60.32	36.82	40.25	38.48	41.19
Germany	73.56 *	68.36	71.44	75.98	61.14 ***	39.26	42.53	45.01
India	22.79	42.79	47.21	49.05	31.88	40.17	44.68	46.17
Italy	62.35 **	54.99	63.67	66.24	33.64	51.59	67.94	69.22
Japan	75.21 ***	46.91	49.16	53.48	77.03 ***	59.68	62.6	64.47
Russia	29.18	48.25	51.18	55.09	36.42	70.14	77.46	79.07
South Africa	34.59	54.64	58.02	61.18	25.18	43.53	48.13	49.44
United Kingdom	73.18 ***	49.23	54.43	58.73	64.85 ***	54.96	53.07	59.21
United States of America	69.15 ***	46.18	49.36	52.77	60.92 **	60.89	62.68	65.15

Source: own processing. Note: ***, **, *, respectively, indicate that it is significant at 1%, 5%, and 10% significance levels.

Then, the causality between human development (HDI), and CO_2 emissions (CO) was checked through the Kónya [38] bootstrap panel Granger causality test, and the findings in Table 9 revealed a bilateral causality between CO and HDI in countries such as Germany, Japan, the United Kingdom, and the United States of America and a unilateral causality from CO to the HDI for Brazil, Canada, China, and France.

5. Conclusions

The 2015 Paris Agreement represented a milestone for the struggle against greenhouse gas emissions. According to the G8 countries' joint declarations, efforts are consistently taken in order to significantly reduce greenhouse gas emissions and to support sustainable energy and human development. However, as most scholars indicate, there is still a gap between political decisions, on the one hand, and the position of different researchers and analysts regarding the need to further and intensify environmental stringency policies, on the other hand.

The study focuses on the following research questions: do environmental stringency policies and human development reduce CO_2 emissions? What does causality analysis indicate about the G7 and BRICS economies, namely Canada, France, Germany, Italy, Japan, the United Kingdom, and the United States of America, Brazil, China, India, Russian Federation, and South Africa?

In order to tackle these interrogations, the cointegrating relationship among environmental stringency policies, human development and CO_2 emissions were examined through the Westerlund and Edgerton [37] LM bootstrap cointegration test and the Kónya [38] bootstrap panel Granger causality test given the presence of heterogeneity and cross-sectional dependency. The causality analysis disclosed a bilateral causality between environmental stringency policies and CO_2 emissions for Germany, Japan, the United Kingdom, and the United States of America and a unilateral causality from CO_2 emissions to the environmental stringency policies for Canada, China, and France. Therefore, the causality analysis revealed that environmental stringency policy was found to be effective, especially in the developed countries, in the short run. Moreover, a bilateral causality between human development and CO_2 emissions was discovered for Germany, Japan, the United Kingdom, and the United States of America, and unilateral causality from CO_2 emissions to human development was discovered for Brazil, Canada, China, and France.

As we already mentioned in the previous sections, the studies focused on the causal link between CO_2 emissions and HDI have not reached a clear consensus regarding the causality between two variables, and this can be resulted from different variables, methods, time periods or countries with different characteristics. However, the feedback link between CO_2 emissions and human development indicates a mutual interaction between two variables, especially in the leading developing countries in the short run, but a unilateral causality from CO_2 emissions to human development indicates that CO_2 emissions have significant effect on human development.

Furthermore, the cointegration analysis revealed that both environmental stringency policies and human development were effective in decreasing CO_2 emissions in the long run. However, both human development and environmental stringency policies were more effective in decreasing the CO_2 emissions, especially in the developed economies such as the United States of America, Canada, Japan, and the United Kingdom. Therefore, we conclude that the effect of human development and environmental stringency policies on the CO_2 emissions raise in parallel with development level.

Our findings indicated that environmental stringency policies and human development are important for environmental sustainability. However, environmental stringency policies can negatively affect economic growth and employment through raising the costs at the beginning. However, the countries offset the negative economic effects of environmental stringency policies through innovation, considering the Porter hypothesis and empirical findings over time. On the other hand, improvements in human development also are effective for environment sustainability.

Consequently, there are no uniform environment policies with which the countries can achieve their environment targets. Therefore, countries should design an environmental policy mix considering their country-specific characteristics. Future studies should explore the environmental policies at country level by conducting comparative research in different categories of countries (developed, emerging, least developed, OECD, EU countries, etc.).

Author Contributions: Conceptualization: F.H.S., Y.B., L.H., M.D.G.; Methodology: F.H.S., Y.B., L.H., M.D.G.; Resources: F.H.S., Y.B., L.H., M.D.G.; Writing: F.H.S., Y.B., L.H., M.D.G. All authors have contributed significantly for this research in all phases and sections. All authors have read and agreed to the published version of the manuscript.

Funding: The publication of this article was supported by the 2020 Development Fund of the Babeș-Bolyai University.

Institutional Review Board Statement: Not applicable.

Informed Consent Statement: Not applicable.

Data Availability Statement: Not applicable.

Conflicts of Interest: The authors declare no conflict of interest.

Appendix A

Table A1. Descriptive statistics for analyzed countries.

Country	Characteristic	CO	EPS	HDI
	Mean	1.9963	0.4414	0.7042
	Median	1.876441	0.420000	0.700000
	Maximum	2.631290	0.630000	0.756000
Brazil	Minimum	1.594542	0.380000	0.651000
	Std. Dev.	0.281403	0.069013	0.031227
	Skewness	0.983730	1.594038	0.121792
	Kurtosis	2.827510	4.354830	2.148359
	Mean	16.42349	2.138500	0.889150
	Median	16.67965	1.875000	0.895000
	Maximum	17.56134	3.850000	0.921000
Canada	Minimum	14.79888	0.460000	0.861000
	Std. Dev.	0.951222	1.260419	0.019329
	Skewness	−0.402700	−0.014637	−0.071062
	Kurtosis	1.672255	1.275895	1.775377
	Mean	4.834477	1.019500	0.647200
	Median	4.751746	0.830000	0.646500
	Maximum	7.557211	2.160000	0.739000
China	Minimum	2.648649	0.520000	0.554000
	Std. Dev.	1.926844	0.582955	0.060970
	Skewness	0.228806	1.064472	−0.003893
	Kurtosis	1.494350	2.611028	1.623914
	Mean	5.679318	2.480500	0.866250
	Median	5.884539	2.785000	0.869000
	Maximum	6.280954	3.700000	0.895000
France	Minimum	4.550000	1.150000	0.837000
	Std. Dev.	0.548513	1.001790	0.018038
	Skewness	−0.834628	−0.113231	0.014689
	Kurtosis	2.454713	1.301456	1.724320
	Mean	9.714977	2.621500	0.903850
	Median	9.780996	2.670000	0.913000
	Maximum	10.86023	3.140000	0.938000
Germany	Minimum	8.797642	1.850000	0.846000
	Std. Dev.	0.590215	0.473834	0.030465
	Skewness	0.102840	−0.443551	−0.589654
	Kurtosis	2.052530	1.613958	1.990459
	Mean	1.213358	0.832500	0.541700
	Median	1.091958	0.630000	0.541000
	Maximum	1.784334	1.820000	0.624000
India	Minimum	0.898163	0.460000	0.468000
	Std. Dev.	0.295116	0.398641	0.049592
	Skewness	0.613705	0.938410	0.106093
	Kurtosis	1.937756	2.736987	1.739943

Table A1. *Cont.*

Country	Characteristic	CO	EPS	HDI
Italy	Mean	7.256863	2.198000	0.859600
	Median	7.668207	2.280000	0.867500
	Maximum	8.216487	3.280000	0.883000
	Minimum	5.140000	1.350000	0.814000
	Std. Dev.	0.981292	0.719778	0.022892
	Skewness	−1.025494	0.115076	−0.654713
	Kurtosis	2.748522	1.389011	2.044067
Japan	Mean	9.485325	2.001500	0.875600
	Median	9.547599	1.680000	0.877000
	Maximum	9.880903	3.500000	0.908000
	Minimum	8.632100	1.330000	0.847000
	Std. Dev.	0.289335	0.711701	0.019422
	Skewness	−1.305026	1.032466	0.104473
	Kurtosis	4.822582	2.437024	1.877646
Russian Federation	Mean	11.31108	0.616000	0.755750
	Median	11.18706	0.600000	0.756500
	Maximum	12.62027	0.920000	0.809000
	Minimum	10.12729	0.330000	0.703000
	Std. Dev.	0.723075	0.134962	0.036007
	Skewness	0.084680	0.553416	−0.046385
	Kurtosis	1.894299	3.630978	1.711029
South Africa	Mean	8.877895	0.695000	0.646700
	Median	8.751202	0.500000	0.643000
	Maximum	9.979458	1.750000	0.701000
	Minimum	7.727642	0.400000	0.611000
	Std. Dev.	0.575199	0.430037	0.026492
	Skewness	0.276880	1.776961	0.604032
	Kurtosis	2.732339	4.541542	2.290577
United Kingdom	Mean	8.332668	2.141500	0.891950
	Median	8.901417	2.090000	0.896500
	Maximum	9.480231	3.830000	0.925000
	Minimum	6.220240	0.810000	0.851000
	Std. Dev.	1.007337	1.146906	0.021982
	Skewness	−0.821794	0.217461	−0.295253
	Kurtosis	2.259445	1.569621	2.108925
United States of America	Mean	18.49938	1.896500	0.902600
	Median	19.35696	1.715000	0.901500
	Maximum	20.17875	3.170000	0.921000
	Minimum	15.98987	1.050000	0.884000
	Std. Dev.	1.467676	0.764339	0.013520
	Skewness	−0.621797	0.194185	0.080692
	Kurtosis	1.684749	1.329994	1.428619

Source: own processing.

References

1. Singh, R.L.; Singh, P.K. Global environmental problems. In *Principles and Applications of Environmental Biotechnology for a Sustainable Future*; Singh, R., Ed.; Applied Environmental Science and Engineering for a Sustainable Future; Springer: Singapore, 2017. [CrossRef]
2. Sofroniu, A. *United Nations Organization Assessment: Global Politics, Relations & Functions*; Lulu.com: Morrisville, NC, USA, 2018.
3. Manisalidis, I.; Stavropoulou, E.; Stavropoulos, A.; Bezirtzoglou, E. Environmental and Health Impacts of Air Pollution: A Review. *Front Public Health* **2020**, *8*. [CrossRef]
4. Mannucci, P.M.; Franchini, M. Health Effects of Ambient Air Pollution in Developing Countries. *Int. J. Environ. Res. Public Health* **2017**, *14*, 1048. [CrossRef]
5. Greenspan, B.R.; Russell, C. Environmental Policy for Developing Countries. *Issues Sci. Technol.* **2002**, *18*, 63–70. Available online: https://issues.org/greenspan-environmental-policy-developing-countries/ (accessed on 8 June 2021).
6. World Bank. CO$_2$ Emissions (Metric Tons per Capita) 2021. Available online: https://data.worldbank.org/indicator/EN.ATM.CO2E.PC (accessed on 20 February 2021).

7.   Hansen, U.; Nygaard, I.; Romijn, H.; Wieczorek, A.; Kamp, L.; Klerkx, L. Sustainability transitions in developing countries: Stocktaking, new contributions and a research agenda. *Environ. Sci. Policy* **2018**, *84*, 198–203. [CrossRef]

8.   Onoda, T. IEA policies—G8 recommendations and an afterwards. *Energy Policy* **2009**, *37*, 3823–3831. [CrossRef]

9.   Sandalow, D.; Volfson, H. *G8 Summit Leaders' Statements Climate Change Language 1990-2004*; Brookings Institution: Washington, DC, USA, 2005; Available online: https://www.brookings.edu/wp-content/uploads/2012/04/g8climatelanguage.pdf (accessed on 8 June 2021).

10.   Armstrong, K. The Lisbon Strategy and Europe 2020. In *The EU's Lisbon Strategy. Palgrave Studies in European Union Politics*; Copeland, P., Papadimitriou, D., Eds.; Palgrave Macmillan: London, UK, 2012.

11.   Eurostat. Renewable Energy Statistics. 2020. Available online: https://ec.europa.eu/eurostat/statistics-explained/index.php?title=Renewable_energy_statistics (accessed on 14 May 2021).

12.   European Commission (EC). Supporting Climate Action through the EU Budget. 2020. Available online: https://ec.europa.eu/clima/policies/budget/mainstreaming_en (accessed on 14 May 2021).

13.   UNDP. *Human Development Report*; Oxford University Press: Oxford, UK; New York, NY, USA, 1990.

14.   Tudorache, M.D. Examining the Drivers of Human Development in European Union. In Proceedings of the 35th IBIMA Conference, Seville, Spain, 1–2 April 2020; IBIMA Publishing: King of Prussia, PA, USA, 2020. ISBN 978-0-9998551-4-0. Available online: https://ibima.org/accepted-paper/examining-the-drivers-of-human-development-in-europeanunion/ (accessed on 22 May 2021).

15.   Union of Concerned Scientists. Each Country's Share of CO2 Emissions. 2020. Available online: https://www.ucsusa.org/resources/each-countrys-share-co2-emissions (accessed on 21 May 2021).

16.   Ritchie, H.; Roser, M. CO_2 and Greenhouse Gas Emissions. 2020. Published Online at OurWorldInData.org. Available online: https://ourworldindata.org/co2-and-other-greenhouse-gas-emissions (accessed on 21 May 2021).

17.   Wolde-Rufael, Y.; Mulat-Weldemeskel, E. Do environmental taxes and environmental stringency policies reduce CO2 emissions? Evidence from 7 emerging economies. *Environ. Sci. Pollut. Res. Int.* **2021**, *28*, 22392–22408. [CrossRef]

18.   OECD. Environmental Policy Stringency Index 2021. Available online: https://doi.org/10.1787/2bc0bb80-en (accessed on 22 May 2021).

19.   Botta, E.; Kózluk, T. *Measuring Environmental Policy Stringency in OECD Countries: A Composite Index Approach*; Economics; OECD Economics Department Working Papers, No. 1177; OECD Publishing: Paris, France, 2014. [CrossRef]

20.   Ahmed, K.; Ahmed, S. A predictive analysis of CO_2 emissions, environmental policy stringency, and economic growth in China. *Environ. Sci. Pollut. Res. Int.* **2018**, *25*, 16091–16100. [CrossRef] [PubMed]

21.   Ahmed, K. Environmental policy stringency, related technological change and emissions inventory in 20 OECD countries. *J. Environ. Manag.* **2020**, *274*. [CrossRef] [PubMed]

22.   Wang, K.; Yan, M.; Wang, Y.; Chang, C.P. The impact of environmental policy stringency on air quality. *Atmos. Environ.* **2020**, *231*. [CrossRef]

23.   Wolde-Rufael, Y.; Mulat-Weldemeskel, E. Environmental policy stringency, renewable energy consumption and CO_2 emissions: Panel cointegration analysis for BRIICTS countries. *Int. J. Green Energy* **2020**, *17*, 568–582. [CrossRef]

24.   Albrizio, S.; Botta, E.; Kózluk, T.; Zipperer, V. *Do Environmental Policies Matter for Productivity Growth*; OECD Economics Department Working Papers No. 1176; OECD Publishing: Paris, France, 2014. [CrossRef]

25.   Porter, M. America's green strategy. *Sci. Am.* **1991**, *264*, 168. [CrossRef]

26.   Albrizio, S.; Kózluk, T.; Zipperer, V. Environmental policies and productivity growth: Evidence across industries and firms. *J. Environ. Econ. Manag.* **2017**, *81*, 209–226. [CrossRef]

27.   de Vries, F.P.; Withagen, C. *Innovation and Environmental Stringency: The Case of Sulphur Dioxide Abatement*; Discussion Papers/CentER for Economic Research 2005-18; Tilburg University, School of Economics and Management: Tilburg, The Netherlands, 2005.

28.   Kara, E.; Ribaudo, M.; Johansson, R.C. On how environmental stringency influences adoption of best management practices in agriculture. *J. Environ. Manag.* **2008**, *88*, 1530–1537. [CrossRef]

29.   Bieth, R.C.E. The influence of gross domestic product and human development index on CO_2 emissions. *J. Phys. Conf. Ser.* **2021**, *1808*, 012034. [CrossRef]

30.   Azomahou, T.; Laisney, F.; Van, P.N. Economic development and CO_2 emissions: A nonparametric panel approach. *J. Public Econ.* **2006**, *90*, 1347–1363. [CrossRef]

31.   Knight, K.W.; Schor, J.B. Economic growth and climate change: A cross-national analysis of territorial and consumption-based carbon emissions in high-income countries. *Sustainability* **2014**, *6*, 3722–3731. [CrossRef]

32.   Jardón, A.; Kuik, O.; Tol, R.S.J. Economic growth and carbon dioxide emissions: An analysis of Latin America and the Caribbean. *Atmósfera* **2017**, *30*, 87–100. [CrossRef]

33.   Osobajo, O.A.; Otitoju, A.; Otitoju, M.A.; Oke, A. The impact of energy consumption and economic growth on carbon dioxide emissions. *Sustainability* **2020**, *12*, 7965. [CrossRef]

34.   UNCTAD. Human Development Index (HDI) 2021. Available online: http://hdr.undp.org/en/indicators/137506# (accessed on 20 February 2021).

35. United Nations Statistics Division. Millennium Development Goals Indicators. Available online: https://unstats.un.org/unsd/mdg/Metadata.aspx?IndicatorId=0&SeriesId=776 (accessed on 20 February 2021).
36. Pesaran, M.H. A simple panel unit root test in the presence of cross-section dependence. *J. Appl. Econ.* **2007**, *22*, 265–312. [CrossRef]
37. Westerlund, J.; Edgerton, D.L. A panel bootstrap cointegration test. *Econ. Lett.* **2007**, *97*, 185–190. [CrossRef]
38. Kónya, L. Exports and Growth: Granger Causality Analysis on OECD Countries with a Panel Data Approach. *Econ. Model.* **2006**, *23*, 978–992. [CrossRef]
39. Pesaran, M.H.; Ullah, A.; Yamagata, T. A bias-adjusted LM test of error cross-section ndependence. *Econ. J.* **2008**, *11*, 105–127.
40. Pesaran, M.H. *General Diagnostic Tests for Cross Section Dependence in Panels*; IZA Discussion Papers, No:1240; Institute of Labor Economics: Bonn, Germany, 2004.
41. Breusch, T.S.; Pagan, A.R. The lagrange multiplier test and its applications to model specification in econometrics. *Rev. Econ. Stud.* **1980**, *47*, 239–253. [CrossRef]
42. Pesaran, M.H.; Yamagata, T. Testing slope homogeneity in large panels. *J. Econ.* **2008**, *142*, 50–93. [CrossRef]

International Journal of **Environmental Research** **and Public Health**

Article

Comparison of Joint Effect of Acute and Chronic Toxicity for Combined Assessment of Heavy Metals on *Photobacterium* sp.NAA-MIE

Nur Adila Adnan [1], Mohd Izuan Effendi Halmi [1,*], Siti Salwa Abd Gani [2], Uswatun Hasanah Zaidan [3] and Mohd Yunus Abd Shukor [3]

[1] Department of Land Management, Faculty of Agriculture, University Putra Malaysia, Serdang 43400, Selangor, Malaysia; nuradilaadnan09@gmail.com
[2] Department of Agricultural Technology, Faculty of Agriculture, University Putra Malaysia, Serdang 43400, Selangor, Malaysia; ssalwaag@upm.edu.my
[3] Department of Biochemistry, Faculty of Biotechnology and Biomolecular Sciences, University Putra Malaysia, Serdang 43400, Selangor, Malaysia; uswatun@upm.edu.my (U.H.Z.); mohdyunus@upm.edu.my (M.Y.A.S.)
* Correspondence: m_izuaneffendi@upm.edu.my

Abstract: Predicting the crucial effect of single metal pollutants against the aquatic ecosystem has been highly debatable for decades. However, dealing with complex metal mixtures management in toxicological studies creates a challenge, as heavy metals may evoke greater toxicity on interactions with other constituents rather than individually low acting concentrations. Moreover, the toxicity mechanisms are different between short term and long term exposure of the metal toxicant. In this study, acute and chronic toxicity based on luminescence inhibition assay using newly isolated *Photobacterium* sp.NAA-MIE as the indicator are presented. *Photobacterium* sp.NAA-MIE was exposed to the mixture at a predetermined ratio of 1:1. TU (Toxicity Unit) and MTI (Mixture Toxic Index) approach presented the mixture toxicity of $Hg^{2+} + Ag^+$, $Hg^{2+} + Cu^{2+}$, $Ag^+ + Cu^{2+}$, $Hg^{2+} + Ag^+ + Cu^{2+}$, and $Cd^{2+} + Cu^{2+}$ showed antagonistic effect over acute and chronic test. Binary mixture of $Cu^{2+} + Zn^{2+}$ was observed to show additive effect at acute test and antagonistic effect at chronic test while mixture of $Ni^{2+} + Zn^{2+}$ showing antagonistic effect during acute test and synergistic effect during chronic test. Thus, the strain is suitable and their use as bioassay to predict the risk assessment of heavy metal under acute toxicity without abandoning the advantage of chronic toxicity extrapolation.

Keywords: toxicity assessment; luminescent bacteria; acute and chronic toxicity

Citation: Adnan, N.A.; Halmi, M.I.E.; Abd Gani, S.S.; Zaidan, U.H.; Abd Shukor, M.Y. Comparison of Joint Effect of Acute and Chronic Toxicity for Combined Assessment of Heavy Metals on *Photobacterium* sp.NAA-MIE. *Int. J. Environ. Res. Public Health* **2021**, *18*, 6644. https://doi.org/10.3390/ijerph18126644

Academic Editors: Pasquale Avino, Massimiliano Errico, Aristide Giuliano and Hamid Salehi

Received: 24 April 2021
Accepted: 8 June 2021
Published: 21 June 2021

Publisher's Note: MDPI stays neutral with regard to jurisdictional claims in published maps and institutional affiliations.

1. Introduction

The victory of urbanization, the agricultural and pharmaceutical industries, and other economic developments make important contributions to our health and high living standard. Nevertheless, their application is often correlated with persistent problems of heavy metals in solid and liquid wastes and contributes to the pollution of the environment [1,2]. Bioaccumulation properties of metals can drive high biotic tissue concentrations in consumers, predominantly through food webs even less than the range of acute toxic concentrations [3,4]. Globally, improper management of municipal solid waste incineration and sewage is giving rise to mercury (Hg) increasing from 0.5 to 9.0 μg/L in the aquatic environment of China [5–7]. In other cases, Cu and As contaminated areas of the Rhône River (France) are constantly increasing due to mining activities and pesticides in organic agriculture [8,9]. In Malaysia itself, as an important international port, in three subsidiary ports in Port Klang, concentrations of metals As, Cd and Pb were comparatively higher than the background values in the coastal sediment [10].

Unfortunately, the analysis of the total amount of heavy metals is unable to represent the full environmental behaviors and ecological effects of heavy metals [11]. Numerous studies deal with risk effects of single metal types, while in reality the aquatic organisms are often exposed to mixtures of metals [12,13]. Thus, the effects of co-existing multiple chemicals cannot be ignored and interactions of components in a mixture might cause significant changes in the properties of its constituents [14]. Indeed, organisms are susceptible to multiple mixtures of metals in ecosystems, which may have antagonistic, synergistic or additive effects, and complex formations may critically lead to an imbalanced ecosystem [15,16]. In the event two or more chemicals are found concurrently in organisms, the joint effect may bring in the addition of the toxic effect of one chemical to the other, known as additive interaction. Furthermore, antagonistic interaction occurs if the toxic effects created by the mixture are lower than the sum of the toxic effects of the individual elements, and vice versa, synergistic interaction takes place if toxic effects of metal present in the mixture go beyond the total effects of the individual components [17,18].

In current regulatory approaches, bioassays evaluate the physiological or behavioral changes exhibited by living organisms attributed to metabolic disruption caused by toxic compounds [19,20]. Existing standard toxicity tests both on prokaryotic and eukaryotic species, such as algae, fish and daphnid, are expensive; these tests give slow responses and entail a long life cycle to reproducibility [21,22]. Taken for example, Nwanyanwu and co-workers studied the toxicity of a binary mixture of Zn and Cd based on the inhibition of dehydrogenase activity of three bacteria consortium. However, the time taken for the toxicant to react with these bacteria was 24 h [15].

In this regard, it is mandatory to determine the interactive effects of the mixtures of pollutant on environmental microbial associated mixtures of pollutants with the demand of an ecotoxicity test that is cost-effective, rapid response (5 min–30 min), sensitive and reliable. Bioluminescence inhibition assay promises further advantage of easy use, low cost and high reproducibility based on reduction of the light output emitted by naturally luminous bacteria such as *Vibrio fischeri* and *Photobacterium phosphoreum* in the presence of toxicants due to the inhibition of enzyme luciferase [19,23,24]. This bioassay has been accepted as a quick method for chemical toxicity assessment and commercialized as Microtox® (Azur Environmental, California, USA), BioTox™ (Aboatox, Masku, Finland), ToxAlert® (Merck, Darmstadt, Germany) test for ecotoxicological monitoring and chemical testing of wastewater effluents, sediment extracts and contaminated groundwater [19,25]. Even so, this system needs to maintain optimum assay temperatures, which range from 15 to 25 °C. Therefore, the system is not practical in order to conduct an assay in a tropical country like Malaysia, which displays a broad variation of regular temperatures up to 34 °C in the mid-day [26]. A slight change out of this range of temperature could extremely affect the luminescence activity [24]. Besides, the system would depend on a refrigerated water bath to operate, which is not constructive, expensive and difficult to maintain, and instrument-dependent for field applications in the tropical regions [27]. Hence, a cost-effective bioluminescence inhibition assay using local isolate without the application of expensive luminometer and instruments, and, a higher sensitivity towards heavy metals is urgently needed in tropical regions.

At the moment, bioluminescence inhibition assay has been widely used to test acute toxicities of metals. Based on previous studies, *Vibrio fischeri* bioluminescence inhibition bioassay (VFBIA) only provides information about the overall acute metal toxicity [11,17,19]. Direct assessment of the potential chronic effects of metals by applying naturally bioluminescence species is currently lacking. Blasco and Del [28] and Wang et al., [29] recommended that chronic tests could be more applicable to signify a threat to public health. It can be argued that feedback at the chronic joint may be more crucial, because of the occurrence of compounds in the environment for an extended period of time. Thus, it is necessary to study these chronic joint effects. In the prior work, luminescence activity of luminescent bacterium identified as *Photobacterium* sp.NAA-MIE isolated from local marine fish *Selar crumenophthalmus* has exhibited temperature and pH stable in the tropical environment and sensitive towards heavy metals [30]. In this study, acute and chronic metal toxicity for mixture of heavy metals were conducted using this strain without the application of expensive instrumentation. The toxicity results of *Photobacterium* sp.NAA-MIE will be helpful in the future for assessing the individual and joint toxicity during the treatment of industrial wastewater.

2. Materials and Methods

The steps of the experimental procedure are summarized in Figure 1.

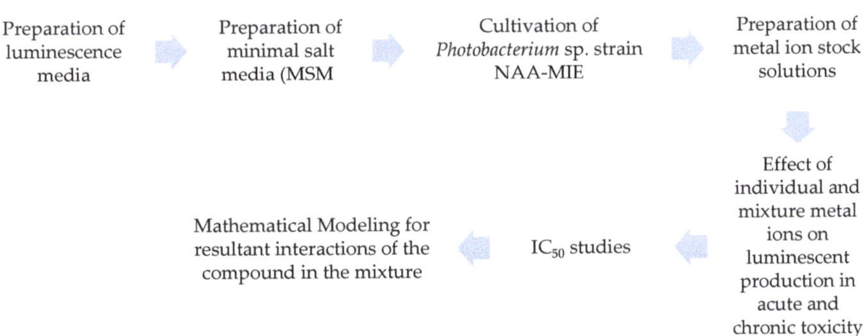

Figure 1. Steps in the experimental procedure for the determination of interaction existing between compounds in acute and chronic toxicity of *Photobacterium* sp.NAA-MIE against 6 heavy metal.

2.1. Preparation of Bacterial Culture and Media for Toxicity Assay

Luminescent bacterium *Photobacterium* sp.NAA-MIE newly isolated from local marine fish *Selar crumenophthalmus* [28] was used in this study. 1.0% (*v/v*) of bacterial culture (OD_{600} = 0.7–0.8) was inoculated into luminescence broth medium by dissolving 10 g of NaCl, 10 g of peptone, 3 mL of glycerol and 3 g of yeast in 1 L of distilled water [20] and grown at room temperature for 12 h on rotary shaker (100 rpm).The culture was then harvested by centrifugation (10,000× *g*, 10 min) and the pellet was diluted using 1 L of minimal salt media (MSM) (12.8 g of Na_2HPO_4 $7H_2O$, 3.1 g of KH_2PO_4, 17 g of NaCl, 1 g NH_4Cl, 0.5 g $MgSO_4$, and 3 mL glycerol) as substitute media for toxicity assay. The luminescence and sensitivity bioassay can be limited by a high density of the bacterial culture. To overcome this, the culture was further diluted 1000 times until the emission was within the range of 40,000 to 90,000 RLU before assaying [22,31,32].

2.2. Preparation of Heavy Metal Solutions

For individual metal toxicity test seven heavy metals were used: Hg^{2+}, Cu^{2+}, Ag^+, Pb^{2+}, Cd^{2+}, Ni^{2+} and Zn^{2+}. Standard procedure was adopted to prepare metal stock solutions ranging over (0.5, 1, 10, 100 and 1000 ppm or mg/L) by dissolving the AAS grade metal in deionized water. The solutions were freshly prepared [13]. Ten concentrations of the tested metals were prepared from stock solutions to carry out a preliminary experiment in triplicate to determine the suitable metal concentration array for determining the half maximal inhibitory concentration IC_{50} of each tested metal. All analytical-grade chemicals used were purchased from Oxoid (North Shore, England), and Merck (Damstardt, Germany).

2.3. Acute and Chronic Toxicity Assay of Single Metals

The assay was conducted by mixing 10 μL of toxicant solutions with 190 μL of bacterial cultures. Deionized water was used as a control to replace the toxicants in this study. The IC_{50} values were determined at total time per assay, approximately 30 min for acute toxicity and 6 h incubation for chronic toxicity, at room temperature (26 °C). The luminescence was determined with a portable luminometer (Lumitester PD-30 by Kikkoman, Tokyo, Japan) and reported as Relative Luminescent Unit (RLU). A total of 200 μL of bacterial culture was inoculated into the LuciPac pen tube before the readings were taken. Each assay was performed at least in triplicate.

2.4. Acute and Chronic Toxicity Assay of Metal Mixture

Concentration for each selected metal mixture was prepared in an equitoxic ratio of 1:1 and 1:1:1 based on concentrations of each toxicant producing 50% light reduction when being tested individually [33]. In this study, Cu + Cd, Cu + Zn, Ni + Zn, Hg + Cu, Hg + Ag, Ag + Cu and Hg + Ag + Cu mixtures were proposed to evaluate possible effects for each combination. The tests for mixtures were conducted similarly to those for the individual metals for both acute and chronic toxicity at 30 min and 6 h exposure time.

2.5. Data Analysis

2.5.1. Half Maximal Inhibitory Concentration (IC_{50}) Study of Metals Toxicant on Luminescence Production

The bioluminescence inhibition study allows identifying the concentration (IC_{50}) of the toxicant (mg/L) that gives a 50% reduction in light [22,34]. The luminescence activity of bacteria samples was inhibited by a series of the studied metals at different ranges of concentrations (mg/L) after exposure for 30 min (acute) and 6 h (chronic). The variation in light reduction and concentration of the toxicant produces a response relationship. IC_{50} calculations according to previous works was outlined by [35,36]. The percentage of luminescence inhibition was determined and IC_{50} was a nonlinear regression One Phase Decay model for one-phase binding and four-parameter logistic models software calculated using Graphpad Prism version 7.04 (California, USA). Means and standard errors were determined according to at least three independent replicates of each test.

2.5.2. Measurement of the Luminescence Inhibition

According to [37] intensity of luminescence produced by luminescence bacteria is inhibited in the presence of the toxicant samples. In this study, the inhibition rate of bacteria for the exposure to a tested metal was calculated by the following equation:

$$\text{Luminescence inhibition } (\%) = \frac{(L_o - L_i)}{L_o} \times 100 \tag{1}$$

where L_o represents luminescence (RLU) of the control sample, and L_i defines the luminescence (RLU) of the sample exposed to metal or mixtures of metals.

2.5.3. Mathematical Modeling for Assessment of Combined Heavy Metals Toxicity

The combined effect of two or three metals eventuates into three types of joint action. They were designated into additive effect, synergistic effect and antagonistic effect [38,39]. In this study, the joint action of each combined metals was determined by Toxicity Unit and Mixture Toxic Index approaches [40–42].

Toxicity Unit (TU)

In this method, emerging ICs of the individual compounds are determined. Toxic Unit (TU) can be defined as the incipient IC_{50} for each compound. Toxic Unit (TU) of each chemical in a mixture is calculated by Equations (2) and (3), respectively

$$TU_i = \frac{C_i}{IC_{50i}} \tag{2}$$

$$M = \sum_{i=1}^{n} TU_i = \left(\frac{C_1}{IC_{50\ 1}} + \frac{C_2}{IC_{50\ 2}} \cdots + \frac{C_n}{IC_{50\ n}} \right) \tag{3}$$

whereby C_i represents the concentration of ith compound when the mixture is at its IC_{50}, and $IC_{50}i$ is the concentration that elicits median inhibition concentration when the compound is individually tested. TU_i is the toxic unit of ith component in the mixture. In Equation (3), M is the sum of the toxic units TU_i. The effect of joint toxicity calculated by M value is characterized in Table 1.

Table 1. Sum of Toxicity Unit values representing the interaction existing between mixture compound [40,41].

M (Sum of Toxic Unit)	Interaction
M < 0.8	Synergism
M between 0.8–1.2	Additive
M > 1.2	Antagonism

Mixture Toxic Index (MTI)

The other approach used to assess the joint toxicity effect was Mixture Toxic Index (MTI), which can be calculated by the Equations (4) and (5)

$$M_o = \frac{M}{TU_i max} \tag{4}$$

$$M_o = \frac{(1 - \log M_o)}{M_o} \tag{5}$$

In Equation (4), M_0 represents the ratio of M and max (TU_i), the maximum value of toxic units TU_i of the single components in the mixture. Table 2 summarizes the values calculated by using the MTI approach and the resultant interactions of the compound in the mixture [33,42].

Table 2. Mixture toxicity indices and interaction existing between compounds [42].

MTI (Mixture Toxicity Index)	Interaction
MTI < 0	Antagonistic
MTI = 0	No addition
0 < MTI < 1	Partial additive
MTI = 1	Additive
MTI > 1	Synergistic (Supra-additive)

3. Results

3.1. Acute Toxicity of Combined Heavy Metals Toxicity

Table 3 show IC_{50}s over acute and chronic toxicity of six metals against the bacteria before carry out mixture metals toxicity in this study. Figures 2a, 3a, 4a, 5a, 6a, 7a, 8a, 9a, 10a, 11a, 12a, 13a, and 14a show the Dose-Inhibition response curve of both individual and mixture metals studied in the acute test respectively constructed by nonlinear regression models in GraphPad Prism 7.04 in terms of IC_{50}s. TU (Toxicity Unit) and MTI (Mixture Toxicity Index) approaches were used for predicting the mixture toxicity between two and three metals in this study. As documented in (TU) concept, toxic units of a single metal tested in this study, and the sum of toxic units in a mixture (M) were computed as listed in (Table 4) in acute test. Afterward, data of M and MTI were compared as proven in (Table 5) and similar action was predicted by the two models.

The results suggested that the acute effect combination of $Hg^{2+} + Ag^+$, $Hg^{2+} + Cu^{2+}$, $Ag^+ + Cu^{2+}$, $Hg^{2+} + Ag^+ + Cu^{2+}$ were antagonistic, as IC_{50}s are higher than the TU values as M > 1 and MTI < 1 (Table 1&2). The sigmoidal curve in (Figure 8a) indicates the inhibition increase at mixed concentrations of Hg^{2+} and Ag^+ checked from 0.0816 mg/L until 189 16.73 mg/L.

Table 3. IC$_{50}$s over acute and chronic toxicity of six metals against *Photobacterium* sp.NAA-MIE.

Metal	*Photobacterium* sp.NAA-MIE IC 50 30 min (mg/L)	*R2*	*Photobacterium* sp.NAA-MIE IC 50 6 h (mg/L)	*R2*
Hg^{2+}	0.0603	0.911	0.05714	0.8853
Ag^+	0.5788	0.9096	1.657	0.9427
Cu^{2+}	2.943	0.9574	0.2421	0.8142
Cd^{2+}	251.9	0.9348	48.52	0.9816
Ni^{2+}	27.24	0.9279	8.343	0.9388
Zn^{2+}	80.57	0.9832	60.03	0.9487

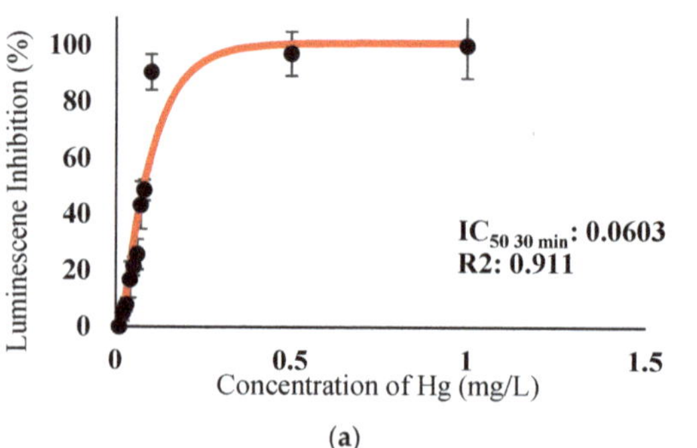

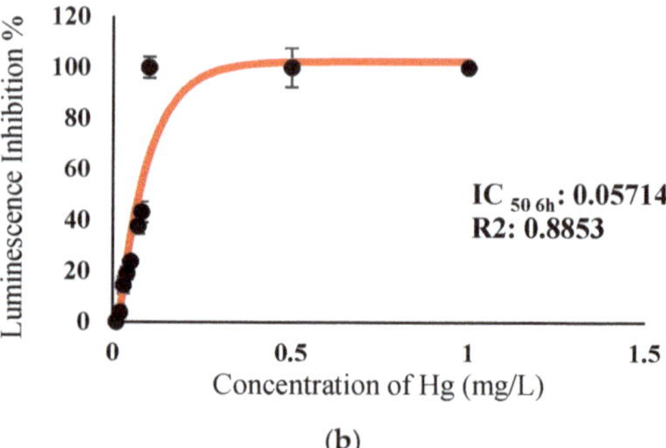

Figure 2. (**a**) Acute and (**b**) Chronic toxicity of Hg^2 on luminescence inhibition of *Photobacterium* sp.NAA-MIE. Data represent mean $\pm$ SD, $n = 3$.

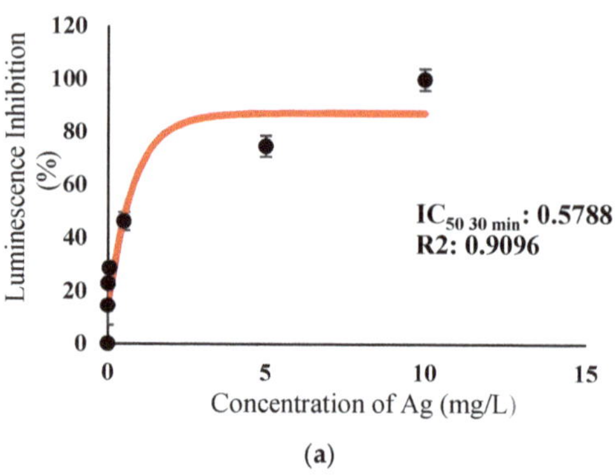

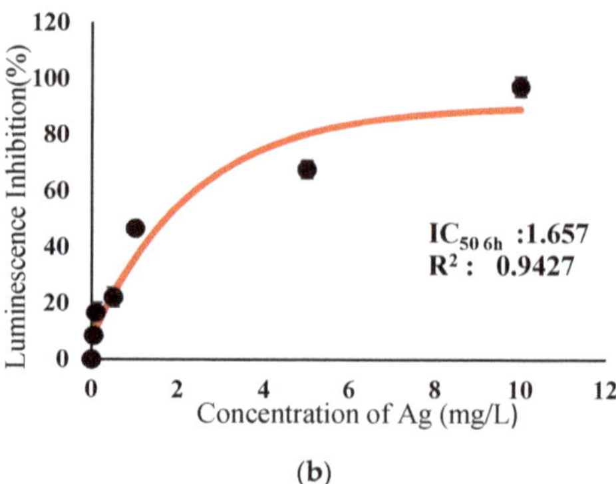

Figure 3. (**a**) Acute and (**b**) Chronic toxicity of Ag^{2+} on luminescence inhibition of *Photobacterium* sp.NAA-MIE. Data represent mean $\pm$ SD, $n = 3$.

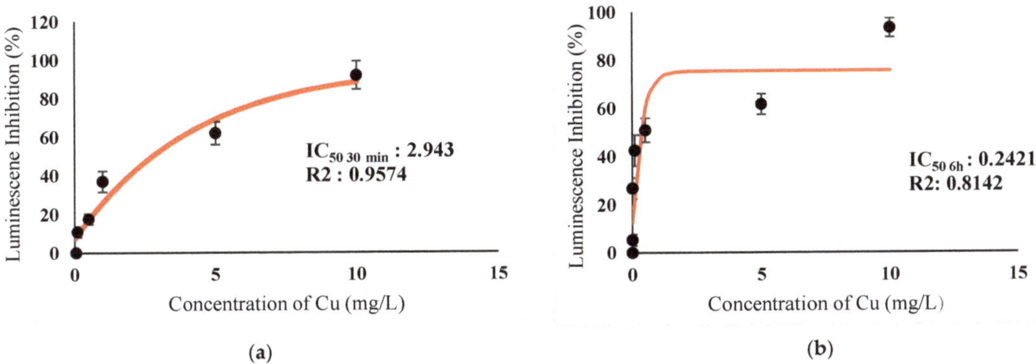

Figure 4. (**a**) Acute and (**b**) Chronic toxicity of Cu^{2+} on luminescence inhibition of *Photobacterium* sp.NAA-MIE. Data represent mean $\pm$ SD, $n = 3$.

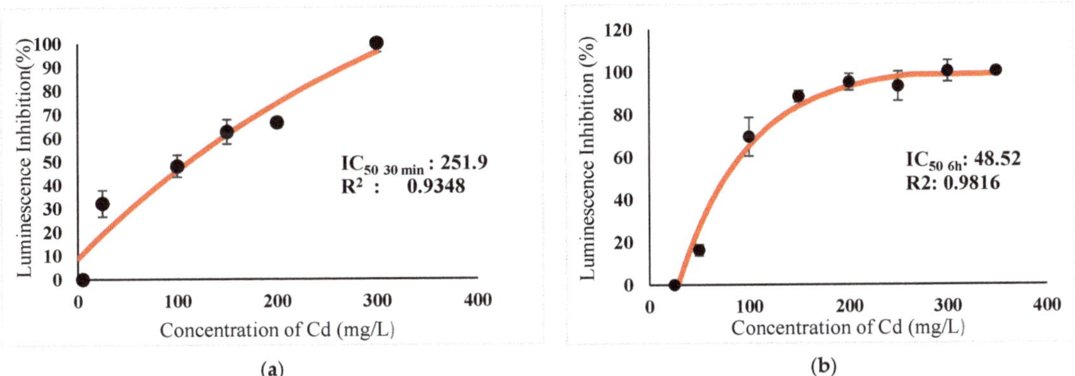

Figure 5. (**a**) Acute and (**b**) Chronic toxicity of Cd^{2+} on luminescence inhibition of *Photobacterium* sp.NAA-MIE. Data represent mean $\pm$ SD, $n = 3$.

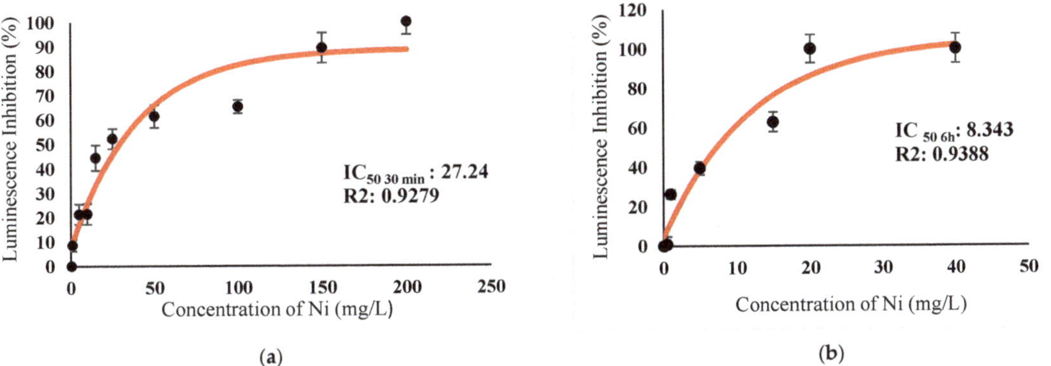

Figure 6. (**a**) Acute and (**b**) Chronic toxicity of Ni^{2+} on luminescence inhibition of *Photobacterium* sp.NAA-MIE. Data represent mean $\pm$ SD, $n = 3$.

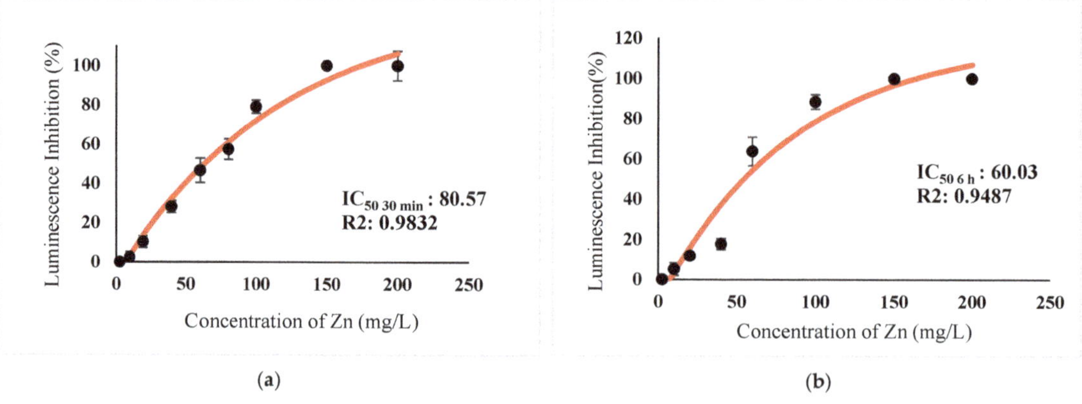

Figure 7. (**a**) Acute and (**b**) Chronic toxicity of Zn^{2+} on luminescence inhibition of *Photobacterium* sp.NAA-MIE. Data represent mean $\pm$ SD, $n = 3$.

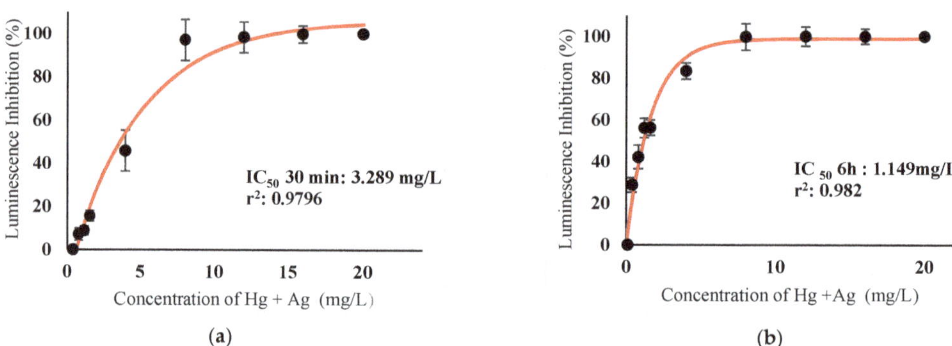

Figure 8. (**a**) Acute and (**b**) Chronic toxicity of mixture $Hg^{2+} + Ag^{+}$ in concentration ratio of 1:1 on luminescence inhibition of *Photobacterium* sp.NAA-MIE. Data represent mean $\pm$ SD, $n = 3$.

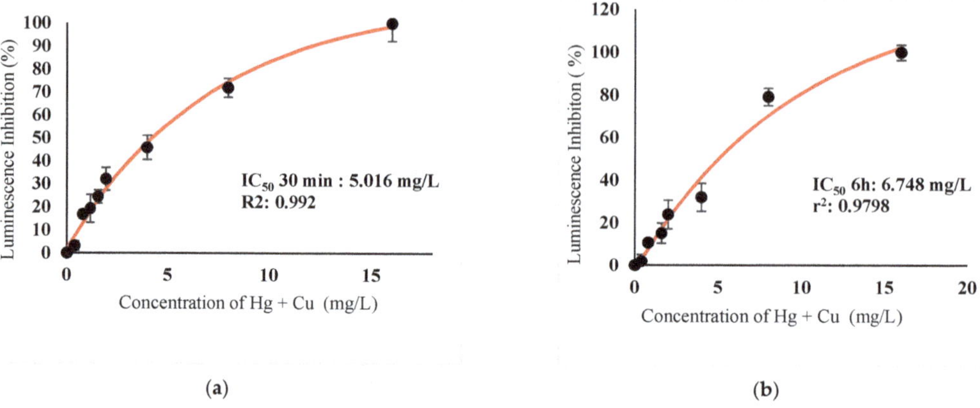

Figure 9. (**a**) Acute and (**b**) Chronic toxicity of mixture $Hg^{2+} + Cu^{2+}$ in concentration ratio of 1:1 on luminescence inhibition of *Photobacterium* sp.NAA-MIE. Data represent mean $\pm$ SD, $n = 3$.

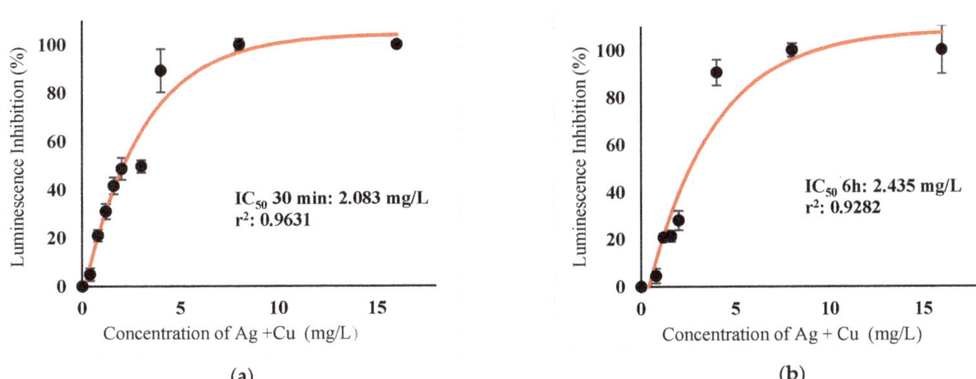

Figure 10. (**a**) Acute and (**b**) Chronic toxicity of mixture $Ag^+ + Cu^{2+}$ in the concentration ratio of 1:1 on luminescence inhibition of *Photobacterium* sp.NAA-MIE. Data represent mean $\pm$ SD, $n = 3$.

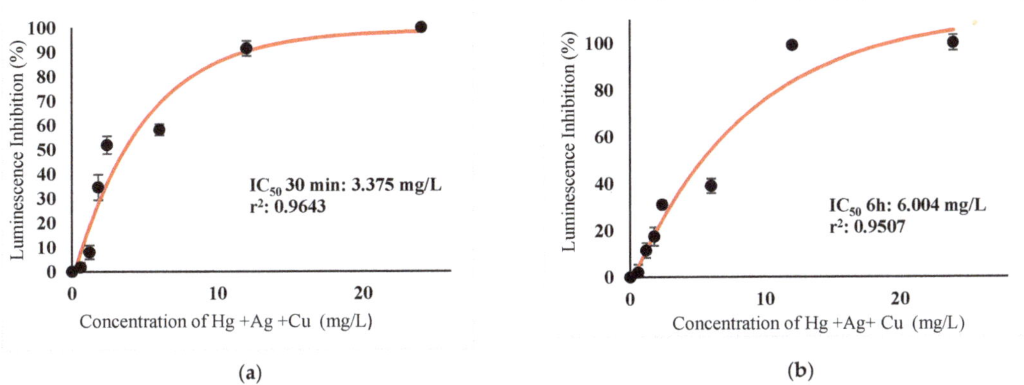

Figure 11. (**a**) Acute and (**b**) Chronic toxicity of mixture $Hg^{2+} + Ag^+ + Cu^{2+}$ in the concentration ratio of 1:1:1 on luminescence inhibition of *Photobacterium* sp.NAA-MIE. Data represent mean $\pm$ SD, $n = 3$.

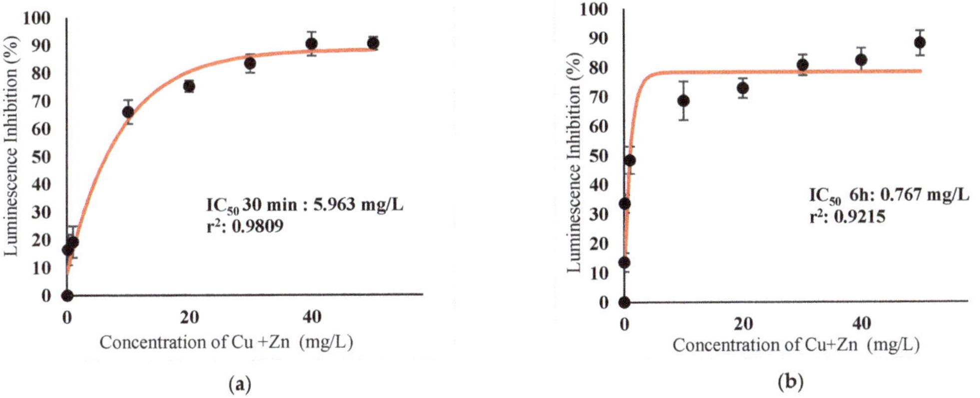

Figure 12. (**a**) Acute and (**b**) Chronic toxicity of mixture $Cu^{2+} + Zn^{2+}$ in the concentration ratio of 1:1 on luminescence inhibition of *Photobacterium* sp.NAA-MIE. Data represent mean $\pm$ SD, $n = 3$.

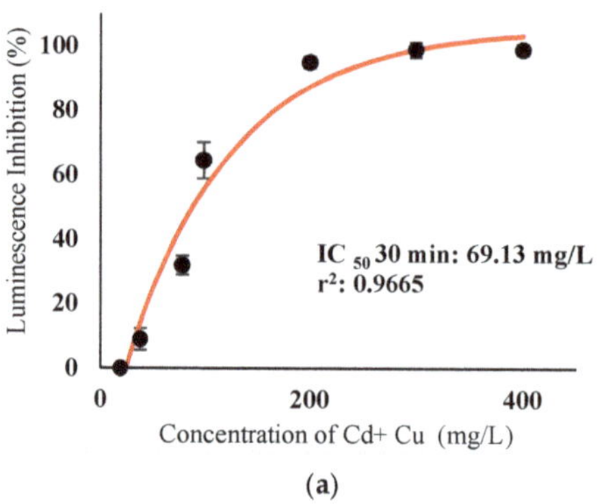
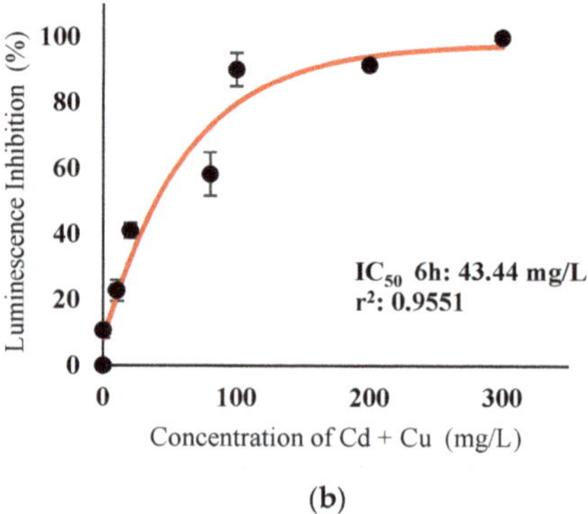

(a)

(b)

Figure 13. (**a**) Acute and (**b**) Chronic toxicity of mixture $Cd^{2+} + Cu^{2+}$ in the concentration ratio of 1:1 on luminescence inhibition of *Photobacterium* sp.NAA-MIE. Data represent mean $\pm$ SD, $n = 3$.

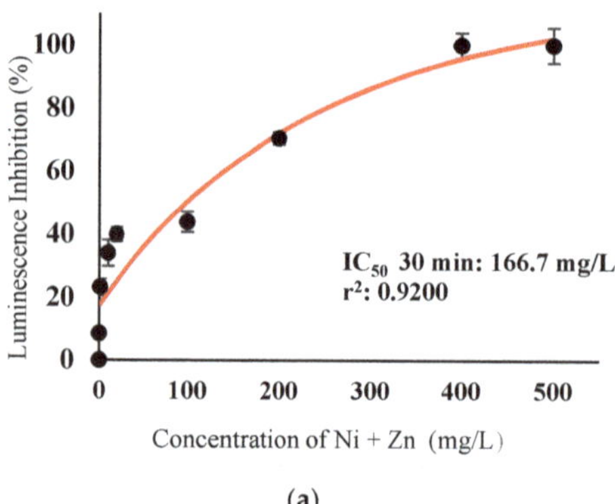
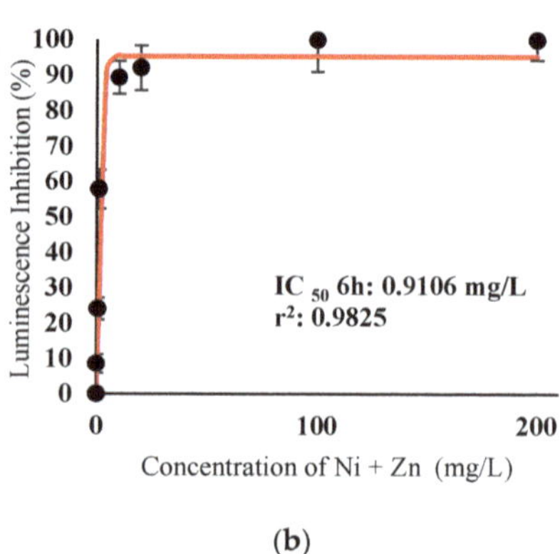

(a)

(b)

Figure 14. (**a**) Acute and (**b**) Chronic toxicity of mixture $Ni^{2+} + Zn^{2+}$ in the concentration ratio of 1:1 on luminescence inhibition of *Photobacterium* sp.NAA-MIE. Data represent mean $\pm$ SD, $n = 3$.

In particular, as was the case with individual metals, the inhibition was less abrupt on the impact of binary mixture compared with individual metal Hg^{2+} and Ag^+ tested at high concentration (Figures 2a and 3a). In the same way, the antagonistic effect of mixture metal tested Hg^{2+} and Cu^{2+} on the bacteria was observed with increased concentrations from 0.327 mg/L until a high concentration 15.67 mg/L was reached (Figure 9a). Meanwhile, (Figure 10a) illustrates the dose-inhibition response curve acute toxicity of metal mixture Ag^+ and Cu^{2+}. The responses produced a sigmoid curve as inhibition increase with mixed concentration from 0.326 mg/L up to 10.12 mg/L. Likewise, acute toxicity of a tertiary mixture of Hg^{2+}, Ag^+ and Cu^{2+} on the bacteria was observed to increase the luminescence inhibition from 0.489 mg/L to 18.61 mg/L (Figure 11a). Generally, the inhibition of those metals in the tertiary mixture was less gradual when metals are combined together at high concentration. Above all, as it was the case with individual metals associating with three high toxicity metals Hg^{2+}, Ag^+ and Cu^{2+} on the bacteria tested in this study as demonstrated in Table 3, the inhibition was less abrupt on the impact in studying mixture metal rather than individually tested metal.

Table 4. List of Toxic Unit (TU) for each heavy metals tested acutely and a total of the toxic units in two and three joint mixtures of heavy metals on *Photobacterium* sp.NAA-MIE.

Mixture	IC_{50} (30 min)	Toxic Unit of Individual Heavy Metals						Sum of Toxic Units in a Mixture (M)
		TU_{Hg}	TU_{Ag}	TU_{Cu}	TU_{Cd}	TU_{Zn}	TU_{Ni}	
$Hg^{2+} + Ag^{+}$	3.289	27.272	2.840	-	-	-	-	30.112
$Hg^{2+} + Cu^{2+}$	5.016	41.592	-	0.8522	-	-	-	42.444
$Ag^{+} + Cu^{2+}$	2.083	-	1.799	0.3539	-	-	-	2.153
$Hg^{2+} + Ag^{+} + Cu^{2+}$	3.375	18.657	1.944	0.3823	-	-	-	20.983
$Cu^{2+} + Zn^{2+}$	5.963	-	-	1.013	-	0.037	-	1.050
$Cd^{2+} + Cu^{2+}$	69.13	-	-	11.745	0.1372	-	-	11.882
$Ni^{2+} + Zn^{2+}$	166.7	-	-	-	-	1.035	3.060	4.095

Table 5. MTI approach to determine the acute effects (30 min) of two and three combined heavy metals on *Photobacterium* sp.NAA-MIE.

Mixture	Acute Toxicity			
	M	Interactive Effect	MTI	Interactive Effect
$Hg^{2+} + Ag^{+}$	30.112	Antagonistic	−33.4	Antagonistic
$Hg^{2+} + Cu^{2+}$	42.444	Antagonistic	−188.25	Antagonistic
$Ag^{+} + Cu^{2+}$	2.153	Antagonistic	−3.265	Antagonistic
$Hg^{2+} + Ag^{+} + Cu^{2+}$	20.983	Antagonistic	−24.843	Antagonistic
$Cu^{2+} + Zn^{2+}$	1.050	Additive	1	Additive
$Cd^{2+} + Cu^{2+}$	11.882	Antagonistic	−206.473	Antagonistic
$Ni^{2+} + Zn^{2+}$	4.095	Antagonistic	−3.842	Antagonistic

Acute exposure to Cu^{2+} and Zn^{2+} mixture showed an additive effect as the sum of toxicity units, M, showed a value equal to 1 and MTI equal to 1. The influence of acute toxicity of the joint metal Cu^{2+} and Zn^{2+} produced a dose-inhibition response curve presented in (Figure 12a). The inhibition of metal Zn^{2+} in the binary mixture at low concentration was more gradual when combined with Cu^{2+} if compared to individually tested Zn^{2+}. Similarly, antagonism was observed for $Cd^{2+} + Cu^{2+}$ in acute test on the bacteria as M > 1 and MTI < 1. Figure 13a demonstrates the inhibition rapidly occurring between 24.49 mg/L and 302.04 mg/L of both Cd^{2+} and Cu^{2+}. The action of metal Cd^{2+} on the luminescence of the bacteria was more gradual at low concentration when combined with Cu^{2+} compared with Cd^{2+} alone in this study. On the other hand, assessment of joint toxicity of $Ni^{2+} + Zn^{2+}$ demonstrated that the acute test of the metal shows antagonistic effect in both approaches, TU and MTI. The impact of Ni^{2+} and Zn^{2+} was increased concentrations of both metals at mixed concentrations from 10.21 mg/L until a high concentration of 489.79 mg/L (Figure 14a).

3.2. Chronic Toxicity of Combined Heavy Metals Toxicity

In the multi-component mixtures, the toxicity of any compounds might change for the extended period of time they mix with the other metals because they would interact in some way. Additional studies with longer exposure times (6 h) were carried out to predict the interaction between metals exposed to chronic action.

(Figures 2b, 3b, 4b, 5b, 6b, and 7b) and (Figures 8b, 9b, 10b, 11b, 12b, 13b, and 14b) show the Dose-Inhibition response curve of both individual and mixture metals studied in the chronic test respectively. The results were used to calculate the combined metal effects over 6 h exposure on the bacteria with the approaches of TU and MTI already discussed for chronic toxicity tests (Tables 6 and 7). Antagonistic effects were observed for Hg^{2+} + Ag^+, Hg^{2+} + Cu^{2+}, Ag^+ + Cu^{2+}, Hg^{2+} + Ag^+ + Cu^{2+}, Cu^{2+} + Zn^{2+} and Cd^{2+} + Cu^{2+} according to M value > 1 and MTI lower than 0. The sigmoidal dose-inhibition response curve shown in (Figure 8b) presents the inhibition increase with the mixed concentration of Hg^{2+} and Ag^+ for the range of concentration checked from 0.408 mg/L to 6.12 mg/L. Furthermore, it can be seen that chronic toxicity of a mixture Hg^{2+} and Cu^{2+} in this study demonstrated an inhibitory effect that gradually increased at mixed concentration range from 0.324 mg/L to 15.67 mg/L at which a slope in (Figure 9b) reaches maximum percentage inhibition. The impact of combined metal Ag^+ and Cu^{2+} on the bacteria under chronic toxicity produces a sigmoidal dose-inhibition response curve in (Figure 10b). The responses of the tertiary mixture of metals Hg^{2+}, Ag^+ and Cu^{2+} are shown in (Figure 11b). The mixtures deliberately inhibited the luminescence activity as the concentration increased from 0.489 mg/L and seemingly reaching saturation by a maximum inhibition at 24 mg/L. In the same way, the sigmoidal curve in (Figure 12b) indicates the impact of joint metals Cu^{2+} and Zn^{2+} under chronic toxicity. Precisely, the inhibition increased rapidly at range of mixed concentration from 1.02 mg/L to 5.10 mg/L. (Figure 13b) illustrates the dose-response relationship curve of metals Cd^{2+} and Cu^{2+} in this study. The sigmoidal curve shows 97.04% inhibition at 275.51 mg/L. In the case of interaction between Ni^{2+} + Zn^{2+}, synergistic effect arose under chronic exposure according to TU value < 1 and MTI greater than 1. The toxicity assessment of metals Ni^{2+} and Zn^{2+} under chronic test produced responses in (Figure 14b). The sigmoidal curve shows complete inhibition on the bacteria as the slope falls off at flat plateau at 12.24 mg/L.

Table 6. List of Toxic Unit (TU) for each heavy metal in chronic toxicity test and sum of the toxic units, M in mixtures of heavy metals on *Photobacterium* sp.NAA-MIE.

Mixture	IC$_{50}$ (6 h)	Toxic Unit of Individual Heavy Metals						Sum of Toxic Units in a Mixture (M)
		TU $_{Hg}$	TU $_{Ag}$	TU $_{Cu}$	TU $_{Cd}$	TU $_{Zn}$	TU $_{Ni}$	
Hg^{2+} + Ag^+	1.149	10.061	0.3467	-	-	-	-	10.408
Hg^{2+} + Cu^{2+}	6.748	59.048	-	13.936	-	-	-	72.984
Ag^+ + Cu^{2+}	2.435	-	0.7351	5.031	-	-	-	5.766
Hg^{2+} + Ag^+ + Cu^{2+}	6.004	35.019	1.208	8.265	-	-	-	44.492
Cu^{2+} + Zn^{2+}	0.7675	-	-	1.585	-	0.006	-	1.591
Cd^{2+} + Cu^{2+}	43.44	-	-	89.715	0.447	-	-	90.162
Ni^{2+} + Zn^{2+}	0.9106	-	-	-	-	0.008	0.055	0.063

Table 7. TU and MTI approaches to determine the interaction of two and three combined heavy metal effects in chronic (6 h) exposure tests mixture on *Photobacterium* sp.NAA-MIE.

Mixture	Chronic Toxicity			
	M	Interactive Effect	MTI	Interactive Effect
Hg^{2+} + Ag^+	10.408	Antagonistic	−22.666	Antagonistic
Hg^{2+} + Cu^{2+}	72.984	Antagonistic	−19.429	Antagonistic
Ag^+ + Cu^{2+}	5.766	Antagonistic	−11.856	Antagonistic
Hg^{2+} + Ag^+ + Cu^{2+}	44.492	Antagonistic	−14.827	Antagonistic

Table 7. *Cont.*

Mixture	Chronic Toxicity			
	M	**Interactive Effect**	**MTI**	**Interactive Effect**
$Cu^{2+} + Zn^{2+}$	1.591	Antagonistic	−115.32	Antagonistic
$Cd^{2+} + Cu^{2+}$	90.162	Antagonistic	−901.58	Antagonistic
$Ni^{2+} + Zn^{2+}$	0.063	Synergistic	22.342	Synergistic

4. Discussion

4.1. Acute and Chronic Action for Individual Metals on Photobacterium sp.NAA-MIE

At present, the highest metal toxicities on the bacteria are Hg^{2+} in both acute and chronic test, and the lowest toxicities are demonstrated by metal Cd^{2+} in acute test and Zn^{2+} in chronic test (Table 3). Mercury is known as the most toxic heavy metal [43, 44]. It enters biomembranes easily due to its lipophilic properties. After its discharge into the environment, inorganic mercury is incited by bacteria, thereby accelerating to form methylmercury, (MeHg) and has the ability to bioaccumulate in fish and other animal tissues [43,45]. Whereas, numerous study have reported, Gram-negative bacteria is less sensitive towards Cd^{2+}. The outer layer of the bacteria membrane is made of exo-polysaccharides which are able to adsorb and trap Cadmium [46–48]. Data in the previous study further supported the least metal toxicities of Cd^{2+} in this study. This may prevent the interaction between Cd^{2+} and key enzymes, and luciferase, which is responsible for bioluminescence production, and ultimately decrease the toxicity of Cd towards the bacteria. In addition, based on previous literature, some metals have been proven as a detrimental element for living organisms and the main integral of metabolic enzymes, such as Zn and Cu [49–51]. These can be toxic to organisms when exceeding threshold levels. In this study, the bacteria might profit at low concentration when Cu^{2+} and Zn^{2+} were acutely exposed, whereas high concentrations at more than IC_{50} value of 2.943 mg/L and 80.57 mg/L for Cu and Zn respectively (Table 3) were observed to increase the percentage of luminescence inhibition.

4.2. Comparison of Acute and Chronic Action for Metal Mixtures on Photobacterium sp.NAA- MIE

The degree of toxicity of pollutants is pertained not just to the exposure concentration, but also to the exposure time, which is a significant influencing factor. The effect of the acute and chronic joint is different thereby toxicity data regarding the mixtures in both 30 min and 6 h exposures were compared and are given in (Table 8).

Antagonistic effect of combination of the $Hg^{2+} + Ag^+$, $Hg^{2+} + Cu^{2+}$, $Ag^+ + Cu^{2+}$, $Hg^{2+} + Ag^+ + Cu^{2+}$ arise from both acute and chronic action on luminescence. The previous section has shown that metals Hg^{2+}, Ag^{2+} and Cu^{2+} were most toxic to bacteria when tested individually as listed in (Table 3). In combination toxicology, the primary way to specify the joint action of the components in the mixture is to conduct experimental studies correlating the effect of the mixture to the effect of the individual compounds. However, when those metal were mixed, a lower toxicity than expected was exhibited. In this event, the interaction effect of $Hg^{2+} + Ag^+$, $Hg^{2+} + Cu^{2+}$, $Ag^+ + Cu^{2+}$, $Hg^{2+} + Ag^+ + Cu^{2+}$ resulted in a weaker effect. These interactions may decrease the toxicity of the active compound, thus their mixture exerts a weaker effect than predicted [52]. Authors of [53] argue that antagonistic interaction occurs due to the competition for binding sites in the biological interface. In this study, the reference metals Hg^{2+}, Ag^+ and Cu^{2+} are three pollutants that exhibited high sensitivity when tested individually to the bacteria for both acute and chronic tests. It can be hypothesized that the metals Hg^{2+}, Ag^+ and Cu^{2+} respectively displayed in the mixtures directly aim at luciferase, which is important in chemical reactions for luminescence production in both acute and chronic actions. Thus resulting in the fierce competition between them, which may impede their interaction with the proteins and ultimately reduce luminescence inhibition. Another toxicity study involving antagonistic interaction of mercury with selenium suggested selenium to lower the effect and depress mercury in fish tissues [54].

Table 8. TU and MTI approaches to determine the interaction of two and three combined heavy metal effects over Acute (30 min) and Chronic (6 h) on *Photobacterium* sp.NAA-MIE.

Mixture	Acute Toxicity				Chronic Toxicity			
	M	Interactive Effect	MTI	Interactive Effect	M	Interactive Effect	MTI	Interactive Effect
$Hg^{2+} + Ag^+$	30.112	Antagonistic	−33.4	Antagonistic	10.408	Antagonistic	−22.666	Antagonistic
$Hg^{2+} + Cu^{2+}$	42.444	Antagonistic	−188.25	Antagonistic	72.984	Antagonistic	−19.429	Antagonistic
$Ag^+ + Cu^{2+}$	2.153	Antagonistic	−3.265	Antagonistic	5.766	Antagonistic	−11.856	Antagonistic
$Hg^{2+} + Ag^+ + Cu^{2+}$	20.983	Antagonistic	−24.843	Antagonistic	44.492	Antagonistic	−14.827	Antagonistic
$Cu^{2+} + Zn^{2+}$	1.050	Additive	1	Additive	1.591	Antagonistic	−115.32	Antagonistic
$Cd^{2+} + Cu^{2+}$	11.882	Antagonistic	−206.472	Antagonistic	90.162	Antagonistic	−901.58	Antagonistic
$Ni^{2+} + Zn^{2+}$	4.095	Antagonistic	−3.842	Antagonistic	0.063	Synergistic	22.342	Synergistic

Similarly, an antagonistic effect was observed for the mixture of metal Cd^{2+} and Cu^{2+} from acute to chronic action on *Photobacterium* sp. NAA-MIE as TU shows M > 1 and MTI < 0. In this study, Cu^{2+} was observed to be more toxic than Cd^{2+} when tested individually. Capability of Cd^{2+} to antagonize the toxic effects of Cu^{2+} was observed. According to [52], the possibility of a partial agonist and full agonist may occur if two substances present in the mixture in a manner of low efficacy of one substance compete with the high efficacy of second substance respectively. Antagonistic interaction similarly observed in studies made by [13,21] based on inhibition determined by Microtox ® assay, under acute toxicity of Cd + Cu against *P. phosphoreum* T3S with (15 min IC_{50} 8.581 mg/L) and *V. fischeri* respectively after 15-min exposure (Table 9). However, most previous studies regarding the luminescence inhibition test have only showed the interactive effect of the combined metals from 5 min to 30 min. Limited available data reported on the chronic effect. At present the disfigurement of the conventional short-term concerning substances with a slowed impact creates interest essentiality for chronic toxicity.

In the case of toxicity action of $Cu^{2+} + Zn^{2+}$, the toxic action changes from additive action in 30 min to antagonistic action in 6 h. The additive effect of the two combined heavy metals was due to the same action mechanism, differing only in their potencies. Evaluation of the combined heavy metals toxicity may not be similar in various studies. Other authors (Table 9) reported that acute toxicity of $Cu^{2+} + Zn^{2+}$ on *V. fischeri* manifested synergistic interaction as 15 min EC_{50} for Zinc in presence of 0.125 mg/L of copper, and was 0.14 mg/L [55]. Surprisingly, after chronic tests, both mixtures of Cu^{2+} and Zn^{2+} showed antagonistic interaction suggesting that Cu may enhance the Zinc toxic effect after 6 h exposure as Zn showed lower sensitivity than Cu to *Photobacterium* sp.NAA-MIE. Results in the present study further supported the idea of cellular uptake of Cu^{2+} as greater than that of Zn^{2+}, which affected the luciferase in the luminescence bioprocess. The majority of published articles regarding the toxicity of Cu and Zn mixtures have been reported using different organisms [33,49,55]. Taken for example, acute toxicity of Zinc pyrithione (ZnPT) and Cu against three marine organisms, based on algal growth inhibition test of *Thalassiosira pseudonana* under 96-h exposure time and mortality rate of polychaete larvae *Hydroides elegans*, and amphipod *Elasmopus rapax* under 48-h and 24-h exposure time respectively presented synergistic interaction [49] in Table 9. Evaluation of mixture toxicants may not give similar interaction for different organisms due to low reproducibility, consequently it takes a long time to give a response rather than applying luminescence inhibition tests that gives fast results. Additionally, the use of luminescence inhibition bioassay has a shorter time advantage compared to other methods. According to [58], among various species (algae, crustaceans, rotifers, bacteria and protozoan), the bacterial bioluminescence assays showed the highest sensitivity in acute toxicity measurement for most of the samples.

Table 9. Comparative metals mixture toxicity studies on *Photobacterium* sp.NAA-MIE with those reported by other investigators among different species.

Mixture	30-min Acute Toxicity *Photobacterium* sp.NAA-MIE	6 h-Chronic Toxicity *Photobacterium* sp.NAA-MIE	Other Studies	
	Interactive Effect	**Interactive Effect**	**Organism**	**Interactive Effect**
Hg + Ag	Antagonistic	Antagonistic	Not applicable	-
Hg + Cu	Antagonistic	Antagonistic	Not applicable	-
Ag + Cu	Antagonistic	Antagonistic	Not applicable	-
Hg + Ag + Cu	Antagonistic	Antagonistic	Not applicable	-
Cu + Zn	Additive	Antagonistic	*Clariasgariepinus* [33] *Thalassiosira pseudonana,* *Hydroides elegan,* and *Elasmopus rapax* [49] *Vibrio fischeri* [55]	Antagonistic [a] Synergistic [a] Synergistic [a]
Cd + Cu	Antagonistic	Antagonistic	*P. phosphoreum* [13] *Vibrio fischeri* [21]	Antagonistic [a] Antagonistic [a]
Ni + Zn	Antagonistic	Synergistic	*Pimephales promelas* [56] *Daphnia magna* [57]	Synergistic [a] Antagonistic [b]

[a] acute toxicity test [b] chronic toxicity test.

Furthermore, [33] in Table 9 reported that the mixture of zinc and copper on *Clarias gariepinus* under 96-h exposure time showed a different reaction, for example antagonism occurred when the metals were mixed at the ratio of 1:1 and synergism at the ratio of 1:2. This study suggested that it is important to take into consideration the effect of low and high concentrations of toxicant. At present, there are limited data on different concentration ratios of constituent metal components in a mixture which influences metal toxicity to *Photobacterium* sp.NAA-MIE, thus additional studies are required in this field.

The interaction between Ni^{2+} and Zn^{2+} metals became more complex over chronic action as synergistic effect was observed. Synergistic interactions of chemicals in mixtures boost the action of other chemicals, so that they allow to jointly strive for a stronger effect than expected [59]. Even though metal Ni^{2+} and Zn^{2+} appeared to be slightly toxic when tested individually, the information theoretically proves the fact that metals forming complexes have better penetrability with regards to the bacterial cell compared to a single metal acting solely. On that account, the eventual toxicity of the mixture brings to bear a stronger effect rather than the degree of toxicity of the single metal. Besides, the concentration for both metals tested is low in mixture compared to being tested individually. It is assumed in this study that particular concentrations of heavy metals regarded as being low and of minimal effect possibly draw out more degree of toxicity in mixture with other metals. Similarly, synergism was observed by Zn and Ni mixtures on *Pimephales promelas* under standard 96-h acute tests for a 4 times concentration of Zn and a 3 times concentration of Ni, as 67% rate of mortality of the organism was predicted [58]. Other studies did not correspond well with the trend in this work. For instance, the mixture chronic toxicity of zinc and nickel to *D. magna* reproduction toxicity test for 21-days exposure time is antagonistic at a low concentration however, it becomes synergistic at the high concentration as predicted by global statistical analysis [59]. Besides, different mixture interactions of metal studied also show that the toxicity mixture present in binary or tertiary joint relies on the tested organisms [53]. Different mixture interactions were observed especially when the metals $Cu^{2+} + Zn^{2+}$ and $Ni^{2+} + Zn^{2+}$ were acutely and chronically exposed in this study.The author cited that mixture effect under chronic tests was dissimilar to those from acute tests as short term exposures tend not to represent metal interactions occurring with longer-term exposures.

Even so, studies on mixture have declared that interactions might be inconsistent throughout various experiments. The interactions count on concentration, types of chemicals mixed and organisms tested [60,61]. Interactions between metals mostly affect some modes that are essential for the differential responses of test organisms to mixtures of metals towards luminescence production, such as bioavailability, internal transportation, binding at the target site and interactions of the toxic elements with luciferase complex in charge for the luminescence. Furthermore, the biological makeup of the receiving organism and physicochemical nature of the metal decide the bioavailability of the metal that will act upon in the organism [62].

An approach for threat identification and risk evaluation of a complex mixture such as in medication therapies interested in vitro experiments of many antibiotics and their combinations or appropriate drug combinations will be governed to assess a public health problem.These drug combinations have derived into additive, synergistic or antagonistic interactions [63,64]. In such a way, the study on mixture compounds is beneficial to heal, delay the progression, or decrease the symptoms of diseases. Another good example that demonstrates this trend is the information regarding toxic action by antibiotic agents on *V. fischeri* from acute synergism to chronic antagonism, which suggests that the time intermission of medication should be reduced to prevent chronic antagonism [17].

5. Conclusions

In the current work, the tropical luminescent bacterium *Photobacterium* sp.NAA-MIE luminescent bacteria test allows the prediction of the joint effect of metal mixture according to the simultaneous analysis of two toxicological endpoints which are acute luminescence inhibition after 30 min, and chronic luminescence inhibition after 6 h. The application of two different mathematical Toxicity Unit (TU) and Mixture Toxicity Index (MTI) observes mixture toxicity of Hg + Ag, Hg + Cu, Ag + Cu, Hg + Ag + Cu and Cd + Cu, showing an antagonistic effect over acute and chronic tests. A binary mixture of Cu + Zn was observed to show additive effect at the acute test and antagonistic effect at chronic test, while the mixture of Ni + Zn showed antagonistic effect during the acute test and synergistic effect during the chronic test. The similarities and differences in prediction of the joint effect of metal mixtures with other standard bioassay in this study indicate possible different toxicity/inhibition mechanisms for metals in different organism under acute to chronic test. In future, additional studies should be conducted to quantify the antagonistic and synergistic effect of metals in the mixture to understand the mechanism of action and potency of each compound.

However, the advantage of the bioluminescence test in this study is applicable in terms of the sensitivity of the bacteria in much shorter time than other assays, is cost-effective, and may be used to counter to the possible environmental risks. Thus, the toxicity result of this strain may be effective to take into theoretical consideration toxicity levels of mixture of pollutants that occur together in ecosystems and in the same way contribute extrapolation studies under acute to chronic exposures for mixture metal toxicity in deriving safe limits and standards aimed at protecting organisms in the environment.

Author Contributions: Conceptualization, M.I.E.H., N.A.A.; data curation, M.I.E.H., N.A.A.; methodology, M.I.E.H.; supervision, M.I.E.H., S.S.A.G., U.H.Z., M.Y.A.S.; writing—Original draft, N.A.A.; writing—Review and editing, M.I.E.H., N.A.A. All authors have read and agreed to the published version of the manuscript.

Funding: This project was financed by funds from Putra Grant (GP-IPM/2017/9532800), Yayasan Pak Rashid Grant UPM (6300893-10201) and Fundamental Research Grant Scheme-FRGS (5540085).

Institutional Review Board Statement: Not applicable.

Informed Consent Statement: Not applicable.

Data Availability Statement: Not applicable.

Conflicts of Interest: The authors declare no conflict of interest.

References

1. Ali, H.; Khan, E.; Ilahi, K. Environmental Chemistry and Ecotoxicology of Hazardous Heavy Metals: Environmental Persistence, Toxicity, and Bioaccumulation. *J. Chem.* **2019**, *2019*, 6730305. [CrossRef]
2. Leizou, K.E.; Horsfall, M.J.; Spiff, A.I. Speciation of some heavy metals in sediments of the Pennington River, Bayelsa State, Nigeria. *Am. Chem. Sci. J.* **2015**, *5*, 238–246.
3. Davis, J.A.; Looker, R.E.; Yee, D.; Marvin-Di Pasquale, M.; Grenier, J.L.; Austin, C.M.; McKee, L.J.; Greenfield, B.K.; Brodberg, R.; Blum, J.D. Reducing methylmercury accumulation in the food webs of San Francisco Bay and its local watersheds. *Environ. Resour.* **2012**, *119*, 3–26. [CrossRef] [PubMed]
4. Foster, K.L.; Stern, G.A.; Pazerniuk, M.A.; Hickie, B.; Walkusz, W.; Wang, F.; Macdonald, R.W. Mercury biomagnification in marine zooplankton food webs in Hudson Bay. *Environ. Sci. Technol.* **2012**, *46*, 12952–12959. [CrossRef] [PubMed]
5. Wang, X.; Qu, R.J.; Wei, Z.B.; Yang, X.; Wang, Z.Y. Effect of water quality on mercury toxicity to *Photobacterium phosphoreum*: Model development and its application in natural waters. *Ecotoxicol. Environ. Saf.* **2014**, *104*, 231–238. [CrossRef] [PubMed]
6. Cheng, H.F.; Hu, Y.A. China needs to control mercury emissions from municipal solid waste (MSW) incineration. *Environ. Sci. Technol.* **2010**, *44*, 7994–7995. [CrossRef]
7. Nansai, K.; Oguchi, M.; Suzuki, N.; Kida, A.; Nataami, T.; Tanaka, C.; Haga, M. High Resolution inventory of Japanese anthropogenic mercury emissions. *Environ. Sci. Technol.* **2012**, *46*, 4933–4940. [CrossRef]
8. Ahmed, A.M.; Lyautey, E.; Bonnineau, C.; Dabrin, A.; Pesce, S. Environmental concentrations of copper, alone or in mixture with arsenic, can impact river sediment microbial community structure and functions. *Metals Effects Sediment Microorg.* **2018**, *9*, 1852.
9. Bereswill, R.; Golla, B.; Streloke, M.; Schulz, R. Entry and toxicity of organic pesticides and copper in vineyard streams: Erosion rills jeopardise the efficiency of riparian buffer strips. *Agric. Ecosyst. Environ.* **2013**, *172*, 49–50. [CrossRef]
10. Sany, S.E.T.; Salleh, A.; Rezayi, M.; Saadati, N.; Narimany, L.; Tehrani, G.M. Distribution and Contamination of Heavy Metal in the Coastal Sediments of Port Klang, Selangor, Malaysia. *Water Air Soil Pollut.* **2013**, *224*, 1476. [CrossRef]
11. Yang, X.P.; Yan, J.; Wang, F.F.; Xu, J.; Liu, X.Z.; Ma, K.; Hu, X.M.; Ye, J.B. Comparison of organics and heavy metals acute toxicities to *Vibrio fischeri*. *J. Serb. Chem. Soc.* **2016**, *81*, 697–705. [CrossRef]
12. Nweke, C.O.; Ike, C.C.; Ibegbulem, C.O. Toxicity of quaternary mixtures of phenolic compounds and formulated glyphosate to microbial community of river water. *Ecotoxicol. Environ. Contam.* **2016**, *11*, 63–71. [CrossRef]
13. Zeb, B.S.; Ping, Z.; Mahmood, Q.; Lin, Q.; Pervez, A.; Irshad, M.; Bilal, M.; Bhatti, Z.A.; Shaheen, S. Assessment of combined toxicity of heavy metals from industrial wastewaters on *Photobacterium phosphoreum* T3S. *Appl. Water Sci.* **2016**, *7*, 2043–2050. [CrossRef]
14. Cobbina, S.J.; Duwiejuah, A.B.; Quansah, R.; Obiri, S.; Bakobie, N. Comparative Assessment of Heavy Metals in Drinking Water Sources in Two Small-Scale Mining Communities in Northern Ghana. *Int. J. Environ. Res. Public Health* **2015**, *12*, 1062–1063. [CrossRef] [PubMed]
15. Nwanyanwu, C.E.; Adieze, I.E.; Nweke, C.O.; Nzeh, B.C. Combined effects of metals and chlorophenols on dehydrogenase activity of bacterial consortium. *Int. Resour. J. Biol. Sci.* **2017**, *6*, 10–20.
16. Lopes, S.; Pinheiro, C.; Soares, A.M.V.M.; Louriero, S. Joint toxicity prediction of nanoparticles and ionic counterparts: Simulating toxicity under a fate scenario. *J. Hazard. Mater.* **2016**, *320*, 1–9. [CrossRef] [PubMed]
17. Wang, T.; Liu, Y.W.; Wang, D.; Lin, Z.; An, Q.Q.; Yin, C.H.; Liu, Y. The joint effects of sulfonamides and quorum sensing inhibitors on *Vibrio fischeri*: Differences between the acute and chronic mixed toxicity mechanisms. *J. Hazard. Mater.* **2016**, *310*, 56–67. [CrossRef]
18. Pablos, M.V.; García-Hortigüela, P.; Fernández, C. Acute and chronic toxicity of emerging contaminants, alone or in combination, in *Chlorella vulgaris* and *Daphnia Magna*. *Environ. Sci. Pollut. Resour.* **2015**, *22*, 5417–5424. [CrossRef]
19. Abbas, M.; Adi, M.; Haque, S.E.; Munir, B.; Yameen, M.; Ghaffar, A.; Shar, G.A.; Tahir, M.A.; Iqbal, M. *Vibrio fischeri* biolumines-cence inhibition assay for ecotoxicity assessment: A review. *Sci. Total Environ.* **2018**, *626*, 1295–1309. [CrossRef]
20. Chen, S.S.; Sun, Y.; Tsang, D.C.; Graham, N.J.; Ok, Y.S.; Feng, Y. Potential impact of flowback water from hydraulic fracturing on agricultural soil quality: Metal/metalloid bioaccessibility, Microtox bioassay, and enzyme activities. *Sci. Total Environ.* **2017**, *579*, 1419–1426. [CrossRef]
21. Yamamuro, Y. Social behavior in laboratory rats: Applications for psycho-neuroethology studies. *Anim. Sci. J.* **2006**, *77*, 386–394. [CrossRef]

22. Cheng, C.Y.; Kuo, J.-T.; Yu, C.; Lin, Y.-C.; Liao, Y.-R.; Chung, Y.-C. Comparisons of *Vibrio fischeri*, *Photobacterium phosphoreum*, and recombinant luminescent using *Escherichia coli* as BOD measurement. *J. Environ. Sci. Health* **2010**, *45*, 233–238. [CrossRef]
23. Mansour, S.A.; Abdel-Hamid, A.A.; Ibrahim, A.W.; Mahmoud, N.H.; Moselhy, W.A. Toxicity of Some Pesticides, Heavy Metals and Their Mixtures to *Vibrio fischeri* Bacteria and *Daphnia magna*: Comparative Study. *J. Biol. Life Sci.* **2015**, *6*, 221–240. [CrossRef]
24. Halmi, M.I.E.; Jirangon, H.; Johari, W.L.W.; Abdul Rachman, A.R.; Shukor, M.Y.; Syed, M.A. Comparison of Microtox and Xenoassay light as a near real time river monitoring assay for heavy metals. *Sci. World J.* **2014**, *2014*, 834202. [CrossRef]
25. Wong, C.L.; Liew, J.; Yusop, Z.; Ismail, T.; Venneker, R.; Uhlenbrook, S. Rainfall characteristics and regionalization in Peninsular Malaysia based on a high resolution gridded data set. *Water* **2016**, *8*, 500. [CrossRef]
26. Dewhurst, R.; Wheeler, J.; Chummun, K.; Mather, J.; Callaghan, A.; Crane, M. The comparison of rapid bioassays for the assessment of urban groundwater quality. *Chemosphere* **2002**, *47*, 547–554. [CrossRef]
27. Halmi, M.I.E.; Kassim, A.; Shukor, M.Y. Assessment of heavy metal toxicity using a luminescent bacterial test based on *Photobacterium* sp. strain MIE. *Rendi Fis. Acc. Lincei* **2019**, *30*, 589–601. [CrossRef]
28. Blasco, J.; Del Valls, A. Impact of emergent contaminant in the environment: Environmental Risk Assessment. In *Emerging Contaminants from Industrial and Municipal Waste*; Springer: Berlin/ Heidelberg, Germany, 2008; Volume 5, pp. 169–188.
29. Wang, D.; Gu, Y.; Zheng, M.; Zhang, W.; Lin, Z.; Liu, Y. A Mechanism-based QSTR Model for Acute to Chronic Toxicity Extrapolation: A Case Study of Antibiotics on Luminous Bacteria. *Sci. Rep.* **2017**, *7*, 6022. [CrossRef] [PubMed]
30. Adnan, N.A.; Halmi, M.I.E.; Shukor, M.Y.; Gani, S.S.A.; Zaidan, U.H.; Radziah, O. Statistical Modeling for the Optimization of Bioluminescence Production by Newly Isolated *Photobacterium* sp. NAA-MIE. *Proc. Natl. Acad. Sci. India Sect. B Biol. Sci.* **2020**, *90*, 797–810. [CrossRef]
31. Chun, U.-H.; Simonov, N.; Chen, Y.; Britz, M.L. Continuous pollution monitoring using *Photobacterium phosphoreum*, *Resour. Conserv. Recycl.* **1996**, *18*, 25–40. [CrossRef]
32. Hong, Y.; Chen, Z.; Zhang, B.; Zhai, Q. Isolation of *Photobacterium* sp. LuB-1 and its application in rapid assays for chemical toxicants in water. *Lett. Appl. Microbiol.* **2010**, *51*, 308–312. [CrossRef] [PubMed]
33. Damian, E.C.; Obinna, O.M.; Ndidi, O.C. Predictive Modelling of Heavy Metal-Metal Interactions in Environmental Setting: Laboratory Simulative Approach. *Environ. Ecol. Res.* **2014**, *2*, 248–252.
34. Villaescusa, I.; Martinez, M.; Murat, J.C.; Costa, C. Cadmium species toxicity on luminescent bacteria, Fresen. *J. Anal. Chem.* **1996**, *354*, 566–570.
35. Shukor, Y.; Baharom, N.A.; Rahman, F.A.; Abdullah, M.P.; Shamaan, N.A.; Syed, M.A. Development of a heavy metals enzymatic-based assay using papain. *Anal. Chim. Acta* **2006**, *566*, 283–289. [CrossRef]
36. Girotti, S.; Bolelli, L.; Roda, A.; Gentilomi, G.; Musiani, M. Improved detection of toxic chemicals using bioluminescent bacteria. *Anal. Chim. Acta* **2002**, *471*, 113–120. [CrossRef]
37. Girotti, S.; Ferri, E.N.; Fumo, M.G.; Maiolini, E. Monitoring of environmental pollutants by bioluminescent bacteria. *Anal. Chim. Acta* **2008**, *608*, 2–29. [CrossRef] [PubMed]
38. Calabrese, E.J. *Multiple Chemical Interactions*; Lewis Publishers: Chelsea, MI, USA, 1991.
39. Zheng, P.; Feng, X.S. *Biotechnology for Waste Treatment*; Higher Education Press: Beijing, China, 2006.
40. Marking, L.; Dawson, V.K. Method of assessment of toxicity or efficacy of mixture of chemical. *Investig. Fish Control* **1975**, *67*, 1–8.
41. Broderius, S.J.; Kahl, M.D.; Hoglund, M.D. Use of joint toxic response to define the primary mode of toxic action for diverse industrial organic chemicals. *Environ. Toxicol. Chem.* **1995**, *14*, 1591–1605. [CrossRef]
42. Konemann, H. Fish toxicity tests with mixtures of more than 2 chemicals-a proposal for a quantitative approach and experimental results. *Toxicology* **1981**, *19*, 229–238. [CrossRef]
43. Jan, A.T.; Azam, M.; Siddiqui, K.; Ali, A.; Choi, I.; Haq, Q.M.R. Heavy Metals and Human Health: Mechanistic Insight into Toxicity and Counter Defense System of Antioxidants. *Int. Mol. Sci.* **2015**, *16*, 29592–29630. [CrossRef] [PubMed]
44. Chen, C.W.; Chen, C.F.; Dong, C.D. Distribution and Accumulation of Mercury in Sediments of Kaohsiung River Mouth, Taiwan. *APCBEE Procedia* **2012**, *1*, 153–158. [CrossRef]
45. Hintelmann, H. Organomercurials: Their formation and pathways in the environment. *Met. Ions Life Sci.* **2010**, *7*, 365–401. [PubMed]
46. Bauda, P.; Block, J.C. Role of envelops of Gram negative bacteria in cadmium binding and toxicity. *Toxic. Assess.* **1990**, *5*, 47–60. [CrossRef]
47. Fulladosa, E.; Murat, J.C.; Martınez, M.; Villaescusa, I. Patterns of metals and arsenic poisoning in *Vibrio fischeri* bacteria. *Chemistry* **2005**, *60*, 43–48. [CrossRef]
48. Kungolos, A.; Hadjispyrou, S.; Petala, M.; Tsiridis, V.; Samaras, P.; Sakellaropoulos, G.P. Toxic properties of metals and organotin compounds and their interactions on *Daphnia magna* and *Vibrio fischeri*. *Water Air Soil Pollut.* **2004**, *4*, 101–110. [CrossRef]
49. Bao, V.W.W.; Leung, K.M.A.; Kwok, K.W.H.; Zhang, A.Q.; Lui, G.C. Synergistic toxic effects of zinc pyrithione and copper to three marine species: Implications on setting appropriate water quality criteria, *Mar. Pollut. Bull.* **2008**, *57*, 616–623. [CrossRef]
50. Sfakianakis, D.G.; Renieri, E.; Kentouri, M.; Tsatsakis, A.M. Effect of heavy metals on fish larvae deformities: A review. *Environ. Res.* **2005**, *137*, 246–255. [CrossRef]
51. Li, L.; Tian, X.L.; Yu, X.; Dong, S.L. Effects of Acute and Chronic Heavy Metal (Cu, Cd, and Zn) Exposure on Sea Cucumbers (*Apostichopus japonicus*). *Biomed. Res. Int.* **2016**, *13*. [CrossRef]

52. Singh, N.; Yeh, P. Suppressive drug combinations and their potential to combat antibiotic resistance. *J. Antibiot.* **2017**, *70*, 1033–1042. [CrossRef]
53. Bryan, G.W.; Gibbs, P.E. Heavy metals in the Fal estuary, Cornwall: A study of long term contamination by mining waste and its effects on estuarine organisms, Occasional Publications. *Mar. Biol. Assoc. U. K.* **1983**.
54. Paulson, K.; Lunderberg, K. Treatment of mercury contaminated fish by selenium Addition. *WaterAir Soil Pollut.* **1991**, *56*, 833–841. [CrossRef]
55. Utkigar, V.P.; Chaudhary, N.; Koeniger, A.; Tabak, H.H.; Haines, J.R.; Govind, R. Toxicity of metals and metal mixtures: Analysis of concentration and time dependence for zinc and copper. *Water Resour.* **2004**, *38*, 3651–3658.
56. Lynch, N.R.; Hoang, T.C.; O'Brien, T.E. Acute Toxicity of Binary-Metal mixtures of Copper, Zinc and Nickel to *Pimephales Promelas*: Evidence of More than Additive Effect. *Environ. Toxicol. Chem.* **2016**, *35*, 446–457. [CrossRef] [PubMed]
57. Nys, C.; Asselman, J.; Hochmuth, J.D.; Janssen, C.R.; Blust, R.; Smolders, E.; De Schamphelaere, K.A.C. Mixture toxicity of nickel and zinc to *Daphnia magna* is non interactive at low effect sizes but becomes synergistic at high effect sizes: Interactive effects in a binary Ni-Zn mixture. *Environ. Toxicol. Chem.* **2015**, *5*, 1091–1102. [CrossRef] [PubMed]
58. Padrtova, R.R.; Marsalek, B.; Holoubek, I. Evaluation of alternative and standard toxicity assays for screening of environ-mental samples: Selection of an optimal test battery. *Chemosphere* **1998**, *37*, 495–507. [CrossRef]
59. Cedergreen, N. Quantifying Synergy: A Systematic Review of Mixture Toxicity Studies within Environmental Toxicology. *PLoS ONE* **2014**, *9*, e96580. [CrossRef]
60. Gao, Y.F.; Feng, J.F.; Kang, L.; Xu, X.; Zhu, L. Concentration addition and independent action model: Which is better in predicting the toxicity for metal mixtures on zebrafish larvae. *Sci. Total Environ.* **2018**, *610*, 442–450. [CrossRef]
61. Anyanwu, B.O.; Ezejiofor, A.N.; Igweze, Z.N.; Orisakwe, O.E. Heavy Metal Mixture Exposure and Effects in Developing Nations: An Update. *Toxics* **2018**, *6*, 65. [CrossRef]
62. Alsop, D.; Wood, C. Metal uptake and acute toxicity in zebrafish: Common mechanisms across multiple metals. *Aquat. Toxicol.* **2011**, *105*, 385–393. [CrossRef]
63. Olajuyigbe, O.O.; Animashaun, T. Synergistic Activities of Amoxicillin and Erythromycin against Bacteria of Medical Importance. *Pharmacologia* **2012**, *3*, 450–455. [CrossRef]
64. Loewe, S. The problem of synergism and antagonism of combined drugs. *Arznei-mittel-Forsc Drug Resour.* **1953**, *3*, 285–290.

International Journal of
Environmental Research and Public Health

MDPI

Article

An Innovative Approach to Determining the Contribution of Saharan Dust to Pollution

Nicoletta Lotrecchiano [1,2], Vincenzo Capozzi [2,3] and Daniele Sofia [1,2,*]

1   DIIN-Department of Industrial Engineering, University of Salerno, Via Giovanni Paolo II, 132, 84084 Fisciano, Italy; nlotrecchiano@unisa.it
2   Department of Science and Technology, University of Naples "Parthenope", Centro Direzionale di Napoli, Isola C4, 80134 Naples, Italy; vincenzo.capozzi@uniparthenope.it
3   Research Department, Sense Square Srl, 84084 Salerno, Italy
*   Correspondence: dsofia@unisa.it

Abstract: Air quality is one of the hot topics of today, and many people are interested in it due to the harmful effects that environmental pollution has on human health. For this reason, in recent years, measurement systems based on advanced technology have been implemented to integrate national air quality networks. This study aimed to analyze the air quality data of the monitoring network of the regional agency for environmental protection of the Campania region (Italy), integrated with a monitoring station based on IoT technology to highlight criticalities in the levels of pollution. The data used was from the month of February 2021 and measured in a medium-large city in southern Italy. In-depth analyses showed that two events related to Saharan dust occurred, which led to an increase in the measured PM10 values.

Keywords: air quality; Saharan dust; environmental pollution; mineral dust

Citation: Lotrecchiano, N.; Capozzi, V.; Sofia, D. An Innovative Approach to Determining the Contribution of Saharan Dust to Pollution. *Int. J. Environ. Res. Public Health* **2021**, *18*, 6100. https://doi.org/10.3390/ijerph18116100

Academic Editors: Pasquale Avino, Massimiliano Errico, Aristide Giuliano and Hamid Salehi

Received: 29 April 2021
Accepted: 3 June 2021
Published: 5 June 2021

Publisher's Note: MDPI stays neutral with regard to jurisdictional claims in published maps and institutional affiliations.

1. Introduction

Europe's industrialized regions are continuous sources of anthropogenic particulate matter. In addition to the gaseous emissions, these sources also produce fine powders. Particles can be inhaled while breathing, therefore high concentrations of atmospheric particles can have dangerous effects on human health [1]. However, not all particles are harmful, only those with small grain sizes.

Particulate material (PM) is a mixture of different components that vary both locally and regionally. PM can have various origins; it can be generated by natural phenomena, such as soil erosion or, more commonly, from vehicle combustion or industrial plant emissions. PM10 and PM2.5 are the two parts into which particulate matter is divided in particles with an aerodynamic diameter of $\leq$10 μm and $\leq$2.5 μm, respectively. Particles with a diameter less than 10 μm constitute the inhalable fraction, able to reach the broncho-tracheal area, while particles with a diameter less than 2.5 μm, which constitute the breathable fraction, can reach the alveoli lungs, conveying the substances of which they are composed into the body. PM10 is partly of the primary type, being entered directly into the atmosphere, and partly of the secondary type, being produced by physico-chemical transformations involving various substances such as SO_x, NO_x, VOC, and NH_3, which determine its production and/or removal.

Particulate matter of anthropogenic origin largely belongs to the fraction with a diameter less than 2.5 μm, while desert clouds of dust contain a significant fraction with larger particle size; a considerable decrease in the PM2.5/PM10 ratio is, therefore, a contributing index of natural particulate matter of desert origin [2]. Saharan episodes are characterized by a large increase in the PM10 concentration, not followed by that of PM2.5 and PM1, with a consequent decrease in the PM2.5/PM10 ratio, an increase of the crustal elements (Al, Si, Ca, Ti, Fe) in the PM10 fraction, and an increase in the Si/Fe, Al/Fe, and Ti/Fe ratios [3,4]. Moreover, the effects of these Saharan events have repercussions, not only for outdoor air quality, but also for indoor air quality, which is affected by the high concentrations of aero-dispersed dust [5].

From a climatic point of view, the Mediterranean atmosphere (MED), as defined by Jeftic et al. [6], is characterized by rainy winter seasons and hot and dry summer seasons, which affect the continental areas that surround it. It is therefore full of particulates of anthropogenic origin, coming from the vast industrialized European regions, and of natural particulates of crustal origin, coming from the extensive arid and semi-arid areas of Africa and the Middle East, and of aerosols of marine origin, generated by the surface of the MED, and of volcanic dust, emitted by the main volcanoes of the basin, such as Etna and Stromboli [7]. From the studies carried out on the influence of Saharan dust on a large scale, it is highlighted that latitude is a very important factor that must be taken into account in studies concerning PM10 concentrations. This same difference can be seen between the number of Saharan transport events that take place in Spain and those that take place in Italy or England [8]. The possibility of there occurring an interference of the powder transport in the PM10 concentration is higher in Spain than in northern Italy, and much greater than in England. Various studies have involved Italy in recent years, highlighting how, depending on the areas investigated, the days of transport of dust increase from north to south [2]. Therefore, a need is highlighted, concerning European countries located at higher latitudes, to look for these signals, both in PM10 and PM2.5. The problem of dust can be analyzed using integrated approaches to define the link between the natural and the urban dust within the environment [9].

From this perspective, air quality monitoring systems have been introduced. These systems have evolved over the years using increasingly innovative technologies based on the principles of the Internet of things (IoT). The new smart monitoring devices [10] integrate the traditional monitoring stations implemented by the institution to create a monitoring network with high spatial and temporal resolution. The new monitoring networks, given their characteristics and small dimensions, allow their installation in many points of the urban or extra-urban context, optimizing their position [11]. The IoT-based sensors include a data protection system based on the blockchain principle that ensures data integrity [12]. Recently, the technology has developed to the point of making the monitoring stations dynamic, i.e., housed on a moving support vehicle that allows on-the-road real-time monitoring [13]. This new type of sensor can also be portable and worn by users so that they can directly assess their exposure to indoor and outdoor pollution [14]. Knowledge of airborne pollutant concentrations is a fundamental part of the definition of strategies for pollution reduction [15,16]. The pollutants being analyzed are commonly the average particulate PM10 and PM2.5, to which PM1 is sometimes added, as well as gases such as NO_2, SO_2, CO, H_2S, O_3, and VOC (volatile organic compounds). The measured air quality data, albeit in large quantities, are discontinuous in the territory and need to be integrated into modeling systems [17] that allow their analysis, forecasting, and spatialization. In particular, the in-depth analysis of air quality data allows the study of accidental events such as fires [18], the analysis of the effects of pollutant dispersion produced by industrial activities [19], and the implementation of forecasting models that can provide detailed information on air quality over time [20] and space. Knowing the behavior of pollutants, and in particular of airborne dust, is also useful for defining the levels of healthiness in closed environments, such as homes or workplaces. [21]. Unlike the mineral aerosol which is of natural origin, the PM10 produced in the urban environment comes from different anthropogenic sources. The major particulate matter sources in urban and extra-urban areas are represented by industrial plants, domestic heating systems, emissions due to vehicular traffic, tire breaking, agriculture, and livestock. In detail, the present study aims to evaluate the contribution of Saharan dust to the environmental pollution measured by the air quality monitoring stations in an urban context.

The Case Study

The area investigated in this work is Avellino, a medium-sized city in Campania situated in a plain surrounded by mountains. The city of Avellino is one of the most polluted in central-southern Italy and seventh in the national ranking, with 78 days of overruns compared to the legal limit of 50 µg/m^3 during 2020. The harsh winters and obsolete domestic heating technologies, added to the various industrial emissions and intense vehicular traffic, mean that the measured average levels of particulate matter PM10 and PM2.5 are very high. Moreover, the complex mountain topography does not facilitate the natural dispersion of pollutants. For all these reasons, the city is interested in monitoring air quality, and great attention is paid to the analysis of data from the regional monitoring stations.

2. Materials and Methods

2.1. Air Quality Sources

The control of the air quality parameters according to the regulatory provisions of Legislative Decree 155/2010 represents one of the main institutional activities of the Regional Agency for Environmental Protection of Campania region (ARPAC). The agency manages the monitoring network determined according to regional specifications for the identification and management of air quality monitoring stations, to assess the air quality and polluting emissions spread over the area. The configuration of the network in the Campania region (a southern Italian region) includes 36 fixed monitoring stations and 5 mobile laboratories (Figure 1).

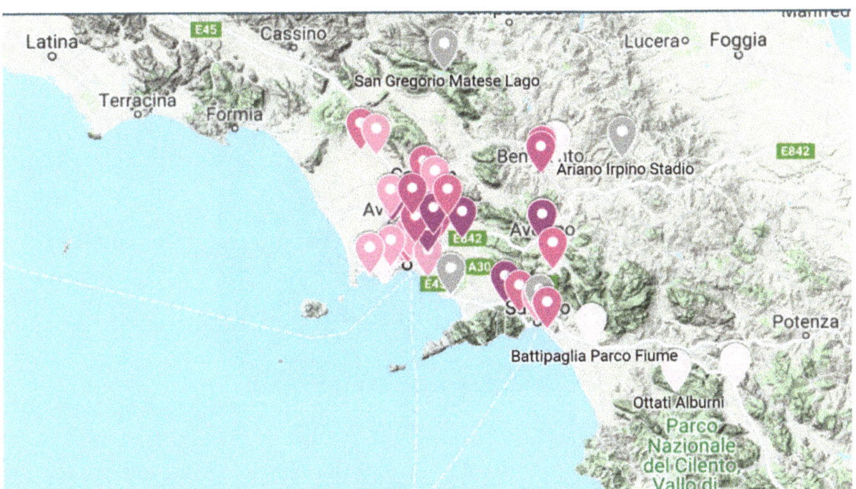

Figure 1. ARPA air quality monitoring network located in the Campania region.

The monitoring stations are located in sensitive areas, following the classification of the regional territory. There are also 10 additional fixed monitoring stations installed near the waste treatment plants ("STIR" network) that, although not part of the regional network, provide additional and support measures for the interpretation of the evolutionary phenomena of air quality on a regional basis.

The technology on which the measuring instruments of the ARPAC network are based comply with the technical standard UNI EN 12341: 2014. The measurement principle is gravimetry, and the reference method for the determination of PM10 particulate material is based on the collection of the PM10 fraction on a special filter and the subsequent determination of its mass by gravimetric methods in the laboratory, after conditioning the filter in controlled conditions of temperature (20 °C ± 1) and humidity (50 ± 5%).

In Avellino, the ARPA air quality monitoring network is composed of two monitoring stations, V° Circolo (VC) and Scuola Alighieri (SA), classified as suburban background and urban traffic, respectively. Both monitoring stations provide for PM10 and PM2.5 particulate matter measurements with a daily resolution using gravimetric devices and according to the legal specifications. The VC station also provides hourly particulate matter analysis by using laser scatter technology. Both stations also provide NO_2 measurements, and SA adds measurements for CO and benzene, while VC adds only O_3 data. The main characteristics of the stations considered in this work are summarized in Table 1. The SA station is located on one of the most congested roads of the city, while the VC station is located near a school that represents a sensitive point (see Figure 2). For this work, data for February 2021 were investigated.

Table 1. Characteristics of the monitoring stations.

Station	Acronym	Type	Measurements Technique	
V° Circolo	VC	Suburban background	Gravimetric Laser scattering	ARPAC
Scuola Alighieri	SA	Urban traffic	Gravimetric	ARPAC
Mt. Vergine Observatory	MV		Laser scattering	MVOBSV

Figure 2. Map of measuring stations located in Avellino.

In the city considered there is also an extra air quality monitoring station, part of the meteorological network managed by the non-profit organization MVOBSV. The sensor used is a PMSA003 Digital Laser Dust Sensor (MV) placed on a palace rooftop (18 m) in the city center (see Figure 2). The sensor is based on laser scattering technology with a time resolution of 10 min. The sensor uses the principle of laser dispersion, that is, by irradiating the suspension of particles in the air, it collects the scattered light at a certain angle and obtains a light dispersion curve that varies over time. The wavelength of the laser is 1100 μm. The particles of equivalent diameter and the number of particles with a different diameter per unit of volume can be calculated in real time by a microprocessor based on the MIE theory. In this case the sensor provides for PM10 and PM2.5 concentrations.

Since the ARPA data were compared with MVOBSV data that are based on the technology of laser scattering, a comparison with gravimetry was mandatory. The comparison between the two measurement techniques allowed for data validation.

Figure 3 reports the daily average PM10 and PM2.5 concentrations measured by the VC station with the two measurement techniques. From inspection of Figure 3, it is clear that the values measured with gravimetry and laser scattering are almost the same, with an average difference of 0.23% for PM10 and 0.5% for PM2.5. The good agreement of the data measured means that the laser scattering measurements are reliable for further analysis.

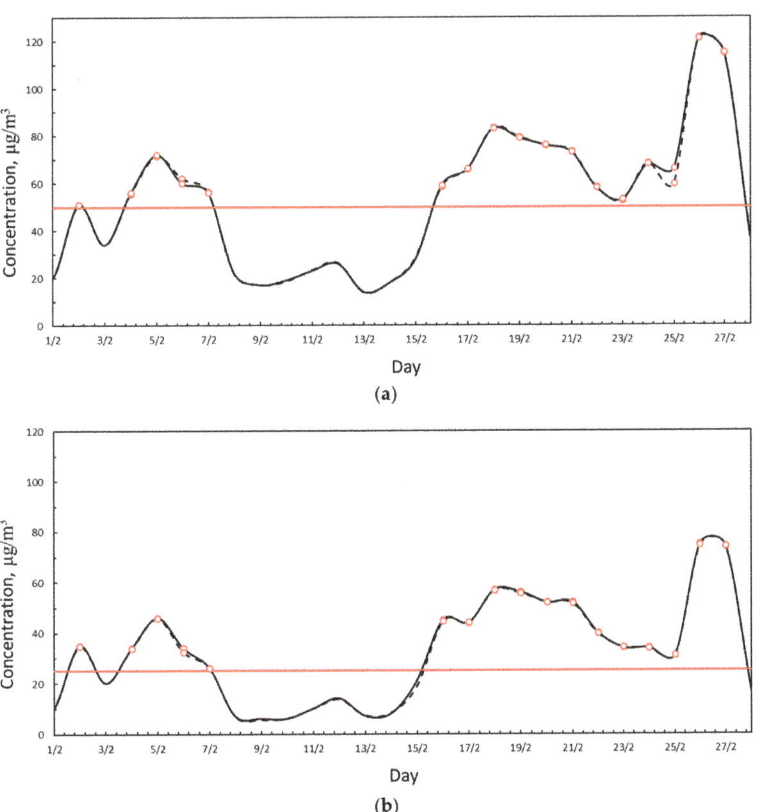

Figure 3. (**a**) PM10 daily average concentration measured at the VC station with gravimetry (solid black line) and with laser scattering (dashed black line). Red line represents the legal limit concentration according to D.Lgs. 155/2010. Red dots represent exceedance of the legal limit. (**b**) PM2.5 daily average concentration measured at the VC station with gravimetry (solid black line) and with laser scattering (dashed black line). Red line represents the legal limit concentration according to D.Lgs. 155/2010. Red dots represent exceedance of the legal limit.

The MV station is placed on the roof 18 m from the ground, above the urban canyon, which acts as a mixing cell. This height should ensure that Saharan dust is not directly measured since it is mainly made up of PM10 and naturally tends to stagnate on the ground. The proposed approach consisted of the following fundamental steps:

- identification of the ARPA monitoring device with the same technological characteristics of the monitoring station to be compared, located at a distance, not exceeding 1000 m, and placed in the same urban context,
- validation of experimental data of basic pollution conditions through the analysis of comparison graphs,

- identification of Saharan events,
- comparison of the experimental daily values obtained by the measuring device with the ARPA values, and subsequently investigating in more depth the days with greater relevance using the hourly averages.

In this case study, the monitoring station analyzed was represented by MV and the ARPAC monitoring station used for the comparison was VC. The two stations considered (MV and VC) are both based on laser scattering technology and are located at a distance of 600 m in a context of medium urbanization.

2.2. Meteorological Data

Meteorological data are fundamental for pollution dispersion analysis. The data used in this study were provided by an automatic weather station (Davis Vantage Pro 2) co-located with an air quality sensor. This station is part of the meteorological network managed by the non-profit organization MVOBSV and includes sensors for the measurement of essential atmospheric parameters, i.e., air temperature, relative humidity, air pressure, rainfall, and wind speed and direction. The data are stored with a temporal resolution of 10 min and are fully available for the analyzed period.

3. Results

In the first analysis, the daily data provided by the MV station was compared with those of the station VC which is closest (Figure 4). Comparing the PM10 and PM2.5 trends of the stations showed a good agreement between the measurements. The measured data aligned around the central trend line, with small deviations, especially for PM2.5.

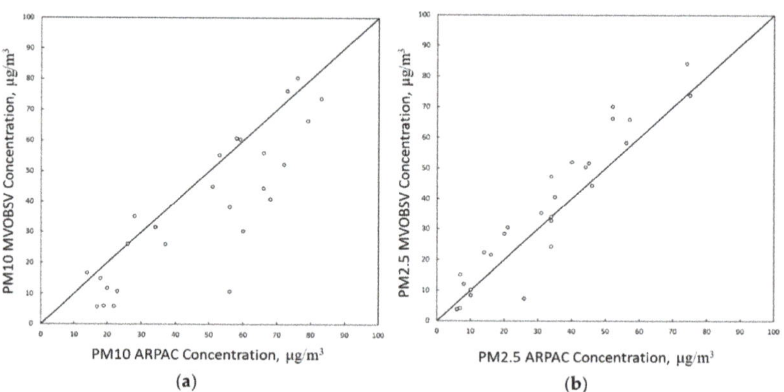

Figure 4. (**a**) Comparison of PM10 daily average concentrations measured at the MV and VC stations. (**b**) Comparison of PM2.5 daily average concentrations measured at the MV and VC stations.

From the analysis of the PM10 and PM2.5 trends (Figure 5) it can be seen that for many days in February, both the recorded PM10 and PM2.5 exceeded the legal limit. In particular, the days exceeding the limit threshold of 50 µg/m³ for PM10 were 16 at the SA station and 17 at the VC station. For PM2.5, the daily limit of 25 µg/m³, similarly to PM10, was exceeded 16 times at the SA station and 17 times at the VC station. In February, the days when the pollution levels were low coincided with favorable weather conditions for dispersion. The three data trends have the same behavior, and differences can be attributed to the different positions in the urban context and, therefore, to the local conditions.

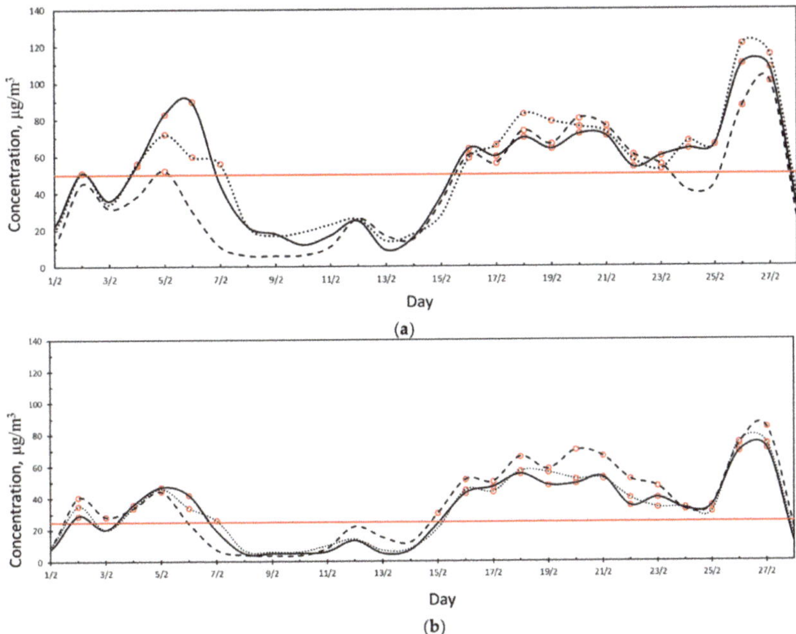

Figure 5. (**a**) PM10 daily average concentration measured at the SA station (solid black line), VC station (dotted black line), and MV station (dashed black line). Red line represents the legal limit concentration according to D.Lgs. 155/2010. Red dots represent exceedance with respect to the legal limit. (**b**) PM2.5 daily average concentration measured at the SA station (solid black line), VC station (dotted black line), and MV station (dashed black line). Red line represents the legal limit concentration according to D.Lgs. 155/2010. Red dots represent exceedance of the legal limit.

Subsequently, the trends of PM10, PM2.5, and their ratio were analyzed for each measuring station (Figure 6). An inspection of Figure 6 reveals that in the periods 5–7 and 23–28 of February, the difference between PM10 and PM2.5 concentrations was high for the SA and VC stations (Figure 6a,b); moreover, in such periods, the ratio between the particulate matter was low. This means that the contribution of PM10 to pollution was higher than PM2.5, and in particular, an external contribution represented by Saharan dust was added to the PM10 produced in the urban context. This external contribution was represented by the Saharan dust that in these days was dispersed over Italian territory.

The transported particulate is of desert and sandy origin, mainly composed of siliceous materials; unlike PM10 of urban origin which is characterized by the presence of carbon and metal-based particles. The analysis confirmed that the contribution of the mineral aerosol to the PM10 values can be very important, favoring the breaking of the legislative limit.

Looking in depth at the two periods selected, it is possible to analyze the hourly trends of pollutants (Figures 7 and 8). In the first Saharan event, PM10 concentration reaches a peak of 180 μg/m^3 measured at the VC station. As highlighted in Figure 7, the PM10/PM2.5 ratio was extremely variable at the VC station (Figure 7a), while it was slightly variable at the MV station (Figure 7b). Moreover, at the MV station there was no evidence of Saharan dust contributing to PM10 concentrations, which remained proportional to PM2.5.

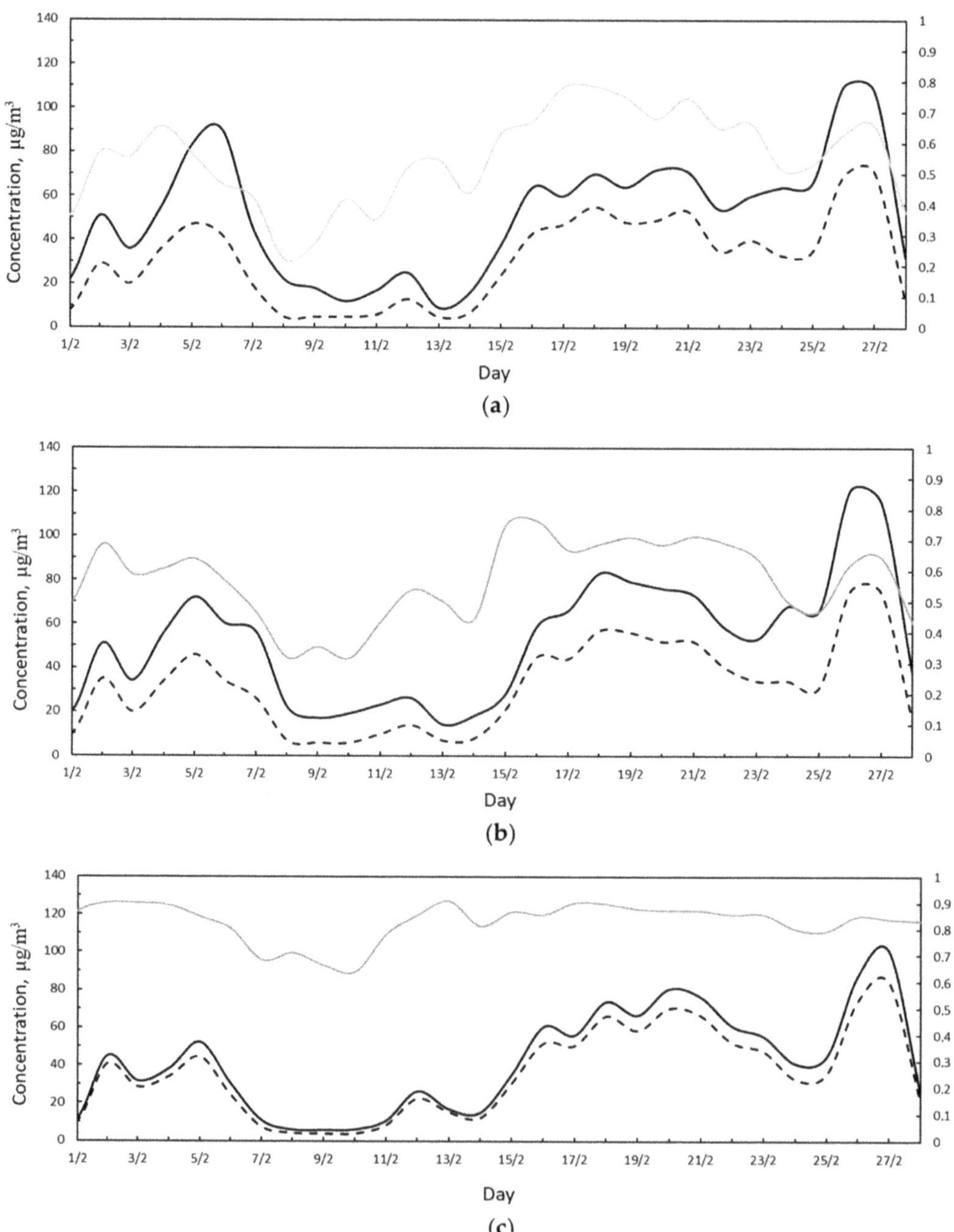

Figure 6. (**a**) Daily PM10 (black solid line), PM2.5 (black dashed line), and PM2.5/PM10 concentrations (grey solid line) measured during February 2021 at the SA station. (**b**) Daily PM10 (black solid line), PM2.5 (black dashed line), and PM2.5/PM10 concentrations (grey solid line) measured during February 2021 at the VC station. (**c**) Daily PM10 (black solid line), PM2.5 (black dashed line), and PM2.5/PM10 concentrations (grey solid line) measured during February 2021 at the MV station.

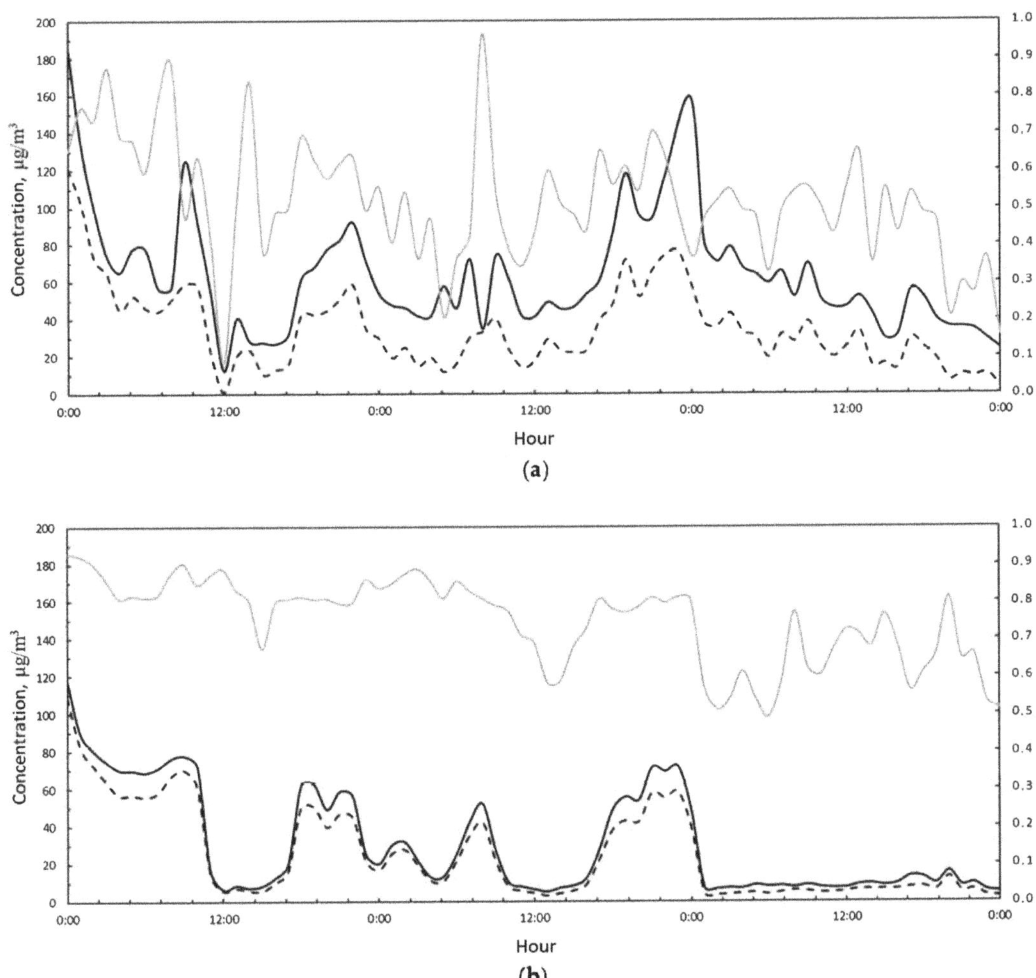

Figure 7. (**a**) PM10 (black solid line), PM2.5 (black dashed line), and PM2.5/PM10 hourly concentrations (grey solid line) measured during the period 5–7 February 2021 at the VC station. (**b**) PM10 (black solid line), PM2.5 (black dashed line), and PM2.5/PM10 hourly concentrations (grey solid line) measured during the period 5–7 February 2021 at the MV station.

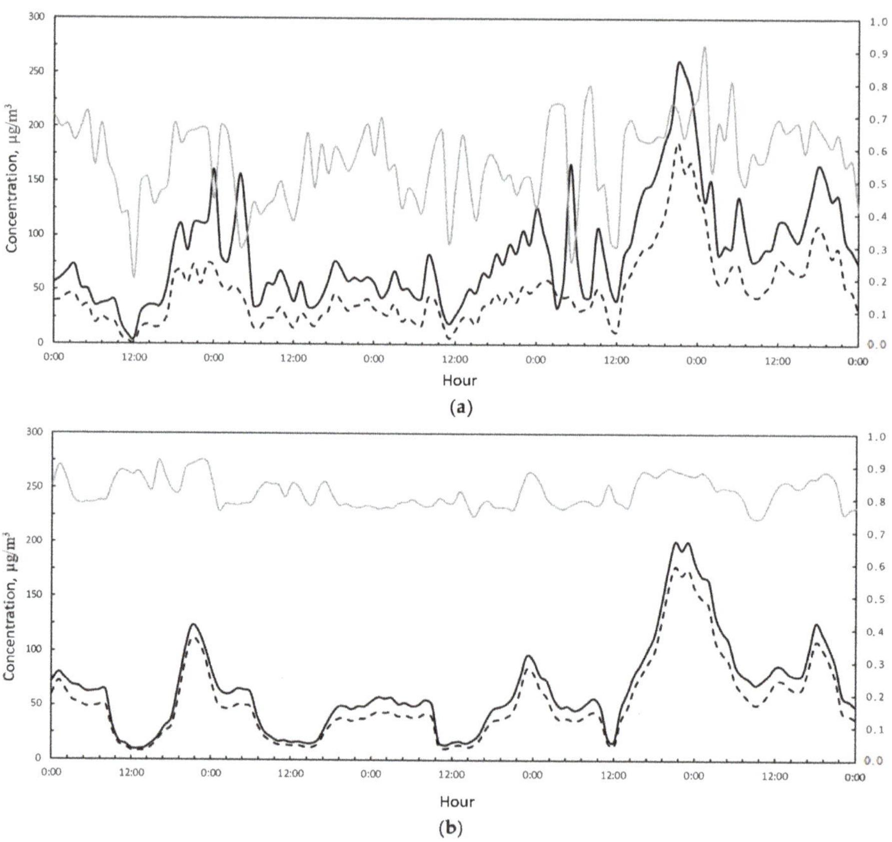

Figure 8. (**a**) PM10 (black solid line), PM2.5 (black dashed line), and PM2.5/PM10 hourly concentrations (grey solid line) measured during the period 23–28 February 2021 at the VC station. (**b**) PM10 (black solid line), PM2.5 (black dashed line), and PM2.5/PM10 hourly concentrations (grey solid line) measured during the period 23–28 February 2021 at the MV station.

Considering the second Saharan event, during 23–28 February and reported in Figure 8, the PM10 concentration reached an hourly peak of 260 µg/m³ measured by the VC station (Figure 8a). As in the previous case, the MV measuring station did not show any evidence of an external PM10 concentration contribution, in fact, the PM10/PM2.5 ratio was slightly variable and the pollutant trends were similar (Figure 8b). There was only a slight difference between PM10 and PM2.5 on days 26–27. The absence of evidence of a Saharan influence phenomenon at the MV station is probably due to the different height at which it is positioned compared to the ARPAC monitoring stations. The MV station is positioned on the roof of a building, while the two ARPAC stations are located near the street. The open-field exposure of the MV station does not allow the detection of the Saharan phenomenon, since the dust is more dispersed and diluted in the atmosphere at that height.

Further analysis included the daily pollution inspection, to compare the trends measured by the VC and MV stations. In particular, some of most interesting days were more deeply analyzed, and also including the meteorological parameters.

In Figure 9, the hourly measured concentration trends at the VC and MV stations appear to be in good agreement on most of the days analyzed. In detail, the measurements were also in agreement in their values and not only in their trend in 25% of cases.

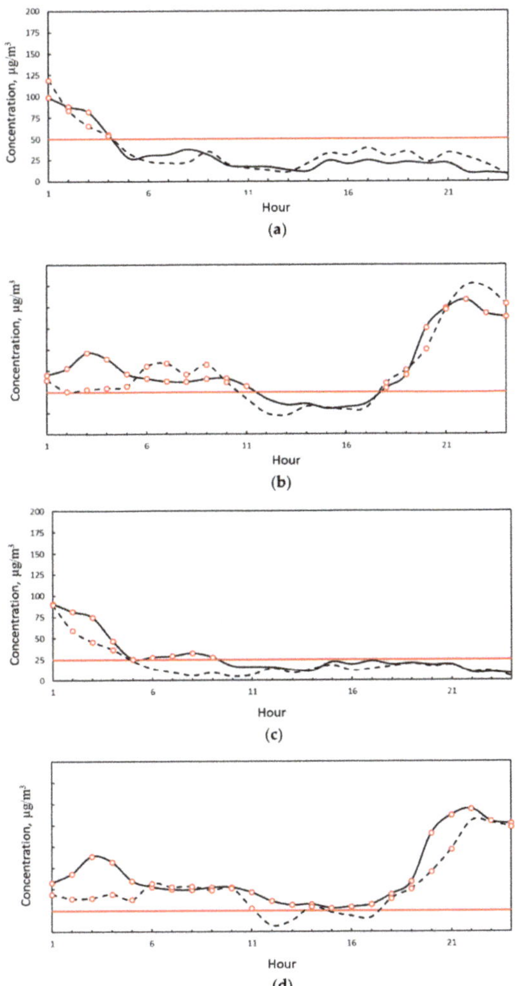

Figure 9. (**a**) PM10 hourly average concentration measured at the VC station (solid black line) and MV station (dashed black line) 3 February 2021. Red line represents the legal limit concentration according to D.Lgs. 155/2010. Red dots represent exceedance of the legal limit. (**b**) PM10 hourly average concentration measured at the VC station (solid black line) and MV station (dashed black line) 21 February 2021. Red line represents the legal limit concentration according to D.Lgs. 155/2010. Red dots represent exceedance of the legal limit. (**c**) PM10 hourly average concentration measured at the VC station (solid black line) and MV station (dashed black line) 3 February 2021. Red line represents the legal limit concentration according to D.Lgs. 155/2010. Red dots represent exceedance of the legal limit. (**d**) PM2.5 hourly average concentration measured at the VC station (solid black line) and MV station (dashed black line) 21 February 2021. Red line represents the legal limit concentration according to D.Lgs. 155/2010. Red dots represent exceedance of the legal limit.

On both days, at around 7 and 9 a.m. the pollutant levels increased. Particulate matter tends to settle on the ground due to gravity, both when airborne and as a result of rain. It remains on the asphalt if there is no movement of vehicles, such as during the night. When city activities start early in the morning and vehicles start moving on the roads, the particulate matter is dislodged from the road surface and released back into the atmosphere, increasing particulate concentrations. This phenomenon was more evident at the MV station than the VC station because it is located in the more congested city center.

Figure 10 shows the 500-hPa and sea level pressure fields for a typical scenario in which low pollutants concentrations were detected in the area of Avellino due to the simultaneous action of rain and winds. More specifically, the left panel of Figure 10 (Figure 10a) shows the synoptic scenario at 00:00 UTC on 11 February 2021, when the Italian peninsula was affected by a low-pressure area, well-structured at both mid and low tropospheric levels. This cyclonic system caused light rainfall in the study area (the AWS involved in this study registered a 24-h accumulated precipitation value of 1.5 mm), associated with moderate south-western winds. In the subsequent 12 h, the low-pressure area moved towards the Balkan peninsula (Figure 10b); this caused an improvement in weather conditions in Avellino. However, a moderate breeze from the west was still observed and, therefore, favorable pollutant dispersion renewed.

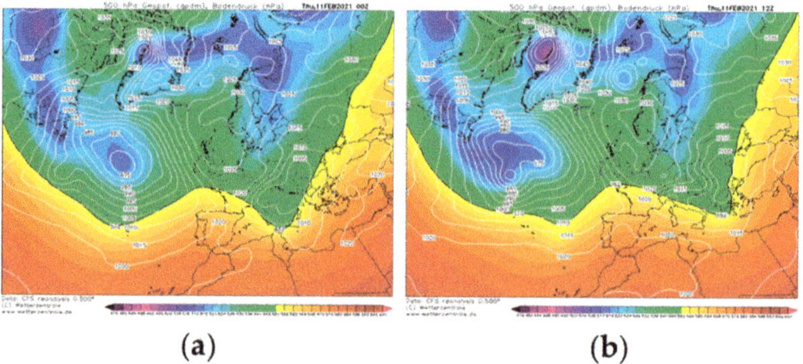

(a)   (b)

Figure 10. (a) Geopotential height at 500 hPa (shaded color) and sea level pressure (white solid line) fields at 00:00 UTC on 11 February 2021. (b) Geopotential height at 500 hPa (shaded color) and sea level pressure (white solid line) fields at 12:00 UTC on 11 February 2021 and 12:00 UTC. Both fields were retrieved from Climate Forecast System Renalysis of the National Centers for Environmental Prediction.

The Campania region was affected by two Saharan events in the periods 5–7 and 23–28 of February. Here, we provide a brief meteorological analysis of such events, and to better visualize the influence of Saharan dust on the Italian peninsula, synoptic maps were created.

Saharan event occurring on 5–7 February 2021.

On 5–7 February 2021, the European synoptic scenario resembled one of the most favorite patterns for the incoming Saharan air masses in the central Mediterranean area. The analysis of the 500-hPa geopotential height field revealed that, on 5 February (Figure 11a), the presence of a trough over Western Europe extended from the British Isles to the western sector of Morocco. This trough was the result of a relevant oscillation of the polar front that occurred in the East Atlantic between 2 and 4 February. The sea-level pressure distribution (also depicted in Figure 11a) showed a low-pressure area downstream of the trough axis, near the Gibraltar strait. A strong ridge, synonymous with stable weather conditions, instead modulated the atmospheric scenario in the Central and Eastern Mediterranean basins. In the subsequent 24 h, the trough moved eastwards (Figure 11b) and caused a drop of atmospheric pressure on the western side of the Italian peninsula. The consequence of this pressure pattern was a dry and warm meridional flow, which advected on the central Mediterranean area subtropical continental air masses coming from the Sahara region. In this respect, the 850-hPa equivalent potential temperature field presented in Figure 11c gives clear evidence of the warm tongue extending from the inland sectors of North Africa up to the Italian peninsula. On 7 February, the cyclonic system moved further to the east, causing a strengthening of warm air advection on southern Italy in the morning hours, and, consequently, a massive transport of Saharan dust. The meteorological conditions abruptly changed in the evening hours of 7 February, when the subtropical continental air was replaced by colder and moist air masses coming from the west.

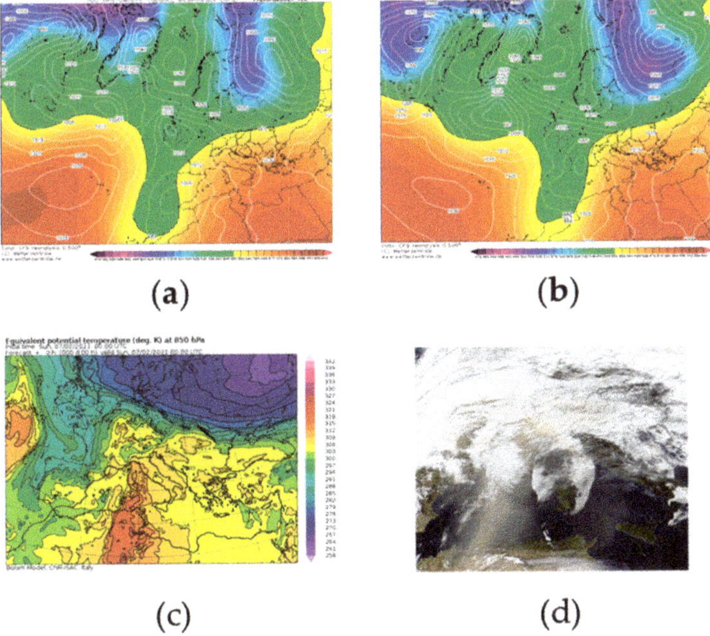

Figure 11. (**a**) Geopotential height at 500 hPa (shaded color) and sea level pressure (white solid line) fields at 12:00 UTC of 5 February 2021. (**b**) Geopotential height at 500 hPa (shaded color) and sea level pressure (white solid line) fields at 12:00 UTC 6 February 2021. Both fields were retrieved from Climate Forecast System Renalysis of National Centers for Environmental Prediction. (**c**) 850 mb equivalent potential temperature (shaded color) at 00:00 UTC 7 February 2021 from the BOLAM model of the Institute of Atmospheric Science and Climate of the Italian National Research Council. (**d**) Image of the Saharan dust event acquired on 6 February 2021 by the Moderate Resolution Imaging Spectroradiometer (MODIS) onboard NASA's Terra satellite. Images in panels (**a**) and (**b**) are courtesy of http://www.wetterzentrale.de, (accessed on 23 February 2021) whereas the image in (**c**) was retrieved from https://www.isac.cnr.it/dinamica/projects/forecasts/index.html. (accessed on 23 February 2021) The picture in panel (**d**) is courtesy of the NASA Worldview Snapshots application (https://wvs.earthdata.nasa.gov) (accessed on 23 February 2021).

Figure 11d offers a true-color image of the central and western Mediterranean basins, collected by The Moderate Resolution Imaging Spectroradiometer (MODIS) onboard NASA's Terra satellite on 6 February 2021. This picture highlights the Saharan dust cloud rising from North Africa and crossing the Mediterranean Sea.

Saharan event occurring on 23–26 February 2021.

A second relevant Saharan event took place in the study area at the end of February 2021 (days 23rd–26th). The meteorological dynamics that forced the incoming of desert dust in the central Mediterranean area had some similarities with those involved in the previous event. To shed light on these dynamics, it is necessary to analyze the 500-hPa geopotential height field on 22 February at 12 UTC (Figure 12a), which reveals the presence of a cut-off low over the Alboran Sea, generated by a Rossby wave that affected the Iberian Peninsula and Morocco during 20 and 21 February. This upper-level circulation was associated with a low-level cyclonic area, located inland of Algeria, as clearly highlighted by the sea-level pressure field depicted in Figure 12a. On the Italian peninsula, stable weather conditions prevailed, due to the presence of a ridge. This synoptic scenario promoted an advection of warm sub-tropical continental air, which can be easily identified by means of the 850-hPa equivalent potential temperature field (see Figure 12b). The latter clearly shows the pattern followed by warm air, which reached central Europe after crossing the Sardinian Sea. On 23 February 2021, the ridge located over Italy strengthened (see Figure 12c), obstructing the natural eastward movement of the low-pressure area located over the western Mediterranean basins. This evolution led to the translation of the cyclonic system inland of Algeria and, therefore, to the end of the warm advection in the central Mediterranean area. The previously advected sub-tropical air (rich in Saharan dust), following a clockwise motion imposed by the anticyclone, spread out over the Italian peninsula and Balkan regions. Owing to the persistence of the ridge in the subsequent days, the Saharan dust cloud continued to flow around Italy although it had been inevitably subjected to a gradual dilution process.

Satellite evidence of Saharan dust affecting central and northern Italy is provided by the image in Figure 12d, acquired by MODIS on 23 February 2021.

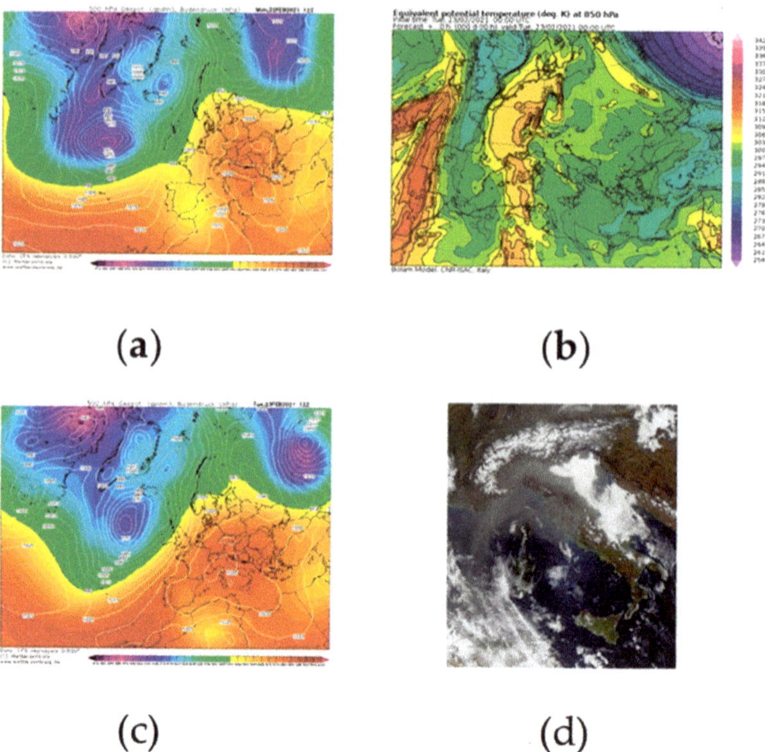

Figure 12. (a) Geopotential height at 500 hPa (shaded color) and sea level pressure (white solid line) fields at 12:00 UTC of 22 February 2021. Fields were retrieved from Climate Forecast System Renalysis of National Centers for Environmental

Prediction. (**b**) 850 mb equivalent potential temperature (shaded color) at 00:00 UTC of 23 February 2021 from the BOLAM model of the Institute of Atmospheric Science and Climate of the Italian National Research Council. (**c**) Geopotential height at 500 hPa (shaded color) and sea level pressure (white solid line) fields at 12:00 UTC of 23 February 2021. Fields were retrieved from Climate Forecast System Renalysis of National Centers for Environmental Prediction. (**d**) Image of the Saharan dust event acquired on 23 February 2021 by the Moderate Resolution Imaging Spectroradiometer (MODIS) onboard NASA's Terra satellite. Images in panels (**a**) and (**b**) are courtesy of http://www.wetterzentrale.de (accessed on 23 February 2021, whereas the image in (**c**) was retrieved from https://www.isac.cnr.it/dinamica/projects/forecasts/index.html (accessed on 23 February 2021). The picture in panel (**d**) is courtesy of the NASA Worldview Snapshots application (https://wvs.earthdata.nasa.gov) (accessed on 23 February 2021).

4. Discussion

The analysis of Saharan dust contribution to PM10 in urban areas is defined by the high levels of PM10 measured not followed by a simultaneous increase in PM2.5 concentrations. By applying to this case study the method proposed by Escudero et al. [22], it was possible to define the contribution of PM10 measured on the days in which the Saharan events occurred, using the 30th percentile of the monthly concentrations. By applying this procedure to the background VC station, we obtained the PM10 value in the absence of mineral aerosol. Comparing the values obtained after the application of the procedure suggested by Escudero et al., PM10 values similar to those measured at the MV station were obtained. As an example for the VC station, a daily value of 60 μg/m^3 was measured, and the monthly 30th percentile value was 27 for this month. Therefore, the net dust contribution at this point, for this day, was 33 μg/m^3. For the same day considered, the MV station recorded a PM10 daily value of 31 μg/m^3. Comparing this last value with the previous one obtained without considering the mineral dust contribution to the PM10, it is clear that they are similar. This confirms that the Saharan dust phenomenon was not measured from the station located at a height of 15/20 m, as suggested by the method proposed in this study. With this approach it is possible to also evaluate the exceedance in PM10 concentrations due to the mineral dust to the urban background.

This type of approach can be used in any city with orographic characteristics, urbanization, industrial supplies, population density, type of heating, and degree of traffic similar to the city in the case study. The peculiarity of being in a basin, surrounded by mountains is a frequent feature of Italian and European cities, which makes the case study adaptable to many real situations, such as the Po valley in Lombardy (Italy).

5. Conclusions

The occurrence of natural events such as the transport of Saharan dust can lead to a significant increase in the PM10 concentrations measured in urban areas. The correct analysis of the air quality data, including the evaluation of the influence of these natural phenomena, is necessary to define the real levels of urban pollution. The Saharan phenomena bring the levels of PM10 to very high values, as in this case study to 260 μg/m^3, which if interpreted incorrectly can generate unnecessary alarmism in the population and institutions. In fact, due to their nature, the events that transport Saharan dust are completely unpredictable and therefore unmanageable by man. What can be done to lower pollution levels is the reduction in the hours of ignition of domestic heating and the use of eco-friendly fuels, a significant reduction in traffic emissions, and the control of emissions from agriculture. Surely, in a territory that is based on an economy including agriculture, the elimination of phenomena such as weed burning would have a positive influence on the reduction of pollution. Furthermore, it must be considered that the territory of this case study, as previously mentioned, has an orography that is unfavorable to pollutant dispersion, therefore it is necessary to reduce the levels of pollution, which tend to rise due to the low natural dispersion. The implementation within the urban context of an air quality monitoring network is of fundamental importance for the definition of the city pollution levels. The regional networks of environmental protection agencies (ARPA) must be integrated with additional monitoring stations to obtain a greater spatial resolution of the data. Furthermore, the implementation of further measurement stations based on IoT technology allows real-time knowledge of the levels of airborne pollutants. The spatial and temporal detail of the measured data is the starting point for having a clear vision of the pollution phenomena, which can be analyzed in detail, also highlighting natural phenomena such as Saharan dust. From this perspective, the MV air quality monitoring device of the Mt.Vergine Observatory is placed such that, with the latest generation technology, it provides data to support traditional technologies.

Author Contributions: Conceptualization, D.S., N.L., V.C.; methodology, D.S., N.L., V.C.; software, D.S., N.L., V.C.; validation, D.S., N.L., V.C.; formal analysis, D.S., N.L., V.C..; investigation, D.S., N.L., V.C.; resources, D.S., N.L., V.C.; data curation, D.S., N.L., V.C.; writing—original draft preparation, D.S., N.L., V.C.; writing—review and editing, D.S., N.L., V.C.; visualization, D.S., N.L., V.C.; supervision, D.S., N.L., V.C.; project administration, D.S., N.L., V.C.; funding acquisition, D.S., V.C. All authors have read and agreed to the published version of the manuscript.

Funding: This research received external funding from Sense Square and the no profit organization MVOBSV.

Institutional Review Board Statement: Not applicable.

Informed Consent Statement: Not applicable.

Data Availability Statement: Data are available at https://www.arpacampania.it/aria.

Conflicts of Interest: The authors declare no conflict of interest.

References

1. Hasegawa, K.; Toubou, H.; Tsukahara, T.; Nomiyama, T. Short-Term Associations of Ambient Fine Particulate Matter (PM$_{2.5}$) with All-Cause Hospital Admissions and Total Charges in 12 Japanese Cities. *Int. J. Environ. Res. Public Health* **2021**, *18*, 4116. [CrossRef] [PubMed]
2. Di Menno di Bucchianico, A.; Catambrone, M.; Perrino, C.; Passariello, B.; Uqresima, S. Eventi di trasporto di sabbie sahariane nell'Italia centrale. In Proceedings of the Convegno Nazionale La Valutazione del Materiale Particellare Alla Luce del DM 2/4/2002, Marina di Carrara, Italy, 2 April 2002.
3. Chiari, M.; Lucarelli, A.; Nava, S.; Paperetti, L.; D'Alessandro, A.; Mazzei, F.; Prati, P.; Zucchiatti, A.; Marcazzan, G.; Valli, G.; et al. Aerosol transport episodes in italian urban environments detected by PIXE analysis. In Proceedings of the 10th International Conference on Particle Induced X-ray Emission and Its Analytical Applications PIXE, Portoroz, Slovenia, 4–8 June 2004.
4. Chiari, M.; Del Carmine, P.; Lucarelli, F.; Paperetti, L.; Nava, S.; Prati, P.; Valli, G.; Vecchi, R. Particulate matter characterization in an industrial district near Florence, by PIXE and PESA. In Proceedings of the 10th International Conference on Particle Induced X-ray Emission and Its Analytical Applications PIXE, Portoroz, Slovenia, 4–8 June 2004.
5. Katra, I.; Krasnov, H. Exposure Assessment of Indoor PM Levels During Extreme Dust Episodes. *Int. J. Environ. Res. Public Health* **2020**, *17*, 1625. [CrossRef] [PubMed]
6. Kenworthy, J.M.; Jeftic, L.; Milliman, J.D.; Sestini, G. *Climatic Change and the Mediterranean: Environmental and Societal Impacts of Climatic Change and Sea-Level Rise in the Mediterranean Region*; Arnold Publisher: London, UK, 1992.
7. Arnold, M.; Seghaier, A.; Martin, D.; Buat-Ménard, P.; Chesselet, R. Géochimie de l'aérosol marin de la Méditerranée Occidentale. In Proceedings of the Mediterranean Science Commission-CIESM. VI Journees Etud. Pollution, Cannes, France, 2–4 December 1982.
8. Ryall, D.; Derwent, R.; Manning, A.; Redington, A.; Corden, J.; Millington, W.; Simmonds, P.; O'Doherty, S.; Carslaw, N.; Fuller, G. The origin of high particulate concentrations over the United Kingdom, March 2000. *Atmos. Environ.* **2002**, *36*, 1363–1378. [CrossRef]
9. Doronzo, D.M.; Al-Dousari, A. Preface to Dust Events in the Environment. *Sustainability* **2019**, *11*, 628. [CrossRef]
10. Sofia, D.; Giuliano, A.; Gioiella, F. Air quality monitoring network for tracking pollutants: The case study of Salerno city center. *Chem. Eng. Trans.* **2018**, *68*, 67–72. [CrossRef]
11. Sofia, D.; Lotrecchiano, N.; Giuliano, A.; Barletta, D.; Poletto, M. Optimization of number and location of sampling points of an air quality monitoring network in an urban contest. *Chem. Eng. Trans.* **2019**, *74*, 277–282. [CrossRef]
12. Sofia, D.; Lotrecchiano, N.; Trucillo, P.; Giuliano, A.; Terrone, L. Novel Air Pollution Measurement System Based on Ethereum Blockchain. *J. Sens. Actuator Netw.* **2020**, *9*, 49. [CrossRef]
13. Lotrecchiano, N.; Sofia, D.; Giuliano, A.; Barletta, D.; Poletto, M. Real-time on-road monitoring network of air quality. *Chem. Eng. Trans.* **2019**, *74*, 241–246. [CrossRef]

14. Agrawaal, H.; Jones, C.; Thompson, J. Personal Exposure Estimates via Portable and Wireless Sensing and Reporting of Particulate Pollution. *Int. J. Environ. Res. Public Health* **2020**, *17*, 843. [CrossRef] [PubMed]
15. Sofia, D.; Gioiella, F.; Lotrecchiano, N.; Giuliano, A. Mitigation strategies for reducing air pollution. *Environ. Sci. Pollut. Res.* **2020**, *27*, 19226–19235. [CrossRef] [PubMed]
16. Sofia, D.; Gioiella, F.; Lotrecchiano, N.; Giuliano, A. Cost-benefit analysis to support decarbonization scenario for 2030: A case study in Italy. *Energy Policy* **2020**, *137*, 111137. [CrossRef]
17. Shao, Y.; Wyrwoll, K.-H.; Chappell, A.; Huang, J.; Lin, Z.; McTainsh, G.H.; Mikami, M.; Tanaka, T.Y.; Wang, X.; Yoon, S. Dust cycle: An emerging core theme in Earth system science. *Aeolian Res.* **2011**, *2*, 181–204. [CrossRef]
18. Lotrecchiano, N.; Sofia, D.; Giuliano, A.; Barletta, D.; Poletto, M. Pollution Dispersion from a Fire Using a Gaussian Plume Model. *Int. J. Saf. Secur. Eng.* **2020**, *10*, 431–439. [CrossRef]
19. Sofia, D.; Lotrecchiano, N.; Cirillo, D.; Villetta, M.L. NO_2 Dispersion model of emissions of a 20 kwe biomass gasifier. *Chem. Eng. Trans.* **2020**, *82*, 451–456. [CrossRef]
20. Lotrecchiano, N.; Gioiella, F.; Giuliano, A.; Sofia, D. Forecasting Model Validation of Particulate Air Pollution by Low Cost Sensors Data. *J. Model. Optim.* **2019**, *11*, 63–68. [CrossRef]
21. Salehi, H.; Sofia, D.; Barletta, D.; Poletto, M. Dust Generation in Vibrated Cohesive Powders. *Chem. Eng. Trans.* **2015**, *43*, 769–774. [CrossRef]
22. Escudero, M.; Querol, X.; Peya, J.; Alastuey, A.; Pérez, N.; Ferreira, F.; Alonso, S.; Rodríguez, S.; Cuevas, E. A methodology for the quantification of the net African dust load in air quality monitoring networks. *Atmos. Environ.* **2007**, *41*, 5516–5524. [CrossRef]

International Journal of
Environmental Research and Public Health

Article

Purification of Wastewater from Biomass-Derived Syngas Scrubber Using Biochar and Activated Carbons

Enrico Catizzone [1,*], Corradino Sposato [1], Assunta Romanelli [1], Donatella Barisano [1], Giacinto Cornacchia [1], Luigi Marsico [2], Daniela Cozza [2] and Massimo Migliori [2]

1   ENEA-Italian National Agency for New Technologies, Energy and Sustainable Economic Development, Trisaia Research Center, Department of Energy Technologies and Renewable Sources, I-75026 Rotondella, Italy; corradino.sposato@enea.it (C.S.); assunta.romanelli@enea.it (A.R.); donatella.barisano@enea.it (D.B.); giacinto.cornacchia@enea.it (G.C.)
2   Department of Environmental and Chemical Engineering, University of Calabria, via P. Bucci, 44a, I-87036 Rende, Italy; luigi.marsico92@gmail.com (L.M.); daniela.cozza@unical.it (D.C.); massimo.migliori@unical.it (M.M.)
*   Correspondence: enrico.catizzone@enea.it

Abstract: Phenol is a major component in the scrubber wastewater used for syngas purification in biomass-based gasification plants. Adsorption is a common strategy for wastewater purification, and carbon materials, such as activated carbons and biochar, may be used for its remediation. In this work, we compare the adsorption behavior towards phenol of two biochar samples, produced by pyrolysis and gasification of lignocellulose biomass, with two commercial activated carbons. Obtained data were also used to assess the effect of textural properties (i.e., surface area) on phenol removal. Continuous tests in lab-scale columns were also carried out and the obtained data were processed with literature models in order to obtain design parameters for scale-up. Results clearly indicate the superiority of activated carbons due to the higher pore volume, although biomass-derived char may be more suitable from an economic and environmental point of view. The phenol adsorption capacity increases from about 65 m/g for gasification biochar to about 270 mg/g for the commercial activated carbon. Correspondingly, service time of commercial activated carbons was found to be about six times higher than that of gasification biochar. Finally, results indicate that phenol may be used as a model for characterizing the adsorption capacity of the investigated carbon materials, but in the case of real waste water the carbon usage rate should be considered at least 1.5 times higher than that calculated for phenol.

Keywords: biomass; syngas scrubber wastewater; environmental pollution; pollutant abatement technologies; biochar; adsorption

Citation: Catizzone, E.; Sposato, C.; Romanelli, A.; Barisano, D.; Cornacchia, G.; Marsico, L.; Cozza, D.; Migliori, M. Purification of Wastewater from Biomass-Derived Syngas Scrubber Using Biochar and Activated Carbons. *Int. J. Environ. Res. Public Health* **2021**, *18*, 4247. https://doi.org/10.3390/ijerph18084247

Academic Editor: Pasquale Avino

Received: 26 February 2021
Accepted: 13 April 2021
Published: 16 April 2021

Publisher's Note: MDPI stays neutral with regard to jurisdictional claims in published maps and institutional affiliations.

1. Introduction

Phenolic compounds attract great attention in the international scientific community because they are chemicals that are capable of persisting in the environment for long periods of time and can exert toxic effects on humans and animals [1]. Moreover, phenols cause negative effects on drinking water and environment. Some phenolic compounds are abundant in nature and are associated with the colors of flowers and fruits. Others are synthesized and they are the basic ingredient for many synthetic organic compounds, but the high toxicity may affect aquatic life, causing ecological imbalance. Phenol can be absorbed by the human body through the respiratory organs, skin and the alimentary canal [2].

Wastewaters containing phenol may be produced in biomass-based industry [3]. Biomass gasification is the thermochemical process used to convert solid biomass into syngas, a gaseous mixture consisting of hydrogen, carbon monoxide and carbon dioxide, which can find different uses for energy purposes [4,5]. One of the main issues related to biomass gasification is the presence of tars, a group of organic compounds [6] whose presence could make this technology unsuccessful from a commercial point of view [7]. In order to overcome this problem, two strategies may be adopted, namely the optimization of gasification operation conditions with addition of sorbents and catalysts directly in the gasification reactor (primary methods) or syngas cleaning, by means of systems and processes implemented downstream of the gasifier (secondary methods) [8–10]. In this regard, several technologies have been developed with the aim to push towards an efficient and economic/environmental sustainable process [11]. Tar is a complex mixture of condensable hydrocarbons, including oxygenated or polycyclic aromatics. Among the several cleaning technologies proposed, tar removal through physical processes are the most used at demonstrative scale. Physical processes may be dry or wet. Adsorbents, activated carbons and catalytic filters may be used as dry technologies for tar removal. On the contrary, the absorption tower (scrubber) is the most used wet technology. The advantage of a scrubber unit is related to the possibility to use the same unit as both quencher and absorber. For instance, after the gasifier, a wet scrubber with an organic liquid (e.g., bio-diesel) may be used for quenching the hot syngas and for removing the hydrophobic and heavy tar molecules from the syngas [12]. The exhausted bio-diesel could be used as fuel for energy production [13].

Afterwards, the content of hydrophilic or light tars, such as phenol, can be further reduced by a secondary wet scrubber using water as absorbent liquid. In that case, the produced wastewater has to be properly treated and phenol is usually the main organic tar to be removed [14].

The concentration of phenolic compounds in scrubber wastewater strongly depend on both phenol content in the syngas and process scheme and parameters, e.g., water-to-syngas flowrate ratio in the wet scrubber unit. For instance, Akhlas et al. report that phenol concentration may be in the range 950–4630 mg/L [15], as a function of the process conditions of the biomass gasification plant, although the lower value, i.e., 772 mg/L, is reported by Panchratna et al. [16]. Lower phenol concentration may be found in the case of coal gasification. In particular, Li et al. [17] report values lower than 500 mg/L for a full scale wastewater treatment facility of a coal gasification plant installed in Harbin (China). A phenol concentration up to 1600 mg/L is reported by Wang et al. for a Lurgi coal gasification wastewater, indicating that phenol concentration in scrubber wastewater strongly depends on the characteristics of the plant also for the coal gasification process [18].

In order to effectively eliminate phenolic compounds from wastewater, different processes are commonly used such as photocatalytic degradation, ozonation, extraction (liquid-liquid or solid phase extraction), biological methods (via microbial or enzymatic methods), membrane-based separation methods, and adsorption [19]. Activated carbons are versatile adsorbents, particularly effective in the adsorption of organic and inorganic pollutants from aqueous solutions [20]. In chemical industry, the most common applications of activated carbon are in fixed bed reactors or fluidized bed reactors in which the passage of a gas or liquid stream permit the removal of a large number of contaminants. Carbon characteristics, such as the large percentage of micropores, high specific surface area, high pore volumes, good mechanical strength, etc., provide a framework type with chemical and physical characteristics useful for absorption application [21]. Activated carbon is commonly used in powder and granular form. Commercially, both forms are sold and the most appropriate is chosen according to industrial applications; other forms such as felts, fibers and cloths are being studied in the scientific research [22].

In the literature, many works have been done on the aromatic compounds' absorption in aqueous solution and, in particular, phenols attract great attention [23–28].

Mukherjee et al. worked on phenol adsorption efficiency of different adsorbents including bagasse ash, activated carbon and charcoal from wastewater [29]. The monitored parameters for evaluating adsorption efficiency was the influence of pH, concentration of ethylenediaminetetraacetic acid (EDTA), anions and adsorbent dose. Their result showed

98, 90 and 90% phenol removal efficiencies by activated carbon, wood charcoal and bagasse ash systems, respectively. Removal efficiency was observed to increase with a decrease in the pH of the system. Effects of EDTA and nitrate ion content of the solution were identified as the factors that influenced the adsorption process.

Akl et al. [30] used sugarcane bagasse-based activated carbons to remove phenol form aqueous solution. They found that the proposed adsorbents are capable of phenol elimination from water. In particular, the process depends on pollutant concentration, pH solution and temperature.

Alhamed [31] studied the kinetics of phenol adsorption on activated carbon produced from waste dates' stones using four different solid particle sizes (1.47 to 0.225 mm) and initial concentrations of phenol of 200 and 400 ppm. He found that the initial rate of adsorption, predicted from the pseudo-second order model, decreased with increasing particle diameter as a result of higher interfacial area provided by particles with smaller diameter. Breakthrough curves for phenol removal using a packed bed of activated carbon were well predicted using an axial dispersion model.

Daifullah et al. [32] studied the removal of phenol and its derivate by adsorption on activated carbons prepared from H_3PO_4-impregnated powdered apricot stone shells and carbonized at 300–500 °C. Uptake of phenols increases with the respective increase in molecular dimensions and acidity of the organic compound and decrease in solubility of the sorbates.

In this work, the phenol removal from aqueous media was carried on by using residual biochars. For comparison of the treatment effectiveness, commercial activated carbons were included. The latter were kindly provided by Sicav S.p.A., both in granular and powder form; these materials were considered as representative of high performance materials with elevated BET specific surface area and pore volume. The considered biochars were agroindustrial biomass residues produced in pyrolysis and gasification processes of concern at the ENEA—Trisaia Research Centre (south of Italy). In this context, in the specific plant configurations, sections for gas cleaning based on combined biodiesel and water scrubbing are under consideration. In fact, the effect of carbon amount on treatment performances of real scrubber water was also assessed.

Therefore, the obtained data were analyzed to evaluate the performance of the experimental biochars in wastewater treatment for possible reuse options of the two residues, with a view to closing the production cycle of the thermochemical processes involved, reducing wastewater and solid process residues to limit the environmental impact and increase the sustainability of the processes.

2. Materials and Methods

2.1. Materials

Two activated carbons with specific surface areas of 800 m^2/g (SP800) and 1000 m^2/g (SP1000) were kindly provided by Sicav S.p.A (Gissi, Italy) in powder form with sizes < 75 μm and grain form with sizes of 1–1.5 mm. Both samples were used as-received with no additional treatments. The two biomass-derived carbons (biochar) were residual materials produced during processes of pyrolysis (SPBCP) or gasification (SPBCG). The SPBCP sample originated from wood chips subjected to slow pyrolysis carried out at 550 °C for 16 h, while the SPBCG came from a biomass gasification plant. The biomass gasification plant consists of a 120-kWth downdraft gasifier utilizing air as the gasification agent. The produced syngas is then purified in a cleaning section which consists of four units: cyclones, wet scrubbers, demisters and a chips filter. Wet scrubbers consist of two units in series. Heavy tars (mainly hydrophobic molecules) are removed in the primary biodiesel scrubber. Residual light tars, mainly phenolic molecules, are then removed in the secondary water scrubber. In that last unit, wastewater is produced and requires purification. Almond shells were used as feedstock and the gasifier was operated at a temperature of about 850 °C.

Before use, both the biochar samples were ground in a blade mill and then sieved; the powder fraction with size < 75 μm was used for the adsorption tests. To facilitate the grinding, amounts of the two samples were first kept at 70 °C for three days to reduce the moisture content. In order to carry out continuous tests in the column, the produced biochars were also sieved in the range 1–1.5 mm.

2.2. Characterization of the Investigated Carbons

All the investigated carbons were analyzed by thermogravimetric analysis in the range 30–850 °C, in air flow, and with a heating rate of 10 °C/min (SDT 650, TA Instruments, New Castle, DE, USA). The produced biochars (namely SPBCP and SPBCG) were also characterized in terms of textural properties. In this regard, the specific surface area (SSA) of the produced samples were obtained by adopting the Brunauer–Emmett–Teller model (B.E.T.) to nitrogen adsorption isotherms performed at 77 K with an ASAP 2020 Micromeritics instrument. In particular, the B.E.T. was applied in the $P/P°$ range 0.01–0.1 in order to obtain a positive C-value, which is exponentially related to the adsorption energy of the monolayer [33]. Total pore volume was calculated at $P/P° = 0.99$. Pore size distribution was determined by Non-Local Density Functional Theory (NLDFT) by using MicroActive Version 4.06 software (Micromeritics, Norcross, GA, USA) and by adopting an available model for carbon slit-shape pores. Before the analysis, the investigated samples were degassed, by heating at 300 °C, at a rate of 5 °C min^{-1}, under a high vacuum ($<10^{-8}$ mbar) for 12 h. Scanning electron microscope (SEM) images of the investigated samples were collected by a scanning electron microscope (FEI model Inspect, ThermoFisher Scientific, Hillsboro, OR, USA).

2.3. Batch Adsorption Tests

The batch phenol adsorption tests were carried out in a glass laboratory bottle using 50 mL of an aqueous solution containing 5 g/L of phenol (PhOH). The mass of adsorbent (activated carbons or biochars) was varied between 0.5 g and 4 g. The adsorption tests were carried out under vigorous stirring at 25 °C and the effect of contact time was investigated in the range 0–24 h. Details are reported in the preliminary study [1].

Batch adsorption tests were also carried out by using scrubber wastewater produced in the demonstrative biomass gasification plant described above.

After the adsorption test, the solid was separated from the liquid by centrifugation at 4000 rpm for 10 min. Any carbon residues suspended in the liquid were carefully eliminated by vacuum filtration followed by a filtration step through a 0.45 PTFE filter before the HPLC-UV-Vis analysis. UV-Vis analysis was performed by an Agilent 1100 HPLC system coupled with a DAD detector (Eluent: H20/ACN 80/20, supplied in isocratic mode; flow: 1 mL/min; column: C18; λ: 254 nm). The amount of adsorbed phenol as a function of time was fitted with two adsorption kinetic models, i.e., the pseudo-first and -second order models [34], while equilibrium values were modelled with Langmuir, Freundlich and Temkin equations [34,35]. Scrubber wastewater treatment was assessed by analyzing the treated samples with a Varian Cary 300 Scan UV-Vis spectrophotometer (Agilent Technologies, Inc., Santa Clara, CA, USA) in the wavelength range 190–400 nm.

2.4. Continuous Adsorption Tests

Continuous adsorption tests were carried out in the lab-scale experimental set-up reported in Figure 1.

The system consisted of a calibrated peristaltic pump that fed the phenol solution to the bottom of a glass cylindrical column containing the carbon.

The column had an internal diameter of 29 mm and a total length of 60 cm. The liquid flowrate was set to 55 mL/min to obtain an empty bed liquid velocity equal to 5 m/h. Different amounts of carbons (namely 20, 30 and 50 g) were used to study the effect of the length of the adsorption bed on the breakthrough curve. Before each test, the carbon bed was purged with water in order to eliminate air contained inside the pores. Furthermore, this purge also allowed the elutriation of the residual fine particulate from the carbon grains. From the top of the column the out-stream was sampled every one minute. The recovered samples were stored at 4 °C in closed bottles and analyzed by the HPLC technique as previously described.

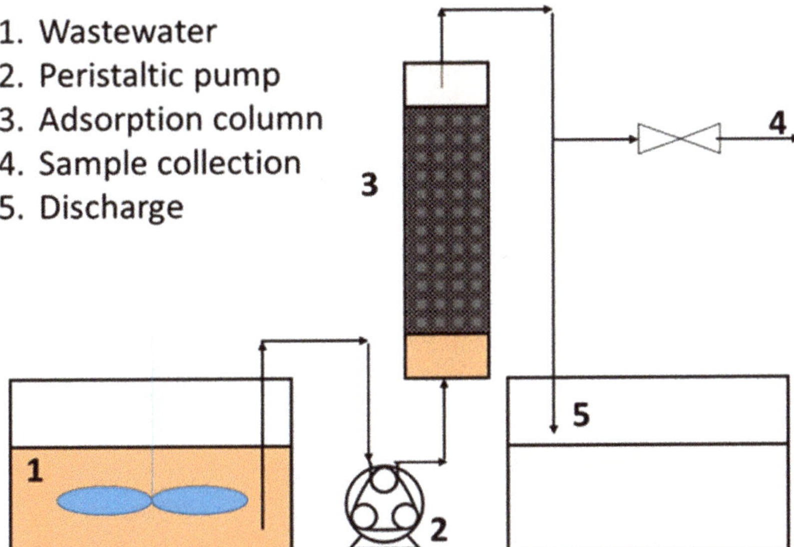

Figure 1. Scheme of the experimental set-up used for continuous adsorption tests.

The obtained breakthrough curves were modelled with the Thomas model [36]:

$$\frac{C_t}{C_0} = \frac{1}{1 + exp\left(\frac{k_{Th}}{Q}(q_{Th}m - C_0Qt)\right)} \tag{1}$$

where C_t is the phenol concentration (unit: mg/mL) in the column outlet at the time t (unit: min), C_0 is the initial phenol concentration (unit: mg/mL) in the feed, k_{Th} is the Thomas constant (unit: mL/min/mg), q_{Th} is the maximum adsorption capacity (unit: mg/g), Q is the wastewater flowrate (unit: mL/min) and m is amount of carbon used for the experiment (unit: mg). The Thomas model assumes Langmuir isotherms for equilibrium and pseudo-second-order reaction kinetics [36]. These assumptions agree with the equilibrium and kinetics insights found for the investigated process, as reported in a previous study [1]. One of the main weaknesses of the Thomas model is that its derivation assumes that the adsorption process is limited by kinetics, while it is often controlled by interphase mass transfer. Nevertheless, Thomas is widely used to describe the adsorption process in packed bed columns [37–41].

Furthermore, experimental data were also modelled with the Modified Dose Response (MDR) model [42]:

$$\frac{C_t}{C_0} = 1 - \frac{1}{1 + \left(\frac{Q \cdot C_0 \cdot t}{q_{MDR} \cdot M}\right)^a} \tag{2}$$

where C_t is the phenol concentration (unit: mg/mL) in the column outlet at the time t (unit: min), C_0 is the initial phenol concentration (unit: mg/mL) in the feed, q_{MDR} is the maximum adsorption capacity (unit: mg/g), "a" is a MDR parameter (unit: none), Q is the wastewater flowrate (unit: mL/min) and M is amount of carbon used for the experiment (unit: g). MDR is an empirical model usually adopted to minimize the error that results from use of the Thomas model, especially with lower and higher breakthrough curve times.

The obtained data were also used for scaling up by adopting the Bed Depth Service Time (BDST) approach [43].

3. Results

3.1. Characterization of the Samples

The B.E.T specific surface area (SSA), B.E.T. C-value, and total pore volume of the investigated samples are reported in Table 1, while pore size distribution is reported in Figure 2. In the literature it is known that, in the case of biomass belonging to similar types, the pyrolysis biochar has a typically specific surface area lower than biochar from gasification. It is also known that the characteristics of the char produced are also strongly influenced by the type of biomass and by the specific morphology [44,45]. These aspects apply well to the case of the chars used in this work, which in fact come from different biomasses, wood chips and almond shells, respectively known for having different textural characteristics (higher porosity in the former, low in the latter) [46–48]. Pore size distribution reveals that commercial activated carbons possess pores with an average pore width of about 11–12 Å (Å is Ångström, 10^{-10} m). Smaller pores are not excluded. On the contrary, lager pores, i.e., >15 Å can be observed for the investigated biochar. The pores are suitable for the inclusion of phenol, which has a molecular diameter of 7.46 Å [49].

Table 1. B.E.T. specific surface area and pore volume of the investigated samples.

Sample	B.E.T. SSA (m^2/g)	B.E.T C-Value $(-)$	Total Pore Volume (cm^3/g)
SP1000	1281	6289	0.643
SP800	698	6155	0.413
SPBCP	211	81	0.126
SPBCG	63	844	0.069

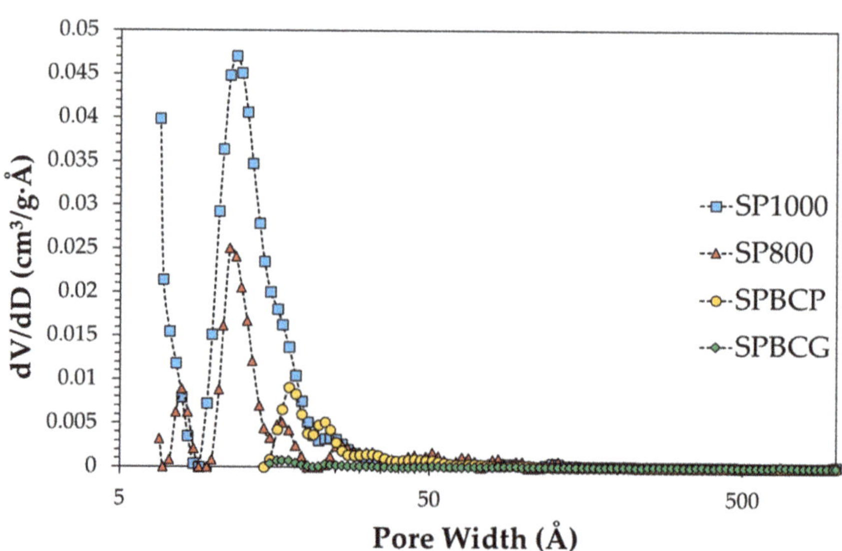

Figure 2. Pore size distribution of the investigated samples.

Thermogravimetry analysis carried out in air flow revealed that the combustion of both the activated carbons occurs in the temperature range 400–600 °C with a maximum heat flow at about 580 °C, indicating a similar structure. On the contrary, the combustion of the biomass-derived chars start at about 250 °C with a maximum heat flow in the range 400–500 °C. After complete combustion, the residual matter was about 5% for SP1000, about 18% for SP800 and about 20% for SPBCG, while it was below 1% for SPBCP. Such differences may be related to both the thermal preparation process and the original feedstock that affects the carbon content and inorganic matter in the produced solid.

Figure 3 shows the SEM images of the investigated samples at two different magnifications. All the samples exhibit a not well-defined shape with macropores that may be observed for the SP1000 sample. The presence of some large pores on the external surface may be also observed for the SPBCG sample. On the contrary, no large pores are observed on the SPBCP sample probably due to the milder conditions adopted for its production (i.e., slow pyrolysis), with respect to the SPBCG sample which was produced in more severe conditions (i.e., high temperature gasification).

Figure 3. SEM images of the investigated SP1000, SP800, SPBCG and SPBCP samples. Subfigures are SEM images at higher magnification.

3.2. Adsorption Tests under Batch Conditions

Batch tests for phenol removal were described and discussed in a previous work [1]. Briefly, the trend of phenol adsorption as a function of time was modelled with both pseudo-first and pseudo-second order kinetics, and F-tests indicate the latter as the best fitting model. In particular, the model also indicates that the SP1000 sample shows an adsorption rate four times higher than the SP800 sample, which displays a kinetic constant one order of magnitude higher than biochars. Concerning biochars, a higher kinetic constant is calculated for the SPBCG sample, despite a lower equilibrium capacity.

The Langmuir model was found to be the best fitting model for equilibrium data. Details about F-tests and comparison with the other models, i.e., Temkin and Freundlich, are reported elsewhere [1]. The Langmuir model predicted a phenol maximum quantity adsorbed at 25 °C equal, respectively, to 270 mg/g, 233 mg/g, 104 mg/g and 65 mg/g, for SP1000, SP800, SPBCP and SPBCG, while the equilibrium constants were, respectively, 1.7×10^{-2} L/mg, 3.0×10^{-3} L/mg, 5.4×10^{-4} L/mg and 9.4×10^{-4} L/mg.

In particular, the calculated adsorption capacity increases as a function of pore volume as reported in Figure 4, also according to literature data [50–55]. In contrast, no trend was found for the Langmuir equilibrium constants. In fact, phenol adsorption appears to be more favorable on SPBCG than on SPBCP, despite the smaller surface area, indicating that the investigated biochars exhibit different phenol–surface interactions.

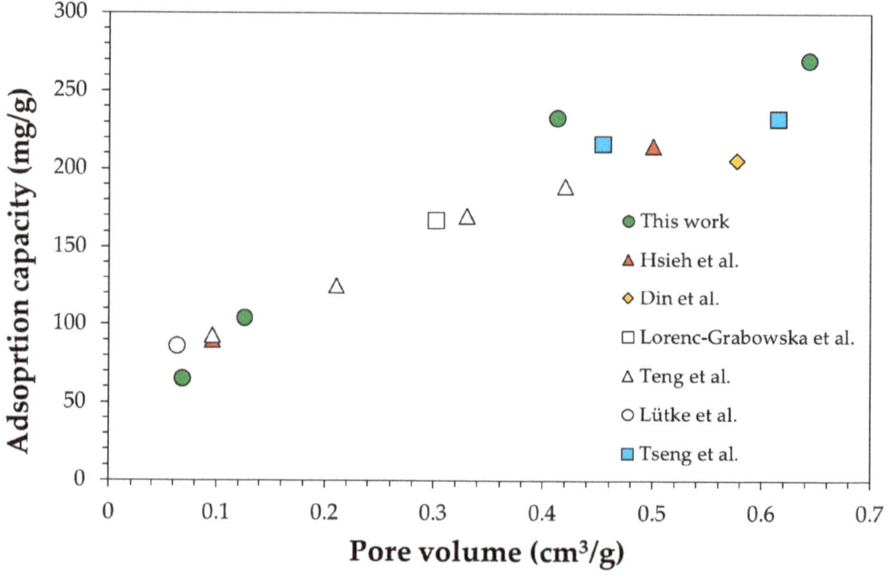

Figure 4. Langmuir adsorption capacity at 25 °C as a function of pore volume. Comparison with literature data [50–55].

Obtained data indicate SP1000 and SPBCP as the best carbons among the investigated activated carbons and biochars, respectively. Nevertheless, the utilization of the SPBCG sample for wastewater treatment may be interesting from a process circularity perspective. In fact, the char produced during the biomass gasification process is considered as a low value material, with very limited uses, to be disposed of as a process waste stream. Therefore, the utilization of biomass gasification char as adsorbent material for the treatment of wastewater produced during syngas wet cleaning may be an interesting strategy for its valorization. Biochar from gasification may be considered as a zero-value product (that may become a negative value in the case it is disposed of as waste), while biochar from pyrolysis usually has a positive production cost, since it is a target product of the process. Therefore, albeit that the SPBCG sample exhibits the lowest capacity of adsorption towards phenol, it was studied as a potential adsorbent for the treatment of real scrubber wastewater produced in a biomass-based gasification plant. For comparison, the SP1000 sample was also studied in terms of capacity of purification of real syngas scrubber wastewater.

UV-Vis profiles of the real syngas scrubber wastewater and phenol solution used in this study are reported in Figure 5.

UV-Vis spectra of the phenol solution exhibit a well-defined absorption band centered at about 285 nm. On the contrary, a broad band is observed for the syngas scrubber wastewater indicating the presence of several compounds. However, also for the investigated wastewater the maximum absorption occurs at about 285 nm, suggesting that phenol may be considered a molecule suitable for modelling.

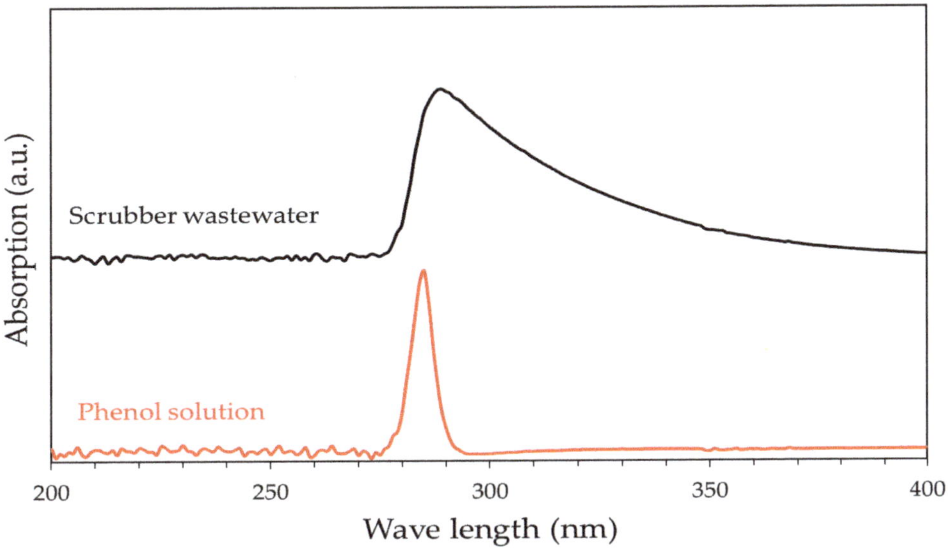

Scrubber wastewater

Phenol solution

Absorption (a.u.)

Wave length (nm)

200   250   300   350   400

Figure 5. UV-Vis profiles of real syngas scrubber wastewater and 5 g/L phenol solution used in this work.

The UV-Vis spectra of the syngas scrubber water as a function of carbon amount after 24 h of treatment are reported in Figure 6. The effect of carbon amount of wastewater treatment level is reported in Figure 7.

Results clearly confirm the superiority of activated carbons over biochar in terms of purification level for both scrubber wastewater and phenol removal.

The purification level of both real scrubber wastewater and phenol model solution increases as the carbon amount increases.

In particular, the purification level of wastewater and phenol solution increases from 13% to 25% and from 31% to 64%, respectively, by increasing the SPBCG amount from 1 g to 4 g. In the case of commercial activated carbon SP1000, the purification level increases from 47% to 92% in the case of real syngas scrubber wastewater, and from 68% to 99% in the case of phenol model solution.

In fact, as an example, with 4 g of carbon, a wastewater purification level of about 92% is achieved for commercial activated carbon SP1000, while only 25% is measured for gasification biochar SPBCG. A similar trend is also observed in terms of phenol removal.

It is important to observe that the purification level is higher in the case of phenol model solution. As an example, with 4 g of carbon, the purification level was 64% for phenol model solution and 25% for scrubber wastewater in the case of SPBCG, and 92% and 96% for phenol model solution and scrubber wastewater, respectively, in the case of commercial activated carbon. This experimental evidence is of paramount importance. In fact, phenol is usually considered as a model for such a kind of investigation but the complexity of the real wastewater stream may have a significant impact on adsorption performances.

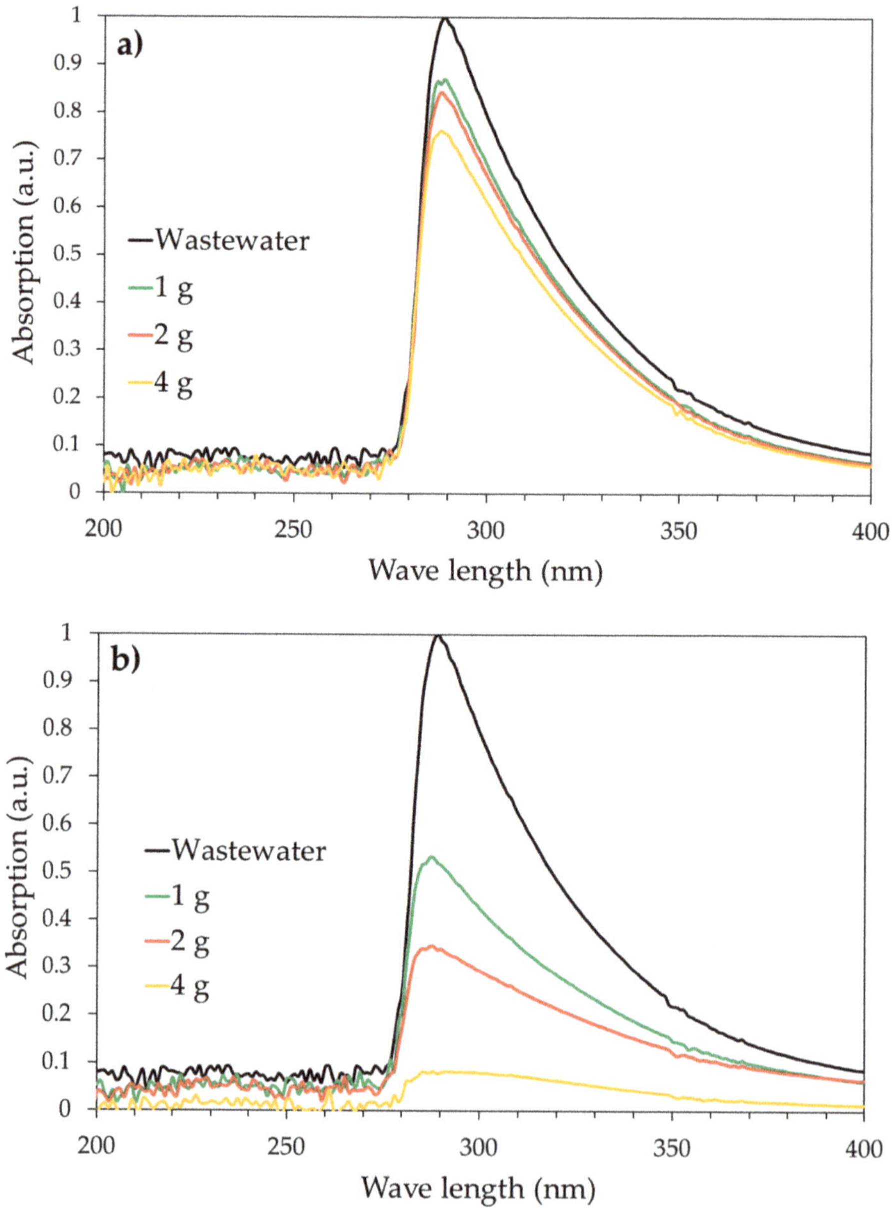

Figure 6. UV-Vis profiles of scrubber wastewater as a function of carbon amount (50 mL of solution for each test) for the SPBCG sample (**a**) and the SP1000 sample (**b**).

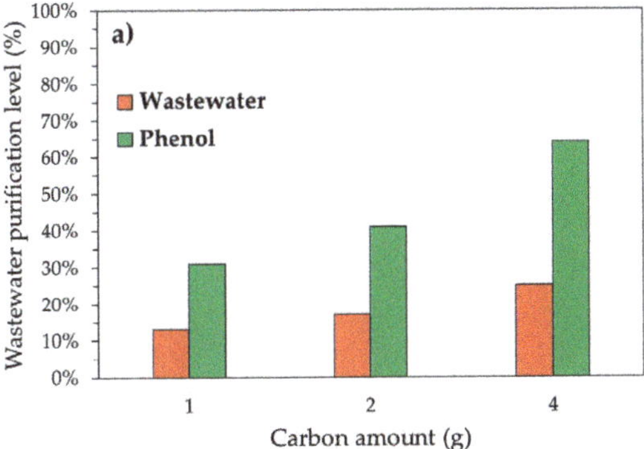
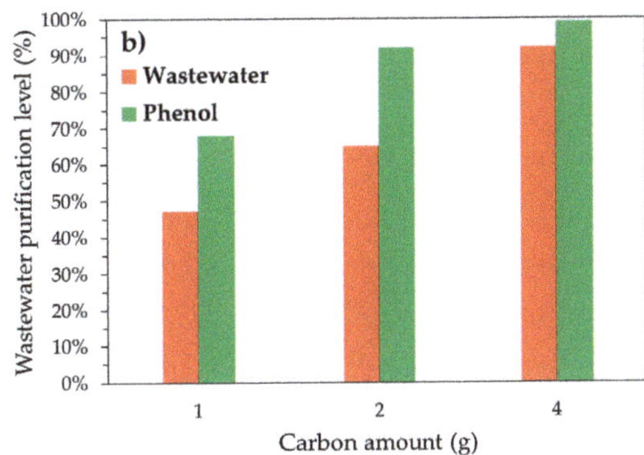

Figure 7. Scrubber wastewater and phenol model solution purification level as a function of carbon amount for the SPBCG (**a**) and SP1000 (**b**) samples, after 24 h treatment.

3.3. Adsorption Tests in Continuous Mode: Experimental, Modelling and Scale-Up

Continuous tests at bench-scale were carried out over the SP1000 and SPBCG samples, available in the size range 1–1.5 mm. As previously observed, SP1000 is the sample with the best adsorption properties in terms of both kinetics and thermodynamics. On the contrary, SPBCG showed the slower adsorption kinetics and the lower adsorption capacity. However, the potential utilization of SPBCG may be also related to economic consideration. For instance, phenol-contaminated wastewater may be produced in biomass-fueled gasification plants. To regenerate a such waste stream, it may be both environmentally and economically beneficial to consider an on-site treatment with the biochar produced as a residue of gasification rather than its disposal.

The breakthrough curves comparing the adsorption activity of commercial activated carbon vs. the investigated gasification biochar are reported in Figure 8. Assuming a C/C0 value at the breakpoint equal to 0.1, under the conditions adopted the collected data allow estimation of service times of approximately 12 min for the SP1000 sample and 2–3 min for the SPBCG sample. Therefore, with the same fluid dynamic conditions, pollutant load, and service time, these outputs indicate that the quantity of biochar required is at least four times the quantity of commercial activated carbon. This difference may increase for effluents with lower phenol concentrations, due to the differences observed in the adsorption isotherms. In particular, commercial activated carbon exhibits an almost irreversible isotherm, while gasification biochar shows a much less favorable adsorption, with a less pronounced increase in the quantity adsorbed with the solution concentration.

The effect of adsorption bed weight on breakthrough curve was studied for the SP1000 sample and results are reported in Figure 9.

From the mass balance it is possible to calculate the characteristic parameters reported in Table 2.

The reported data show how the abatement level increases as the height of the bed increases. The total adsorbed quantity increases from 143 to 177 mg/g, demonstrating that column is not operating in equilibrium conditions, the equilibrium adsorption capacity being equal to about 270 mg/g. As the height of the bed increases, the phenol adsorption capacity at the breakpoint also increases.

As previously reported, the obtained breakthrough curves were modelled with both the Thomas and MDR models.

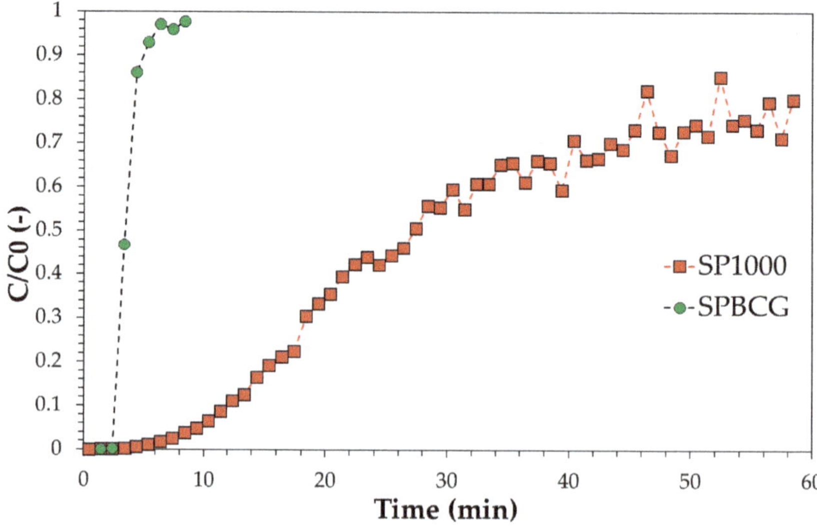

Figure 8. Breakthrough curve of the SP1000 and SPBCG samples for phenol model solution. Solution initial concentration: 5000 mg/g, flowrate: 55 mL/min, carbon amount: 50 g.

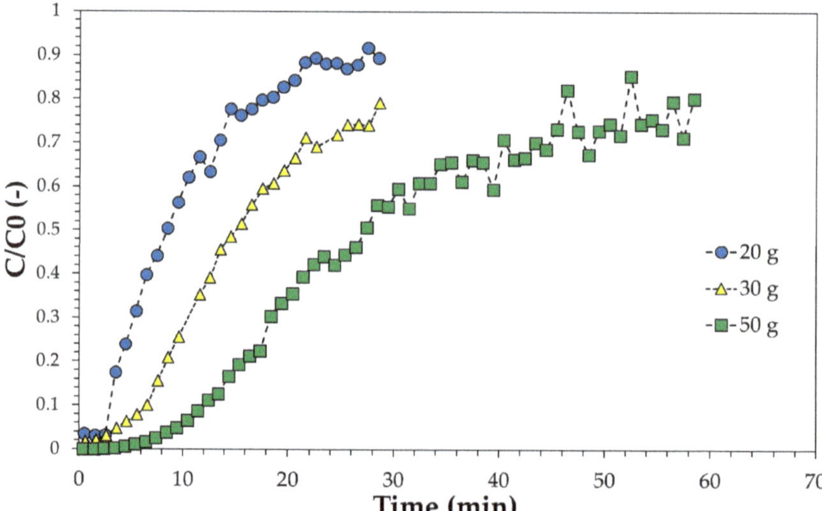

Figure 9. Breakthrough curve of SP1000 as a function of carbon amount for the phenol model solution. Solution initial concentration: 5000 mg/g, flowrate: 55 mL/min.

Figure 10 shows the linear plot of the investigated models, and the estimated parameters are reported in Table 3. The comparison between models and experimental data is reported in Figure 11. Results clearly show that the MDR model satisfactorily fits the experimental data in the whole time range and for each carbon amount. On the contrary, the Thomas model diverges from experimental data at low time values and for higher carbon amounts. It may be then concluded that MDR is better than the Thomas model. Both Thomas and MDR models were also applied for describing the breakthrough curve of the SPBCG sample and results are reported in Figure 12. The model parameters estimated by linear regression are reported in Table 3. As in the case of SP1000, the MDR model may be considered the best fitting mathematical model of the breakthrough curve of the SPBCG sample.

Table 2. Parameters estimated from the breakthrough curve of SP1000 for phenol solution.

Carbon Amount (g)	Bed Height (cm)	Total Adsorption Capacity (mg/g)	Breakthrough Time [a] (min)	Breakthrough Adsorption Capacity [b] (mg/g)	Stoichiometric Time [c] (min)
20	5.9	133	3	40	11
30	8.9	152	6	49	21
50	14.9	177	12	71	37

[a] Calculated at $C/C_0 = 0.1$; [b] adsorption capacity at $C/C_0 = 0.1$; [c] calculated in accordance with [56].

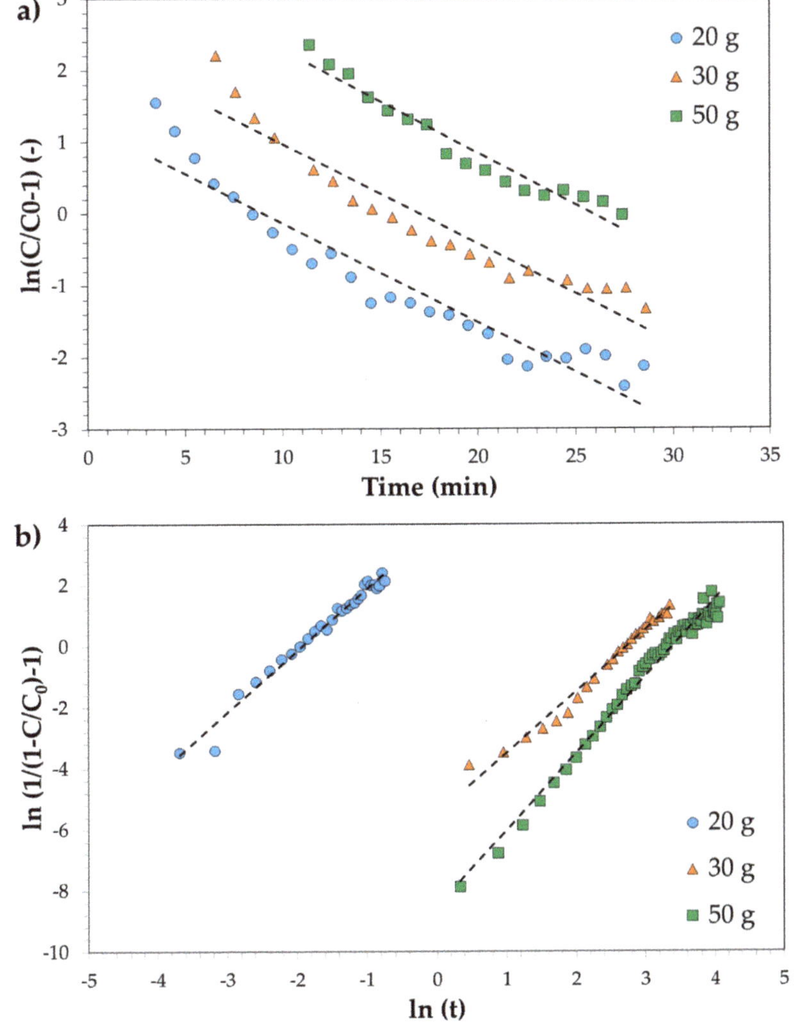

Figure 10. Linear plot of the Thomas (**a**) and MDR (**b**) models. Dashed lines refer to the fitted linear regression model.

Table 3. Parameters estimation of the Thomas and MDR models for the SP1000 and SPBCG samples.

Sample	Carbon Amount (g)	Thomas			MDR		
		$K_{th} \cdot 10^2$ (min/mL/mg)	q_{th} (mg/g)	R^2	a (−)	q_{MDR} (mg/g)	R^2
SP1000	20	2.76	118	0.959	1.8	109	0.998
	30	2.78	155	0.987	2.0	144	0.997
	50	2.88	142	0.985	2.5	168	0.984
SPBCG	50	14.01	13.1	0.949	5.6	19.5	0.995

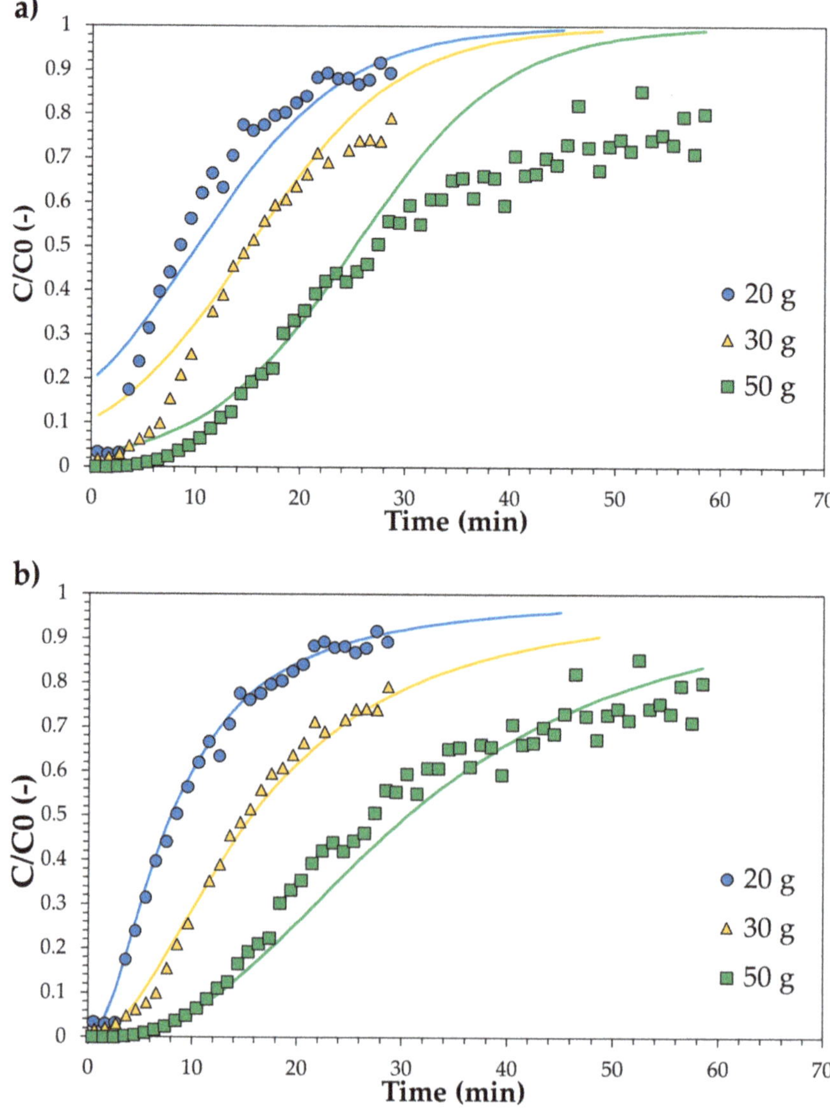

Figure 11. Comparison with experimental data of the Thomas (**a**) and MDR (**b**) models. Continuous lines refer to the fitted linear regression models.

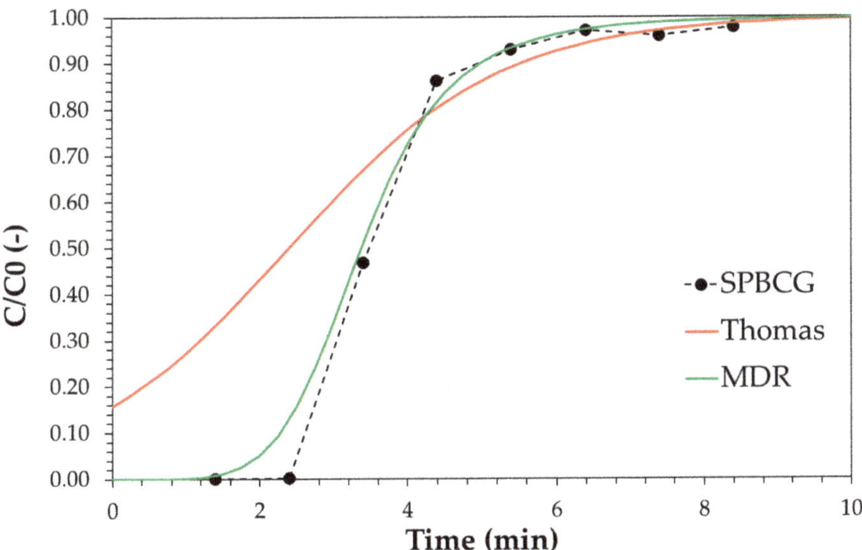

Figure 12. Comparison with experimental data of the Thomas (red line) and MDR (green line) models for the biochar SPBCG sample.

On the basis of the obtained breakthrough curves, scale-up may be carried out by the Bed Depth Service Time (BDST) model, which may be expressed as follows:

$$t = \frac{N \cdot H}{C_0 \cdot v} - \frac{1}{k \cdot C_0} ln\left(\frac{C_0}{C_b} - 1\right) \tag{3}$$

where C_0 is the initial concentration of phenol, C_b is the breakpoint concentration, v is the liquid empty bed velocity, t is the service time, H is the bed height, N is the adsorption capacity of the bed and k is a kinetic parameter. By plotting the service time obtained experimentally vs. bed height it is possible to obtain N and k values, which are the scale-up parameters. Figure 13 shows the plot of the BDST equation applied to the obtained experimental data.

The value of the BDST model are N = 45,438 g/m^3 and k = 7.2 × 10^{-3} m^3/h/g. Therefore, the obtained parameters may be used to estimate the service time as a function of bed height, superficial velocity, and desired breakpoint concentration.

Another practical tool to estimate the service time is to consider a design adsorption capacity (W_{design}) as 50% lower than that calculated from the isotherm ($W_{theoretical}$) [57]:

$$W_{design} = 0.5 \, W_{theoretical} \tag{4}$$

On the basis of obtained results, Langmuir isotherms may be used to calculate $W_{theoretical}$ as a function of initial phenol concentration. The carbon usage rate (CUR) may be then estimated as follows:

$$CUR = \frac{Q \cdot C_0 \cdot \left(1 - \frac{C_b}{C_0}\right)}{W_{design}} \tag{5}$$

where Q is the wastewater flowrate (m^3/h), C_0 is the phenol concentration in the wastewater (g/m^3), C_b is the maximum admissible concentration in the treated stream, i.e., breakpoint concentration, and W_{design} is the design adsorption capacity (g$_{PhOH}$/kg$_{carbon}$). As an example, Table 4 shows the CUR (kg$_{carbon}$/h) of the investigated samples estimated for treating 1 m^3/h of wastewater contaminated with phenol for several initial concentrations.

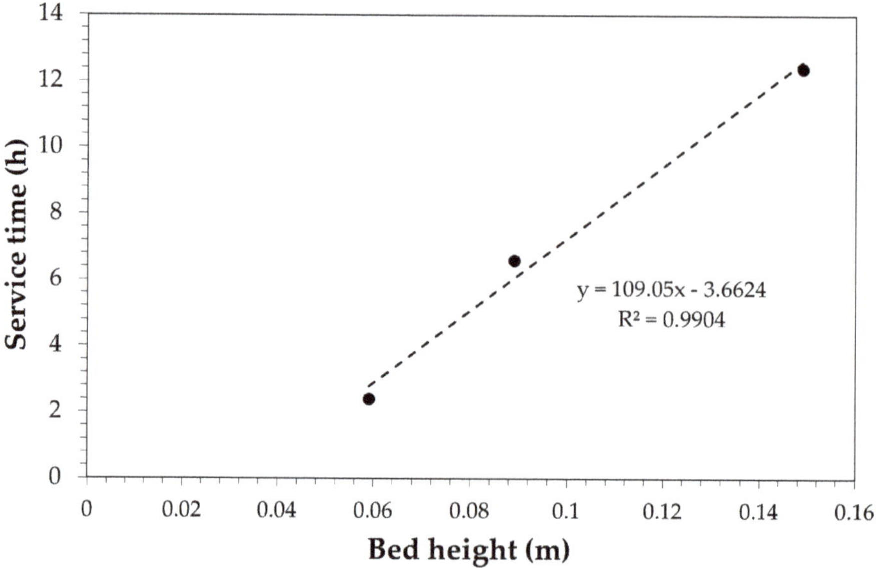

Figure 13. Linear plot of the BDST model. In the linear regression model, y refers to the service time (h) and x refers to the bed height (m).

Table 4. Carbon usage rate (CUR) for the SP1000, SP800, SPBCG and SPNCP samples as a function of phenol concentration. Wastewater flowrate: 1 m^3/h.

CUR (kg/h)	SP1000	SP800	SPBCG	SPBCP
C0 = 5000 ppm	41	46	187	132
C0 = 2500 ppm	21	24	110	84
C0 = 1000 ppm	8	11	64	55
C0 = 500 ppm	4	7	48	45
C0 = 100 ppm	1	4	36	38

By comparing the results, performance ratios ranging from a minimum value of about 3 (SPBCP vs. SP800) to a maximum value of around 40 (SPBCP vs. SP1000) can be estimated, depending on the specific materials and the concentration of phenol in the solution. Concerning the two biochars, no significant differences may be observed for low phenol concentration values, while a lower carbon usage rate is calculated for pyrolysis biochar with respect to gasification biochar at high phenol concentration.

A continuous test was also carried out for real scrubber wastewater over commercial activated carbon in order to obtain insights about the suitability of phenol as a model molecule for wastewater purification. Results are reported in Figure 14.

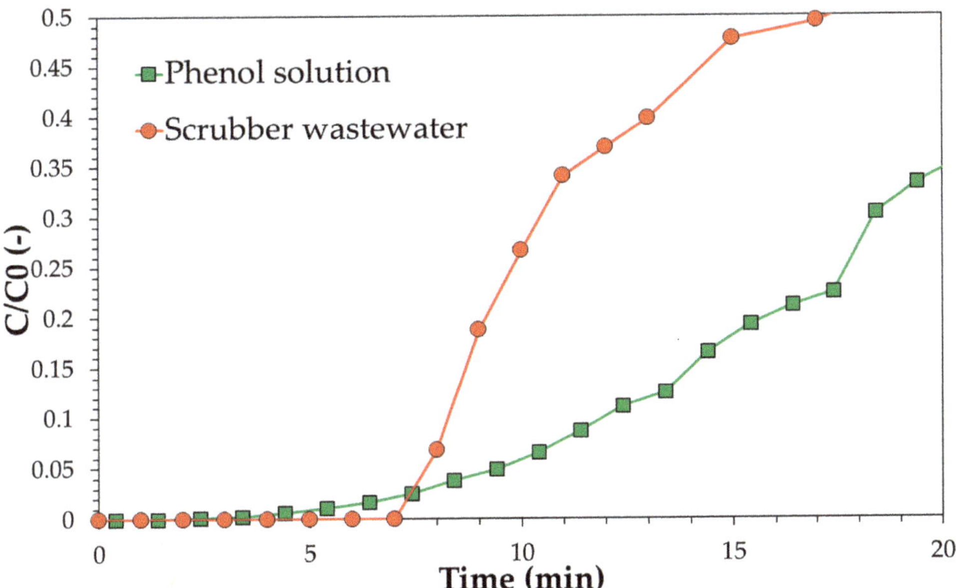

Figure 14. Breakthrough curve of the SP1000 sample for the phenol model solution and real scrubber wastewater. Solution initial phenol concentration: 5000 mg/g, flowrate: 55 mL/min, carbon amount: 50 g.

Results of continuous tests confirm data obtained during batch adsorption tests. The level of stream purification is lower in the case of real scrubber wastewater. In particular, if $C/C_0 = 0.1$ is assumed as the breakpoint value, the service time is reduced from 12 min to about 8 min. This may be also considered as a design parameter. Therefore, based on the collected data, the CUR estimated for phenol should be increased by a factor $12/8 = 1.5$ in the case of a real scrubber wastewater stream. GC-MS analysis was carried out on both fresh and treated wastewater in order to assess the effect of other molecules on phenol removal. Wastewater consists of phenol as the main hydrophilic tar, with the presence of indene, a hydrophobic tar. The presence of indene may be responsible for the lower purification level observed in the case of real wastewater, especially for materials with low adsorption capacity. Indeed, in the case of the SPBCG sample, after 24 h treatment under batch conditions with 4 g of carbon, indene was completely removed, while phenol was still found in the residual solution, indicating that the adsorption of hydrophobic tars, e.g., indene hinders the adsorption of hydrophilic tars, e.g., phenol. Therefore, in the case of a syngas cleaning system consisting of a water scrubber downstream of a biodiesel scrubber, the latter unit should be suitably designed in order to avoid the presence of hydrophobic tars in the syngas, with a negative effect on the purification of wastewater produced in the secondary scrubber.

The obtained results may be exploited for both technical and economic assessments. For instance, the utilisation of biochar produced from a gasification well agrees with the concept of the circular economy. In fact, the biochar produced during gasification is considered a residual stream usually treated as a waste. Therefore, the estimated carbon usage rate gives information about the possibility to reuse the produced biochar within the same gasification plant before its disposal. In that scenario, it is important to note that, in the case of a biomass gasification plant fuelled with commonly used lignocellulose feedstocks such as poplar wood, almond shell, pine wood, eucalyptus, wheat straw, poultry litter, maize cobs, grape marc, miscanthus, and switchgrass, the biochar yield is usually lower than 20% with respect to the initial biomass flowrate to the gasifier [58]. Furthermore, both the type of gasifier and the operation conditions also strongly affect the formation of tars, with an effect on the amount and the characteristics of wastewater to be treated; this aspect should also be taken into account for the assessment. For instance, tar content in the syngas may be up to 100–150 g/Nm^3 in the case of an updraft gasifier, and lower than 10 g/Nm^3 in the case of a downdraft gasifier, and phenolic compounds represent usually about 10% of the total tar content [59–62]. Future research should focus on technical and economic assessment of the possible utilization of biochar as a partial or complete substitute to commercial activated carbons for the purification of wastewater contaminated with phenols at pilot or demonstrative scale.

4. Conclusions

In this work, two commercial activated carbons and two residual biochars obtained from pyrolysis and gasification processes were assessed as potential adsorbents for the purification of wastewater produced in a syngas wet scrubber unit of a biomass gasification plant of concern at the ENEA Trisaia Research Centre. Phenol solution was used as a model solution for the investigations. Data obtained from both batch and continuous tests were used to compare the investigated samples in terms of kinetics and adsorption capacity. On the basis of results obtained at bench-scale, design parameters were also estimated. Experimental tests indicate that the phenol adsorption capacity estimated by the Langmuir model depend on the pore volume of the carbons. Continuous tests give evidence of the lower efficacy of the considered residual biochars, compared to the commercial activated carbons. Design parameters were estimated by applying the BDST model. Furthermore, a practical tool was also used for estimating the carbon usage rate (CUR) and a case study was assessed. Based on the collected results, the two investigated biochars exhibit a similar CUR at low phenol concentration, while a lower CUR is estimated for pyrolysis biochar with respect to gasification biochar at a high phenol concentration level.

Overall, the collected results indicate that the performance of the biochars are significantly lower than commercial materials; nevertheless, the residual chars under examination have proven a certain degree of effectiveness as activated carbons that can be used for the regeneration of water contaminated by phenols. This achievement therefore provides an incentive for further evaluation, both technical and economic, with respect to their use for the regeneration of water contaminated by phenols and the design of a related plant unit.

Author Contributions: Conceptualization, E.C., C.S.; methodology, C.S., A.R.; investigation, E.C., A.R., L.M., D.C.; data curation, E.C., L.M.; writing—original draft preparation, E.C.; writing—review and editing, E.C., D.B.; supervision, G.C., M.M. All authors have read and agreed to the published version of the manuscript.

Funding: This research received no external funding.

Institutional Review Board Statement: Not applicable.

Informed Consent Statement: Not applicable.

Data Availability Statement: All data are reported in the paper.

Conflicts of Interest: The authors declare no conflict of interest.

References

1. Sposato, C.; Catizzone, E.; Romanelli, A.; Marsico, L.; Barisano, D.; Migliori, M. Phenol Removal from Water with Carbons: An Experimental Investigation. *Tech. Ital. Ital. J. Eng. Sci.* **2020**, *64*, 143–148. [CrossRef]
2. Available online: https://echa.europa.eu/substance-information/-/substanceinfo/100.003.303 (accessed on 15 November 2020).
3. Galanopoulus, C.; Giuliano, A.; Barletta, D.; Zondervan, E. An Integrated Methodology for the Economic and Environmental Assessment of a Biorafinery Supply Chain. *Chem. Eng. Res. Des.* **2020**, *160*, 199–215. [CrossRef]
4. Giuliano, A.; Catizzone, E.; Barisano, D.; Nanna, F.; Villone, A.; De Bari, I.; Cornacchia, G.; Braccio, G. Towards Methanol Economy: A Techno-environmental Assessment for a Bio-methanol OFMSW/Biomass/Carbon Capture-based Integrated Plant. *Int. J. Heat Technol.* **2019**, *37*, 665–674. [CrossRef]

5. Giuliano, A.; Catizzone, E.; Freda, C.; Cornacchia, G. Valorization of OFMSW Digestate-Derived Syngas toward Methanol, Hydrogen, or Electricity: Process Simulation and Carbon Footprint Calculation. *Processes* **2020**, *8*, 526. [CrossRef]

6. Milne, T.; Evans, R.J.; Abatzoglou, N. Biomass Gasifier "Tars": Their Nature, Formation, and Conversion. NREL/TP-570-25357. Available online: https://www.nrel.gov/docs/fy99osti/25357.pdf (accessed on 15 November 2020).

7. Luo, X.; Wu, T.; Shi, K.; Song, M.; Rao, Y. Chapter 1–Bionmass Gasification: An Overview of Technological Barriers and Socio-Environmental Impact. In *Gasification of Low Grade Feedstock*; Yun, Y., Ed.; IntechOpen: London, UK, 2018.

8. Zwart, R.W.R. Gas Cleaning Downstream Biomass Gasification–Status Report 2009. ECN-E–08-078. Available online: www.ieatask33.org (accessed on 15 November 2020).

9. Singh, R.N.; Singh, S.P.; Balwanshi, J.B. Tar Removal from Producer Gas: A Review. *Res. J. Eng. Sci.* **2014**, *3*, 16–22.

10. Devi, L.; Ptasinski, K.J.; Janssen, F.J.J.G. A Review of the Primary Measures for Tar Elimination in Biomass Gasification Processes. *Biomass Bioenerg.* **2003**, *24*, 125–140. [CrossRef]

11. Heidenreich, S.; Müller, M.; Foscolo, P.U. *Advanced Biomass Gasification New Concepts for Efficiency Increase and Product Flexibility*; Academic Press: Cambridge, MA, USA, 2016; pp. 1–134.

12. Madav, V.; Das, D.; Kumar, M.; Surwade, M.; Parikh, P.P.; Sethi, V. Studies for Removal of Tar from Producer Gas in Small Scale Biomass Gasifiers Using Biodiesel. *Biomass Bioenerg.* **2019**, *123*, 123–133. [CrossRef]

13. Hrbek, J. *Status Report on Thermal Biomass Gasification in Countries Participating in IEA Bioenergy Task 33*; University of Technology: Viena, Austria, April 2016. Available online: http://www.ieatask33.org/app/webroot/files/file/2016/Status_report.pdf (accessed on 15 November 2020).

14. De Bari, I.; Barisano, D.; Cardinale, M.; Matera, D.; Nanna, F.; Viggiano, D. Air Gasification of Biomass in a Downdraft Fixed Bed: A Comparative Study of the Inorganic and Organic Products Distribution. *Energy Fuels* **2000**, *14*, 889–898. [CrossRef]

15. Akhlas, J.; Bertucco, A.; Ruggeri, F.; Collodi, G. Treatment of Wastewater from Syngas Wet Scrubbing: Model-based Comparison of Phenol Biodegradation Basin Configurations. *Canad. J. Chem. Eng.* **2017**, *95*, 1652–1660. [CrossRef]

16. Panchratna, D.; Sethi, V.; Mukherji, S. Sorption of Simulated Biomass Gasification Wastewater on Rice Husk Char and Activated Carbon. *Int. J. Environ. Eng. Manag.* **2013**, *4*, 489–496.

17. Li, H.-Q.; Han, H.-j.; Du, M.-A.; Wang, W. Removal of Phenols, Thiocyanate and Ammonium from Coal Wastewater Using Moving Bed Biofilm Reactor. *Biores. Technol.* **2011**, *102*, 4667–4673. [CrossRef] [PubMed]

18. Wang, Z.; Xu, X.; Gong, Z.; Yang, F. Removal of COD, Phenols and Ammonium from Lurgi Coal Gasification Wastewater Using A^2O-MBR System. *J. Hazard. Mater.* **2012**, *235-236*, 78–84. [CrossRef] [PubMed]

19. Bianco, L.; D'Amico, E.; Villone, A.; Nanna, F.; Barisano, D. Bioremediation of Wastewater Stream from Syngas Cleaning via Wet Scrubbing. *Chem. Eng. Trans.* **2020**, *80*, 31–36.

20. Wong, S.; Ngadi, N.; Inuwa, I.M.; Hassan, O. Recent Advances in Applications of Activated Carbon from Biowaste for Wastewater Treatment: A Short Review. *J. Clean. Prod.* **2018**, *175*, 361–375. [CrossRef]

21. Khandaker, T.; Hossain, M.S.; Dhar, P.K.; Rahman, M.S.; Hossain, M.A.; Ahmed, M.B. Efficacies of Carbon-Based Adsorbents for Carbon Dioxide Capture. *Processes* **2020**, *8*, 654. [CrossRef]

22. Marsh, H. Chapter 2–Activated Carbon: Structure Characterization Preparation and Applications. In *Introduction to Carbon Technologies*; Rodriguez-Reinoso, F., Ed.; University of Alicante: Alicante, Spain, 1997; p. 35.

23. Kilic, M.; Apaydin-Varol, E.; Putun, A.E. Adsorptive Removal of Phenol from Aqueous Solutions on Activated Carbon Prepared from Tobacco Residues: Equilibrium, Kinetics and Thermodynamics. *J. Hazard. Mater.* **2011**, *189*, 397–403. [CrossRef]

24. El-Naas, M.; Al-Zuhair, S.; Alhaija, M. Removal of Phenol from Petroleum Refinery Wastewater through Adsorption on Date-pit Activated Carbon. *Chem. Eng. J.* **2010**, *162*, 997–1005. [CrossRef]

25. Rodrigues, L.A.; Pinto da Silva, M.L.C.; Alvarez-Mendes, M.O.; dos reis Coutinho, A.; Thim, G.P. Phenol Removal from Aqueous Solution by Activated Carbon Produced from Avocado Kernel Seeds. *Chem. Eng. J.* **2011**, *174*, 49–57. [CrossRef]

26. Yang, G.; Tang, L.; Zeng, G.; Cai, Y.; Tang, J.; Pang, Y.; Zhou, Y.; Liu, Y.; Wang, J.; Zhang, S.; et al. Simultaneous Removal of Lead and Phenol Contamination from Water by Nitrogen-functionalized Magnetic Ordered Mesoporous Carbon. *Chem. Eng. J.* **2015**, *259*, 854–864. [CrossRef]

27. Beker, U.; Ganbold, B.; Dertlu, H.; Gulbayir, D.D. Adsorption of Phenol by Activated Carbon: Influence of Activation Methods and Solution pH. *Energ. Conv. Manag.* **2010**, *51*, 235–240. [CrossRef]

28. Gundogdo, A.; Duran, C.; Senturk, H.B.; Soylak, M.; Ozdes, D.; Serencam, H.; Imamoglu, M. Adsorption of Phenol from Aqueous Solution on a Low-Cost Activated Carbon Produced from Tea Industry Waste: Equilibrium, Kinetic, and Thermodynamic Study. *J. Chem. Eng. Data* **2012**, *57*, 2733–2743. [CrossRef]

29. Mukherjee, S.; Kumar, S.; Misra, A.K.; Fan, M. Removal of Phenols from Water Environment by Activated Carbon, Bagasse Ash and Wood Charcoal. *Chem. Eng. J.* **2007**, *129*, 133–142. [CrossRef]

30. Akl, M.A.; Dawy, M.B.; Serage, A.A. Efficient Removal of Phenol from Water Samples Using Sugarcane Bagasse Based Activated Carbon. *J. Anal. Bioanal. Techn.* **2014**, *5*, 1–12. [CrossRef]

31. Alhamed Yahia, A. Adsorption Kinetics and Performance of Packed Bed Adsorber for Phenol Removal Using Activated Carbon from Dates' Stones. *J. Hazard. Mater.* **2009**, *170*, 763–770. [CrossRef]

32. Daifullah, A.A.M.; Girgis, B.S. Removal of Some Substituted Phenols by Activated Carbon Obtained from Agricultural Waste. *Water Res.* **1998**, *32*, 1169–1177. [CrossRef]

33. Thommes, M.; Kaneko, K.; Neimark, A.V.; Olivier, J.P.; Rodriguez-Reinoso, F.; Rouquerol, J.; Sing, K.S.W. Physisorption of Gases, with Special Reference to the Evaluation of Surface Area and Pore Size Distribution (IUPAC Technical Report). *Pure Appl. Chem.* **2015**, *87*, 1051–1069. [CrossRef]

34. Vuono, D.; Catizzone, E.; Aloise, A.; Policicchio, A.; Agostino, R.G.; Migliori, M.; Giordano, G. Modelling of Adsorption of Textile Dyes over Multi-walled Carbon Nanotubes: Equilibrium and Kinetic. *Chin. J. Chem. Eng.* **2017**, *25*, 523–532. [CrossRef]

35. Al-Ghouti, M.A.; Daana, D.A. Guidelines for the Use and Interpretation of Adsorption Isotherm Models: A Review. *J. Hazard. Mater.* **2020**, *393*, 122383. [CrossRef]

36. Benmahdi, F.; Semra, S.; Haddad, D.; Mandin, P.; Kolli, M.; Bouhelassa, M. Breakthrough Curves Analysis and Statistical Design of Phenol Adsorption on Activated Carbon. *Chem. Eng. Technol.* **2019**, *42*, 355–369. [CrossRef]

37. Hasan, S.H.; Ranjan, D.; Talat, M. Agro-industrial Waste "Wheat Bran" for the Biosorptive Remediation of Selenium through Continuous Up-flow Fixed-bed Column. *J. Hazard. Mater.* **2010**, *181*, 1134–1142. [CrossRef]

38. Girish, C.R.; Ramachandra, M.V. Removal of Phenol from Wastewater in Packed Bed and Fluidized Bed Columns: A Review. *Int. Res. J. Environ. Sci.* **2013**, *2*, 96–100.

39. Srivastava, V.C.; Prasad, B.; Mishra, I.M.; Mall, I.D.; Swamy, M.M. Prediction of Breakthrough Curves for Sorptive Removal of Phenol by Bagasse Fly Ash Packed Bed. *Ind. Eng. Chem. Res.* **2008**, *47*, 1603–1613. [CrossRef]

40. El-Naas, M.H.; Alhaija, M.A.; Al-Zuhair, S. Evaluation of an Activated Carbon Packed Bed for the Adsorption of Phenols from Petroleum Refinery Wastewater. *Environ. Sci. Pollut. Res.* **2017**, *24*, 7511–7520. [CrossRef] [PubMed]

41. Negrea, A.; Mihailescu, M.; Mosoarca, G.; Ciopec, M.; Duteanu, N.; Negrea, P.; Minzatu, V. Estimation of Fixed-bed Column Paramters Pf Breakthrough Behaviours for Gold Recovery by Adsorption onto Modifies/Functionalized Amberlite XAD7. *Int. J. Environ. Res. Public Health* **2020**, *17*, 6868. [CrossRef] [PubMed]

42. Su, Y.; Xiao, W.; Han, R. Adsorption Behavior of Light Green Anionic Dye Using Cationic Surfactant-modified Wheat Straw in Batch and Column Mode. *Environ. Sci. Pollut. Res.* **2013**, *20*, 5558–5568. [CrossRef]

43. Walker, G.M.; Weatherley, L.R. Adsorption of Acid Dyes on to Granular Activated Carbon in Fixed Beds. *Water Res.* **1997**, *31*, 2093–2101. [CrossRef]

44. Fryda, L.; Visserr, R. Biochar for Soil Improvement: Evaluation of Biochar from Gasification and Slow Pyrolysis. *Agriculture* **2015**, *5*, 1076–1115. [CrossRef]

45. Dias, D.; Lapa, N.; Bernardo, M.; Godinho, D.; Fonseca, I.; Miranda, M.; Pinto, F.; Lemos, F. Properties of Chars from the Gasification and Pyrolysis of Rice Waste Streams Towards Their Valorisation as Adsorbent Materials. *Waste Manag.* **2017**, *65*, 186–194. [CrossRef]

46. Bordbar, M. Biosynthesis of Ag/Almond Shell Nanocomposite as a Cost-effective and Efficient Catalyst for Degradation of 4-nitrophenol and Organic Dyes. *RSC Adv.* **2017**, *7*, 180. [CrossRef]

47. Xue, J.; Wu, Y.; Li, M.; Sun, X.; Wang, H.; Liu, B. Characteristic Assessment of Diesel-degrading Bacteria Immobilized on Natural Organic Carriers in Marine Environment: The Degradation Activity and Nutrient. *Sci. Rep.* **2017**, *7*, 8635–8644. [CrossRef]

48. Singhal, A.; Jaiswal, P.K.; Jha, P.K.; Thapliyal, A.; Thakur, I.S. Assessment of *Cryptococcus albidus* for Biopulping of Eucalyptus. *Biochem. Biotechnol.* **2013**, *43*, 735–749.

49. Lorenc-Grabowska, E.; Diez, M.A.; Gryglewicz, G. Influence of Pore Size Distribution on the Adsorption of Phenol on PET-baserd Activated Carbons. *J. Coll. Interf. Sci.* **2016**, *469*, 205–212. [CrossRef] [PubMed]

50. Hsieh, C.-T.; Teng, H. Liquid-phase Adsorption of Phenol onto Activated Carbons Prepared with Different Activation Levels. *J. Coll. Interf. Sci.* **2000**, *230*, 171–175. [CrossRef] [PubMed]

51. Din, A.T.M.; Hameed, B.H.; Ahmad, A.L. Batch Adsorption of Phenol onto Physciochemical-activated Coconut Schell. *J. Hazard. Mat.* **2009**, *161*, 1522–1529.

52. Lorenc-Grabowska, E.; Gryglewicz, G.; Diez, M.A. Kinetics and Equilibrium Study of Phenol Adsorption on Nitrogen-enriched Activated Carbons. *Fuel* **2013**, *114*, 235–243. [CrossRef]

53. Teng, H.; Hsieh, C.-T. Influence of Surface Characteristics on Liquid-phase Adsorption of Phenol by Activated Carbons Prepared from Bituminous Coal. *Ind. Eng. Chem. Res.* **1998**, *37*, 3618–3624. [CrossRef]

54. Lütke, S.F.; Igansi, A.V.; Pegoraro, L.; Dotto, G.L.; Pinto, L.A.A.; Cadaval Jr, T.R.S. Preparation of Activated Carbon from Black Wattle Bark Waste and Its Application for Phenol Adsorption. *J. Environ. Chem. Eng.* **2019**, *7*, 103396–103406. [CrossRef]

55. Tseng, R.-L.; Wu, F.-C.; Juang, R.-S. Liquid-phase Adsorption of Dyes and Phenols Using Pinewood-based Activated Carbons. *Carbon* **2003**, *41*, 487–495. [CrossRef]

56. Lukchis, G.M. Adsorption Systems, Part I-Design by Mass-Transfer-Zone Concept. *Chem. Eng.* **1973**, *30*, 111–116.

57. Kuo, J. *Practical Design Calculations for Groundwater and Soil Remediation*, 2nd ed.; CRC Press: Boca Raton, FL, USA, 2014.

58. You, S.; Ok, Y.S.; Chen, S.S.; Tsang, D.C.W.; Kwon, E.E.; Lee, J.; Wang, C.-H. A Critical Review on Sustainable Biochar System through Gasification: Energy and Environmental Applications. *Bioresour. Technol.* **2017**, *246*, 242–253. [CrossRef]

59. Asudullah, M. Biomass Gasification Gas Cleaning for Downstream Applications: A Comparative Critical Review. *Renew. Sustain. Energy Rev.* **2014**, *40*, 118–132. [CrossRef]

60. Yu, H.; Zhang, Z.; Li, Z.; Chen, D. Characteristics of Tar Formation during Cellulose, Hemicellulose and Lignin Gasification. *Fuel* **2014**, *118*, 250–256. [CrossRef]

Int. J. Environ. Res. Public Health **2021**, *18*, 4247

61. Coll, R.; Salvadò, J.; Farriol, X.; Montané, D. Steam Reforming Model Compounds of Biomass Gasification Tars: Conversion at Different Operating Conditions and Tendency Towards Coke Formation. *Fuel Process. Technol.* **2001**, *74*, 19–31. [CrossRef]
62. Wolfesberger, U.; Aigner, I.; Hofbauer, H. Tar Content and Composition in Produced Gas of Fluidized Bed Gasification of Wood–influence of Temperature and Pressure. *Environ. Prog. Sustain. Energy* **2009**, *28*, 372–379. [CrossRef]

International Journal of
*Environmental Research
and Public Health*

Article

Environmental Compliance and Enterprise Innovation: Empirical Evidence from Chinese Manufacturing Enterprises

Meng Liu, Yun Liu * and Yongliang Zhao *

College of Economics, Jinan University, Guangzhou 510632, China; maxlou1991@163.com
* Correspondence: 18236955609@163.com (Y.L.); ericjue@sohu.com (Y.Z.)

Abstract: This paper embeds environmental compliance factor and compliance cost factor into the M-O monopolistic competition and multi-product firm model to construct a theoretical model applicable to environmental compliance and enterprise innovation. In addition, we also construct a new environmental compliance index. We use the random-effects Tobit model and the double hurdle model to empirically test the micro-data from the Database of China Industrial Enterprises from 1998 to 2013, then we use the Generalized Propensity Score Matching (GPSM) to conduct a robustness test. The robustness conclusion is that environmental compliance has a significant U-shaped relationship with enterprise innovation, which means, environmental compliance will inhibit enterprise innovation on the left of the inflection point of environmental compliance (0.669), while environmental compliance on the right of the inflection point will promote enterprise innovation. The sub-sample regressions show that, enhanced environmental compliance of state-owned enterprises, mature enterprises, core area enterprises and export enterprises with low level of the environmental compliance, makes the greater inhibition to enterprise innovation, and enhanced environmental compliance of above enterprises with high level of the environmental compliance, makes the greater contribution to enterprise innovation. To this end, the government should adopt the policy, from the shallower to the deeper, to promote the construction of environmental compliance, and identify the inflection point of the environmental compliance on enterprise innovation to stimulate the role of environmental compliance in promoting enterprise innovation.

Keywords: environmental compliance; enterprise innovation; Chinese manufacturing enterprises; U-shaped relationship

Citation: Liu, M.; Liu, Y.; Zhao, Y. Environmental Compliance and Enterprise Innovation: Empirical Evidence from Chinese Manufacturing Enterprises. *Int. J. Environ. Res. Public Health* **2021**, *18*, 1924. https://doi.org/10.3390/ijerph18041924

Academic Editors: Pasquale Avino, Andrew S. Hursthouse, Massimiliano Errico, Aristide Giuliano and Hamid Salehi

Received: 14 December 2020
Accepted: 11 February 2021
Published: 17 February 2021

Publisher's Note: MDPI stays neutral with regard to jurisdictional claims in published maps and institutional affiliations.

1. Introduction

The sixth "Global Environment Outlook" (GEO-6) put forward by the United Nations in March 2019 claimed that environmental pollution has become a main issue restricting the steady rise of economic and social development. It is difficult to solve the problem of environmental pollution effectively by means of market regulation, the reason is that the environment has the characteristics of public goods. Although environmental regulation can make up for part of the market failure and increase the profitability of enterprises [1], it would inhibit enterprise innovation [2]. Some scholars (e.g., [3]) believe that technological innovation has become the high priority for the market sector to balance environmental protection and economic benefits. Since the reform and opening up, China has experienced a transition period from rapid economic growth to high-quality economic growth and environmental pollution has more and more inhibitory effects in economic development in China. Therefore, it is an urgent task for China to enter into building an innovative country in an all-round way to evaluate comprehensively and reflect the effectiveness of environmental compliance and implement effective adjustment and reform. Hence, whether environmental compliance promotes enterprise innovation or not is the topic of this paper, by aiming at solving the problem that it is difficult for enterprises to balance environmental protection and economic benefits.

According to much scholars, with the strength of environmental regulation increasing, the enterprise innovation tends to rise, decline, decline first and then rise, or insignificant. The first category is the promoting effect. The Porter hypothesis claimed that environmental protection policies will promote enterprises' technological innovation and rise enterprises' competitiveness. The positive effects of environmental regulation on the macroeconomy are mainly manifested through the innovative behavior of enterprises [4,5]. Then, some scholars found that the improvement of environmental regulation intensity could significantly promote technological innovation (e.g., [6–10]). The second category is the inhibitory effect. Neoclassical economics believes that environmental protection policies will lead to an increase in the production cost of enterprises, and the positive effects of environmental protection policies on the society will all be offset by the increased production cost of enterprises, on the contrary, are not conducive to enterprise innovation and economic development. This proposition is in sharp contrast to the Porter hypothesis. Meanwhile, other scholars also showed that environmental protection policies would inhibit technological innovation. For example, both Zhao and Sun and Shi et al. claimed that environmental regulation would inhibit the efficiency of technological innovation [11,12]. Chen claimed that environmental regulation would inhibit the R&D investment (including the investment of research institutions, the investment of technology introduction and the investment of internal R&D) of industries with different pollution levels [13]. The third category is nonlinear effect. Some studies have showed that environmental regulation has a significant U-shaped relationship with enterprise innovation. With the strength of environmental regulation increasing, the enterprise innovation tends to decline first and then rise (e.g., [14,15]). Su and Zhou claimed that China's formal environmental regulations would promote the innovation performance of firms with higher environmental standards, but inhibit the innovation performance of firms with lower environmental standards [16]. Long and Wan found that environmental regulation would promote the profit margin improvement of large-scale enterprises with low compliance costs and reduce the profit margin of small-scale enterprises with high compliance costs [17]. The fourth category is that there is no significant correlation between environmental policy and enterprise innovation. Kang et al. found that there was no significant correlation between incentive-based environmental regulation policies and enterprise innovation [18]. Sharif used the micro-enterprise data from 1983 to 1992 in the United States to study and found that there was no clear evidence to prove that environmental regulations would promote enterprise innovation [19].

According to the above analysis, much scholars used traditional indicators such as single pollutant discharge per unit output or pollution control expenditure to measure the intensity of environmental regulation. Different from previous studies, we use a new method to measure the environmental compliance index of enterprises by referring to the logic of accounting in measuring abnormal financial indicators. Specifically, the extent to which an enterprise complies with the government's environmental laws and regulations is used to measure the intensity of environmental compliance. The larger the difference between the government's environmental prescribed parameters and the enterprise's environmental indicators, the smaller the enterprise's environmental compliance index will be, and vice versa. The new measurement method of environmental compliance index constructed in this paper is an innovation in this research field.

In order to study the impact of environmental compliance on enterprise innovation, we build a mathematical model to theoretically analyze the impact of environmental compliance on enterprise innovation. According to the results of theoretical analysis, we propose not only hypothesis H_1: Environmental compliance has a significant U-shaped relationship of first suppression and then promotion with enterprise innovation, but hypothesis H_2: In terms of the overall effect, the promotion effect may play a leading role in the effect of environmental compliance on enterprise innovation in China's industrial enterprises. In order to verify the hypothesis based on the theoretical analysis results, we use the random-effects Tobit model and the double hurdle model to empirically test the micro-data from China Industrial Enterprise Database from 1998 to 2013. Hence, by aiming at the improvement of the above points, we will have a certain degree of innovation. Firstly, based on the M-O theoretical model, we relax the original producer behavior hypothesis and embed environmental compliance factors and cost factors into the multi-product enterprise model, then, we construct a theoretical model applicable to environmental compliance and enterprise innovation. Secondly, based on previous studies, this paper refers to the logic of accounting to measure abnormal financial indicators and uses a new method to measure the environmental compliance index of enterprises. Finally, in terms of research methods, we use the random-effects Tobit model and the double hurdle model to empirically test the micro-data from China Industrial Enterprise Database from 1998 to 2013. Then, by using the Generalized Propensity Score Matching (GPSM) method to conduct robustness test. The robustness conclusion is drawn that the impact of environmental compliance on enterprise innovation shows a U-shaped relationship.

The structure of the remaining paper is organized as follows: Section 2 presents the theoretical model. Section 3 shows methodology and data, which illustrates the research methods and models, variable definition, data source and processing and descriptive statistical analysis. Section 4 describes empirical analysis and results, which illustrates benchmark results analysis, heterogenous texts and robustness test. Finally, Section 5 completes the conclusions and policy implications.

2. Theoretical Model

Different from previous studies on environmental regulations, this paper defines environmental compliance as the degree to which an enterprise's production activities comply with environment-related regulations. Due to the different cognition and norms of different enterprises, much enterprises have different environmental compliance index. Based on M-O monopolistic competition and multi-product enterprise model, we relax the original producer hypothesis and embed the environmental compliance factor and cost factor into the multi-product enterprise equation. The equation is similar in theory to the equation proposed by Melitz and Ottaviano [20], but there are the entry of compliance factor and cost factor and the adjustment of entry cost, expected environmental compliance cost and expected environmental violation cost.

2.1. Consumer Equation

This paper retains the hypothesis of the equation proposed by Melitz and Ottaviano [20], that is, economic activity involves only one factor of production labor and two consumer goods. The first commodity is homogeneous, one unit of labor is needed to produce this commodity, and it is sold in perfect competition, and this commodity is chosen as money, so equilibrium wages are equal to 1. The other kind of commodity is horizontally differentiated (new product), continuously provided by N enterprises with returns on scale and monopolistic competition. Its preference is described by the following utility function:

$$U(q_0; q(i), i \in [0, N]) = \alpha \int_0^N q(i)di - \frac{\beta}{2} \int_0^N [q(i)]^2 di - \frac{\gamma}{2} \left[\int_0^N q(i)di \right]^2 + q_0 \tag{1}$$

where $q(i)$ is the quantity of variety $i \in [0, N]$ and q_0 is the quantity of the homogeneous good. α is a measure of the intensity of preferences for the differentiated good with respect to the numéraire and may therefore be viewed as a measure of the size of the market, where γ expresses the substitutability between varieties: the higher γ, the closer substitutes the varieties. $\beta > 0$ indicates that the representative consumers prefer the diversified consumption of their varieties.

According to the model hypothesis, the price equation is finally obtained:

$$p(i) = \alpha - \beta q(i) - \gamma Q \tag{2}$$

where $Q = \int_0^N q(j)dj$ represents the total market output of the differentiated product. j stands for different product categories.

2.2. Producter Equation

The cost involved in production equation mainly includes entry cost, fixed cost and marginal cost. Referring to the logic of Rocha et al. in measuring the labor compliance [21], this paper assumes that marginal cost can be divided into two parts. One part is that the marginal requirement of labor is assumed to be zero. The other part is that the marginal requirement of environmental compliance is assumed to be $\tau_i(h,t)$, and it consists of two parts. And the first part is the penalty cost faced by environmental non-compliance, and it is the increasing function of the possibility of being caught for environmental non-compliance. With the increase of $h(\bullet)$, the probability of being caught for environmental non-compliance and being punished will increase, and the increase of the possibility of being punished will lead to the expansion of environmental compliance costs. The second part is a function of the production time of an firm. In a short period of time, an firm must pay sunk costs if it wants to achieve certain environmental compliance. Hence, it is possible that the revenue brought by environmental compliance will be less than the cost of environmental compliance in the short term. This can explain the conclusion drawn by some scholars that the effect of environmental regulation on enterprise innovation is U-shaped. $h(\bullet)$ denotes the possibility of being caught by the government for environmental non-compliance, where $(\bullet)$ refers to the environmental compliance factor. The bigger $h(\bullet)$ is, the more likely the firm will be caught for environmental noncompliance. If $h(\bullet) = 1$ means that it is an established fact that the enterprise is caught, then the enterprise will lose its production capacity. If $h(\bullet) = 0$, it means that the firm cannot be caught and has full production capacity. We define $1 - h(\bullet)$ as environmental compliance index. The larger $1 - h(\bullet)$ is, the stronger environmental compliance will be, and the smaller expected marginal requirement will be faced by enterprises. We assume that the enterprise is faced with the same entry cost G, the expected marginal cost of production is $\tau_i(h,t)$, and the fixed cost of each type of product is F, then the total production cost of enterprise i is $G + \tau_i(h,t) + F$. To avoid losing the generality of the equation, we readjust the demand intercept α to zero, and the adjusted entry cost G becomes zero. The firm profit equation can be expressed as follows:

$$\Pi(i) = (1 - h(\bullet))p(i)q(i) - \tau_i(h,t)q(i) - F \tag{3}$$

where, $\tau_i(h,t) = C(1-h) + \phi/t(1-h)$ is a function of expected environmental compliance cost $C(1-h)$, expected environmental violation cost $\phi/t(1-h)$ and time t, and ϕ is a constant. Expected environmental compliance cost $C(1-h)$ is an increasing function of environmental compliance $(1-h(\bullet))$. If time T is fixed, the expected marginal cost of a firm is negatively correlated with environmental compliance, that is, with the increase of environmental compliance, the expected cost of environmental violation $\phi/t(1-h)$ decreases.

Combining Equations (2) and (3), we can obtain the output q^* and price p^* under general equilibrium, and then the number of equilibrium producers N^* is given by:

$$N^* = \frac{\sqrt{\beta/F(1-h(\bullet))} \cdot [\alpha(1-h(\bullet)) - \tau_i(h,t)] - 2\beta}{\gamma} \tag{4}$$

Entry and exit are free so that profits are zero in equilibrium. Thus, the optimal q^* and p^* are given by:

$$q^* = \frac{1}{\sqrt{(1-h(\bullet))}}\sqrt{F/\beta}; p^* = \frac{1}{\sqrt{(1-h(\bullet))}}\sqrt{F \cdot \beta} \tag{5}$$

2.3. In Equilibrium

Combining $Q = Q_i + Q_{-i} = \int_{\omega \in \pi_i} q_i(\omega)d\omega + (M-1)|\pi|q$ and utility are maximized, where, ω represents the product category and M represents the number of enterprises with a given multi-product. Q_i is the total output of firm i's diversified products, and Q_{-i} is the total output of firm i's competitors. $\pi \subseteq R_+$, represents the collection of all product categories produced by enterprise i ($i = 1......M$), we get an equilibrium p^* and q^*:

$$\begin{cases} q^* = \frac{\alpha(1-h(\bullet))-r(1-h)(M-1)|\pi|q-\tau_i(h,t)|\pi|}{2(1-h(\bullet))(\beta+\gamma|\pi_i|)} \\ p^* = \frac{\alpha(1-h(\bullet))-r(1-h)(M-1)|\pi|q+\tau_i(h,t)|\pi|}{2(1-h(\bullet))} \end{cases} \tag{6}$$

This paper assumes that the cost for environmental compliance in the production process of the firm can be ignored, then the product and price functions of the firm are given by:

$$\begin{cases} q^* = \frac{\alpha(1-h(\bullet))-\gamma(1-h)(M-1)|\pi|q}{2(1-h(\bullet))(\beta+\gamma|\pi_i|)} \\ p^* = \frac{\alpha(1-h(\bullet))-\gamma(1-h)(M-1)|\pi|q}{2(1-h(\bullet))} \end{cases} \tag{7}$$

at this point, we have the profit function:

$$\Pi(i) = (1 - h(\bullet))p(i)q(i) - F \tag{8}$$

Using (7) and (8), we have the product range of profit maximization, so that the product range of the M-th enterprise $|\pi_i|^*$ is given by Equation (9) (the specific solution process can be seen in the derivation of the Melitz and Ottaviano equation)

$$|\pi_i|^* = \frac{\alpha\sqrt{(1-h(\bullet))} \cdot \sqrt{\beta/F} - 2\beta}{\gamma(M+1)} \tag{9}$$

According to Schumpeter's innovation theory, the expansion of product range can measure the enhancement of innovation ability. Equation (9) shows that the product range $|\pi_i|^*$ produced by the enterprise will expand with the enhancement of environmental compliance $1 - h(\bullet)$, that is, with the enhancement of environmental compliance, the product range of the firm will expand accordingly. This is also consistent with the connotation of the Porter hypothesis.

In fact, ignoring the cost of environmental compliance is an absolute idealization that does not exist in real life. Therefore, considering $\tau_i(h,t) \neq 0$, the impact of environmental compliance on enterprise innovation depends on the size of expected environmental compliance cost $C(1-h)$ and expected environmental violation cost $\phi/t(1-h)$. If $C(1-h) \geq \phi/t(1-h)$, environmental compliance is not conducive to enterprise innovation, or even inhibits. If $C(1-h) < \phi/t(1-h)$, environmental compliance will promote enterprise innovation. Considering that enterprises need to buy a large amount of machinery and equipment in the early stage of environmental compliance, and the newly formulated punishment standards for environmental violation will not be too strict, the initial cost of environmental compliance of enterprises $C(1-h)$ will be greater than the cost of environmental violation $\phi/t(1-h)$. In the later stage of environmental compliance, enterprises only need to improve their machinery and equipment, and the perfect environmental compliance standard will impose more and more strict penalties on enterprises violating environmental regulations. In this case, the cost of environmental compliance $C(1-h)$ will be less than the cost of environmental violation $\phi/t(1-h)$. Therefore, we propose the first hypothesis:

Hypothesis 1 (H1): *Environmental compliance has a significant U-shaped relationship of first suppression and then promotion with enterprise innovation.*

Given early implementation of environmental compliance regulations, most of China's manufacturing companies may have already crossed the environmental compliance inflection point. In the 1980s, China began to explore the prevention and control of industrial pollution, including China's first environmental protection conference held in 1982 and the subsequent "Trial Implementation of the Environmental Protection Law" and the "Several Decisions on Environmental Protection Work" promulgated in 1984. Therefore, we propose the second hypothesis:

Hypothesis 2 (H2): *In terms of the overall effect, the promotion effect may play a leading role in the effect of environmental compliance on enterprise innovation in China's industrial enterprises.*

3. Methodology and Data

3.1. Research Methods and Models

Considering the characteristics of the data set (in the Database of Chinese Industrial Enterprises, the dependent variable greater than 0 has a wide range of distribution, which can be regarded as a continuous variable whose condition depends on some factors and obeys normal distribution. In addition, the actual dependent variable cannot be negative, so deleting all the observed values of the dependent variable displayed as 0 will lead to biased estimation. Therefore, the Tobit model of random effects panel we chose is valid) and the fact that the Tobit model allows the estimated parameters to be unbiased and consistent, this paper uses the Tobit model to test the impact of environmental compliance on enterprise innovation. It is usually impossible for the nonlinear model of fixed effects to obtain a consistent estimate panel data [22]. Moreover, the preliminary test results of panel data of China' industrial enterprises from 1998 to 2013 in this paper show that the maximum likelihood ratio rejects the original hypothesis that the mixed section Tobit model has no difference. In order to prevent spurious regression, we also conducted panel unit root test and co-integration test, and the results prove that the panel data used in this paper is stable and co-integration. Therefore, we select Tobit model of random effects panel. In view of the above analysis, used with Equation (9), we have the empirical model:

$$Pnov_{c,i,t} = \alpha + \beta_1 H_{c,i,t-1} + \beta_2 H^2_{c,i,t-1} + \gamma CV_{c,i,t-1} + \varepsilon_{c,i,t} \tag{10}$$

where, $Pnov_{c,i,t}$ represents the innovation level of enterprises in city c, industry i and year t, $H_{c,i,t-1}$ refers to 1 period lag environmental compliance. Considering the marginal requirement of environmental compliance $\tau_i(h,t) \neq 0$ and the non-linear relationship between environmental compliance and enterprise innovation confirmed by some scholars (e.g., [23]), we control the second term of 1 period lag environmental compliance $H^2_{c,i,t-1}$. $CV_{c,i,t-1}$ represents 1 period lag control variable, and $\varepsilon_{c,i,t}$ represents the random error term.

3.2. Variable Definition

3.2.1. Enterprise Innovation

According to existing studies, financial indicators related to new products, success rate of scientific research subjects or intellectual property data were chosen as indicators of enterprise innovation performance. In this paper, enterprise innovation is measured for the proportion of sales revenue of new products in the sales revenue of enterprises, which according to the first-level index of the evaluation index system of enterprise independent innovation ability put forward by The National Bureau of Statistics of China. In order to ensure the robustness of the empirical model, we also use enterprises' R&D investment as a substitute variable for enterprise innovation ($Pnov_{c,i,t}$).

3.2.2. Environmental Compliance Index

Environmental compliance refers to the degree to which an enterprise complies with the government's environmental laws and regulations. The larger the difference between the government's environmental regulation and the enterprise's environmental indicators, the smaller the enterprise's environmental compliance will be, and vice versa. This research assumes that the environmental expenses of a firm remain unchanged in the same region, industry and year. The mean environmental expenses of the region, industry and year are taken as the compliance standard, and the grouping scope is limited to the same region, industry and year. The main reason is that there are significant differences in the dimension of environmental costs of different firms. Although the assumption that "environmental costs of enterprises remain unchanged in the same region, industry and year" is strong, it conforms to the general equilibrium principle in reality [24]. Therefore, environmental compliance is measured by the difference between the proportion of sewage charge in product sales revenue of the enterprise and the average proportion of sewage charge in product sales revenue of the industry. According to the definition of environmental compliance $(1 - h(\bullet))$ in this research, if the difference between the actual average environmental cost of an enterprise and the actual average environmental cost of the industry is positive, it means that the probability of an enterprise being caught for non-compliance is zero, and then the compliance index $1 - h(\bullet)$ is 1. If the difference between the two is negative, it means that the probability of being caught for environmental noncompliance is greater than zero, and the smaller the negative number is, the more noncompliance the enterprise environment is. To this end, we use the logic of enterprise abnormal investment measurement () to standardize the data [25]. The specific formula of environmental compliance index after treatment can be written as follows:

$$\widetilde{H}_{k,i} = \begin{cases} 1, H_{k,i} - \overline{H}_i \geq 0 \\ 1 + \left\langle H_{k,i} - \overline{H}_i \right\rangle^{standard}, H_{k,i} - \overline{H}_i < 0 \end{cases} \tag{11}$$

where $\widetilde{H}_{k,i}$ represents the environmental compliance index, which is expressed as the difference between the proportion of pollutant discharge fee in product sales revenue of the firm and the proportion of pollutant discharge fee in product sales revenue of the industry (due to the lack of uniform parameters stipulated by the government, this research measures environmental compliance with the difference between the proportion of pollutant discharge fee in product sales revenue of an enterprise and the proportion of pollutant discharge fee in product sales revenue of an industry. Although this method has some endogenous problems, it can also reflect the environmental compliance intensity to a certain extent. For example, the proportion of pollutant discharge fee in product sales revenue of enterprise A is lower than the proportion of pollutant discharge fee in product sales revenue of the industry in which enterprise A is located. According to the assumption of this paper, enterprise A may be in violation of environmental compliance behaviors to A certain extent without considering reasonable avoidance of taxes. Moreover, the proportion of pollutant discharge fee in product sales revenue of enterprise A is lower than that of the industry in which enterprise A is located, and the higher the proportion of pollutant discharge fee in product sales revenue of the industry in which enterprise A is located, the greater the possibility that enterprise A has environmental compliance violations. In the empirical process, we used the robust standard error of aggregation at the industry level, and calculated the conventional robust standard error after controlling the industry fixed effect, that is, heteroscedasticity was allowed between groups, and the heteroscedasticity could be corrected to some extent by clustering within each group to reduce the variance). $\overline{H}_{i,j}$ represents the average value of the industry environmental compliance index, and i represents different industries. $\left\langle H_{k,j,i} - \overline{H}_{i,j} \right\rangle^{standard}$ represents the difference between the actual sewage charge of the enterprise after standardized treatment and the mean value of the industry (Table 1).

Table 1. Usage description of main variables.

Variable Types	Variable Code	Variable Name	Explanation	Expected Symbol
Explained variable	Enterprise innovation	$Pnov_{c,i,t}$	The ratio of new product sales revenue to the total product sales revenue.	
Explanatory variables	Environmental compliance index	$H_{c,i,t-1}$	Calculated using the calculation equation in this research. See Equation (11) for details, and 1 period lag.	+
	Enterprise size	$lnSize_{c,i,t-1}$	The total fixed assets take logarithm, and 1 period lag.	+
	Total factor productivity (TFP)	$lnTfp_{c,i,t-1}$	LP method is used to calculate, and 1 period lag.	+
	The age of operation	$lnAge_{c,i,t-1}$	The age of enterprise operation take logarithm, and 1 period lag.	+
Control variables	The number of employees	$lnL_{c,i,t-1}$	The number of employees in an enterprise takes the logarithm, and 1 period lag.	+
	Asset-labor ratio	$lnKL_{c,i,t-1}$	The ratio of assets to labor is logarithmic, and 1 period lag.	+
	Current ratio	$Current_{c,i,t-1}$	The ratio of current assets to current liabilities, and 1 period lag.	$+/-$
	Financing constraints	$Fin_{c,i,t-1}$	The ratio of interest expense to fixed asset net value, and 1 period lag.	+
	Subsidies income	$lnSubsidy_{c,i,t-1}$	The ratio of Subsidy income to sales income ratio, and 1 period lag.	+
	Export intensity	$lnExport_{c,i,t-1}$	Take the logarithm of export delivery value, and 1 period lag.	+

Source: The Database of China Industrial Enterprises (With the total factor productivity estimation method put forward by Berry, Levinsohn and Pakes (LP method for short) [26], LP method uses the intermediate input variable of the firm as the adjustable factor input when the firm is impacted by productivity. The total factor productivity measured by the LP method can be expressed as: $tfp_{it}^{LP} = Y - \hat{\beta}_{it}^{LP} - \hat{\beta}_{l}^{LP}L_{it} - \hat{\beta}_{k}^{LP}K_{it} - \hat{\beta}_{m}^{LP}M_{it} - \eta_{it}$. We use industrial added value to replace the output of enterprises, and use the producer price index of products in different industries for the reduction. Referring to the treatment method of Brandt et al., we use the perpetual inventory method to calculate the capital stock in order to achieve detailed and accurate TFP estimation).

3.3. Data source and Processing

In view of the fact that the empirical research of this research needs to use indicators such as industrial enterprise value-added, product sales income, and industrial enterprise value-added, and in order to follow the completeness and continuity of the data, this paper selects the data of industrial enterprises above a certain scale the Database of China Industrial Enterprises from 1998 to 2013 (although the database of Chinese Industrial Enterprises has been updated to 2014 (the open data of The National Bureau of Statistics of China is only available until 2014), and there is no experts has yet used the 2014 data in view of the inconsistency between the 2014 data and the previous years' data. In addition, the statistical data before 1998 were not complete, and there were many key variables missing about this research. Therefore, we selected the interval data from 1998 to 2013). We refer to intertemporal matching method of Brandt et al. and combine Generally Accepted Accounting Principles (GAAP) [27], the sample is deleted in one of the following situations: the enterprise code is missing; total assets, paid-up capital, net fixed assets, sales are missing or less than 0; the number of employees in the enterprise is less than 8; the total assets are less than current assets or less than the net value of fixed assets; the enterprise age is obviously wrong (the establishment year is behind the reporting year; the month is less than 1 or larger than 12); the enterprises are in abnormal business; the observed value of the main business income is less than 5 million yuan from 1998 to 2010 and the industrial enterprises with the main business income is less than 20 million yuan from 2011 to 2013. We eliminate the extreme value beyond the range of 5~95% of the key index. When there is no industrial output value (e.g., 2004), the calculation is carried out by using the industrial output value estimation formula [28]. We use interpolation method to complete the missing indicators from 1998 to 2013 (the concrete method of interpolation we used is as follows: if the enterprise appears in the previous year, the missing indicators will be supplemented according to the growth rate of product sales revenue. If the enterprise does not appear in the previous year, the missing indicators will be made up according to the growth rate of product sales revenue in the industry where the enterprise is located). Finally, the total number of effective samples obtained was 3,318,020, among which the sample size increased from 67,797 in 1998 to 259,468 in 2013.

In order to capture the unpredictable factors that change with the industry characteristics, we classify state-owned and non-state-owned manufacturing enterprises above designated size into 30 specific manufacturing industries according to the GB/T4754-2011 two-digit classification published by the National Bureau of Statistics of China, with 30 codes ranging from 13 to 42. We classify 30 specific manufacturing industries into three categories of labor-intensive, capital-intensive and technology-intensive, and the classification results are shown in Table 2 below. During the sample period, the corresponding industry code of industrial enterprises above the size has been adjusted for three times and finally adjusted to be the enterprise sample of the two-digit manufacturing industry in the industry classification standard of the national economy in 2011. Although the sample data only reaches 2013, the sample size is huge, and the panel data can reduce the endogeneity problem of empirical regression to some extent.

Table 2. Industry classification by factor intensity (30 industries).

Labor-Intensive Industry	Capital-Intensive Industry	Technology-Intensive Industry
Agricultural and sideline food processing industry (13) Food manufacturing industry (14) Wine, beverage and refined tea manufacturing industry (15) Tobacco Products industry (16) Textile industry (17) Textile clothing, clothing industry (18) Leather, fur, feather and their products and footwear (19) Wood processing and wood, bamboo, rattan, palm, grass products industry (20) Furniture manufacturing (21) Paper and paper products industry (22) Printing and recording media reproduction industry (23) Culture and education, industrial beauty, sports and entertainment manufacturing (24) Rubber and plastic products (29) Other manufacturing industries (41) Comprehensive utilization of waste resources (42)	Petroleum processing, coking and nuclear fuel processing industries (25) Non-metallic mineral products industry (30) Ferrous metal smelting and calendering industry (31) Non-ferrous metal smelting and calendering industry (32) Metal products industry (33) General equipment manufacturing (34) Special equipment manufacturing (35) Manufacturing of railway, shipping, aerospace and other transportation equipment(37) Instrumentation and culture, office machinery manufacturing (40)	Manufacturing of chemical raw materials and chemical products (26) Pharmaceutical manufacturing (27) Chemical fiber manufacturing (28) Automobile manufacturing industry (36) Electrical machinery and equipment manufacturing (38) Manufacturing of computers, communications and other electronic equipment (39)

Source: China Industry Business Performance Data.

3.4. Descriptive Statistical Analysis

In order to capture the distribution of environmental compliance index in detail, we calculated the characteristics of the cleaned samples. The descriptive statistical results of the main variables are shown in Table 3. There were 3,317,975 observed samples of enterprise innovation, with a standard deviation of 0.083, which indicating that there was a small difference in the innovation level of the sample enterprises. The minimum value and maximum value are 0 and 0.339 respectively, which indicates that the innovation level of Chinese manufacturing enterprises is generally low, which also conforms to the practical inspiration of this goal we study to generally improve the innovation level of China's enterprises. The standard deviation of environmental compliance in the first period after standardized treatment is 0.342, indicating that the compliance distribution of sample enterprises is relatively uniform. The minimum and maximum values of environmental compliance are 0 and 1, which indicating that the construction of environmental compliance indicators in this paper is in line with expectations. It is an established fact that enterprises with low environmental compliance (less than 0) will be caught. In this case, enterprises will lose production capacity, namely, $h(\bullet) \to 1$. The standard deviation of the enterprise's operating life and financing constraints is less than 1, and the distribution is relatively uniform. The descriptive statistics of the remaining variables are similar to previous studies, so it will not be repeated here.

Table 3. Descriptive statistics of main variables.

Variables	Observations	Mean	Standard Deviation	Minimum	Maximum
$Pnov_{c,i,t}$	3,317,975	0.027	0.083	0	0.339
$H_{c,i,t-1}$	2,586,771	0.736	0.342	0	1
$lnAge_{c,i,t-1}$	2,506,420	1.869	0.880	0	7.602
$lnSize_{c,i,t-1}$	2,574,927	8.695	1.708	-0.095	18.94
$lnTfp_{c,i,t-1}$	2,563,855	8.154	1.033	2.957	16.54
$lnL_{c,i,t-1}$	2,585,580	4.930	1.108	0	12.29
$lnKL_{c,i,t-1}$	2,563,874	5.164	1.678	-9.879	9.245
$Current_{c,i,t-1}$	2,537,542	2.157	5.624	0.001	64.09
$Fin_{c,i,t-1}$	2,486,844	0.066	0.167	-0.056	1.633
$lnSubsidy_{c,i,t-1}$	2,255,394	0.787	2.070	0	15.39
$lnExport_{c,i,t-1}$	2,274,359	2.974	4.575	0	19.04

Note: Except the dependent variable, all the other variables 1 period lag.

4. Empirical Analysis and Findings

4.1. Benchmark Results Analysis

In order to test the rationality of random-effects Tobit model we constructed, we use the LR test to determine which estimation method is more suitable for the mixed Tobit model and the random-effects Tobit model. The test of LR rejects the null hypothesis (p value is 0), so there is an individual effect in enterprise innovation. The Tobit model of random effects panel adopted in this paper is more effective, and the basic regression results are shown in Table 4. We find that the empirical results of the innovation level decision-making of the two-column model in column (8) are basically the same as those in column (5), which confirms that the empirical model constructed in this paper is robust. According to the theory proposed by Scherer [29], we realize that compliance may have a nonlinear relationship with enterprise innovation, so columns (5) and (6) are regression results after controlling the quadratic term of the environment compliance ($H^2_{c,i,t-1}$). Therefore, column (5) has a more reasonable test result for controlling the second term of environment compliance. On the one hand, the regression coefficient of the quadratic term of compliance is relatively positive and significant at the 1% significance level. On the other hand, the 95% confidence interval of the inflection point calculated according to the columns (1) and (2) or the columns (5) and (6) is within the range of the independent variable. Therefore, the regression coefficient of the quadratic term of environmental compliance satisfies the assumption of the non-linear relationship between environmental compliance and enterprise innovation [30], and the impact of environmental compliance on enterprise innovation has a non-linear relationship. In order to further confirm the non-linear relationship of the impact of environmental compliance on corporate innovation, we use the three statistics of LM, LMF, and LRT to perform non-linear tests on columns (5) and (6), and all the results reject the null hypothesis that there is no non-linear relationship. According to the above analysis, environmental compliance has a significant non-linear relationship with corporate innovation, so there is no bias in the empirical model.

Table 4. Environmental compliance and enterprise innovation: Benchmark results.

Variable	Random-Effect Tobit Model						Double Hurdle	
	(1)	**(2)**	**(3)**	**(4)**	**(5)**	**(6)**	**Willingness**	**Level**
$H_{c,i,t-1}$	−0.114 ***	−0.119 ***	−0.103 ***	−0.104 ***	−0.122 ***	−0.128 ***	−1.448 ***	−0.194 ***
	(0.001)	(0.001)	(0.000)	(0.000)	(0.001)	(0.001)	(0.025)	(0.008)
$H_{c,i,t-1}^2$	0.109 ***	0.113 ***			0.091 ***	0.096 ***	1.251 ***	0.145 ***
	(0.001)	(0.001)			(0.001)	(0.001)	(0.021)	(0.006)
$lnAge_{c,i,t-1}$			0.001 ***	0.001 ***	0.001 ***	0.001 ***	0.144 ***	−0.001
			($\approx$0)	($\approx$0)	($\approx$0)	($\approx$0)	(0.002)	(0.001)
$lnSize_{c,i,t-1}$			0.002 ***	0.002 ***	0.002 ***	0.002 ***	0.043 ***	−0.006 ***
			($\approx$0)	($\approx$0)	($\approx$0)	($\approx$0)	(0.001)	(0.000)
$lnTfp_{c,i,t-1}$			−0.004 ***	−0.004 ***	−0.004 ***	−0.004 ***	−0.088 ***	−0.002 ***
			($\approx$0)	($\approx$0)	($\approx$0)	($\approx$0)	(0.0017)	(0.001)
$lnL_{c,i,t-1}$			0.002 ***	0.002 ***	0.002 ***	0.002 ***	0.176 ***	0.001 *
			($\approx$0)	($\approx$0)	($\approx$0)	($\approx$0)	(0.002)	(0.001)
$lnKL_{c,i,t-1}$			0.011 ***	0.011 ***	0.011 ***	0.011 ***	0.100 ***	0.012 ***
			($\approx$0)	($\approx$0)	($\approx$0)	($\approx$0)	(0.001)	(0.000)
$Current_{c,i,t-1}$			0.000 ***	0.000 ***	0.000 ***	0.000 ***	−0.005 ***	−0.000 **
			($\approx$0)	($\approx$0)	($\approx$0)	($\approx$0)	(0.000)	($\approx$0)
$Fin_{c,i,t-1}$			−0.033 ***	−0.034 ***	−0.034 ***	−0.034 ***	0.239* **	−0.03 8***
			(0.000)	(0.000)	(0.000)	(0.000)	(0.009)	(0.003)
$lnSubsidy_{c,i,t-1}$			0.001 ***	0.001 ***	0.001 ***	0.001 ***	0.041 ***	0.002 ***
			($\approx$0)	($\approx$0)	($\approx$0)	($\approx$0)	(0.001)	(0.000)
$lnExport_{c,i,t-1}$			0.001 ***	0.001 ***	0.001 ***	0.001 ***	0.022 ***	0.000
			($\approx$0)	($\approx$0)	($\approx$0)	($\approx$0)	(0.000)	($\approx$0)
Constant	0.011 ***	0.020 ***	0.013 ***	0.021 ***	0.016 ***	0.024 ***	−2.459 ***	0.224 ***
	(0.000)	(0.000)	(0.001)	(0.001)	(0.001)	(0.001)	(0.0130)	(0.004)
Industry characteristics	Control 1	Control 2	Control 1	Control 2	Control 1	Control 2	Control 1	Control 2
/sigma_u	0.050 ***	0.050 ***	0.055 ***	0.055 ***	0.055 ***	0.055 ***		
	($\approx$0)	($\approx$0)	($\approx$0)	($\approx$0)	($\approx$0)	($\approx$0)		
/sigma_e	0.047 ***	0.047 ***	0.049 ***	0.049 ***	0.049 ***	0.049 ***		
	($\approx$0)	($\approx$0)	($\approx$0)	($\approx$0)	($\approx$0)	($\approx$0)		
sigma								0.151 ***
								(0.0003)
ρ	0.534	0.532	0.558	0.556	0.558	0.555		
Log-L	3,707,020.7	3,623,543.7	2,785,268.5	2,721,770.2	2,785,418.5	2,721,995.2	3,163,286.3	3,164,526.4
Observations	2,586,729	2,531,584	2,023,259	1,979,583	2,023,259	1,979,583	2,418,112	2,418,112

Note: Both explanatory variables and control variables 1 period lag, where $H_{c,i,t-1}^2$ represent the second-order 1 period lag environmental compliance items. This research adopts the robustness standard errors aggregated at the industry level. ***, **, * indicate significant at 1%, 5%, and 10% respectively, and the robust standard error is reported in parentheses, the same below.

It is easy to find that the regression coefficient of 1 period lag operating life ($lnAge_{c,i,t-1}$) is 0.001 and is significant at the 1% significance level with the observation in column (5). The reason seems that enterprises with long operating years are at a relatively mature stage of enterprise innovation, and their marginal cost of innovation is relatively low, so operating years are positively correlated with corporate innovation. The size of the firm ($lnSize_{c,i,t-1}$) 1 period lag is significantly positively correlated with firm innovation, which indicates that the larger the firm, the higher its level of innovation. This is consistent with the conclusion of competition inhibition innovation under Schumpeter's monopoly market research, that is, the larger the enterprise scale, the weaker the market competition effect, and the stronger the enterprise's innovation ability. To our surprise, the regression coefficient of 1 period lag total factor productivity ($lnTfp_{c,i,t-1}$) is significantly negative. The reason seems like factor market distortions lead to the inhibitory effect of total factor productivity on enterprise innovation, that is, market rent-seeking brought about by factor market distortions may cause enterprises with higher total factor productivity to invest in market rent-seeking activities rather than enterprise innovation activities. We suppose that total factor productivity has a nonlinear relationship with enterprise innovation. Therefore, we introduce the quadratic term of total factor productivity into the empirical model, and the empirical results confirm our conjecture that the influence of total factor productivity on enterprise innovation has a nonlinear relationship (where control 1 refers to the control of classifying state-owned and non-state-owned manufacturing enterprises above designated size into 30 specific manufacturing industries according to the two-digit classification published by the China's National Bureau of Statistics. Control 2 refers to the control of labor-intensive, capital-intensive and technology-intensive three industry categories). The regression coefficient of 1 period lag current ratio ($Current_{c,i,t-1}$) is significantly negative, and the possible reason is that excessive financing constraints lead to a greater debt pressure on enterprises and its lack of innovation motivation. The remaining control variables are in line with the expectations of this research, that is, the number of employees, per capita capital, export intensity, and subsidy income will all promote enterprise innovation.

The comparison results of column (5) and column (6) show that controlling the characteristics of different industries has little effect on the overall regression results (where control 1 refers to the control of classifying state-owned and non-state-owned manufacturing enterprises above designated size into 30 specific manufacturing industries according to the two-digit classification published by the China's National Bureau of Statistics. Control 2 refers to the control of labor-intensive, capital-intensive and technology-intensive three industry categories), and the regression model is relatively robust. The empirical results of column (5) in Table 4 indicate that the regression coefficient of the first-stage environmental compliance item is significantly negative, and the regression coefficient of the second-order environmental compliance item is significantly positive. The inflection point calculated according to column (5) is 0.669, and its 95% confidence interval is within the range of the independent variable. This indicates that environmental compliance has a U-shaped relationship with enterprise innovation. The environmental compliance will inhibit corporate innovation, when the enterprise environmental compliance index is lower than 0.669.

Taking the empirical results of column (5) in Table 4 as an example, we conducted a more in-depth analysis of the distribution characteristics of specific enterprises and industries around the inflection point of environmental compliance. Firstly, the sample of companies on the left of the inflection point of environmental compliance is 925,268, accounting for 26.53% of the total number of companies, and the sample of companies on the right of the inflection point of environmental compliance is 2,562,702, accounting for 73.47% of the total number of companies. Therefore, in terms of the effect of environmental compliance of China's manufacturing enterprises above designated size on enterprise innovation, environmental compliance has a promoting effect on enterprise innovation in a relatively large number of enterprise samples, and it has an inhibitory effect on corporate innovation in a relatively small sample of industries. To some extent, this can be interpreted as, from the overall effect, the promotion effect may play a leading role in the impact effect of environmental compliance on enterprise innovation of industrial enterprises above a large scale in China. Secondly, the proportion of enterprises on both sides of the inflection point of environmental compliance in the total number of enterprises in the industry in Figure A1 (Appendix A) shows that the sample industries in which environmental compliance has a promoting effect on enterprise innovation are mainly distributed in labor-intensive and capital-intensive industries. However, there is a polarization in technology-intensive industries. For instance, the distribution of manufacturing of chemical raw materials and chemical products, pharmaceutical manufacturing and chemical fiber manufacturing and other technology-intensive industries is relatively large. In contrast, the sample industries in which environmental compliance has a inhibition effect on enterprise innovation are mainly distributed in automobile manufacturing, electrical machinery and equipment manufacturing, and computer, communications and other electronic equipment manufacturing and other technology-intensive industries, and relatively less in labor-intensive and capital-intensive industries. The comparison results further indicate that from the point of view of China's current development facts, China's industrial sector is in the process of transformation and upgrading as the technology-intensive manufacturing industry develops and grows. Although governments at all levels will give various guidance and incentive policies to technology-intensive manufacturing enterprises to some extent, half of technology-intensive industries, including automobile manufacturing, electrical machinery and equipment manufacturing, and computer, communications and other electronic equipment manufacturing, still face the low environmental compliance index trap.

4.2. Heterogenous Texts

In order to investigate the impact of environmental compliance on enterprise innovation in heterogeneous enterprises, this research divides enterprise samples into state-owned enterprises and non-state-owned enterprises according to the structure of enterprise property rights. According to the enterprise's operating years, whether the enterprise is located in the core area and whether the enterprise exports, the enterprise samples are divided into mature enterprises and non-mature enterprises, core enterprises and non-core enterprises, and export enterprises and non-export enterprises. The results of heterogeneity test about environmental compliance affecting enterprise innovation are shown in Table 5.

Table 5. Environmental compliance and enterprise innovation: Heterogeneity test results.

Variables	(1) State-Owned	(2) Non-State-Owned	(3) Mature	(4) Immature	(5) Core Areas	(6) Non-Core Areas	(7) Export	(8) Non-Export
$H_{c,i,t-1}$	−0.123 ***	−0.114 ***	−0.123 ***	−0.114 ***	−0.122 ***	−0.121 ***	−0.133 ***	−0.118 ***
	(0.001)	(0.002)	(0.001)	(0.002)	(0.001)	(0.003)	(0.002)	(0.001)
$H_{c,i,t-1}^2$	0.117 ***	0.112 ***	0.117 ***	0.112 ***	0.117 ***	0.116 ***	0.123 ***	0.116 ***
	(0.001)	(0.002)	(0.001)	(0.002)	(0.001)	(0.002)	(0.002)	(0.001)
Constant	0.014 ***	−0.005 ***	0.014 ***	−0.005 ***	0.020 ***	0.002	0.002	0.019 ***
	(0.001)	(0.001)	(0.001)	(0.001)	(0.001)	(0.002)	(0.002)	(0.001)
Control variables	Control	Control	Control	Control	Control	Control	Control	Control
Industry characteristics	Control 1	Control 1	Control 1	Control 1	Control 1	Control 1	Control 1	Control 1
sigma_u	0.055 ***	0.054 ***	0.055 ***	0.054 ***	0.054 ***	0.058 ***	0.064 ***	0.055 ***
	($\approx$0)	(0.0001)	($\approx$0)	(0.000)	($\approx$0)	(0.000)	(0.000)	($\approx$0)
sigma_e	0.050 ***	0.041 ***	0.050 ***	0.041 ***	0.048 ***	0.051 ***	0.059 ***	0.041 ***
	($\approx$0)	($\approx$0)	($\approx$0)	($\approx$0)	($\approx$0)	($\approx$0)	($\approx$0)	($\approx$0)
ρ	0.552	0.632	0.552	0.632	0.552	0.567	0.536	0.641
Log-L	2,082,950.3	684,766.71	2,082,950.3	684,766.71	2,173,377.8	615,479.08	744,259.07	1,963,919.6
Observations	1,535,020	488,239	1,535,020	488,239	1,560,682	462,577	621,838	1,323,456

*** indicate significant at 1%.

The primary coefficients of environmental compliance of different types of enterprises are all significantly negative, and the secondary coefficients of environmental compliance are all significantly positive. Specifically, the primary environmental compliance coefficients of state-owned enterprises, mature enterprises, core area enterprises, and export enterprises are all lower than the primary environmental compliance coefficients of non-state-owned enterprises, immature enterprises, non-core area enterprises, and non-export enterprises. This means that among companies with lower levels of environmental compliance, environmental compliance has a greater inhibitory effect on the innovation of state-owned enterprises, mature enterprises, core area enterprises and export enterprises. The possible reason is that compared with non-state-owned enterprises, immature enterprises, non-core area enterprises and non-export enterprises, state-owned enterprises, mature enterprises, core area enterprises and export enterprises tend to have larger scale and more sectors. Therefore, these types of enterprises below the inflection point of environmental compliance often need to spend huge costs to purchase large-scale machinery and equipment to reduce environmental pollution. Therefore, the environmental compliance cost $C(1 - h)$ is relatively high, and environmental compliance has a greater inhibitory effect on these types of enterprise innovation.

Compared with the non-state-owned enterprises, immature, enterprises in non-core areas and non-export enterprises, state-owned enterprises, mature, enterprises in core areas (according to China's "Seventh Five-year Plan" adopted at the fourth Session of the sixth National People's Congress, the core areas refer to the 11 provinces (cities) in the eastern region, including Beijing, Tianjin, Hebei, Liaoning, Shanghai, Jiangsu, Zhejiang, Fujian, Shandong, Guangdong and Hainan, while the non-core areas refer to the provinces (cities) in the central and western regions) and export enterprises of environmental compliance quadratic term coefficient is larger, which means environmental compliance had a greater promoting effect on innovation about state-owned enterprises, mature enterprises, enterprises in core areas and export enterprises over the inflection point of enterprise environmental compliance. The possible reasons are as follows: First, compared with state-owned enterprises, it is difficult for non-state-owned enterprises to overcome the bottleneck of enterprise innovation financing, which may lead to the lack of available innovation capital for state-owned enterprises, and the low innovation efficiency leads to the small promoting effect of environmental compliance on enterprise innovation [31,32]. Therefore, beyond the inflection point of environmental compliance, environmental compliance of state-owned enterprises has a great promotion effect on enterprise innovation. Second, compared with immature enterprises, mature enterprises are already in the relatively mature stage of enterprise innovation, and their innovation marginal cost is relatively low. Therefore, under the condition of constant marginal rate of return, the optimal innovation investment increases, and more funds are invested in existing innovation projects, so that the contribution of environmental compliance investment to the innovation of mature enterprises is greater than its contribution to non-mature enterprises. Third, compared with enterprises in non-core areas, enterprises in core areas are faced with higher marketization level, stricter market access and negative list, and their environmental violation costs will be higher than those of enterprises in non-core areas. Therefore, beyond the inflection point of environmental compliance, the promoting effect of environmental compliance on enterprise innovation in core areas is greater than that of enterprises in non-core areas. Fourth, on the one hand, according to the export self-selection mechanism of Melitz [33], the enterprises that take the lead in export have higher productivity and yield rate. The export enterprises obtain the research and development (R&D) funds from operating profit are greater than those non-export enterprises, and it is easier for export enterprises to obtain various innovation resources from the outside. On the other hand, export enterprises face stricter and more severe international environmental compliance rules, and the cost of enterprise violation is far greater than the cost of enterprise environmental compliance construction. Therefore, the environmental compliance of export enterprises with high environmental compliance contributes more to enterprise innovation.

4.3. Robustness Test

To further prove the robustness of the empirical model, we incorporated 1 period lag enterprise innovation ($Pnov_{c,i,t-1}$) into the random-effects Tobit model, and used 1 period lag ratio of enterprises' R&D investment to product sales revenue ($RD_{c,i,t-1}$) as an alternative variable of enterprise innovation. To reflect the causal impact of environmental compliance on enterprise innovation better and eliminate measurement errors caused by heterogeneity between the treatment group and the control group before receiving policy treatment, this research uses a Generalized Propensity Score Matching method (GPSM) developed by Hirano and Imbens [34], which is based on continuous treatment of variables, to further test the impact of different environmental compliance levels on enterprise innovation.

The empirical results of 1 period lag enterprise innovation ($Pnov_{c,i,t-1}$) are shown in column (3) in Table 6. The empirical results of introducing 1 period lag enterprise innovation into column (3) are consistent with the random-effects Tobit model of column (1) and the double hurdle model of column (2) after controlling the control variables and industry characteristics. We use the 1 period lag ratio of R&D investment to product sales revenue ($RD_{c,i,t-1}$) as an alternative variable of enterprise innovation in column (4), and the empirical results are basically consistent with the column (1) and column (2), and it indicates that the empirical results of this research are relatively robustness.

Table 6. Environmental compliance and enterprise innovation: Robustness.

Variables	(1) Random-Effect Tobit Model	(2) Double Hurdle Model	(3) Random-Effect Tobit Model	(4) Random-Effect Tobit Model	(5) GPSM
$Pnov_{c,i,t-1}$			0.564 ***		
			(0.001)		
$H_{c,i,t-1}$	−0.122 ***	−0.194 ***	−0.140 ***	−0.995 ***	
	(0.001)	(0.008)	(0.001)	(0.022)	
$H_{c,i,t-1}^2$	0.091***	0.145 ***	0.10 5***	0.151 ***	
	(0.001)	(0.006)	(0.001)	(0.019)	
$Pscore_{c,i,t-1}$					−0.121 ***
					(0.001)
$Pscore_{c,i,t-1}^2$					0.118 ***
					(0.001)
Constant	0.224 ***	0.224 ***	−0.016 ***	0.361 ***	−0.005 ***
	(0.004)	(0.004)	(0.000)	(0.016)	(0.001)
Control variables	Control	Control	Control	Control	Control
Industry characteristics	Control 1	Control 1	Control 1	Control 1	Control 1
/sigma_u	0.055 ***		0.0137 ***	3.075 ***	0.056 ***
	($\approx$0)		(0.000)	(0.003)	($\approx$0)
/sigma_e	0.049 ***		0.052 ***	0.791 ***	0.049 ***
	($\approx$0)		($\approx$0)	(0.000)	($\approx$0)
sigma		0.151***			
		(0.0003)			
ρ	0.558		0.065	0.940	0.492
Log-L	2,785,418.5	3,164,526.4	−3,511,470.2	−3,443,592	1,894,004.5
Observations	2,023,259	2,418,112	2,023,259	1,991,203	1,389,267

Note: Where $Pscore_{c,i,t-1}$ is the environmental compliance agent variable matched by GPSM method, and $Pscore_{c,i,t-1}^2$ is the quadratic terms of environmental compliance matched by GPSM method. *** indicate significant at 1%.

5. Conclusions and Policy Implications

5.1. Conclusions

This paper embeds environmental compliance factors and compliance cost factors into the M-O monopolistic competition and multi-product enterprise model to construct a theoretical model applicable to environmental compliance and enterprise innovation. In addition, we also construct a new environmental compliance index. We use the random-effect Tobit model and double hurdle model to empirically test the micro-data of China's manufacturing enterprises from 1998 to 2013, then we use the GPSM to conduct a robustness test. The robustness conclusion is that environmental compliance has a significant U-shaped relationship with enterprise innovation, that is, environmental compliance will inhibit enterprise innovation on the left of the inflection point of environmental compliance (0.669), while environmental compliance on the right of the inflection point will promote enterprise innovation. This is an important finding that is different from the existing researches, which reveals the complex and nonlinear impact of environmental compliance on enterprise innovation. The unique meaning of this nonlinear effect is that only when the environmental compliance index reaches a certain level can it have a promoting effect on enterprise innovation. Otherwise, the environmental compliance behavior of enterprises will have a significant crowding out effect on enterprise innovation. We also conduct empirical tests based on further consideration of enterprise heterogeneity factors. It is found that environmental compliance has greater inhibitory effects on innovation of state-owned enterprises, mature enterprises, core areas enterprises and export enterprises among enterprises with a lower level of environmental compliance, but environmental compliance has a greater promoting effect on state-owned enterprises, mature enterprises, core areas enterprises and export enterprises, when it beyond the inflection point of enterprise environmental compliance.

5.2. Policy Implications

The research conclusion provides the following policy implications. Firstly, the environmental compliance measurement system should be established. Due to the environmental compliance is a legal issue regulated through "yes" or "no" condition, the government should focus its supervision on "no" enterprises, i.e., those with an environmental compliance less than 1. Therefore, based on the environmental compliance measurement index constructed in this research, it is necessary to further improve the measurement system of environmental compliance and incorporate more environmental pollution indicators into the measurement system. The government focuses on regulating various types of companies whose indicators exceed those permitted by the license of that particular firm. On the other hand, firms need to make adjustments according to various indicators of the environmental compliance in order to fulfill their obligations of environmental laws and regulations. Secondly, the government should implement environmental compliance construction step by step. The government should adopt progressive policies to promote environmental compliance construction, and search the inflection point of the impact of current environmental compliance on corporate innovation, so as to stimulate the role of environmental compliance in promoting micro-enterprise innovation. And the government should subsidy the technology-intensive enterprises at low level of environmental compliance and support the innovation of technology-intensive enterprises in breakthrough of key core technologies. Thirdly, the government should strengthen environmental compliance supervision of labor-intensive and capital-intensive enterprises. Because of increasing the punishment for violation can further improve the environmental compliance index of labor-intensive and capital-intensive enterprises by increasing the cost of environmental violation, hence, expanding the transformation and upgrading of labor-intensive and capital-intensive enterprises. Finally, the government should encourage heterogeneous enterprises to enhance compliance construction. The environmental compliance has greater promoting effects on state-owned enterprises, mature enterprises, core enterprises and export enterprises beyond the inflection point of enterprise environmental compliance. Therefore, when innovative resources are completely allocated by the market, there will inevitably be market failure in non-state-owned enterprises, non-mature enterprises, enterprises in non-core areas and non-export enterprises without innovation resources. Therefore, the government should give certain innovative preferential policies to non-state-owned enterprises, non-mature enterprises, enterprises in non-core areas and non-export enterprises.

Author Contributions: M.L. established a theoretical model commonly used to study the environmental compliance and enterprise innovation and a new set of environmental compliance index; Y.L. processed enterprise-level micro panel data obtained from the Database of China Industrial Enterprises from 1998 to 2013, sorted out some literatures and translated the articles into English; M.L. and Y.L. analyzed and explained the empirical results and made relevant suggestions together, and they are the major contributors in writing the manuscript. Y.Z. provided study materials, reagents, materials, patients, laboratory samples, animals, instrumentation, computing resources, or other analysis tools. All authors have read and agreed to the published version of the manuscript.

Funding: This research received no external funding.

Institutional Review Board Statement: Not applicable.

Informed Consent Statement: Not applicable.

Data Availability Statement: The datasets used and/or analysis during the current study are available from the corresponding author on reasonable request.

Conflicts of Interest: The authors declare that they have no competing interests.

Appendix A

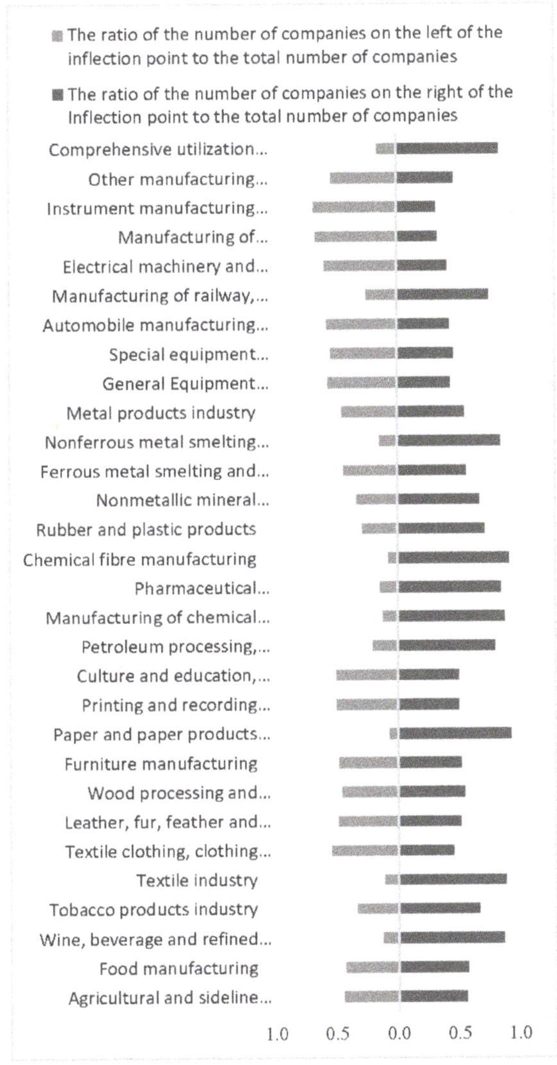

Figure A1. The ratio of the number of enterprises on both sides of the environmental compliance inflection point to the total number of enterprises in China's binary code industry.

References

1. Shabbir, M.S.; Wisdom, O. The Relationship Between Corporate Social Responsibility, Environmental Investments and Financial Performance: Evidence from Manufacturing Companies. *Environ. Sci. Pollut. Res.* **2020**, *27*, 39946–39957. [CrossRef]
2. Bitat, A. Environmental regulation & eco-innovation: The Porter hypothesis refined. *Eurasian Bus. Rev.* **2018**, *8*, 299–321.
3. Nordhaus, W. Designing A Friendly Space for Technological Change to Slow Global Warming. *Energy Econ.* **2011**, *33*, 665–673. [CrossRef]
4. Porter, M.E. America's green strategy. *Sci. Am.* **1991**, *264*, 168. [CrossRef]
5. Porter, M.E.; Van der Linde, C. Toward A New Conception of the Environment-Competitiveness Relationship. *J. Econ. Perspect.* **1995**, *9*, 97–118. [CrossRef]

6. Fadhilah, M.A. Strategic Implementation of Environmentally Friendly Innovation of Small and Medium-Sized Enterprises in Indonesia. *Eur. Res. Stud. J.* **2017**, *20*, 134–148. [CrossRef]
7. Hashmi, R.; Alam, K. Dynamic Relationship among Environmental Regulation, Innovation, CO_2 Emissions, Population, and Economic Growth in OECD Countries: A Panel Investigation. *J. Clean. Prod.* **2019**, *231*, 1100–1109. [CrossRef]
8. Weiss, J.; Stephan, A.; Anisimova, T. Well-Designed Environmental Regulation and Firm Performance: Swedish Evidence on the Porter Hypothesis and the Effect of Regulatory Time Strategies. *J. Environ. Plan. Manag.* **2019**, *62*, 342–363. [CrossRef]
9. Liu, H.Y.; Owen, K.A.; Yang, K.; Zhang, C.H. Pollution Abatement Costs and Technical Changes Under Different Environmental Regulations. *China Econ. Rev.* **2020**, *62*, 2–13. [CrossRef]
10. Li, S.H.; Xu, B.C. Environmental Regulation and Technological Innovation: Empirical Evidence from China's Prefecture-level Cities. *Mod. Econ. Res.* **2020**, *11*, 31–40, (In China CSSCI).
11. Zhao, X.; Sun, B. The Influence of Chinese Environmental Regulation on Corporation Innovation and Competitiveness. *J. Clean. Prod.* **2016**, *112*, 1528–1536. [CrossRef]
12. Shi, B.; Qiu, M.; Feng, C.; Ekeland, A. Innovation Suppression and Migration Effect: The Unintentional Consequences of Environmental Regulation. *China Econ. Rev.* **2018**, *49*, 1–23. [CrossRef]
13. Chen, L. How Does Environmental Regulation Affect Different Approaches of Technical Progress? Evidence from China's industrial sectors from 2005 to 2015. *Clean. Prod.* **2019**, *209*, 572–580.
14. Yang, L.X.; Liu, Y.C. Environmental Regulation and Regional Innovation Efficiency: Quasi-natural Experimental Evidence based on Carbon Emission Trading Pilot. *Commer. Res.* **2020**, *9*, 11–24, (In China CSSCI).
15. Chi, C.J.; Wu, Y.J. An Empirical Study on The Influence of Environmental Regulation on Process Innovation and Product Innovation. *Stat. Decis.* **2020**, *36*, 174–178, (In China CSSCI).
16. Su, X.; Zhou, S.S. Dual Environmental Regulation, Government Subsidy and Enterprise Innovation Output. *China Popul. Resour. Environ.* **2019**, *29*, 31–39, (In China CSSCI).
17. Long, X.N.; Wan, W. Environmental Regulation, Corporate Profit Margins and Compliance Cost Heterogeneity of Different Scale Enterprises. *China Ind. Econ.* **2017**, *6*, 155–174, (In China CSSCI).
18. Kang, Z.Y.; Tang, X.L.; Liu, X. Environmental Regulation, Enterprise Innovation and Export of Chinese Enterprises Retest Based on Porter Hypothesis. *J. Int. Trade* **2020**, *2*, 125–141, (In China CSSCI).
19. Sharif, H. Panel Estimation for CO_2 Emissions, Energy Consumption, Economic Growth, Trade Openness and Urbanization of Newly Industrialized Countries. *Energy Policy* **2011**, *29*, 6991–6999. [CrossRef]
20. Melitz, M.J.; Ottaviano, G.I.P. Market Size, Trade, and Productivity. *Rev. Econ. Stud.* **2008**, *75*, 295–316. [CrossRef]
21. Rocha, R.; Ulyssea, G.; Rachter, L. Do Lower Taxes Reduce Informality? *Evidence from Brazil. J. Dev. Econ.* **2018**, *4*, 134.
22. Heckman, J.; Macurdy, T. A Life Cycle Model of Female Labor Supply. *Rev. Econ. Stud.* **1980**, *47*, 47–74. [CrossRef]
23. Ramanathan, R.; Black, A.; Nath, P.; Muyldermans, L. Impact of Environmental Regulations on Innovation and Performance in the UK Industrial Sector. *Manag. Decis.* **2010**, *48*, 1493–1513. [CrossRef]
24. Fan, Z.Y.; Wang, Q. The Inefficiency of Financial Subsidies: From the perspective of Tax Surplus. *China Ind. Econ.* **2019**, *12*, 23–41, (In China CSSCI).
25. Richardson, S. Overinvestment of Free Cash Flow. *Rev. Account. Stud.* **2006**, *11*, 159–189. [CrossRef]
26. Berry, S.; Levinsohn, J.; Pakes, A. Differentiated Products Demand Systems from a Combination of Micro and Macro Data: The New Car Market. *Political Econ.* **2004**, *112*, 68–105. [CrossRef]
27. Brandt, L.; Van Biesebroeck, J.; Zhang, Y. Creative Accounting or Creative Destruction? Firm-Level Productivity Growth in Chinese Manufacturing. *Dev. Econ.* **2012**, *97*, 339–351. [CrossRef]
28. Liu, X.; Li, S. Determinants of the Relative Efficiency of China's Manufacturing Enterprises (2000–2004). *China Econ. Q.* **2008**, *3*, 843–868, (In China CSSCI).
29. Scherer, F. Firm Size, Market Structure Opportunity and The Output of Patented. *Am. Econ. Rev.* **1965**, *55*, 1097–1125.
30. Lind, J.; Mehlum, H. With or without U? The Appropriate Test for A U-Shaped Relationship. *Oxf. Bull. Econ. Stat.* **2010**, *72*, 109–118. [CrossRef]
31. Peters, M.E. Heterogeneous Mark-Ups, Growth and Endogenous Misallocation. *Econometrica* **2020**, *88*, 2037–2073. [CrossRef]
32. Midrifan, V.; Xu, D.Y. Finance and Misallocation: Evidence from Plant-Level Data. *Am. Econ. Rev.* **2014**, *104*, 422–458. [CrossRef]
33. Melitz, M. The Impact of Trade on Intra-Industry Reallocations and Aggregate Industry Productivity. *Econometrica* **2003**, *71*, 1695–1725. [CrossRef]
34. Hirano, K.; Imbens, G.W. The propensity score with continuous treatments. *Appl. Bayesian Model. Causal Inference Incomplete-Data Perspect.* **2004**, *226164*, 73–84.

International Journal of
Environmental Research and Public Health

Review

Treatment of Manure and Digestate Liquid Fractions Using Membranes: Opportunities and Challenges

Maria Salud Camilleri-Rumbau [1,2,*], Kelly Briceño [1], Lene Fjerbæk Søtoft [1], Knud Villy Christensen [1], Maria Cinta Roda-Serrat [1], Massimiliano Errico [1] and Birgir Norddahl [1]

[1] Department of Green Technology, Faculty of Engineering, University of Southern Denmark, Campusvej 55, 5230 Odense, Denmark; kellybm2004@gmail.com (K.B.); lfj@kbm.sdu.dk (L.F.S.); kvc@igt.sdu.dk (K.V.C.); mcs@igt.sdu.dk (M.C.R.-S.); maer@igt.sdu.dk (M.E.); bno@igt.sdu.dk (B.N.)

[2] Aquaporin A/S, Nymøllevej 78, 2800 Kongens Lyngby, Denmark

* Correspondence: macarum@gmail.com

Citation: Camilleri-Rumbau, M.S.; Briceño, K.; Fjerbæk Søtoft, L.; Christensen, K.V.; Roda-Serrat, M.C.; Errico, M.; Norddahl, B. Treatment of Manure and Digestate Liquid Fractions Using Membranes: Opportunities and Challenges. *Int. J. Environ. Res. Public Health* **2021**, *18*, 3107. https://doi.org/10.3390/ijerph18063107

Academic Editor: Paul B. Tchounwou

Received: 22 February 2021
Accepted: 13 March 2021
Published: 17 March 2021

Publisher's Note: MDPI stays neutral with regard to jurisdictional claims in published maps and institutional affiliations.

Abstract: Manure and digestate liquid fractions are nutrient-rich effluents that can be fractionated and concentrated using membranes. However, these membranes tend to foul due to organic matter, solids, colloids, and inorganic compounds including calcium, ammonium, sodium, sulfur, potassium, phosphorus, and magnesium contained in the feed. This review paper is intended as a theoretical and practical tool for the decision-making process during design of membrane-based systems aiming at processing manure liquid fractions. Firstly, this review paper gives an overview of the main physico-chemical characteristics of manure and digestates. Furthermore, solid-liquid separation technologies are described and the complexity of the physico-chemical variables affecting the separation process is discussed. The main factors influencing membrane fouling mechanisms, morphology and characteristics are described, as well as techniques covering membrane inspection and foulant analysis. Secondly, the effects of the feed characteristics, membrane operating conditions (pressure, cross-flow velocity, temperature), pH, flocculation-coagulation and membrane cleaning on fouling and membrane performance are presented. Finally, a summary of techniques for specific recovery of ammonia-nitrogen, phosphorus and removal of heavy metals for farm effluents is also presented.

Keywords: anaerobic digestion; digestate; liquid-solid separation; membrane separation; nutrient recovery; membrane fouling

1. Introduction

Animal manure was defined by Shobert and Maguire as the solid, semisolid, and liquid by-product generated by animals grown to produce meat, milk, eggs, and other agricultural products for human use and consumption [1]. Over a billion tonnes of manure are annually produced in the Unites States and 1.4 billion tonnes in Europe [2,3].

Animal manure represents a valuable fertilizer. It is estimated that globally livestock manure provided over 115 million tonnes of nitrogen (N) as input to agricultural soils in 2017 [4]. Nevertheless, this resource needs to be carefully managed to avoid ammonia emissions and nutrient losses to water recipients. In the last decades, changes in animal production have resulted in increased production of wastewater volumes, pollution of air, aquifers, surface waters and soil [5–7]. The intensive livestock production has led to challenges in the management, treatment and distribution of the manure nutrients, increasing difficulties in planning solution and investing in technologies to effectively valorize the manure produced. Manure is responsible for 7% of both agriculture CH_4 and N_2O emissions, being the second largest source of greenhouse gas emissions on a dairy farm [8]. It is estimated that methane emissions resulting from manure management have increased by 21% between 1990 and 2020, reaching 500 million tonnes of CO_2 equivalents [9]. In this context, the development and implementation of technologies for manure valorization appear of paramount importance.

Anaerobic digestion, which is a common practice in some European regions with intensive farming, is an efficient biomass treatment to reduce greenhouse gas emissions and produce energy [10–12]. During anaerobic digestion, the organic matter is converted into biogas containing primarily methane (CH_4) and carbon dioxide (CO_2). The obtained digestate is rich in primary nutrients and can improve soil structure when applied in agriculture, by helping soil particles to bind together into aggregates [13]. This improves soil nutrient and water holding capacity preventing erosion [14]. Moreover, the use of efficient separation techniques for the digestate could be beneficial in the overall management of animal waste.

The separation of the digestate into a liquid and a solid fraction is recommended for reducing animal waste volumes and the costs associated to transportation. Solid-liquid separation is also beneficial for producing a concentrated and ready-to-use agricultural fertilizer [5,15,16]. The organic matter contained in the digestate solid fraction provides a readily available carbon source improving the biological, chemical and physical soil characteristics as a soil amendment [13,17,18]. However, solid-liquid separation does not guarantee high recovery of the nutrients still available in the obtained liquid fraction. Therefore, if a further post-treatment of the still diluted liquid fraction is required, membrane filtration offers an efficient technical solution able to re-distribute the unbalanced nutrient concentration on the resulting liquid fraction [19–22]. Membrane technologies, including microfiltration (MF), ultrafiltration (UF), nanofiltration (NF) and reverse osmosis (RO), are extensively used for water and wastewater treatment and have been used before for separation of animal waste sources. Alternative technologies such as forward osmosis (FO), electrodialysis (ED) and membrane distillation (MD) have also raised attention for recovery of nutrients from farm effluents [19,20,23–29].

However, membrane technologies have been used to a limited extent during separation of farm effluents. This limited application is mainly due to the challenges related in using membrane technologies. One of the main limitations is fouling, which leads to permeate flux decay and ultimately to the loss of membrane performance [30,31]. Despite membrane fouling being the main challenge during wastewater filtration, it can be controlled by establishing an efficient operation and membrane pretreatment. In this regard, Pontie et al. [32] suggested that using membrane systems for RO pretreatment in seawater feed is by far the best available technology due to ease of operation, low footprint and lower chemical usage compared to conventional pretreatment systems. During effluent nutrient recovery, MF and UF membranes perform as very efficient solid-liquid separators that can reject nutrients associated with particles such as phosphorus [33,34], whereas NF, RO, FO, MD and ED can be used for the separation and concentration of nitrogen compounds and potassium [20,28,29,34].

In this review, nutrient recovery from raw and digested manure using membranes is discussed, with focus on the state-of-the-art on membrane separation applied to raw manure and anaerobic digestate liquid fractions. Factors affecting the membrane nutrient separation performances such as feed composition, membrane pretreatment and operation, membrane fouling, membrane cleaning strategies and membrane ageing are also described. This review guides the reader in understanding the application of membrane technology to farm effluents, considering opportunities and challenges.

2. Methodology

The methodology followed is a state-of-the-art review approach, with an angle towards a critical review on manure and digestate post-treatment using membranes [35,36]. The authors focused on reviewing studies from recent years, while incorporating a combined retrospective by including some the most relevant membrane studies in the field conducted since the last decades. The review focused on an article search per topic using keywords. The authors collected, tabulated, and compared results from literature regarding state-of-the-art solid-liquid separation systems, feed characteristics for raw and digested manure, commercial membranes applied during separation of raw and digested manure, nutrient recovery methods using membranes and drawbacks of membrane application during treatment of farm effluents, among others. Additionally, the authors envisioned areas which need further investigations for implementing novel membranes and building sustainable processes during manure and digestates treatment. Data grouping, and comparisons were done when relevant, based on statistical analysis for average values and standard deviations, using the Microsoft Excel® statistical analysis package.

3. Manure Treatment Processes

Manure treatment strategies have been developed to reduce water and air pollution from animal wastes. By doing so, nutrient recovery becomes possible and adds value to the entire manure management chain. The main manure treatment approaches consist of aerobic and anaerobic processes. In aerobic processes, bio-available compounds and nitrogenous compounds are oxidized decreasing the ammonia emissions [37]. Bacteria, protozoa and fungi are the main microorganisms degrading the organic matter and releasing CO_2, H_2O, and biomass as final products [38]. Approaches such as autothermal thermophilic aerobic digestion of liquid swine manure have also been reported [39], where the main benefits of this technology are the process simplicity, its robustness, the high reaction rate and the consequent small equipment, the conservation of N and the possibility of heat recovery. However, the applicability of this technology is limited. Anaerobic digestion (AD) is in fact recognized as the most sustainable and cost-effective technology for waste stabilization and production of valuable by-products like fertilizers and biogas [40,41].

3.1. Anaerobic Digestion

Anaerobic digestion has been successfully applied to different wastes [42,43] and manure sources, like poultry manure [44], pig manure [45], dairy manure [46], cow paunch manure [47], cattle manure [48], horse manure [49]. In general, AD follows four successive stages: hydrolysis, acidogenesis, acetogenesis, and methanogenesis with the overall process depending on the interaction of the microorganisms responsible for the different stages [50]. Digestate effluents are known as the material remaining after anaerobic digestion of organic wastes.

Compared to raw animal manure, the digestate has higher total ammoniacal-nitrogen (TAN) ratio, decreased organic matter content, decreased total and organic carbon, reduced biological oxygen demand, elevated pH, smaller carbon (C) to N ratio (C/N), and reduced viscosity [51]. From an agricultural point of view, parameters like pH, salinity, mineral N, especially in the form of ammonium, macro- and micronutrients, organic matter and concentration of heavy metals are important parameters when considering the application of digestates in agriculture [52].

Optimum C/N ratios for anaerobic digestion range between 20 and 30, with an optimal ratio of 25/1 for anaerobic bacterial growth in an AD system [53,54]. Improper C/N ratios could result in high TAN release and/or high accumulation of volatile fatty acids (VFA) in the digester, which would ultimately decrease the methanogen bacterial activity and cause failure of the AD process.

In the case of unbalanced C/N feedstocks, co-digestion of manure with different substrates increases the biogas production rate, improves the fertilizer value of the digestate and mitigates the greenhouse gas emissions [55,56].

The optimum biogas production during anaerobic digestion is achieved when the pH value of the ingestate is between 6 and 7. When the methane production level is stabilized, the pH range remains buffered between 7.2 and 8.3 [57]. Hartmann et al. [58], showed that during co-digestion of the organic fraction of municipal solid waste with cow manure, the pH rose to a value of 8 and the reactor showed stable performance with high biogas yield and low VFA levels.

Anaerobic digestion positively increases the agronomic value of the biomass treated by production of a digestate with a higher proportion of mineral N and less decomposable organic matter [59]. Masse et al. found that the ratio between TAN and the Total Kjeldahl Nitrogen (TKN) increased from 74% in raw manure to 85% in manure digestate. This was possibly due to the mineralization of organic nitrogen during anaerobic treatment [45]. Phosphorus (P) concentration in digestate represented 42% of that in raw manure [45,60]. However, the P availability results for AD are contradictory. Massé et al. [61] reported positive effects of AD in P availability, whereas Möller and Müller [51] maintained a more neutral position. Other authors found that co-digestates might be poorer in nutrients due to dilution [62], especially if the organic waste used for the co-digestion is low in nutrients or simply because of the addition of water into the anaerobic reactor to maintain a proper total solid content.

In relation to the reduction of particle size during AD, Møller et al. [63] found that the relative amount of swine manure particles smaller than 1.6 µm decreased from 27% to 10% during a seven-month storage period at 20 °C under anaerobic conditions. Masse et al. [45] also found that anaerobic treatment reduced swine manure solids concentration by 70% and significantly reduced the particle size. It was found that particles smaller than 10 µm represented 84% of dry matter (DM) for digested manure, whereas for raw swine manure these represented 64% of DM. Among particles smaller than 10 µm for anaerobic digestates, there were no particles larger than 2 µm detected. This suggests that, if a microfiltration step is intended for digested swine manure, the removal of particles larger than 10 µm may be sufficient as a pretreatment [64].

3.2. Solid-Liquid Separation (SLS)

When coupled with AD, manure SLS has been reported to add environmental benefits and increase the flexibility in manure management [65]. Manure separation processes aim at achieving a high separation efficiency in terms of concentration of suspended and colloidal particles in the solid phase, which is beneficial for further processing of the liquid fraction. Among the established treatment practices, solid-liquid separation of manure generates a solid fraction rich in P and a liquid fraction rich in N and K and, to lesser extent, in P [5].

Normally, centrifugation, gravity drainage and pressure filtration are used as manure pretreatment. Low-tech separation implies however a more difficult treatment of the liquid fraction due to the high amount of residual colloidal particles in the liquid phase. Most commercial solid-liquid separators can remove a considerable fraction of raw manure DM, but except for the decanter centrifuge, they are not efficient in terms of nutrient and heavy metal separation [66]. SLS can be performed before or after the AD. The separation before the AD process has the advantage of removing the material that hinders pumping and mixing. Nevertheless, part of the volatile DM is lost in the solid fraction decreasing the recoverable energy [67].

An overview of the main SLS techniques applied to manure separation with the DM from feed and obtained liquid fractions are summarized in Table 1. If the liquid fraction is further concentrated using membrane filtration, particles larger than 10 µm in anaerobic digestates would be removed during pretreatment of anaerobic digestates, as indicated previously by Masse et al. [45].

Table 1. Main manure SLS systems with characteristics of obtained liquid fractions.

Solid-Liquid Separation (SLS) System	Characteristics	Type of Feed	DM [%]		Reference
			Feed	**Obtained Liquid Fraction after SLS**	
Decanter centrifuge	Alfa Laval, NX 309B-31, Denmark)	Pig manure co-digestate	4.8 ± 0.2	2.1 ± 0.09	[66]
	AD-1220, GEA, Germany	Pig, cattle and chicken manure co-digestate	NA	2.7 (MF feed)	[26]
	Gennaretti centrifuge GHT VF, Italy	Pig slurry co-digestate	4.56	2.2	[19,68]
	Centrifuge, Bauer GmbH	Cow co-digestate	NA	4	[20]
Screw press	FAN separator, max. feed rate 6.5 m^3·h^{-1}	Sow slurry	1.5–2	NA	[21]
	NA	Dairy manure digestate	NA	4.1 ± 0.3	[69]
Filter press	ID construct (chamber filter press), 0.20–0.35 m^3·h^{-1}, 10–15 bar, polyamide-based filter cloth Rilsan® types R43 and R57	Sow slurry	1.5–2	NA	[21]
Vibration screen	CRETEL, max. feed rate 1 m^3·h^{-1}, 0.5 mm mesh size	Sow slurry	1.5–2	NA	[21]
Liquid cyclone	SRC, 15 m^3·h^{-1}, 5–7 bar, diameters of 7,12 and 18 mm	Sow slurry	1.5–2	NA	[21]
Bag filter	800 μm mesh size	Sow slurry	1.5–2	NA	[21]
Tangential flow separation	System supplied by Gaston County Dyeing Machine Company (GCDMC)	Pig manure (flushed waste)	1.11 ± 0.49	0.96 ± 0.31	[70]
Sedimentation tank	Settling basin	Sow slurry	1.5–2	NA	[21]
	Storage tank	Pig manure	2.0	NA	[71]
Combined: screw press + centrifugation + MF	Screw press: FAN Separator, pore diameter of 0.25 mm; decanter centrifuge: Baby II; capacity of 0.7 m^3·h^{-1}; 7.5 kW (Pieralisi, Italy); MF pore size 0.2 μm (Zenon GmbH, Germany)	Liquid pig manure	5.1 ± 2.2	8.1 ± 1.0 (MF concentrate)	[60]
Combined: straw filter + sedimentation	Uncompressed straw bed on a trenched concrete floor and settling tanks	Liquid pig manure	6.1 ± 1.1	1.9 ± 0	[60]
Combined: nitrification-denitrification + sedimentation	Nitrification/denitrification system and settling basin	Sow liquid manure	5.3 ± 2.9	0.91 ± 0.04	[58]

Values are expressed as average ± standard deviation.

The removal of colloidal, suspended particles and soluble macromolecules from liquid fractions obtained after manure or digestate solid-liquid separation is possible using MF or UF [26,72]. If RO is applied, the recovery of concentrated soluble compounds such as ammonium, phosphates and potassium, apart from other ions in the concentrate is possible as well [27,28]. The advantage of using RO is the relatively large permeate flow achieved with a relatively small concentrate flow, while MF can be used as a pretreatment for RO [21,57]. The mineral concentrates obtained after membrane separation can be further post treated, as for example [5]:

- Precipitation of struvite ($MgKNH_4PO_4$) by increasing pH to over 9 through addition of $Ca(OH)_2$, MgO or $MgCl_2$ depending on which metal in the concentrate is limiting.

- Recovery of potassium by precipitating potassium struvite (KMgPO$_4$), which occurs at high pH if the concentration of TAN is low and potassium, magnesium and PO$_4^{3-}$ are present in equimolar amounts.
- Evaporation of ammonia from the struvite precipitate by heating it and recovery of magnesium hydrogen phosphate for reuse
- Ammonia stripping by absorption in an inorganic acid or water
- Distillation and concentration of ammonia in a water phase
- Precipitation/crystallization of PO$_4^{3-}$ as calcium phosphate or calcium hydroxyl phosphate with the addition of Ca(OH)$_2$

3.3. Combined Manure Treatment and Separation Strategies

The synthesis of manure treatment processes implies a sequence of unit operations focusing on organic matter removal, nutrient recovery, solid separation, etc. As an example, Figure 1 shows the optimal solution proposed by Camilleri-Rumbau et al. [73] obtained by comparing different membrane technologies, physico-chemical operations and SLS techniques to recover nutrient-rich fractions from biogas digestate. The process was predicted to have a low energy consumption during solid-liquid separation and a low chemical consumption, thus proving the advantages of using membrane techniques to concentrate the nutrients present in digestate liquid fractions.

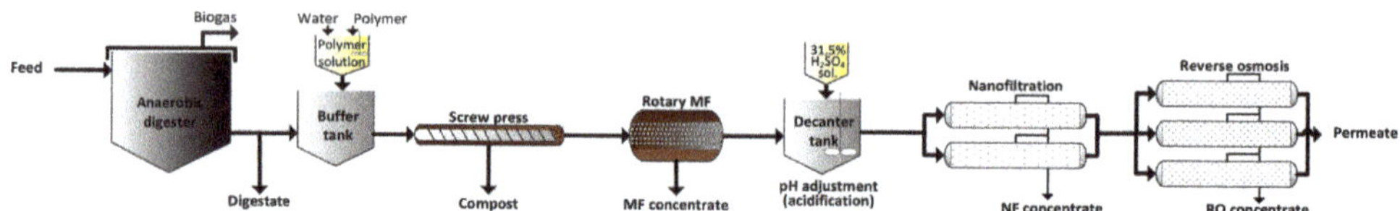

Figure 1. Post-treatment of digestate manure after solid liquid separation using screw press and different membrane technologies.

Table 2 gathers some essential literature about different manure treatment processes.

Table 2. Combined manure treatment systems.

Combined Manure Treatment Systems	Operation Units	Reference
	Ceramic MF membranes and polymeric RO membranes membrane technologies. RO was fed with the microfiltrate in sow slurry separation to obtain a high-quality liquid fraction	[21]
PIGMAN concept	Decanter separator, stirred tank and up flow anaerobic sludge blanket reactors, post digestion, partial oxidation and oxygen-limited autotrophic (nitrification-denitrification) (OLAND)	[66]
AnMBR + UF	Anaerobic membrane bioreactor with ultrafiltration	[72]
Several case studies on digestate post-treatment technologies	AD, ammonia stripping and membrane separation based on MF, NF and RO	[73]
BIOREK® concept	AD, ammonia stripping and membrane separation based on UF and RO	[74]
ADEPT (AD Elutriated Phased Treatment) -SHARON (Single reactor system High Ammonium Removal Over Nitrite)-ANAMMOX (Anaerobic Ammonium Oxidation)	hydrolysis/acidification reactor, methanogenic reactor, SHARON-ANAMOX reactor	[75]
Several case studies on manure treatment technologies	Combined, among other techniques, AD, centrifuge separation with flocculation, acidification, nitrification and de-nitrification, combination of anaerobic digestion–evaporation and drying composting, etc.	[76]

4. Raw Manure and Digestate Composition

Table 3 reports the composition and characteristics of pig manure as raw slurry and as separated liquid fraction. For the same manure, Table 4 summarizes the characteristics of the digestate before and after SLS. The raw manure composition gives fundamental information about the biogas potential achievable during AD, possibilities in land application or further treatment using membrane technologies for the liquid fractions. Similar information for other types of raw manures are provided in the Supplementary Materials Table S1.

4.1. Pig Slurry Composition

Pig slurry is a highly charged stream source rich in H_2O, organic compounds (present in both suspended and colloidal particles); nitrogen compounds (mainly in ammonia/ammonium form), P, major cations and anions (K, Ca, Mg, Na, Cl and sulphate); heavy metals (mainly Cu, Zn and Cd) and organic pollutants such as weeds, pathogens, medicine residues and residues of pesticides, herbicides and fungicides [60].

According to Table 3, the pH of raw pig slurry varies between 6.7 and 8.35 depending on the storage period, application of buffers and temperature [77]. As expected, total solids (TS) concentration is reduced mainly after manure solid-liquid separation treatment. Decanter centrifugation presents a higher TS separation efficiency than the one observed for screw press (approximately 67% and 38%, respectively) [5,26,45,60,73]. TS can also be reduced by about 70% using straw filtration [60]. In terms of nutrient recovery, approximately 78% of the P contained in the raw slurry can be recovered when applying straw filtration and around 83% by microfiltration. Masse et al. [45] suggested that in digestates and raw manure, approximately 20% of total P is in soluble form, whereas another 50% is associated with particles between 0.45 and 10 µm. They reported that only 30% P is linked to particles larger than 10 µm. About 80% P in swine manure is linked to suspended solids mostly attached to particles within 0.45–250 µm diameter.

In relation with DM content, Westerman et al. [70] showed that there was little difference between the concentration values of flushed wastes from finishing pigs and the screened liquid. Concretely, they showed that after separating manure by screw press, fine screening and dewatering the resulting fraction, the separated solids still contained almost 40% of DM and a high nutrient value (about 18 mg N, 10.4 mg P and 4.6 mg K per gram of DM). While total suspended solids (TSS) and TS decreased by about 20% and 13.5%, respectively after screening, while nutrient concentrations in the screened liquid remained almost unchanged.

4.2. Anaerobic Digestate Composition

Table 4 presents the composition and characteristics of anaerobically digested manure. It can be observed that pH values remain between 8.01–8.30 for all digestate sources. However, a lower pH value of 7.2 was detected by Masse et al. [45] during psychrophilic dry anaerobic digestion of dairy cow manure. TS were reduced by about 50% after applying decanter centrifuging to digestates [66]. Ammonia remained between 3.3–5.9 g·kg^{-1} in digestates, with minimal effect from separation [26,67]. K was mainly found as dissolved K^+ remains also practically unchanged after solid-liquid separation with values between 2.0 and 3.0 g·kg^{-1}. P in separated digestates, being mainly bound in particulate matter, decreased from 0.78–1.67 g·kg^{-1} to 0.21–0.67 g·kg^{-1} [26,63] when using decanter centrifuge or screw press. Camilleri-Rumbau et al. [26] further found that microfiltration could recover about 80% of the P present in digestate liquid fractions.

As can be observed from the data presented in Tables 3 and 4, mechanical separation greatly influences manure and digestate composition. As reported by Masse et al. [45], a decanter centrifuge is one of the most efficient solid-liquid separators in terms of nutrient and heavy metal separation. In general, a decanter centrifuge removed all particles > 2 µm [23] while a screw press mainly retained particles > 1 mm [63]. Additionally, mainly the smaller particles are degraded during the anaerobic digestion leading to a relative increase in the proportion of larger particles (>1.4 mm) [78]. A particular separator may be found superior to another based on testing that alters screen size, flow rate, or influent manure DM concentration. However, factors such as power requirements and cost have to be taken into account to evaluate the techno-economic performance of these separators [79].

4.3. Heavy Metal Composition

In the livestock production industry, it is a common practice to use Zn and Cu supplements in animal feed due to their growth-stimulating and antimicrobial effects [80–82]. Lowering the dietary supply of these elements to the livestock would be the most effective way to control heavy metal contents in digested manure slurries [82]. However, the removal of heavy metals from wastewater sources is a major concern mainly due to abrupt interference with the environment, bioaccumulation and related health risks [83]. After thermal treatment of manure, Li et al. [84] found that 75–90% of heavy metals such as Cr, Ni, and Mn are mainly found in the solid-phase, while heavy metals such as Cd, As, Hg, and Pb are found in the aqueous phase and gas phase, accounting for less than 5% of their total concentrations.

Despite the fact that animal manures are likely to contain high levels of heavy metals that pose risks to the environment and to human health, the addition of certain metals to the feed material has also been found to increase biogas production [85]. It has been demonstrated that efficient removal of propionate at high levels of VFA requires supplementation of Ca, Fe, Ni, and Co in a thermophilic non-mixed reactor [86]. Masse et al. [45] reported that approximately 80% of Zn and more than 95% of Cu in anaerobic digestate swine manure were associated with particles between 0.45 and 10 μm. Jin and Chang [82] found that total concentrations of Zn, Cu and As in digested pig slurries were <10, <5 and 0.02–0.1 mg·L^{-1}, respectively. Low concentration of Cu and Zn are also present in screened liquid slurries (8.5 and 11.2 μg·g^{-1} dry basis, respectively) [70]. Leclerc and Laurent [87] presented a global compilation of national release inventories for heavy metals considering 215 countries during a 15-year period. They found that mercury, zinc and copper are mostly responsible for the toxic impacts on human health and freshwater ecosystems resulting from manure application to land. Further information about the composition of manure and digestates in terms of heavy metals is shown in Table 5.

Although it is well-known that heavy metals are present in animal manure and can be concentrated during manure treatment, there is limited literature on membrane technology and the effect of heavy metals concentration [88].

Table 3. Composition of raw pig slurry and separated raw pig slurry.

Raw Pig Slurry									
pH	TVS [g·kg^{-1}]	COD [g L^{-1}]	TS [g·kg^{-1}]	DM [g·L^{-1}]	TKN [g·kg^{-1}]	TAN [g·kg^{-1}]	P [g·kg^{-1}]	K [g·kg^{-1}]	Reference
NA	-	NA	NA	67 ± 26	7.5 ± 2.5	4.5 ± 2.1	2.1 ± 0.8	3.3 ± 1.1	[5]
NA	-	NA	NA	18.4 ± 0.7	NA	2.06 ± 0.19	0.65 ± 0.13	2.32 ± 0.33	[21]
7.7 ± 0.1	-	45.01 ± 3.20	NA	45.5 ± 3.1	6.7 ± 0.5	5.5 ± 0.4	1.6 ± 0.2	2.6 ± 0.2	[28]
7.32 ± 0.16	74.72 ± 12.67	NA	95.84 ± 15.22	NA	10.49 ± 1.56	7.72 ± 1.11	2.49 ± 0.30	4.83 ± 0.80	[45]
7	-	-	NA	73	6.3	NA	1.62	5.98	[57]
8.5 ± 0.2	42 ± 11	NA	61 ± 11	-	7.0 ± 0.5	3.5 ± 0.5	1.8 ± 0.2	5.5 ± 0.5	[60]
7.4 ± 0.2	34 ± 18	NA	51 ± 22	-	5.1 ± 1.2	3.5 ± 0.6	1.2 ± 0.6	4.7 ± 0.7	[60]
7.5	-	45.96	NA	53.2	4.2	3.6	1.26	3.2	[63]
7.09 ± 0.10	36 ± 2	26.00 ± 3.10	48.0 ± 1.8	NA	5.6 ± 0.1	4.8 ± 0.7	1.6 ± 0.05	NA	[66]
6.72	-	NA	NA	34.5	9.35	3.66	0.74	3.62	[77]
8.35 ± 0.23	8.43 ± 6.54	NA	14.87 ± 9.56	NA	NA	3.033 ± 1.12	0.19 ± 0.16	NA	[89]
NA	-	-	26.9	NA	2.12	1.46	1.46	1.36	[90]

Table 3. *Cont.*

				Raw Pig Slurry					
pH	TVS [g·kg⁻¹]	COD [g L⁻¹]	TS [g·kg⁻¹]	DM [g·L⁻¹]	TKN [g·kg⁻¹]	TAN [g·kg⁻¹]	P [g·kg⁻¹]	K [g·kg⁻¹]	Reference
				Separated raw pig slurry					
8.3 ± 0.2	-	40.32 ± 1.93	-	15.6 ± 0.5	NA	5.1 ± 0.5	0.24 ± 0.02	2.5 ± 0.3	[28] [1]
8.4 ± 0.2	8.0 ± 2.6	NA	19 ± 0.0	-	4.7 ± 0.8	2.7 ± 0.4	0.4 ± 0.3	4.9 ± 0.5	[60] [2]
7.8 ± 0.2	8.4 ± 1.8	NA	18 ± 3	-	3.4 ± 0.4	3.1 ± 0.3	<0.2	4.6 ± 0.5	[60] [3]
8.01 ± 0.12	-	9.433 ± 0.472	14.49 ± 0.72	-	1.71 ± 0.068	1.41 ± 0.071	0.20 ± 0.01	NA	[72] [4]
6.76 ± 0.03	-	-	NA	6.10 ± 0.49	NA	1.66 ± 0.07	0.06 ± 0.0	1.09 ± 0.05	[91,92] [5]

Values are expressed as average ± standard deviation. NA: not available. All values in g·kg⁻¹ considering that slurries and digestates have a density of about 1kg·L⁻¹ [5]. [1]: data obtained for vacuum filtration; [2]: data obtained for straw filter made of three-layered plastic foil, total tickness 0.2 mm; [3]: data obtained for the sequence screw press, decanter centrifuge and MF; [4]: data obtained for liquid filtered by a 0.5 mm screen. [5]: data obtained for belt separator, decanted and passed through a 800 μm bag.

Table 4. Composition of anaerobic digestates.

				Digestate before Separation				
pH	TVS [g·kg⁻¹]	TS [g·kg⁻¹]	DM [g·L⁻¹]	TKN [g·kg⁻¹]	TAN [g·kg⁻¹]	P [g·kg⁻¹]	K [g·kg⁻¹]	Reference
8.24 ± 0.23	16.20 ± 3.74	28.29 ± 5.87	NA	6.9 ± 1.08	5.89 ± 0.73	1.04 ± 0.34	2.99 ± 0.31	[45] [1]
8.3	-	NA	56.2	NA	4.2	0.89	2.54	[63] [2]
8.1	-	NA	65.3	NA	5.0	1.67	2.31	[63] [3]
8.1	-	NA	35.5	NA	3.8	1.11	2.71	[63] [4]
8.01 ± 0.11	16.2 ± 1.01	48.1 ± 1.9	NA	5.3 ± 0.08	3.7 ± 0.4	1.5 ± 0.05	NA	[66] [5]
			Effluent after digestate separation					
8.2 ± 0.1	-	NA	27	3.4	3.15	0.46	2.03	[26] [6]
NA	-	NA	52.3	3.80	NA	1.17	NA	[63] [7]
NA	-	NA	19.6–24.4	3.1–5.2	NA	0.21–0.51		[63] [8]
8.09 ± 0.10	-	21.0 ± 0.9	NA	4.3 ± 0.08	3.5 ± 0.4	0.6 ± 0.05	NA	[66] [2]

Values are expressed as average ± standard deviation. NA: not available. [1]: manure from growing finishing swine operation treated in anaerobic sequencing batch reactors at 25 °C; [2]: mix of 75% pig manure and 25% other waste fish-processing waste continuous AD in a thermophilic (53 °C) apparatus; [3]: mix of 75% pig manure and 25% other waste fish-processing waste continuous AD in a mesophilic (35 °C) apparatus; [4]: mix of 98% pig manure and 2% fatty waste continuous AD in a mesophilic (38 °C) apparatus; [5]: mix of 90% pig manure and 10% fish-processing waste continuous AD in a thermophilic (55 °C) plant; [6]: liquid fraction obtained through decanter centrifugation; [7]: obtained with screw press of the digestate of 2; [8]: obtained with decanter centrifuge of the digestate of 3 and 4 respectively.

Table 5. Typical heavy metal concentrations in raw and digestate manure.

Source	Zn [mg/L]	Cu [mg/L]	Fe [mg/L]	As [mg/L]	Solid-Liquid Separation Technique	Reference
Raw pig manure slurry	12.75 ± 1.65	7.54 ± 2.96	-	0.13 ± 0.10	None	[82]
Digested pig slurry	20.66 ± 6.99	16.30 ± 4.1	-	0.26 ± 0.14	None	[82]
Solid fraction of digested pig slurry	477 ± 40.4	204 ± 30	-	2.19 ± 0.88	Sedimentation in anaerobic digester	[82]
Digested swine manure (for particles < 10 μm)	45.90	19.68	71.70	-	None	[45]
Digested swine manure (for particles < 0.45 μm)	2.17	1.07	4.20	-	None	[45]
Manure co-digestate [1]	16.4	6.4	1099	-		[93]
Digested swine manure	64 ± 2.00	9.05 ± 0.1	-	-		[94]
Pig biogas slurry [2]	9.88 ± 2.1	2.74 ± 0.45	-	-		[95]
Co-digestate pig slurry	24.6 ± 4.2	4.8 ± 0.7	-	-	Decanter centrifuge	[68]

Table 5. *Cont.*

Source	Zn [mg/L]	Cu [mg/L]	Fe [mg/L]	As [mg/L]	Solid-Liquid Separation Technique	Reference
Co-digestate liquid fraction of pig slurry	27	5	46	-	Decanter centrifuge	[19]
Co-digestate liquid fraction of cow manure	5	1.4	208	-	Decanter centrifuge	[20]

Values are expressed as average ± standard deviation. [1]: digestate obtained from animal manure, energy maize and food industry residues. [2]: pig manure anaerobically digested. The digestate is stored for 45 days and separated by natural sedimentation in biogas slurry and biogas residue.

5. Membrane Technologies for Post-Treatment of Manure and Digestate Liquid Fractions

After solid-liquid separation of manure or digestate, the obtained liquid fraction can be further concentrated using membranes. One of the main limitations of using membranes is membrane fouling, which represents one of the major problems during processing since it limits the membrane continuous operation. Several parameters can affect membrane fouling severity. Flow conditions, membrane pore size and/or selectivity, ion rejection capacity, membrane material, physico-chemical properties, porosity and morphology of the surface [96], as well as the characteristics of the effluent being treated are the main parameters contributing to fouling. Understanding the fouling mechanisms involved during concentration of livestock manure liquid fractions is one of the main challenges during the application of this technology.

5.1. Membrane Classification and Material Properties

Previous research studies on application of membranes during manure and digestate liquid fractions treatment have mainly used MF, UF, RO, membrane contactors (MC) and ED. Introduction of FO in this field is a relatively new method.

The selection of membrane material plays a very important role in separation performance during farm effluent processing. Membrane performance is mainly linked to physical and chemical interactions between the membrane surface and foulants during processing of water and wastewater sources. Material characteristics such as material type, porosity, hydrophobicity-hydrophilicity, surface charge, membrane polarity, permeability, selectivity, etc. are important factors affecting membrane performance. Table 6 presents some of the most used commercial polymeric membranes for the treatment of farm effluents. Polymeric membranes are typically made of polysulphone (PS), polyethersulphone (PES), polyvinylidene fluoride (PVDF), polypropylene (PP), polytetrafluoroethylene (PTFE), cellulose acetate (CA), cellulose triacetate (CTA), polyamide thin film composite (PA), while inorganic membranes typically are based on aluminum oxide or titanium oxide.

Membrane hydrophobicity or hydrophilicity influences membrane fouling. Hydrophobic membranes are widely reported to be more susceptible to adsorptive fouling by organic particles than hydrophilic ones [97]. Zhang et al. [98] found that the adsorptive fouling degrees were in increasing order PAN < PVDF < PES. Camilleri-Rumbau et al. [26] reported that in the initial stage of the digestate liquid fraction concentration process, PS membranes had a higher fouling tendency than PVDF membranes. However, after cake layer formation, the influence of the membrane material became less relevant and the cake layer controlled the filtration mechanism. Studies from Boerlage et al. [97] also demonstrated that PS membranes had a higher tendency to foul compared to polyacrilonitrile (PAN) membranes, as expected due to the higher hydrophobicity of PS compared to PAN. Furthermore, in their study, a homogeneously permeable surface enhanced particle deposition and formation of a regular cake layer, which was easier to remove with cleaning regardless of the low surface porosity detected [97].

López-Fernández et al. [72] further found that PES had a higher tendency to foul than PVDF. The permeability of PES membranes decreased drastically by 93%, while the permeability on PVDF membranes decreased by around 25%. This was possibly due to the higher affinity of the extracellular polymeric substances to the PES membrane [98]. The same study reported that the selectivity of the UF membranes filtering swine manure liquid fractions was directly related to the membrane rejection and the selected operating conditions. It was found that the increased permeate flux when using PVDF membranes improved the filtration selectivity, with higher COD rejection values for PVDF membranes compared to PES membranes (70% and 55–70%, respectively). Boerlage et al. [97] also reported the importance of the membrane material on the fouling mechanism. It was suggested that fouling observed on PAN membranes was dominated by physical deposition while fouling observed on PS was more likely chemically adsorbed, due to the more hydrophobic nature of PS. During MF of digested swine manure, Camilleri-Rumbau et al. [26] observed that permeate flux decline initially occurs due to a fast fouling formation followed by a filtration period with a slower relatively constant flux decline. This study further showed that foulants were adsorbed stronger to the PS membrane surface than to PVDF membrane surface. Similar conclusions were obtained in other studies using the same type of membrane materials [90,98].

Apart from the influence of membrane material during the filtration process, high porosity of membrane surfaces can also enhance fouling [99], due to the increased flux across the membrane which will drag more foulants towards the membrane surface and hence could provoke pore blocking. The effect of membrane charge and polarity, measured as electrical conductivity and ion exchange capacity, during separation of manure compounds such as NH_3-N is of special relevance when using ED. Mondor et al. [28,100] used AMX anion-exchange and CMB cation-exchange membranes to isolate and concentrate total NH_3-N from swine manure, achieving five-fold concentrations compared to feed manure. However, the main drawbacks of this technique were fouling of the AMX membranes mainly from calcium carbonate and silica, and consequently a loss in stack average current density. Fouling on CMB would however be minimal due mainly to electrostatic repulsion with Ca^{2+} ions and the negatively charged silica colloidal particles.

5.2. Recovery of Total Ammonia Nitrogen (TAN)

Approximately 70% of the N in raw pig slurries is dissolved and present as ammonium while the rest is bound to particles such as organic macromolecules, proteins and inorganic precipitates. At a pH higher than 8, TAN is mostly present in manure as uncharged ammonia, due to the existing equilibrium between ammonium (NH_4^+) and ammonia (NH_3). Contrarily, at low pH, the TAN is mostly present as positively charged ammonium (NH_4^+) [20]. Based on the ammonia-ammonium equilibrium chemistry, there here have been different approaches on ammonia recovery from animal waste in literature.

In digestates, for instance, ammonia recovery can be done by adding sulfuric acid to the liquid fraction obtained after mechanical separation of solids, as a treatment step after phosphorus recovery as calcium phosphate or struvite precipitate. Using concentrated sulfuric acid would volatilize CO_2 and capture ammonium, which would be stable at this pH, as an ammonium sulphate solution [57]. Recovery of nitrogen by ammonia stripping is also possible and it requires however a partial increase of pH in the stripping phase [57]. By neutralizing the stripped liquid fraction, RO can be applied for obtaining a concentrate with low concentration in ammonia and phosphates and a high concentration in K, and possibly some small amount of precipitate [6].

During ammonia recovery using RO membranes, the rejection of ammonium is higher than its uncharged form (ammonia), where rejection depends strongly on the pH [27,100]. RO membranes have proven to be able to recover more than 99% of the TAN present in raw slurry at pH < 6.5 [27]. At a similar pH level, a TAN rejection higher than 95.5% could also be achieved using aquaporin-based forward osmosis membranes on digested manure [20]. However, Li et al. [88] found FO ineffective in retaining N species. This result could be attributed to volatilization from the feed solution or N attachment on the membrane surface.

MD has also been studied as an ammonia recovery technique from raw swine manure. One of the main drawbacks of this technique is membrane wetting due to fouling formation, which provokes a loss in hydrophobicity and hinders the ammonia stripping capabilities of the membrane [101]. However, pretreatment methods such as MF and UF showed to be effective in enhancing the ammonia mass transfer coefficient, concretely by two in the case of PTFE membranes and by four in the case of PP membranes [101].

Ammonia recovery using ED has previously been applied in swine manure [101,102]. Mondor et al. [28] further considered a combination of ED and RO in treating liquid swine manure pretreated by vacuum filtration. The system showed promising results, but the possibility of ammonia volatilization represented a challenge to be taken into account. The authors reported that after 10 ED batches, approximately 17% of NH_3 was volatilized. In order to minimize nitrogen losses during RO processes, pumping, storing, increasing temperature, etc. it is suggested to use a closed system. However, ammonia emissions are more likely to occur during subsequent storing of the concentrate rather than before the concentration process [6].

Garcia-Gonzalez and Vanotti [103] have grouped the technologies for N recovery in five categories:

(1)   RO using high pressure and hydrophilic membranes,
(2)   air-stripping using stripping towers and acid absorption,
(3)   zeolite adsorption through ion exchange,
(4)   co-precipitation with phosphate and magnesium to form struvites,
(5)   gas-permeable membranes at low pressure.

In the first category, Zhou et al. [104] developed a pre-treatment system composed by a sequence of sand filter, steel plate and frame filter, ceramic UF membrane before a RO module. They consider biogas slurry feedstock odtained from chicken manure. Single and double disk tube RO (DTRO) modules were used in combination with seawater RO (SWRO). The authors optimized the slurry pH, operating temperature and pressure to maximize the ammonia recovery. The dual DTRO-SWRO at the optimized conditions (pH 6.1, 5 MPa, 35 °C) reached 99.1% TAN rejection.

In the second category more traditional technologies based on stripping performed in towers are included. Ammonia is stripped from pre-treated manure using air, steam or biogas and then absorbed in an acid solution to produce added-value fertilizers [105–107]. The modelling of this systems represents a challenge as highlighted for an analogous system by Madeddu et al. [108,109].

The third category zeolites can be added to both manure or sewage sludge-based digestate solid to provide active sites for molecular adsorption or exchangable cations for ammonium ions. Both mechanisms can increase the N retention [110]. The formation of struvites, included as forth category is discussed in the Section 4.3 together with the phosphorus recovery. In the last category, NH_3 pass through a microporous hydrophobic membrane and is concentrated in a stripping solution on the other side of the membrane. Garcia-Gonzalez and Vanotti [103] applied this technology to liquid swine manure using submerged tubular expanded polytetrafluoroethylene membranes. A H_2SO_4 solution was used to circulate acid through the membrane. The authors observed that adjusting the manure pH to 9 the TAN removal efficiency reached 88–94%. The same system was used to study the effect of other parameters like the aeration on the removal efficiency [111,112].

5.3. Recovery of Phosphorus

The liquid fraction obtained after mechanical separation of manure is rich in soluble components as well as colloidal and suspended particles. Colloidal and suspended particles contain insoluble organic and inorganic compounds such as phosphate or other P components, Ca and Mg precipitates.

Schoumans et al. [57] reviewed and assessed treatment strategies with regards to P recovery from manure and sewage sludge. Some of these strategies involve recovery of

P_2O_5 in ash after incineration, recovery of phosphate as calcium phosphate or struvite and P-biochar after pyrolysis from which resulting pyrolytic oils (tar) and gases could be used for producing energy during pyrolytic process. Phosphates could be obtained from the sludge liquid fraction by precipitation with aluminum sulphate, although this method seems uneconomical because the entire liquid fraction has to be treated. In this regard, Christensen et al. [113] observed that about 70–90% of the total phosphate in raw pig slurries was found in the particulate fraction while the remainder occurred as dissolved phosphate.

A possible pathway for P recovery is its precipitation in the form of struvite by co-precipitation of ammonium, potassium and phosphate [114]. However, the formation of struvite requires a sufficient amount of phosphate, with the disadvantage that the solubility of struvite in water is relatively high [5]. Struvite precipitation takes place during anaerobic reaction at a pH of 8.3 approximately. The formation of struvite can be enhanced by addition of dissolved magnesium. After centrifugation of the digestate, the solid fraction, with approximately 80% of the total P, can be sent to composting while the liquid fraction could undertake several pathways in order to recover ammonia/ammonium as a product (see Section 4.2. Recovery of TAN) [57]. Karakashev et al. [66] also presented a system for P recovery via struvite precipitation coupled with anaerobic digestion, achieving a phosphate removal of 96%, 7% ammonium removal, a slight decrease in COD, while practically no change in TS and TSS was achieved. Maurer et al. [115] also reported that recovery of P by struvite formation was highly effective (P recovery rate of 90–100%); by applying volume reduction processes such as evaporation, freeze-thawing and electrodialysis prior to the P recovery step.

Phosphorus recovery using membranes has also been well documented. Camilleri-Rumbau et al. [19,26] used MF and UF membranes for treatment of digested manures. They found that more than 80% of the total phosphorus could be rejected using PS and PES membranes. Some authors have used pretreatment strategies of the liquid fraction to achieve precipitation of P and its removal together with colloidal and suspended particles. Christensen et al. [113] suggested that pH control could be used to regulate the concentration of dissolved P in manure and as precipitated struvite. The addition of CaO, MgO or $Ca(OH)_2$ would care an increase in pH and promote precipitation of phosphate as calcium phosphates or calcium carbonates [5], while ammonia could be removed by stripping of the subsequent UF permeate.

Pramanik et al. [116] used a flat sheet FO membrane to pre-concentrate anaerobically treated dairy manure. The authors tested NaCl, $MgCl_2$ and EDTA-2Na as possible draw solutions obtaining in all cases more than 98% PO_4^{3-} rejection. However, supersaturation of different chemical species close to the membrane surface can cause membrane fouling. For this reason, Shi et al. [24] investigated using electrodialysis reversal (EDR) where the polarities of the electrodes are frequently inverted inducing a self-cleaning mechanism. The system was tested with a pre-treated and acidified pig manure digestate using cation and anionic membrane sheets. The authors measured a removal of phosphate up to 84%.

5.4. Rejection of Heavy Metals

Masse et al. [45] reported that in anaerobic digestates from swine manure 80% of Zn, Ca, and Fe and over 95% of Cu approximately are associated with particles between 0.45 and 10 μm. Several treatment strategies for the removal of heavy metals have been reported in literature. The addition of lime is used to promote precipitation of heavy metals, such as Cu and Zn hydroxides that can be efficiently removed from flushed raw swine manure waste [70]. During membrane separation, micellar-enhanced UF (MEUF) has also shown be a viable technique for separating phosphates from heavy metals, such as Cd and Cu, since P was not retained by the micelles and passed through the UF membrane [117]. Rejection coefficients up to 98% were achieved for both metals when no P was present, using sodium dodecyl sulfate (SDS) as a surfactant. Camilleri-Rumbau et al. [19] found that 96.9% Cu, Zn, Fe, Ca, Mg and Al could be rejected during the UF of centrifuged digestate manure, regardless of the use of flocculation-coagulation during the solid-liquid separation by centrifugation. FO aquaporin and conventional polyamide TFC FO membranes were tested by Li et al. [88] for digestate centrate liquid fraction obtained by natural settling. For both membranes the authors observed a rejection of heavy metals higher than 80%. Recently, Fernandes et al. [118] proposed a membrane sequence of MF + UF/NF for processing the digestate from kitchen and food waste. They observed that Al was reduced by 81% after MF and the NF was ineffective in Al rejection. When UF was used after MF, the Al was reduced by 87% compared to the pre-treated digestate. Zn was removed by 97% after MF without further improvement by adding the UF or NF step. Fe was reduced by 63% by the system MF-NF. The authors highlighted NF was the only process that produced a colorless permeate.

Table 6. Classification of the main commercial membranes applied for separation of manure and digestate liquid fractions treatment.

Technique	Pore Size [nm]	Permeability [L·m^{-2}·h^{-1}·bar^{-1}]	Applied Pressure [bar]	Membrane Material	Module Configuration	Feed	Rejection	References
MF	100–10,000	>1000	0.1–2.0	PS, PVDF, Al$_2$O$_3$, monolith ceramic membrane	Flat sheet Submerged capillary	Mainly digestate Raw slurry	Particles	[21,26,60,66,119]
UF	2–100	10–1000	0.1–5	PS, PES, PVDF, coated cellulose, inorganic-silicon carbide, cellulose acetate	Tubular Submerged hollow fiber	Mainly digestate	Suspended solids, particulate phosphorus, nitrogen, COD, 99% coliforms; soluble COD, phosphates or nitrogen require a biological step to be removed; other macromolecules and multivalent ions in a lesser extent Note: Concentration factors between 1.7 and 9 v/v depending on the pretreatment used before the membrane step	[19,27,71,72,119–121]
NF	0.5–2	1.5–30	3–20				Particles, macromolecules, multivalent ions and small organic compounds in a lesser extent	[115]
RO	99% salt rejection	0.05–1.5	5–120	Polyamide thin film-composite Cellulose acetate	Flat sheet Spiral wound	Mainly raw slurry	Particles, macromolecules, monovalent and multivalent ions and small organic compounds	[21,27,91,92,122–128]
FO	99% salt rejection	>7 (in FO mode)	0 (just residual pressure from flow velocity)	Polyamide thin film composite (TFC)	Flat sheet plate and frame Spiral wound Hollow fiber Tubular	Digestate liquid fraction Digestate from municipal wastewater	Particles, macromolecules, monovalent and multivalent ions and small organic compounds	[20,88,116,129]
MC	100–10,000	NA	NA	PP PTFE PVDF	Flat sheet Tubular Hollow fiber	Raw pig slurry and digestate (from municipal waste)	Monovalent and multivalent ions	[29,101,102]
ED	Typically 200 Da apparent pore size	NA	NA	Anionic/ cationic membranes	Electrodialysis cell	Mainly raw slurry (or urine source wastewater)	Monovalent and multivalent ions	[28,100,115,130]

6. Membrane Fouling and Factors Influencing Membrane Performance

Manure is a complex wastewater source with a high membrane fouling propensity. Swine manure, for instance, is a highly complex mixture of inorganic colloids and suspended organics, volatile fatty acids, proteins, bacteria, suspended solids and inorganic elements including Ca, NH_4^+, Na, S, K, P and Mg [33,131,132]. Ruan et al. [133] reported that the main composition of the inorganic fouling found on RO membranes processing digestate slurry was mainly Ca (as $CaCO_3$ and $CaSO_4$), Si (as SiO_2 colloid), O, C, Cl, Na, S, and P as well as organic foulants based on complex components, including hydrocarbons, aliphatic acids and their analogues.

Effluents with high ionic strength have been previously correlated to increased fouling [134,135] and cause more rapid flux decline [136], being membrane scaling and fouling the main factors affecting membrane performance [137].

The importance of membrane fouling during manure processing with membranes has been addressed in the last years. Reverse osmosis membranes has been studied previously for membranes treating swine wastewater [92,122,124,131] and other wastewater sources [23]. Fouling has also been studied in MF and UF systems [19,26,72,120,138,139] treating anaerobically digestated slurries; in membrane contactors for ammonia stripping from pig slurry [101,102] and in ED membranes during concentration and recovery of ammonia from swine wastewater [100]. Studies on fouling of forward osmosis membranes for processing manure wastewater sources are also limited [20].

Several factors have been attributed in the literature to be the main cause for membrane fouling and membrane performance loss. This section gives an overview of literature describing fouling mechanisms, operating conditions, cleaning and ageing of membranes processing mainly raw and digestate manure. An overview of membrane and fouling characterization techniques is available in the Section S1 of the Supplementary Materials.

6.1. Membrane Fouling Mechanisms

Membrane fouling mechanisms are complex and thus difficult to define. Several authors have put efforts in explaining the related mechanisms for specific applications. For instance, Lim and Bai [140] suggested that during MF of activated sludge the fouling mechanism could initially be based on "membrane-limited fouling", followed by "pore blocking", and eventually "cake formation". Other authors explain fouling mechanisms based on an initial concentration polarization, followed by gel layer formation and finally cake layer formation [141].

The use of models to explain the complex mechanisms taking place during membrane fouling is thus of importance to understand the filtration data. The resistance in series model as shown in Equation (1) is the calculation method most often used for evaluation of membrane permeate flux and fouling resistance:

$$J_p = \frac{\Delta P - \Delta \pi}{\mu(T) \cdot R_t} \tag{1}$$

where J_p ($m \cdot s^{-1}$ or $L \cdot m^{-2} \cdot h^{-1}$) is the permeate flux through the membrane, ΔP (Pa or bar) is the applied pressure, $\Delta \pi$ (Pa or bar) is the osmotic pressure, $\mu(T)$ is the viscosity of the fluid at a given temperature and R_t (m^{-1}) is the total resistance of the fouling layer-membrane system which acts as a barrier to the permeate flux.

The total resistance R_t can be divided into a pure membrane resistance and reversible and irreversible fouling contributions. The classification of membrane total resistance can be defined as shown in Equation (2). However, there is not a universal agreement on this classification and authors use different contributions in their definition of R_t and definitions for reversible and irreversible fouling [125,139,142,143]:

$$R_t = R_m + R_s + R_f \tag{2}$$

Camilleri-Rumbau et al. [91,92] classified R_t from RO filtration data from raw swine wastewater, as the sum of R_m, R_s and R_f (Equation (2)). R_m was defined as the intrinsic membrane resistance to water flux, which may increase as the membrane is put into use, due to compaction and irreversible fouling. The substrate resistance (R_s) was attributed to specific membrane-solute interactions which might occur even in the absence of flow. Fouling resistance (R_f) was the result of a variety of phenomena such as concentration polarization, gel layer formation and cake layer formation on the membrane surface. Several authors related the rapid increase in R_f during swine wastewater processing, to the transition between the pressure-dependent flux region and the pressure-independent flux region (i.e., gel-polarized region) [91,92].

Fouling mechanisms for membrane processes using MF/UF/NF/FO/ED/MD are more difficult to be identified. In this regard, authors normally refer to studies on municipal wastewater applications to explain the observations during the filtration of manure effluents.

6.2. Permeate Flux Decay and Pressure Control

Following Darcy's law, pure water fluxes in pressurized systems are proportional to the operating pressure applied, regardless of the variation on cross-flow velocity. However, during membrane processing of highly charged wastewater streams, deviations from linearity occur due to the presence of foulants. For instance, the water flux for clean UF membranes can be between 600–1000 $L \cdot m^{-2} h^{-1}$, while during manure filtration the flux is reduced to 20–40 $L \cdot m^{-2} h^{-1}$. This reduction might be explained by gel polarization and cake formation phenomena [115]. However, turbulence on the membrane surface, although generally helps to remove/lift foulants on the membrane surface, could also promote the inclusion of small particles in the cake layer. This could result in a less porous layer and higher specific resistance resulting in a lower permeate flux [136]. In general, compact smooth biofilms are formed at high shear force, while thick, fluffy biofilms are produced at low shear force [134,144]. Further increasing the applied pressure results in a denser and more compact cake layer which increases the specific cake resistance.

During MF of digestate liquid fractions, Camilleri-Rumbau et al. [26] observed that an increase in the feed cross-flow velocity increased the permeate flux on PS and PVDF membranes. This was attributed to concentration polarization and a reversible fouling layer that could be removed when increasing turbulence at the membrane surface. However, combining a higher pressure with a low cross-flow velocity resulted in the lowest permeate fluxes probably due to a significant increase in the fouling layer resistance caused by increased compression of the fouling layer [115]. Similar observations were obtained during UF of pretreated raw pig slurry by Fugère et al. [71]. They found that the optimal operating parameters during UF of pretreated raw pig slurry remained similar even though manure compositions and pretreatments were different. The authors explained this by the major contribution of fouling by gel-cake layer formation, as gel-cake layer formation is influenced more by the hydrodynamic conditions at the membrane surface than by the solids concentration in each manure type. Waeger et al. [138] also suggested the dominance of cake layer formation over permeate flux decay, during processing of anaerobic digestate liquid fraction. Therefore, increases in cross-flow velocities instead of the applied pressure might sometimes benefit the permeate flux rate [120,138], although higher velocities also imply a higher energy consumption.

6.3. Influence of Manure Composition and Particle Size

As discussed in Section 3, raw and digestate liquid fractions are highly charged wastewater streams rich in particles, proteins, colloids, nitrogen, phosphorus, potassium and metals, among other elements. In particular, the influence of metals during membrane processing cannot be neglected, as non-bound metals can cause severe membrane flux decay. Divalent cations, especially Ca, can form strong complexes with organic matter and polysaccharides, such as extracellular polymeric substances (EPS), resulting in the formation of a compact cake layer highly resistant to hydrodynamic forces [131,145]. Approximately 95% Fe and 80–90% Ca and Mg were found in the particulate fraction in raw pig manure [113], while approximately 80% of Zn, Ca, and Fe and over 95% of Cu were associated with particles between 0.45 and 10 µm in anaerobic pig digestate. As both Ca and Mg may form complexes with dissolved organic matter, the amount of ionic Ca and Mg may be 10–20% of the total concentration in the dissolved fraction [113,146].

In a study comparing different wastewater streams including cattle and pig slurry, Reimann et al. [137] found that the permeability of UF and RO membranes is determined by the concentration of organic matter in the wastewater and is independent of the type of wastewater. They reported a permeability loss over 83% when the COD-concentration increased from 0.63 g·L^{-1} to 42.8 g·L^{-1}. However, Fugère et al. [71] found that COD decreased significantly in raw manure with longer hydraulic retention time since allowed manure biodegradation which resulted in smaller particles. High organic matter concentrations together with the formation of bigger flocks in fresh filtered manure, make manure more difficult to treat, thus leading to a faster membrane flux decrease [71].

Particle size and particle concentration play an important role during membrane operations processing wastewater effluents. Studies show that the higher the feed concentration, the more the permeate flux decline is expected. Additionally, the cake resistance increases with the feed concentration although the specific cake resistance is not influenced and only the cake layer thickness is increased. Masse et al. [45] observed that anaerobic manure treatment of swine manure had an effect on particle size. The smallest particle fraction was between 2 and 10 µm for anaerobic digestates. This would suggest that a pretreatment for removing particles larger than 10 µm would be sufficient as a pretreatment for MF. However, Lee et al. [136] suggested that the smaller the particle size, the bigger the permeate flux decline because small colloids are more likely to aggregate, which would provoke an increase in cake resistance. During manure filtration, Fugère et al. [71] reported that the particle size of the treated manure could influence UF membrane fluxes and possibly define the membrane pore blocking mechanism and cake layer compactness.

6.4. Influence of Temperature

Temperature is an important parameter to take into account during evaluation of membrane performance. An increase in temperature produces a decrease in effluent viscosity and an increase in permeability [72]. This could imply lower membrane costs, a smaller membrane plant footprint and reduced operating costs. However, high temperatures, such as those for thermophilic anaerobic digestion (~55 °C), could also produce irreparable damage to polymeric membranes [6]. It is a good strategy thus to normalize the obtained permeate fluxes by correcting the experimental temperature using the manufacturer's correction factor for the particular membrane [6].

During UF of anaerobically digestated swine manure liquid fraction, increasing the operation temperature decreased the performance and selectivity of PES external tubular membranes compared to PVDF submerged hollow fiber membranes. [72].

Masse and Massé [147] showed that the flocculation performance of high molecular weight cationic polymers during flocculation of high DM swine manure was not affected by the operating temperature (6–25 °C). Furthermore, there was no negative effect from temperature on the removal efficiency of TSS and phosphorus during the process. However, animal wastewater processing at high temperatures implies losing nitrogen due to ammonia volatilization [6]. In the same study, during RO processing of dairy and raw pig slurries, it was found that permeate flux increased by about 50% by increasing the operation temperature from 10 to 20 °C. However, when the volume reduction reached 60%, the flux decreased even though the temperature was still approximately 44 °C. This flux decrease was attributed to the rapid volume reduction with the related fast increase in osmotic pressure.

6.5. Influence of pH

Acidification is a common practice in agricultural manure management to reduce greenhouse gas emissions as well as to increase the fertilizer value of slurry [148]. The influence of the wastewater pH on membrane performance and compound rejection has been previously studied mainly for RO membranes. In general, small amounts of acid are recommended to avoid considerably altering the osmotic pressure of the wastewater for assisting MF-RO steps [21]. However, increases in pH in manure and digestate liquid fractions enhanced the removal of residual colloids [57].

During acidification of swine wastewater separation using RO membranes, it was found that acidification of the pretreated manure to a pH of 6.5, for typical swine manure containing about 3000 $mg \cdot L^{-1}$ of TAN, would ensure high TAN rejection ranging from 97.5% to 99.8% at a high permeate recovery rate of 80% [27]. However, when increasing the pH to 7 the permeate recovery rate had to be reduced to 50% in order to maintain the overall retention of TAN at 97% and concentrations of TAN below 100 $mg \cdot L^{-1}$ in the permeate. Additionally, a pretreatment of the swine wastewater with either NF or a first stage RO further increased feed pH values up to 8.5 due to a decreased alkalinity, which required less acid to lower pH but considerably reduced the TAN rejection [30]. Acidification also enhanced slightly K rejection on RO membranes, achieving a rejection of 71% in urea-source wastewater [115].

In another study, it was observed that during acidification, solubilization of P from the solid manure phase into the liquid phase might occur [148], promoting the dissolution of struvite. Christensen et al. [113] also suggested that pH adjustment could control the amount of dissolved phosphorus in manure. However, during micellar-enhanced UF (MEUF) applied for recovery of P, as pH increased, removal of heavy metals decreased to 80% due to P complex formation that was minimized at lower pH values [118].

6.6. Influence of Flocculants/Coagulants Used during Pretreatment

Physico-chemical methods alone, including coagulation-flocculation, have been found to be effective processes to remove solids and nutrients from animal manure [19,149]. The use of organic and inorganic coagulants and flocculants are recommended to achieve an optimal mechanical separation (e.g., centrifugation, gravitational sedimentation, flotation, gravity drainage, pressure filtration) between the obtained solid and liquid fractions. This results in a solid fraction containing mainly suspended solids, and a liquid fraction, containing most of the dissolved components originally in manure [5,99,150,151]. If nutrient recovery such as phosphorus is intended to be recovered from the obtained fractions to be re-used in agriculture, then Fe-containing coagulants or flocculants must be avoided. Other agents primarily used to precipitate dissolved phosphorus, such as salts of calcium, iron, or aluminum can also form precipitates with dissolved phosphorus. These agents also aid the coagulation/flocculation of suspended solids thus enhancing settling [70].

Electrostatic interactions due to the chemical structure of the flocculant are relevant in the subsequent membrane fouling mechanism. However, flocculation-coagulation can reduce membrane fouling at the optimal dosage by minimizing foulants attachment on the membrane surface due to the formation of larger flocs and restraining the formation of gel layer [152–154]. Furthermore, the addition of coagulants could decrease EPS and TOC concentrations in the effluent, thus minimizing membrane fouling [147]. For instance, membrane pressure control can be done with membrane assisted flocculation-coagulation techniques because total organic carbon (TOC) and EPS are major foulants that require a need of pressure increase [153,155].

Despite the benefits of flocculation-coagulation during manure separation, if dosing is not optimized, flocculation-coagulation could potentially cause severe membrane fouling [99,126]. The fouling capacity of polymer flocculants depends highly on the attraction-repulsion forces towards the membrane surface, flocculant molecular weight and solution chemistry (such as pH, solid concentration and presence of ions in the water solution) [99]. Additionally, pig manure contains highly charged particles suggesting that highly charged, high-molecular-weight cationic polymers can be used [113]. However, it was found that severe fouling could be caused by cationic flocculants, because of the electrostatic attraction with the negatively charged membrane surfaces [156]. Masse et al. [157] found that the amount of cationic polymer added during manure pretreatment for NF and RO membranes, should be closely related to the suspended solids present in solution, to minimize the residual polymer in solution. Residual polymer could play a detrimental role in membrane flux as it could facilitate rapid attachment of organic foulants on the membrane surface.

During the application of flocculants and coagulants on manure, Hjorth et al. [151] also observed that the polymer dose required to obtain large flocs during manure treatment was unaffected by the coagulant addition. It was observed that coagulation was minimal and did not significantly affect the polymer flocculation [158]. However coagulation can disturb the stability of colloidal organic matter, forming flocks by adsorption of dissolved materials and hence reducing the membrane fouling potential [152].

Zhu et al. [152] characterized membrane fouling on a MF ceramic membrane system treating secondary effluent. They showed that coagulation with Al was an effective pretreatment for the control of MF fouling with stable operating time and recovered water volume. However, it was suggested that the potential formation of larger complexes due to the presence of multivalent ions, such as Al and Ca, and organic molecules in the feed water, could also lead to the formation of gel layer or even membrane pore clogging. Similarly, Camilleri-Rumbau et al. [19] observed that digestate centrifugation assisted with flocculation-coagulation could alleviate membrane fouling, thus causing a positive impact on the long-term stability of the subsequent ultrafiltration step.

7. Membrane Cleaning and Ageing

Membranes processing highly charged wastewater effluents such as raw slurries or digestate manure require a frequent and efficient cleaning strategy in order to minimize the permeate production loss and restablish membrane performance. The fouling composition should dictate the cleaning strategy to be performed to maintain a stable membrane operation and ensure a long life-time operation. A proper understanding of membrane fouling could avoid excess of chemical cleanings, which increase operation times and cost. Therefore, it is important to determine the type of fouling interactions to establish better cleaning strategies [159].

A standard cleaning strategy takes into account the nature of the foulants in the wastewater and fouling mechanisms in order to select an optimal cleaning procedure. A cleaning agent can affect fouling in three ways (i) removal of the foulants, (ii) changing morphology of the foulants (i.e., swelling or compaction) and (iii) alteration of surface chemistry modifying membrane surface hydrophobicity or charge [159]. However, the achieved cleaning efficiency depends greatly on the nature of the foulant. Traditionally, membranes are cleaned with bases, acids, chelating agents, surfactants, salts, disinfectants, ozone, etc. in combination with deionized water flushing, high temperature, concentration of and contact time between cleaning agent and membrane [160]. Specifically, for RO and for FO, a limiting factor on the application of this membrane techniques is the precipitation of salts on the membrane surface or in the membrane structure, so-called scaling, occurring during water removal.

There are few studies where cleaning strategies are systematically studied on membranes processing farm effluents. For instance, Masse et al. [128] studied the cleaning efficiency of a combination of EDTA-SDS-NaOH solutions to recover flux and remove proteins and bacteria from RO membranes processing raw swine manure. They observed that the SDS-NaOH solution combination could successfully recover membrane flux, while the addition of EDTA did not improve flux. Additionally, flux recovery using NaOH alone could also be achieved by increasing the cleaning time (from 60 to 120 min) and pH (pH 11/40 °C and pH 12/33 °C). In line with these results, Camilleri-Rumbau et al. [92]

further validated that NaOH alone could recover membrane flux succesffuly in long-term operations and that membrane flushing and soaking are important factors to preserve membrane stability. During UF of digested manure, alkaline and acidic cleaning cycles (NaOH and citric acid) were also effective in removing most of the inorganic foulants accumulated on the polysulfone membrane surface [19]. For UF-membranes cleaning and fouling have been reviewd by Shi et al. [161] to which interested readers are referred.

As stated above, temperature is also an important factor influencing foulant detachment of the membrane surface and structure. Madaeni et al. [162] found that, when cleaning RO membranes fouled with domestic wastewater, a flux recovery of 98% could be achieved by increasing the temperature of the cleaning solution to 35 °C, probably due to an increase in the transport rate and solubility of the cleaning agent and foulants [163]. Conversely, Masse et al. [128] found that the cleaning efficiency of the chemical solutions studied in their research increased significantly when pH was increased to 11–12, even if temperature had to be reduced to 35–40 °C.

Membrane rinsing or flushing with water has also been used successfully during cleaning of membranes fouled with farm effluents [115,127,139]. It was observed that membrane fouling resistance was reduced significantly after flushing [92,139]. All in all, membrane cleaning efficiency is a function of multiple parameters such as hydrodynamic conditions, concentration and temperature of chemical cleaning solution, as well as the order of the steps in the cleaning sequence. Some of the parameters, such as pH and temperature have strong non-linear effects on the cleaning effectiveness [164]. The required cleaning time is also a crucial parameter related to the cleaning efficiency. Generally, by increasing the cleaning time, the flux recovery increases sharply up to a certain limit. This means that the removal of loose deposits takes place during an optimal time after which the continuation of the cleaning process cannot significantly remove the strongly adsorbed fouling materials [126,128].

In addition to chemical cleaning methods, physical or mechanical membrane cleaning methods are also of interest, such as compressed air [165], air scrubbing [166] or ultrasound [167]. Backwashing can also be applied in ceramic membranes [151] and, with care, for polymeric MF and UF [97,166] membranes depending on the manufacturer's recommendation. A more extended review of chemical and mechanical cleaning techniques is reported respectively in Sections S2 and S3 of the Supplementary Materials.

Ageing or membrane degradation has been studied previously in relation to membrane cleaning procedures. Former studies have investigated accelerated ageing protocols to assess the long-term impact of low concentrations of chemical compounds on membranes [168,169]. However, this strategy does not consider a realistic chemical cleaning duration compared to the operating conditions generally applied in the industry. Camilleri-Rumbau et al. [91,92] showed that the effect of the chemical traces even after membrane rinsing and a three-day soaking in permeate water further improved the membrane flux recovery, without apparent increase in membrane degradation during the processing period considered. During cleaning of NF membranes fouled by conventionally-treated surface water, Liikanen et al. [170] found that the ion retention loss was most profound in the cleanings containing only alkaline chelating cleaning agents, while acidic cleaning agents played an important role in preserving membrane ion retention. However, other studies showed that the concentration of the feed solution has the highest contribution in ion rejection, whilst transmembrane pressure and temperature of feed solution have minor contributions [171].

During cleaning, membrane modification can also occur. Nyström and Zhu [172] suggested that cleaning may initially increase the membrane flux partly by ridding the pores of the material that is left from the membrane preparation process, and partly by making the pore surfaces more hydrophilic and charged by the adsorption of the cleaning agent. The increased hydrophilicity makes the chemical bonds between the water molecules and surface groups of the membrane stronger, thus reducing the possibilities of the foulants to displace water molecules and adhere on the membrane. Modification of the membrane can also significantly reduce the contact angles measured. Lower contact angles indicate more hydrophilic membranes and such membranes will potentially show better resistance to fouling by hydrophobic foulants [32]. An increased charge of the membrane increases the electrostatic repulsion between the active sites of the membrane, and thus makes the membrane more open [173]. An increased charge of the membrane also increases the repulsive forces between the membrane and similarly charged foulants. In addition to the increased flux, the membrane modification after cleaning cycles can result in lower ion retention, although this retention is often recovered during membrane operation [174].

8. Environmental Advantages of Membrane Separation Processes

The main advantage of incorporating membrane unit operations in farm effluents processing is that the high-quality membrane permeates are particle- and pathogen-free regardless of the initial quality of the feedstock. This allows for nutrient recovery in the form of solid and liquid fertilizers, and water recycling. The separation itself is a physical procedure that does not require use of chemicals other than the cleaning agents. Furthermore, due to their modular nature, membrane units are packed in compact systems with lower footprint that are easily scalable.

From a life cycle assessment perspective, the environmental implications of membrane technology transcend the target separation application, and does also include factors such as membrane manufacturing, transportation, operation, cleaning, reuse and ultimately membrane end-of-life management [175,176].

Tangsukbul et al. [175] performed life cycle assessment of wastewater treatment by microfiltration at different operating conditions. In general, they observed that the highest contribution to environmental impact was that of the energy consumption, followed by membrane manufacturing, chemical usage, and ultimately transportation, which was not significant. Most of the energy demand in membrane processes is attributed to pumping, due to the need of high shear and cross-flow velocities in order to minimize fouling. Different approaches have been reported to reduce energy demand in membrane processes. Tangsukbul et al. [175] recommended operating at relatively low fluxes to reduce energy consumption, however more membrane modules would be needed to achieve the same production, and that would imply moving the problem elsewhere. In this regard, operation at a competitive flux regime however using renewable energy sources was recommended as the option most benign to the environment. With a completely different approach, Gienau et al. [177] reported that modifying the viscosity and rheological behaviour of the feedstock by means of enzymatic pre-treatment could save up to 45% of the energy demand during ultrafiltration of manure digestate keeping the same flow conditions.

Due to fouling and ageing, the membranes need to be cleaned regularly and substituted periodically. Thus, the lifespan of a membrane is directly affected by the nature and severity of the membrane fouling. When membrane performance decays below an acceptable level, the modules are disposed as waste, mostly landfilled. Landarburu-Aguirre et al. [176] reviewed different end-of-life strategies for desalination RO modules including strategies for reuse and recycling. For instance, in some cases the membranes could be cleaned and re-used as RO membranes in processes requiring less water quality, or as NF or UF membranes [178]. However, reusing highly fouled membranes used for animal effluent treatment may be challenging due to the complexity, heterogeneity and possibly pathogenic nature of the foulants. Especially for the membranes used in the initial stages of the process, use of disinfectants would be indispensable prior to eventual membrane recycling. In the particular application of farm effluent processing, adequate fouling prevention and mitigation seems the most adequate approach towards extending membrane lifetime. The environmental impact of membrane cleaning in terms of type of cleaning agent, chemical concentration and disposal also need to be taken into consideration in the environmental assessment of a given process. In a more general perspective, Nunes et al. [179] outline the challenges and prospects in the future of membrane technology, where innovation and sustainability play key roles on the development and manufacturing of novel membranes and in design of separation processes for a sustainable future.

All in all, the implementation of membrane processes can undoubtedly help reduce the environmental burdens associated to animal waste processing and turn the waste into a valuable resource.

9. Future Research Needs

Membrane technologies offer the potential to separate and concentrate the fertilizers present in animal livestock effluents while at the same time producing clean effluents that can be discharged to the environment. The effective usage of membrane technologies in this respect is challenged by (a) the separation of fertilizers from heavy metals, (b) membrane fouling and the need for fouling treatment and (c) the need of a sustainable production, usage and disposal of membranes.

Phosphates and heavy metals are mainly associated with particulate matter in the livestock effluents and the separation of these, apart from the membrane molecular weight cut-off, is influenced by the membrane fouling layer. A better understanding of fouling layer built-up during operation is needed to address points (a) and (b) above. Most methods used to characterize membrane fouling are destructive methods not allowing real-time three-dimensional characterization of the fouling built-up, nor a true picture of how the fouling layer is influenced by fluid dynamics in the membrane module. A relatively novel technique, X-ray tomographic microscopy soft tissue in situ imaging, might just offer that option [180]. This would also help targeting another issue with fouling, a proper treatment and removal of the fouling layer. As most chemicals used for removal of fouling are not environmental-friendly, a more precise targeting and reduced usage is wanted, and alternative green cleaning agents must be developed.

The need for a sustainable production, usage and disposal of membranes is well recognized [179]. Recycling membranes used in treatment of biological waste streams might be difficult due to the risk of pathogens. Controlled chemical modification through oxidation of polyamide active layers could be an alternative for giving a second life to the resulting converted membranes [181].

Considering the short life span for membranes treating livestock effluents, developing low-cost disposable biodegradable membranes based on biopolymers [182] should be a future must for making membrane technology a viable solution in livestock effluent treatment.

10. Conclusions

Membrane fouling represents one of the major drawbacks when using membrane technologies during farm effluent processing. However, by understanding the related mechanisms, fouling composition and establishing efficient membrane pretreatment and cleaning strategies, membrane technologies can lead to an outstanding performance in terms of volume reduction and nutrient recovery. This paper reviews the raw and digestate general physico-chemical characteristics, technologies used for solid-liquid separation pretreatment at farm level, membrane characteristics requirements for treating liquid fractions of raw and digested manure, as well as membrane fouling mechanisms and membrane cleaning strategies described in the literature. Additionally, this review paper also provides valuable information about recovery of nutrients from manure, such as ammonia-nitrogen and phosphorus as well as heavy metal removal from these effluents using membranes.

Supplementary Materials: The following are available online at https://www.mdpi.com/1660-4601/18/6/3107/s1, Table S1: Composition of different raw manure, Table S2: Classification of membrane-foulant characterization technique, Section S1: Techniques for fouling and membrane characterization, Section S2: Chemical cleaning, Section S3: Mechanical cleaning.

Author Contributions: M.S.C.-R. conceived and designed the content of the article; collected and analyzed the data as well as wrote and reviewed the original draft and the writing. K.B. has designed the content of the article, collected data and reviewed the writing. L.F.S. designed the content of the article, supervised, and administered the project. K.V.C. designed the content of the article, validated the data, supervised, and administered the project as well as reviewed the writing. M.C.R.-S. wrote Section 8 and reviewed the writing. M.E. collected data, reviewed, and edited the writing. B.N. validated the data, reviewed the writing, supervised and administered the project as well as acquired the funding. All authors have read and agreed to the published version of the manuscript.

Funding: The authors acknowledge the funding received from Innovation Fund Denmark and the People Programme (Marie Curie Actions) of the European Union Seventh Framework Programme FP7/2007–2013/under REA grant agreement number (289887). This research work was funded by the European Regional Development Fund as part of the Interreg North Sea Region project 38-2-4–17 BIOCAS, circular BIOmass CAScade to 100%, to whom KVC, ME, and MCRS would like to express their gratitude.

Conflicts of Interest: The authors declare no conflict of interest.

References

1. Shobert, A.L.; Maguire, R.O. *Reference Module in Earth Systems and Environmental Science*; Elsevier: Amsterdam, The Netherlands, 2018.
2. Zhang, H.; Schroder, J. *Animal Manure Production and Utilization in the US*; Springer: Dordrecht, The Netherlands, 2014.
3. Inventory of Manure processing Activities in Europe. Available online: https://op.europa.eu/en/publication-detail/-/publication/d629448f-d26a-4829-a220-136aad51d1d9 (accessed on 18 June 2020).
4. FAO. Environmental Statistics. Available online: http://www.fao.org/economic/ess/environment/data/livestock-manure/en/ (accessed on 18 June 2020).
5. Hjorth, M.; Christensen, K.V.; Christensen, M.L.; Sommer, S.G. Solid-liquid separation of animal slurry in theory and practice. A review. *Agron. Sustain. Dev.* **2010**, *30*, 153–180. [CrossRef]
6. Thorneby, L.; Persson, K.; Tragårdh, G. Treatment of liquid effluents from dairy cattle and pigs using reverse osmosis. *J. Agric. Eng. Res.* **1999**, *73*, 159–170. [CrossRef]
7. Zhen, H.; Jia, L.; Huang, C.; Qiao, Y.; Li, J.; Li, H.; Chen, Q. Long-term effects of intensive application of manure on heavy metal pollution risk in protected-field vegetable production. *Environ. Pollut.* **2020**, *263*, 114552. [CrossRef]
8. Aguirre-Villegas, H.A.; Larson, R.A. Evaluating greenhouses gas emissions from dairy manure management practices using survey data and lifecycle tools. *J. Clean. Prod.* **2017**, *143*, 169–179. [CrossRef]
9. EPA. Global Anthropogenic Non-CO2 Greenhouse Gas Emissions: 1990–2020. Available online: https://nepis.epa.gov/Exe/ZyPDF.cgi/2000ZL5G.PDF?Dockey=2000ZL5G.PDF (accessed on 18 June 2020).
10. Havukainen, J.; Vaisanen, S.; Rantala, T.; Saunila, M.; Ukko, J. Environmental impacts of manure management based on life cycle assessment approach. *J. Clean. Prod.* **2020**, *264*, 121576. [CrossRef]
11. Rajendran, K.; Murthy, G.S. Techno-economic life cycle assessments of anaerobic digestion—A review. *Biocatal. Agric. Biotechnol.* **2019**, *20*, 101207. [CrossRef]
12. Daniyan, I.A.; Daniyan, O.L.; Abiona, O.H.; Mpofu, K. Development and optimization of a smart system for the production of biogas using poultry and pig dung. *Procedia Manuf.* **2019**, *35*, 1190–1195. [CrossRef]
13. Svensson, K.; Odlare, M.; Pell, M. The fertilizing effect of compost and biogas residues from source separated household waste. *J. Agric. Sci.* **2004**, *142*, 461–467. [CrossRef]
14. Odlare, M.; Arthurson, V.; Pell, M.; Svensson, K.; Nehrenheim, E.; Abubaker, J. Land application of organic waste—Effects on the soil ecosystem. *Appl. Energy* **2011**, *88*, 2210–2218. [CrossRef]
15. Jensen, L.S. Animal manure fertilizer value, crop utilization and soil quality impacts. In *Animal Manure Recycling: Treatment and Management*; Wiley: Hoboken, NJ, USA, 2013; pp. 295–328.
16. Vaneeckhaute, C.; Meers, E.; Michels, E.; Buysse, J.; Tack, F.M.G. Ecological and economic benefits of the application of bio-based mineral fertilizers in modern agriculture. *Biomass Bioenergy* **2013**, *49*, 239–248. [CrossRef]
17. Odlare, M.; Pell, M.; Svensson, K. Changes in soil chemical and microbiological properties during 4 years of application of various organic residues. *Waste Manag.* **2008**, *28*, 1246–1253. [CrossRef] [PubMed]
18. Guilayn, F.; Jimenez, J.; Rouez, M.; Crest, M.; Patureau, D. Digestate mechanical separation: Efficiency profiles based on anaerobic digestion feedstock and equipment choice. *Bioresour. Technol.* **2019**, *274*, 180–189. [CrossRef]
19. Camilleri-Rumbau, M.S.; Popovic, O.; Briceno, K.; Errico, M.; Søtoft, L.F.; Christensen, K.V.; Norddahl, B. Ultrafiltration of separated digestate by tubular membranes: Influence of feed pretreatment on hydraulic performance and heavy metal removal. *J. Environ. Manag.* **2019**, *250*, 109404. [CrossRef]

20. Camilleri-Rumbau, M.S.; Soler-Cabezas, J.L.; Christensen, K.V.; Norddahl, B.; Mendoza-Roca, J.A.; Vincent-Vela, M.C. Application of aquaporin-based forward osmosis membranes for processing of digestate liquid fractions. *Chem. Eng.* **2019**, *371*, 583–592. [CrossRef]

21. Pieters, J.G.; Neukermans, G.G.J.; Colanbeen, M.B.A. Farm-scale membrane filtration of sow slurry. *J. Agric. Eng. Res.* **1999**, *73*, 403–409. [CrossRef]

22. Hoeksma, P.; Buisonjé, F.E.; Aarnink, A.A. Full-scale production of mineral concentrates from pig slurry using reverse osmosis. In Proceedings of the International Conference of Agricultural Engineering, Valencia, Spain, 8—12 July 2012; pp. 1–6.

23. Bilstad, T. Nitrogen separation from domestic wastewater by reverse osmosis. *J. Membr. Sci.* **1995**, *102*, 93–102. [CrossRef]

24. Shi, L.; Xie, S.; Hu, Z.; Wu, G.; Morrison, L.; Croot, P.; Hu, H.; Zhan, X. Nutrient recovery from pig manure digestate using electrodialysis reversal: Membrane fouling and feasibility of long-term operation. *J. Membr. Sci.* **2019**, *573*, 560–569. [CrossRef]

25. Adam, G.; Mollet, A.; Lemaigre, S.; Tsachidou, B.; Trouvé, E.; Delfosse, D. Fractionation of anaerobic digestate by dynamic nanofiltration and reverse osmosis: An industrial pilot case evaluation for nutrient recovery. *J. Environ. Chem. Eng.* **2018**, *6*, 6723–6732. [CrossRef]

26. Camilleri-Rumbau, M.S.; Norddahl, B.; Wei, J.; Christensen, K.V.; Søtoft, L.F. Microfiltration and ultrafiltration as a post-treatment of biogas plant digestates for producing concentrated fertilizers. *Desalin. Water Treat.* **2014**, *55*, 1639–1653. [CrossRef]

27. Masse, L.; Masse, D.I.; Pellerin, Y. The effect of pH on the separation of manure nutrients with reverse osmosis membranes. *J. Membr. Sci.* **2008**, *325*, 914–919. [CrossRef]

28. Mondor, M.; Masse, L.; Ippersiel, D.; Lamarche, F.; Masse, D.I. Use of electrodialysis and reverse osmosis for the recovery and concentration of ammonia from swine manure. *Bioresour. Technol.* **2008**, *99*, 7363–7368. [CrossRef] [PubMed]

29. Norddahl, B.; Horn, V.G.; Larsson, M.; du Preez, J.H.; Christensen, K. A membrane contactor for ammonia stripping, pilot scale experience and modeling. *Desalination* **2006**, *199*, 172–174. [CrossRef]

30. Charfi, A.; Ben Amar, N.; Harmand, J. Analysis of fouling mechanisms in anaerobic membrane bioreactors. *Water Res.* **2012**, *46*, 2637–2650. [CrossRef] [PubMed]

31. Zhan, Y.; Dong, H.; Yin, F.; Yue, C. The combined process of paper filtration and ultrafiltration for the pretreatment of the biogas slurry from swine manure. *Int. J. Environ. Res. Public Health* **2018**, *14*, 1894. [CrossRef] [PubMed]

32. Pontie, M.; Rapenne, S.; Thekkedath, A.; Duchesne, J.; Jacquemet, V.; Leparc, J.; Suty, H. Tools for membrane autopsies and antifouling strategies in seawater feeds: A review. *Desalination* **2005**, *181*, 75–90. [CrossRef]

33. Masse, L.; Masse, D.I.; Pellerin, Y. The use of membranes for the treatment of manure: A critical literature review. *Biosyst. Eng.* **2007**, *98*, 371–380. [CrossRef]

34. Camilleri-Rumbau, M.S. Development of Membrane Technology for Production of Concentrated Fertilizer and Clean Water. Ph.D. Thesis, University of Southern Denmark, Esbjerg, Denmark, 2015.

35. Snyder, H. Literature review as a research methodology: An overview and guidelines. *J. Bus. Res.* **2019**, *104*, 333–339. [CrossRef]

36. Grant, M.J.; Booth, A. A typology of reviews: An analysis of 14 review types and associated methodologies. *Health Inf. Libr. J.* **2009**, *26*, 91–108. [CrossRef]

37. Vanotti, M.V.; Szogi, A.A.; Millner, P.D.; Loughrin, J.H. Development of a second-generation environmentally superior technology for treatment of swine manure in the USA. *Bioresour. Technol.* **2009**, *100*, 5406–5416. [CrossRef]

38. Girard, M.; Palacios, J.H.; Belzile, M.; Godbout, S.; Pelletier, F. Biodegradation in animal manure. In *Biodegradation—Engineering and Technology*; Rolando, C., Ed.; IntechOpen: London, UK, 2013; pp. 251–274.

39. Juteau, P. Review of the use of aerobic thermophilic bioprocesses for the treatment of swine waste. *Livest. Sci.* **2006**, *102*, 187–196. [CrossRef]

40. Kothari, R.; Pandey, A.K.; Kumar, S.; Tyagi, V.V.; Tyagi, S.K. Different aspects of dry anaerobic digestion for bio-energy: An overview. *Renew. Sustain. Energy Rev.* **2014**, *39*, 174–195. [CrossRef]

41. Li, Y.; Chen, Y.; Wu, J. Enhancement of methane production in anaerobic process: A review. *Appl. Energy* **2019**, *240*, 120–137. [CrossRef]

42. Gao, M.; Zhang, L.; Liu, Y. High-loading food waste and blackwater anaerobic co-digestion: Miximizing bioenergy recovery. *Chem. Eng. J.* **2020**, *394*, 124911. [CrossRef]

43. Priadi, C.; Wulandari, D.; Rahmatika, I.; Moersidik, S.S. Biogas production in the anaerobic digestion of paper sludge. *APCBEE Procedia* **2014**, *9*, 65–69. [CrossRef]

44. Safley, L.M.; Vetter, R.L.; Smith, D. Operating a full-scale poultry manure anaerobic digester. *Biol. Wastes* **1987**, *19*, 79–90. [CrossRef]

45. Masse, L.; Massé, D.I.; Beaudette, V.; Muir, M. Size distribution and composition of particles in raw and anaerobically digested swine manure. *Trans. Asae* **2005**, *48*, 1943–1949. [CrossRef]

46. Lo, K.V.; Liao, P.H. Two-phase anaerobic digestion of screened dairy manure. *Biomass* **1985**, *8*, 81–90. [CrossRef]

47. Tritt, W.P.; Kang, H. Ultimate biodegradability and decay rates of cow paunch manure under anaerobic conditions. *Bioresour. Technol.* **1991**, *36*, 161–165. [CrossRef]

48. Şenol, H.; Açıkel, Ü.; Demir, S.; Oda, V. Anaerobic digestion of cattle manure, corn silage and sugar beet pulp mixture after thermal pretreatment and kinetic modeling study. *Fuel* **2020**, *263*, 116651. [CrossRef]

49. Hadin, Å.; Eriksson, O. Horse manure as feedstock for anaerobic digestion. *Waste Manag.* **2016**, *56*, 506–518. [CrossRef]

50. Meegoda, J.N.; Li, B.; Patel, K.; Wang, L.B. A review of the processes, parameters and optimization of anaerobic digestion. *Int. J. Environ. Res. Public Health* **2018**, *15*, 2224. [CrossRef]
51. Möller, K.; Müller, T. Effects of anaerobic digestion on digestate nutrient availability and crop growth: A review. *Eng. Life Sci.* **2012**, *12*, 242–257. [CrossRef]
52. Makadi, M.; Tomocsik, A.; Orosz, V. Digestate: A new nutrient source—Review. In *Biogas*; Kumar, S., Ed.; InTech: London, UK, 2012; pp. 295–310.
53. Hills, D.J. Effects of carbon:nitrogen ratio on anaerobic digestion of dairy manure. *Agric. Wastes* **1979**, *1*, 267–278. [CrossRef]
54. Wang, X.; Lu, X.; Li, F.; Yang, G. Effects of temperature and carbon-nitrogen (C/N) ration on the performance of anaerobic co-digestion of dairy manure, chicken manure and rice straw: Focusing on ammonia inhibition. *PLoS ONE* **2014**, *9*, e97265.
55. Atandi, E.; Rahman, S. Prospect of anaerobic co-digestion of dairy manure: A review. *Environ. Technol. Rev.* **2012**, *1*, 127–135. [CrossRef]
56. Søndergaard, M.M.; Fotidis, I.A.; Kovalovszki, A.; Angelidaki, I. Anaerobic co-digestion of agricultural byproducts with manure for enhanced biogas production. *Energy Fuels* **2015**, *29*, 8088–8094. [CrossRef]
57. Schoumans, O.F.; Rulkens, W.H.; Oenema, O.; Ehlert, P.A.I. *Phosphorus Recovery from Animal Manure. Technical Opportunities and Agro-Economical Perspectives*; Alterra: Wageningen, The Netherlands, 2010; p. 2158, ISSN 1566-7197.
58. Hartmann, H.; Ahring, B. Anaerobic digestion of the organic fraction of municipal waste: Influence of co-digestion with manure. *Water Res.* **2005**, *39*, 1543–1552. [CrossRef]
59. Tambone, F.; Scaglia, B.; D'Imporzano, G.; Schievano, A.; Orzi, V.; Salati, S.; Adani, F. Assessing amendment and fertilizing properties of digestate from anaerobic digestion through a comparative study with digestated sludge and compost. *Chemosphere* **2010**, *81*, 577–583. [CrossRef]
60. Melse, R.W.; Verdones, N. Evaluation of four farm-scale systems for the treatment of liquid pig manure. *Biosyst. Eng.* **2005**, *92*, 47–57. [CrossRef]
61. Massé, D.I.; Croteau, F.; Masse, L. The fate of crop nutrients during digestion of swine manure in phychrophilic anaerobic sequencing batch reactors. *Bioresour. Technol.* **2007**, *98*, 2819–2823. [CrossRef] [PubMed]
62. Eriksson, L.; Runevad, D. Evaluating digestate processing methods at Linköping biogas plant. A resource efficient perspective. Master's Thesis, Linköping University, Linköping, Sweden, 2016.
63. Møller, H.B.; Sommer, S.G.; Ahring, B.K. Separation efficiency and particle size distribution in relation to manure type and storage conditions. *Bioresour. Technol.* **2002**, *85*, 189–196. [CrossRef]
64. Khoshnevisan, B.; Duan, N.; Tsapekos, P.; Awasthi, M.K.; Liu, Z.; Mohammadi, A.; Angelidaki, I.; Tsang, D.C.W.; Zhang, Z.; Pan, J.; et al. A critical review on livestock manure biorefinery technologies: Sustainability, challenges, and future perspectives. *Renew. Sustain. Eng. Rev.* **2021**, *135*, 110033. [CrossRef]
65. Aguirre-Villegas, H.A.; Larson, R.A.; Sharara, M.A. Anaerobic digestion, solid-liquid separation, and drying manure: Measuring constituents and modeling emission. *Sci. Total Environ.* **2019**, *696*, 134059. [CrossRef]
66. Karakashev, D.; Schmidt, J.E.; Angelidaki, I. Innovative process scheme for removal of organic matter, phosphorus and nitrogen from pig manure. *Water Res.* **2008**, *42*, 4083–4090. [CrossRef] [PubMed]
67. Rico, C.; Rico, J.L.; Tejero, I.; Munoz, N.; Gomez, B. Anaerobic digestion of the liquid fraction of dairy manure in pilot plant for biogas production: Residual methane yield of digestate. *Waste Manag.* **2011**, *31*, 2167–2173. [CrossRef]
68. Popovic, O.; Gioelli, F.; Dinuccio, E.; Rolle, L.; Balsari, P. Centrifugation of digestate: The effect of chitosan on separation efficiency. *Sustainability* **2017**, *9*, 2302. [CrossRef]
69. Porterfield, K.K.; Faulkner, J.; Roy, E.D. Nutrient recovery from anaerobically digested dairy manure using dissolved air flotation (DAF). *ACS Sustain. Chem. Eng.* **2020**, *8*, 1964–1970. [CrossRef]
70. Westerman, P.W.; Bicudo, J.R. Tangential flow separation and chemical enhancement to recover swine manure solids, nutrients and metals. *Bioresour. Technol.* **2000**, *73*, 1–11. [CrossRef]
71. Fugere, R.; Mameri, N.; Gallot, J.E.; Comeau, Y. Treatment of pig farm effluents by ultrafiltration. *J. Membr. Sci.* **2005**, *255*, 225–231. [CrossRef]
72. Lopez-Fernandez, R.; Aristizabal, C.; Irusta, R. Ultrafiltration as an advanced tertiary treatment of anaerobically digested swine manure liquid fraction: A practical and theoretical study. *J. Membr. Technol.* **2011**, *375*, 268–275. [CrossRef]
73. Camilleri-Rumbau, M.S.; Norddahl, B.; Nielsen, A.K.; Christensen, K.V.; Søtoft, L.F. Comparative techno-economical study between membrane technology systems for obtaining concentrated fertilizers from biogas plan effluents. *Ramiran* **2013**, *8*, 7–10.
74. Du Preez, J.; Norddahl, B.; Christensen, K. The BIOREK® concept: A hybrid membrane bioreactor for very strong wastewater. *Desalination* **2005**, *183*, 407–415. [CrossRef]
75. Hwang, I.S.; Min, K.S.; Yu, Z. Resource recovery and nitrogen removal from piggery waste using the combined anaerobic processes. *Water Sci. Technol.* **2006**, *54*, 229–236. [CrossRef] [PubMed]
76. Foged, H.L.; Flotats, X.; Bonmati Blasi, A.; Shelde, K.M.; Palatsi, J.; Magri, A.; Juznik, Z. *Assessment of Economic Feasibility and Environmental Performance of Manure Processing Technologies*; Technical Report No. IV to the European Commission, Directorate-General Environment, Manure and Processing Activities in Europe—Project Reference: ENV.B1/ETU/2010/0007; European Comission: Brussels, Belgium, 2011.

77. Sommer, S.G.; Hutchings, N.J. Ammonia emission from field applied manure and its reduction. *Eur. J. Agron.* **2001**, *15*, 1–15. [CrossRef]
78. Marcato, C.E.; Pinelli, E.; Pouech, P.; Winterton, P.; Guiresse, M. Particle size and metal distributions in anaerobically digested pig slurry. *Bioresour. Technol.* **2008**, *99*, 2340–2348. [CrossRef]
79. Ford, M.; Fleming, R. *Mechanical Solid-Liquid Separation of Livestock Manure Literature Review*; Ridgetown College—University of Guelph: Guelph, ON, Canada, 2002.
80. Jondreville, C.; Revy, P.S.; Dourmad, J.Y. Dietary means to better control the environmental impact of copper and zinc by pigs from weaning to slaughter. *Livest. Prod. Sci.* **2003**, *84*, 147–156. [CrossRef]
81. Popovic, O.; Jensen, L.S. Storage temperature affects distribution of carbon, VFA, ammonia, phosphorus, copper and zinc in raw pig slurry and its separated liquid. *Water Res.* **2012**, *46*, 3849–3858. [CrossRef]
82. Jin, H.; Chang, Z. Distribution of heavy metal contents and chemical fractions in anaerobically digested manure slurry. *Appl. Biochem. Biotechnol.* **2011**, *164*, 268–282. [CrossRef] [PubMed]
83. Abdullah, N.; Yusof, N.; Lau, W.J.; Jaafar, J.; Ismail, A.F. Recent trends of heavy metal removal from water/wastewater by membrane technologies. *J. Ind. Eng. Chem.* **2019**, *76*, 17–38. [CrossRef]
84. Li, S.; Zou, D.; Li, L.; Wu, L.; Liu, F.; Zeng, X.; Wang, H.; Zhu, Y.; Xiao, Z. Evolution of heavy metals during thermal treatment of manure: A critical review and outlooks. *Chemosphere* **2020**, *247*, 125962. [CrossRef]
85. Ward, A.J.; Hobbs, P.J.; Holliman, P.J.; Jones, D.L. Optimisation of the anaerobic digestion of agricultural resources. *Bioresour. Technol.* **2008**, *99*, 7928–7940. [CrossRef] [PubMed]
86. Kim, M.; Ahn, Y.-H.; Speece, R.E. Comparative process stability and efficiency of anaerobic digestion; mesophilic vs. thermophilic. *Water Res.* **2002**, *36*, 4369–4385. [CrossRef]
87. Leclerc, A.; Laurent, A. Framework for estimating toxic releases from the application of manure on agricultural soil: National release inventory for heavy metals in 2000–2014. *Sci. Total Environ.* **2017**, *590–591*, 452–460. [CrossRef] [PubMed]
88. Li, Y.; Xu, Z.; Xie, M.; Zhang, B.; Li, G.; Luo, W. Resource recovery from digested manure centrate: Comparison between conventional and aquaporin thin-film composite forward osmosis membranes. *J. Membr. Sci.* **2020**, *593*, 117436. [CrossRef]
89. Liu, Y.H.; Kwag, J.-H.; Kim, J.-H.; Ra, C.S. Recovery of nitrogen and phosphorus by struvite crystallization from swine wastewater. *Desalination* **2011**, *277*, 364–369. [CrossRef]
90. Atkinson, S. VSEP vibratory membrane filtration system treats hog manure. *Membr. Technol.* **2005**, *1*, 10–11. [CrossRef]
91. Camilleri-Rumbau, M.S.; Masse, L.; Dubreuil, J.; Mondor, M.; Christensen, K.V.; Norddahl, B. Cleaning strategies for reverse osmosis membrane fouled with swine wastewater. In Proceedings of the 15th Aachener Membrane Kolloquium, Aachen, Germany, 12–13 November 2014; pp. 81–86.
92. Camilleri-Rumbau, M.S.; Masse, L.; Dubreuil, J.; Mondor, M.; Christensen, K.V.; Norddahl, B. Fouling of a spiral-wound reverse osmosis membrane processing swine wastewater: Effect of cleaning procedure on fouling resistance. *Environ. Technol.* **2016**, *37*, 1704–1715. [CrossRef] [PubMed]
93. Vaneeckhaute, C.; Darveau, O.; Meers, E. Fate of micronutrients and heavy metals in digestate processing using vibrating reversed osmosis as resource recovery technology. *Sep. Purif. Technol.* **2019**, *223*, 81–87. [CrossRef]
94. Cestonaro do Amal, A.; Kunz, A.; Steinmetz, R.L.R. Zinc and copper distribution in swine wastewater treated by anaerobic digestion. *J. Environ. Manag.* **2014**, *141*, 132–137. [CrossRef]
95. Tang, Y.; Wang, L.; Carswell, A.; Misselbrook, T.; Shen, J.; Han, J. Fate and transfer of heavy metals following repeated biogas slurry application in a rice-wheat crop rotation. *J. Environ. Manag.* **2020**, *270*, 110938. [CrossRef] [PubMed]
96. Fane, A.G.; Fell, C.J.D. A review of fouling and fouling control in ultrafiltration. *Desalination* **1987**, *62*, 117–136. [CrossRef]
97. Boerlage, S.F.E.; Kennedy, M.D.; Dickson, M.R.; El-Hodali, D.E.Y.; Schippers, J.C. The modified fouling index using ultrafiltration membranes (MFI-UF): Characterization, filtration mechanisms and proposed reference membrane. *J. Membr. Sci.* **2002**, *197*, 1–21. [CrossRef]
98. Zhang, G.; Ji, S.; Gao, X.; Liu, Z. Adsorptive fouling of extracellular polymeric substances with polymeric ultrafiltration membranes. *J. Membr. Sci.* **2008**, *309*, 28–35. [CrossRef]
99. Wang, S.; Liu, C.; Li, Q. Fouling of microfiltration membranes by organic polymer coagulants and flocculants: Controlling factors and mechanisms. *Water Res.* **2011**, *45*, 357–365. [CrossRef] [PubMed]
100. Mondor, M.; Ippersiel, D.I.; Lamarche, F.; Masse, L. Fouling characterization of electrodialysis membranes used for the recovery and concentration of ammonia from swine manure. *Bioresour. Technol.* **2009**, *100*, 566–571. [CrossRef]
101. Zarebska, A.; Romero Nieto, D.; Christensen, K.V.; Norddahl, B. Ammonia recovery from agricultural wastes by membrane distillation: Fouling characterization and mechanism. *Water Res.* **2014**, *56*, 1–10. [CrossRef]
102. Zarebska, A.; Amor, A.C.; Ciurkot, K.; Karring, H.; Thygesen, O.; Prangsgaard Andersen, T.; Hagg, M.-B.; Christensen, K.V.; Norddahl, B. Fouling mitigation in membrane distillation processes during ammonia stripping from pig manure. *J. Membr. Sci.* **2015**, *484*, 119–132. [CrossRef]
103. Garcia-Gonzalez, M.C.; Vanotti, M.B. Recovery of ammonia from swine manure using gas-permeable membranes: Effect of waste strength and pH. *Waste Manag.* **2015**, *28*, 455–461. [CrossRef]
104. Zhou, Z.; Chen, L.; Wu, Q.; Zheng, T.; Yuan, H.; Peng, N.; He, M. The valorization of biogas slurry with a pilot dual stage reverse osmosis membrane process. *Chem. Eng. Res. Des.* **2019**, *142*, 133–142. [CrossRef]

105. Errico, M.; Sotoft, L.F.; Nielsen, A.K.; Norddahl, B. Treatment costs of ammonia recovery from biogas digestate by air stripping analyzed by process simulation. *Clean Technol. Environ. Policy* **2018**, *20*, 1479–1489. [CrossRef]

106. Limoli, A.; Langone, M.; Andreottola, G. Ammonia removal from raw manure digestate by means of a turbulent mixing stripping process. *J. Environ. Manag.* **2016**, *176*, 1–10. [CrossRef]

107. Huang, H.; He, L.; Zhang, Z.; Lei, Z.; Liu, R.; Zheng, W. Enhanced biogasification from ammonia-rich swine manure pretreated by ammonia fermentation and air stripping. *Int. Biodeterior. Biodegrad.* **2019**, *140*, 84–89. [CrossRef]

108. Madeddu, C.; Errico, M.; Baratti, R. Rigorous modelling of a CO_2-MEA stripping system. *Chem. Eng. Trans.* **2017**, *57*, 451–456.

109. Madeddu, C.; Errico, M.; Baratti, R. Process analysis for the carbon dioxide chemical absorption-regeneration system. *Appl. Energy* **2018**, *215*, 532–542. [CrossRef]

110. Liu, J.; de Neergaard, A.; Jensen, L.S. Increased retention of available nitrogen during thermal drying of solids of digested sewage sludge and manure by acid and zeolite addition. *Waste Manag.* **2019**, *100*, 306–317. [CrossRef]

111. Garcia-Gonzalez, M.C.; Vanotti, M.B.; Szogi, A.A. Recovery of ammonia from swine manure using gas-permeable membranes: Effect of aeration. *J. Environ. Manag.* **2015**, *152*, 19–26. [CrossRef] [PubMed]

112. Dube, P.J.; Vanotti, M.B.; Szogi, A.A.; Garcia-Gonzalez, M.C. Enhancing recovery of ammonia from swine manure anaerobic digester effluent using gas-permeable membrane technology. *Waste Manag.* **2016**, *49*, 372–377. [CrossRef] [PubMed]

113. Christensen, M.L.; Hjorth, M.; Keiding, K. Characterization of pig slurry with reference to flocculation and separation. *Water Res.* **2009**, *43*, 773–783. [CrossRef] [PubMed]

114. Tao, W.; Fattah, K.P.; Huchzermeier, M.P. Struvite recovery from anaerobically digested dairy manure: A review of application potential and hindrances. *J. Environ. Manag.* **2016**, *169*, 46–57. [CrossRef]

115. Maurer, M.; Pronk, W.; Larsen, T.A. Treatment process for source-separated urine. *Water Res.* **2006**, *40*, 3151–3166. [CrossRef]

116. Pramanik, B.K.; Hai, F.I.; Ansari, A.J.; Roddick, F.A. Mining phosphorus from anaerobically treated dairy manure by forward osmosis membrane. *J. Ind. Eng. Chem.* **2019**, *78*, 425–432. [CrossRef]

117. Landaburu-Aguirre, J.; Pongracz, E.; Keiski, R. Separation of cadmium and copper from phosphorus rich synthetic waters by micellar-enhanced ultrafiltration. *Sep. Purif. Technol.* **2011**, *81*, 41–48. [CrossRef]

118. Fernandes, F.; Silkina, A.; Fuentes-Grunewald, C.; Wood, E.E.; Ndovela, V.L.S.; Oatley-Radcliffe, D.L.; Lovitt, R.W.; Llewellyn, C.A. Valorising nutrient-rich digestate: Dilution, settlement and membrane filtration processing for optimization as a waste-based media for microalgal cultivation. *Waste Manag.* **2020**, *118*, 197–208. [CrossRef]

119. Gerardo, M.L.; Zacharof, M.P.; Lovitt, R.W. Strategies for the recovery of nutrients and metals from anaerobically digested dairy farm sludge using cross-flow microfiltration. *Water Res.* **2013**, *47*, 4833–4842. [CrossRef] [PubMed]

120. Guo, X.; Jin, X. Treatment of anaerobically digested cattle manure wastewater by tubular ultrafiltration membrane. *Sep. Sci. Technol.* **2013**, *48*, 1023–1029. [CrossRef]

121. Barzee, T.J.; Edalati, A.; El-Meshad, H.; Wang, D.; Scow, K.; Zhang, R. Digestate biofertilizers support similar or higher tomato yields and quality than mineral fertilizer in a subsurface drip fertigation system. *Front. Sustain. Food Syst.* **2019**, *3*, 58. [CrossRef]

122. Masse, L.; Bellemare, G.; Dubreuil, J. RO Membranes filtration of pretreated swine manure with high levels of suspended solids. *Trans. Asabe* **2012**, *55*, 1815–1820. [CrossRef]

123. Bilstad, T.; Madland, M.; Espedal, E.; Hanssen, P.H. Membrane separation of raw and anaerobically digested pig manure. *Water Sci. Technol.* **1992**, *25*, 19–26. [CrossRef]

124. Masse, L.; Mondor, M.; Dubreuil, J. Aging of RO membranes processing swine wastewater. *Trans. Asabe* **2013**, *56*, 1571–1578.

125. Masse, L.; Massé, D.I.; Pellerin, Y.; Dubreuil, J. Osmotic pressure and substrate resistance during the concentration of manure nutrients by reverse osmosis membranes. *J. Membr. Sci.* **2010**, *348*, 28–33. [CrossRef]

126. Pedersen, C.O.; Masse, L.; Hjorth, M. The effect of residual cationic polymers in swine wastewater on the fouling of reverse osmosis membranes. *Environ. Technol.* **2014**, *35*, 1338–1344. [CrossRef] [PubMed]

127. Masse, L.; Mondor, M.; Puig-Bargues, J.; Deschenes, L.; Talbot, G. The efficiency of various chemical solutions to clean reverse osmosis membranes processing swine wastewater. *Water Qual. Res. J.* **2014**, *49*, 295–306. [CrossRef]

128. Masse, L.; Puig-Bargues, J.; Mondor, M.; Deschenes, L.; Talbot, G. Efficiency of EDTA, SDS, and NaOH solutions to clean RO membranes processing swine wastewater. *Sep. Sci. Technol.* **2015**, *50*, 2509–2517. [CrossRef]

129. Holloway, R.W.; Childress, A.E.; Dennett, K.E.; Cath, T.Y. Forward osmosis for concentration of anaerobic digester centrate. *Water Res.* **2007**, *41*, 4005–4014. [CrossRef]

130. Ippersiel, D.; Mondor, M.; Lamarche, F.; Tremblay, F.; Dubreuil, J.; Masse, L. Nitrogen potential recovery and concentration of ammonia from swine manure using electrodialysis coupled with air stripping. *J. Environ. Manag.* **2012**, *95*, S165–S169. [CrossRef] [PubMed]

131. Zhang, R.H.; Yang, P.; Pan, Z.; Wolf, T.D.; Turnbull, J.H. Treatment of swine wastewater with biological conversion, filtration, and reverse osmosis: A laboratory study. *Trans. Asae* **2004**, *47*, 243–250. [CrossRef]

132. Masse, L.; Mondor, M.; Talbot, G.; Deschenes, L.; Drolet, H.; Gagnon, N.; St-Germain, F.; Puig-Bargues, J. Fouling of reverse osmosis membranes processing swine wastewater pretreated by mechanical separation and aerobic biofiltration. *Sep. Sci. Technol.* **2014**, *49*, 1298–1308. [CrossRef]

133. Ruan, H.; Yang, Z.; Lin, J.; Shen, J.; Ji, J.; Gao, C.; Van der Bruggen, B. Biogas slurry concentration hybrid membrane process: Pilot-testing and RO membrane cleaning. *Desalination* **2015**, *368*, 171–180. [CrossRef]

134. Van Loosdrecht, M.C.M.; Eikelboom, D.; Gjaltema, A.; Mulder, A.; Tijhuis, L.; Heijnen, J.J. Biofilm structures. *Water Sci. Technol.* **1995**, *32*, 35–43. [CrossRef]

135. Hong, S.; Elimelech, M. Chemical and physical aspects of natural organic matter (NOM) fouling nanofiltration membranes. *J. Membr. Sci.* **1997**, *132*, 159–181. [CrossRef]

136. Lee, Y.; Clark, M.M. Modeling of flux decline during crossflow ultrafiltration of colloidal suspensions. *J. Membr. Sci.* **1998**, *149*, 181–202. [CrossRef]

137. Reimann, W. Influence of organic matter from wastewater on the permeability of membranes. *Desalination* **1997**, *109*, 51–55. [CrossRef]

138. Waeger, F.; Delhaye, T.; Fuchs, W. The use of ceramic microfiltration and ultrafiltration membranes for particle removal from anaerobic digester effluents. *Sep. Purif. Technol.* **2010**, *73*, 271–278. [CrossRef]

139. Zhang, J.; Padmasiri, S.I.; Fich, M.; Norddahl, B.; Morgenroth, E. Influence of cleaning frequency and membrane history on fouling in an anaerobic membrane bioreactor. *Desalination* **2007**, *207*, 153–166. [CrossRef]

140. Lim, A.L.; Bai, R. Membrane fouling and cleaning in microfiltration of activated sludge wastewater. *J. Membr. Sci.* **2003**, *216*, 279–290. [CrossRef]

141. Bourgeous, K.N.; Darby, J.L.; Tchobanoglous, G. Ultrafiltration of wastewater: Effects of particles, mode of operation, and backwash effectiveness. *Water Res.* **2001**, *35*, 77–90. [CrossRef]

142. Nikolova, J.D.; Islam, M.A. Contribution of adsorbed layer resistance to the flux-decline in an ultrafiltration process. *J. Membr. Sci.* **1998**, *146*, 105–111. [CrossRef]

143. Koltuniewicz, A.; Noworyta, A. Dynamic properties of ultrafiltration systems in light of the surface renewal theory. *Ind. Eng. Chem. Res.* **1994**, *33*, 1771–1779. [CrossRef]

144. Tijhuis, L.; Hijman, B.; Van Loosdrecht, M.C.M.; Heijnen, J.J. Influence of detachment, substrate loading and reactor scale on the formation of biofilms in airlift reactors. *Appl. Microbiol. Biotyechnol.* **1996**, *45*, 7–17. [CrossRef]

145. Herzberg, M.; Kang, S.; Elimelech, M. Role of extracellular polymeric substances (EPS) in biofouling of reverse osmosis membranes. *Environ. Sci. Technol.* **2009**, *43*, 4393–4398. [CrossRef]

146. Bril, J. Chemical composition of animal manure: A modelling approach. *NJAS* **1990**, *38*, 333–351. [CrossRef]

147. Masse, L.; Massé, D.I. The effect of environmental and process parameters on flocculation treatment of high dry matter swine manure with polymers. *Bioresour. Technol.* **2010**, *101*, 6304–6308. [CrossRef] [PubMed]

148. Fangueiro, D.; Hjorth, M.; Gioelli, F. Acidification of animal slurry—A review. *J. Environ. Manag.* **2015**, *149*, 46–56. [CrossRef] [PubMed]

149. Chelme-Ayala, P.; El-Din, M.G.; Smith, R.; Code, K.R.; Leonard, J. Advanced treatment of liquid swine manure using physico-chemical treatment. *J. Hazard. Mater.* **2011**, *186*, 1632–1638. [CrossRef]

150. Vanotti, M.B.; Hunt, P.G. Solids and nutrient removal from flushed swine manure using polyacrylamides. *Trans. Asae* **1999**, *42*, 1833–1840. [CrossRef]

151. Hjorth, M.; Christensen, M.L.; Christensen, P.V. Flocculation, coagulation, and precipitation of manure affecting three separation techniques. *Bioresour. Technol.* **2008**, *99*, 8598–8604. [CrossRef]

152. Zhu, H.; Wen, X.; Huang, X. Characterization of a membrane fouling in a microfiltration ceramic membrane system treating secondary effluent. *Desalination* **2012**, *284*, 324–331. [CrossRef]

153. Wu, B.; An, Y.; Li, Y.; Wong, F.S. Effect of adsorption/coagulation on membrane fouling in microfiltration process post-treating anaerobic effluent. *Desalination* **2009**, *242*, 183–192. [CrossRef]

154. Yao, M.; Nan, J.; Chen, T.; Zhan, D.; Li, Q.; Wang, Z.; Li, H. Influence of flocs breakage process on membrane fouling in coagulation/ultrafiltration process—Effect of additional coagulant of poly-aluminum chloride and polyacrylamide. *J. Membr. Sci.* **2015**, *491*, 63–72. [CrossRef]

155. Huang, X.; Gui, P.; Qian, Y. Effect of sludge retention time on microbial behavior in a submerged membrane bioreactor. *Process Biochem.* **2001**, *36*, 1001–1006. [CrossRef]

156. Lim, V.H.; Yamashita, Y.; Doan, Y.T.H.; Adachi, Y. Inhibition of cationic polymer-induced colloid flocculation by polyacrylic acid. *Water* **2018**, *10*, 1215. [CrossRef]

157. Masse, L.; Mondor, M.; Dubreuil, J. Membrane filtration of the liquid fraction from a solid–liquid separator for swine manure using a cationic polymer as flocculating agent. *Environ. Technol.* **2013**, *34*, 671–677. [CrossRef]

158. Hjorth, M.; Jørgensen, B.U. Polymer flocculation mechanism in animal slurry established by charge neutralization. *Water Res.* **2012**, *46*, 1045–1051. [CrossRef] [PubMed]

159. Guo, W.; Ngo, H.-H.; Li, J. A mini-review on membrane fouling. *Bioresour. Technol.* **2012**, *122*, 27–34. [CrossRef] [PubMed]

160. Mohammadi, T.; Madaeni, S.S.; Moghadam, M.K. Investigation of membrane fouling. *Desalination* **2003**, *153*, 155–160. [CrossRef]

161. Shi, X.; Tal, G.; Hankins, N.P.; Gitis, V. Fouling and cleaning of ultrafiltration membranes: A review. *J. Water Process Eng.* **2014**, *1*, 121–138. [CrossRef]

162. Madaeni, S.S.; Samieirad, S. Chemical cleaning of reverse osmosis membrane fouled by wastewater. *Desalination* **2010**, *257*, 80–86. [CrossRef]

163. Ang, W.S.; Lee, S.; Elimelech, M. Chemical and physical aspects of cleaning of organic-fouled reverse osmosis membranes. *J. Membr. Sci.* **2006**, *272*, 198–210. [CrossRef]

164. Chen, J.P.; Kim, S.L.; Ting, Y.P. Optimization of membrane physical and chemical cleaning by a statistically designed approach. *J. Membr. Sci.* **2003**, *219*, 27–45. [CrossRef]
165. Chakravorty, B.; Layson, A. Ideal feed pretreatment for reverse osmosis by continuous microfiltration. *Desalination* **1997**, *110*, 143–149. [CrossRef]
166. Saravia, F.; Zwiener, C.; Frimmel, F.H. Interactions between membrane surface, dissolved organic substances and ions in submerged membrane filtration. *Desalination* **2006**, *192*, 280–287. [CrossRef]
167. Lujan-Facundo, M.J.; Mendoza-Roca, J.A.; Cuartas-Uribe, B.; Alvarez-Blanco, S. Ultrasonic cleaning of ultrafiltration membranes fouled with BSA solution. *Sep. Purif. Technol.* **2013**, *120*, 275–281. [CrossRef]
168. Antony, A.; Fudianto, R.; Cox, S.; Leslie, G. Assessing the oxidative degradation of polyamide reverse osmosis membrane—Accelerated ageing with hypochlorite exposure. *J. Membr. Sci.* **2010**, *347*, 159–164. [CrossRef]
169. Hajibabania, S.; Antony, A.; Leslie, G.; Le-Clech, P. Relative impact of fouling and cleaning on PVDF membrane hydraulic performances. *Sep. Purif. Technol.* **2012**, *90*, 204–212. [CrossRef]
170. Liikanen, R.; Yli-Kuivila, J.; Laukkanen, R. Efficiency of various chemical cleanings for nanofiltration membrane fouled by conventionally-treated surface water. *J. Membr. Sci.* **2002**, *195*, 265–276. [CrossRef]
171. Madaeni, S.S.; Koocheki, S. Application of taguchi method in the optimization of wastewater treatment using spiral-wound reverse osmosis element. *Chem. Eng. J.* **2006**, *119*, 37–44. [CrossRef]
172. Nystrom, M.; Zhu, H. Characterization of cleaning results using combined flux and streaming potential methods. *J. Membr. Sci.* **1997**, *131*, 195–205. [CrossRef]
173. Braghetta, A.; DiGiano, F.A.; Ball, W.P. NOM accumulation at NF membrane surface: Impact of chemistry and shear. *J. Environ. Eng.* **1998**, *124*, 1087–1098. [CrossRef]
174. Liikanen, R. Nanofiltration as a Refining Phase in Surface Water Treatment. Ph.D. Thesis, Helsinki University of Technology, Helsinki, Finland, 2006.
175. Tangsubkul, N.; Parameshwaran, K.; Lundie, S.; Fane, A.; Waite, T. Environmental life cycle assessment of the microfiltration process. *J. Membr. Sci.* **2006**, *284*, 214–226. [CrossRef]
176. Landaburu-Aguirre, J.; García-Pacheco, R.; Molina, S.; Rodríguez-Sáez, L.; Rabadán, J.; García-Calvo, E. Fouling prevention, preparing for re-use and membrane recycling. Towards circular economy in RO desalination. *Desalination* **2016**, *393*, 16–30. [CrossRef]
177. Gienau, T.; Brüß, U.; Kraume, M.; Rosenberger, S. Nutrient recovery from biogas digestate by optimised membrane treatment. *Waste Biomass Valorization* **2018**, *9*, 2337–2347. [CrossRef]
178. Senán-Salinas, J.; García-Pacheco, R.; Landaburu-Aguirre, J.; García-Calvo, E. Recycling of end-of-life reverse osmosis membranes: Comparative LCA and cost-effectiveness analysis at pilot scale. *Resour. Conser. Recycl.* **2019**, *150*, 104423. [CrossRef]
179. Nunes, S.P.; Culfaz-Emecen, P.Z.; Ramon, G.Z.; Visser, T.; Koops, G.H.; Jin, W.; Ulbricht, M. Thinking the future of membranes: Perspectives for advanced and new membrane materials and manufacturing processes. *J. Membr. Sci.* **2020**, *598*, 117761. [CrossRef]
180. Xu, H.; Bührer, M.; Marone, F.; Schmidt, T.J.; Büchi, F.N.; Eller, J. Optimal Image Denoising for In Situ X-ray Tomographic Microscopy of Liquid Water in Gas Diffusion Layers of Polymer Electrolyte Fuel Cells. *J. Electrochem. Soc.* **2020**, *167*, 104505. [CrossRef]
181. Lawler, W.; Bradford-Hartke, Z.; Cran, M.J.; Duke, M.; Leslie, G.; Ladewig, B.P.; Le-Clech, P. Towards new opportunities for reuse, recycling and disposal of used reverse osmosis membranes. *Desalination* **2012**, *299*, 103–112. [CrossRef]
182. Galiano, F.; Briceño, K.; Marino, T.; Molino, A.; Christensen, K.V.; Figoli, A. Advances in biopolymer-based membrane preparation and applications. *J. Membr. Sci.* **2018**, *564*, 562–586. [CrossRef]

MDPI

St. Alban-Anlage 66

4052 Basel

Switzerland

Tel. +41 61 683 77 34

Fax +41 61 302 89 18

www.mdpi.com

International Journal of Environmental Research and Public Health Editorial Office

E-mail: ijerph@mdpi.com

www.mdpi.com/journal/ijerph